# Wiring Regulations in Brief

# Wiring Regulations in Brief

3rd edition

Ray Tricker

 Routledge
Taylor & Francis Group

LONDON AND NEW YORK

First edition published 2007
by Butterworth Heinemann, an imprint of Elsevier

Second edition published 2008
by Butterworth Heinemann, an imprint of Elsevier

This edition published 2013
by Routledge
2 Park Square, Milton Park, Abingdon, Oxon OX14 4RN

Simultaneously published in the USA and Canada
by Routledge
711 Third Avenue, New York, NY 10017

*Routledge is an imprint of the Taylor & Francis Group, an informa business*

© 2013 Ray Tricker

*British Library Cataloguing in Publication Data*
A catalogue record for this book is available from the British Library

*Library of Congress Cataloging in Publication Data*
Tricker. Ray
   Wiring regulations in brief / Tricker Ray.—3rd ed.
   p. cm
   Includes bibliographical references and Indes.
   1. Electric wiring, Interior—Handbooks, manuals, etc.  2. Electric wiring,
   Interior—Standards.  I. Title.
   TK3271.T75 2012
   621.319'240941—dc23
   2011049090

ISBN: 978-0-415-52687-6 (pbk)
ISBN: 978-0-203-11571-8 (ebk)

Typeset in Goudy
by Swales & Willis Ltd, Exeter, Devon

Printed and bound in Great Britain by
CPI Group (UK) Ltd, Croydon, CR0 4YY

# About the author

**Ray Tricker** (MSc, IEng, FIET, FCMI, FCQI, FIRSE) is the Senior Consultant (Management Systems) of Herne European Consultancy Ltd (a company specialising in Outsourced Back Office Functions) is also an established author, with over 30 titles published. He served with the Royal Corps of Signals (for a total of 37 years) during which time he held various managerial posts culminating in being appointed as the Chief Engineer of NATO's Communication Security Agency (ACE COMSEC).

# Contents

# Preface

The Industrial Revolution during the 1800s was responsible for causing poor living and working conditions in ever expanding, densely populated urban areas. Outbreaks of cholera and other serious diseases (through poor sanitation, damp conditions and lack of ventilation) forced the government to take action. Building control took on the greater role of health and safety through the first Public Health Act in 1875, and this eventually led to the first set of national building standards (i.e. the Building Regulations).

As is the case with most official documents, as soon as the Regulations were published they were almost out of date, and consequently needed revising. So it was not too much of a surprise to learn that the committee responsible for writing the Public Health Act of 1875 had overlooked the increased use of electric power for street lighting and/or domestic purposes. Electricity was beginning to become increasingly popular but, as there were no rules or regulations governing their installation at that time, the companies or person responsible simply dug up the roads and laid the cables as and where they felt like it!

From a health and safety point of view, the government of the day expressed extreme concern at this exceedingly dangerous situation, and so in 1882 *The Electric Lighting Clauses Act* (modelled on the previous 1847 *Gas Act*) was passed by Parliament. This legislation was implemented by *Rules and Regulations for the Prevention of Fire Risks Arising from Electric Lighting*, and it is this document that is the forerunner of today's IEE Wiring Regulations. Since then, this document has seen a succession of amendments, new editions and new titles, and has now become the 17th Edition of the IEE Wiring Regulations (i.e. BS 7671:2008 *Requirements for Electrical Installations*). This has now been republished to incorporate Amendment No. 1:2011, which contains new sections on:

- measures against electromagnetic disturbances;
- devices for protection against overvoltage;
- selection of surge protection devices (SPDs);
- medical locations;
- operating or maintenance gangways;
- new model forms for certification and reporting.

Technical authority for this standard is vested in the Joint IET/BSI Technical Committee JPEL/64. This Joint Technical Committee is responsible for the work previously undertaken by the IEE Wiring Regulations Technical Com-

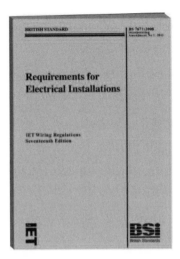

**Figure P1** The 17th edition of the IEE Wiring Regulations incorporating Amendment No. 1:2011.

mittee and the BSI Technical Committee PEL/64. Copyright is held jointly by the IET and BSI.

The latest amendment to the Wiring Regulations (i.e. BS 7671:2008 incorporating Amendment No. 1 *Requirements for Electrical Installations* (the 'Green Book') was issued on 1 July 2011 and came into effect on 1 January 2012. Installations designed after 31 December 2011 are to comply with this latest amendment.

The Regulations apply to the design, erection and verification of electrical installations, and also to additions and alterations to existing installations. Existing installations that have been installed in accordance with earlier editions of the Regulations may not comply with this new edition in every respect, but this does not necessarily mean that they are unsafe for continued use or require upgrading.

The current legislation for all building control is the Building Act 1984, which is implemented by the Building Regulations 2010; these Building Regulations are a set of minimum requirements for the design and construction of buildings, and are designed to secure the health, safety and welfare of people in and around those buildings. They include requirements to ensure that fuel and power are conserved and that facilities are provided for people, including those with disabilities, to get into and move around inside buildings. They are basic performance standards and are supported by a series of documents that correspond to the different areas covered by the regulations. These are called Approved Documents, and they contain practical and technical guidance on ways in which the requirements of the Building Act 1984 can be met.

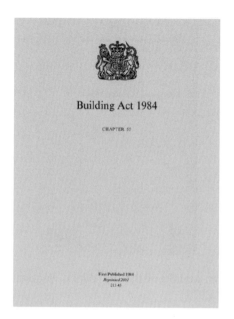

**Figure P2** The Building Act 1984.

Since the introduction of the Public Health Act in 1875 there has always, therefore, been a direct link between electrical installations and building control, and parts of all of the Approved Documents have an effect on these sorts of installation. With the publication of Approved Document P for Electrical Safety in 2005, however, the design, installation, inspection and testing of electrical installations has now become inextricably linked to building control, and the purpose of this book is attempt to draw all the various requirements together.

Over the past 120 years there have been literally hundreds of books written on the subject of electrical installations, but the aim of this 3rd edition of *Wiring Regulations in Brief* is not just to become another book on the library shelf to be looked at occasionally. The intention is that it will provide professional engineers, students and (i.e. to a lesser degree) the unqualified DIY fraternity with an easy to read reference source on the official requirements of BS 7671:2008 for electrical safety and electrical installations.

Although BS 7671:2008 (incorporating Amendment No. 1:2011) is well structured and has separate sections for all the main topics (e.g. safety protection, selection and erection of equipment, and so on), it is not the easiest of standards to get to grips with for a particular situation. Occasionally it can be very confusing, and requires the reader to constantly flick backwards and forwards through the book to find what it is all about.

For example, Regulation 411.4.7 states that:

*"Where a circuit-breaker is used to satisfy the requirements of Regulation 411.3.2.2 or Regulation 411.3.2.3, the maximum value of earth fault loop*

*impedance (Zs) shall be determined by the formula in Regulation 411.4.5. Alternatively, for a nominal voltage (Uo) of 230 V and a disconnection time of 0.4 s in accordance with Regulation 411.3.2.2 or 5 s in accordance with Regulation 411.3.2.3, the values specified in Table 41.3 for the types and ratings of overcurrent devices listed may be used instead of calculation!"*

The intention of *Wiring Regulations in Brief*, therefore, is to peel away some of this confusion, and to 'officialise' and provide the reader with an on-the-job reference source that can be quickly used without having to delve backwards and forwards though the standard.

 Please note, however, that this is only the author's impression of the most important aspects of the Wiring Regulations and their association with the Building Regulations. It should, therefore, only be treated as an *aide memoiré* to the Regulations, and electricians should always consult BS 7671 to satisfy compliance.

## What are the main changes in the First Amendment to the 2008 edition of BS 7671?

In summary, the main changes concern:

- **Conduits** – the requirement for orange as a means of identification of an electrical conduit has been deleted.
- **Definitions** – have been expanded and modified.
- **Diagrams**:
  - the TN–C and IT system diagrams have been deleted; and
  - explanatory diagrams for single-phase 2-wire, single-phase 3-wire, two-phase 3-wire, three-phase 3-wire, three-phase 4-wire a.c. and 2- and 3-wire d.c. have now been included.

- **Electromagnetic disturbances** – a new section has been included (i.e. Section 444) that provides basic requirements and recommendations to enable the avoidance and reduction of electromagnetic disturbances.

 **Notes:**

1. For example, currents due to lightning, switching operations, short-circuits and other electromagnetic phenomena may cause overvoltages and electromagnetic interference.
2. Electromagnetic interference (EMI) may severely disrupt or even damage information technology (IT) equipment/systems as well as equipment with electronic components or circuits.
3. Section 444, therefore, provides basic requirements and recommendations to enable the avoidance and reduction of these potential electromagnetic disturbances.

- **Inspection and testing** – now includes some minor modifications to take account of the change from a periodic inspection report to a condition report.
- **Medical locations** – owing to the increased governmental requirement to ensure safety of patients and medical staff, a new Section 710 has been included. These requirements, in the main, refer to hospitals, private clinics, medical and dental practices, healthcare centres and dedicated medical rooms in the workplace. They also apply to electrical installations in locations designed for medical research.

 **Note:** While these requirements mainly refer to hospitals, private clinics, medical and dental practices, healthcare centres and dedicated medical rooms in the workplace, this section is also relevant to electrical installations in locations designed for medical research.

 It should be noted that the requirements of section 710 do **not** apply to the actual medical electrical equipment.

- **Numbering system** – a new numbering system (i.e. 100) has been introduced for the UK-only Regulations so as to accommodate future IEC changes. This means, for example, that the former Regulation 422.3.14 is now Regulation 422.3.100 and this is then further divided into Regulation 422.3.101 and 422.3.102, etc.

 This only applies to regulations intended solely for the UK.

- **Operating and maintenance gangways** – a new section (Section 729) has been added for the basic protection of authorised persons going into (and when in) restricted-access areas containing switchgear and controlgear assemblies. It includes requirements for the width of gangways and access areas for operational access, emergency access and/or evacuation, and for the transportation of equipment.
- **Special installations or locations** – some minor changes have been introduced that mainly concern PME supplies.
- **Surge protective devices** (SPDs) – a new section (Section 534) deals with the selection and installation of SPDs that are designed to limit transient overvoltages (of atmospheric origin transmitted via the supply distribution system), switching overvoltages and for protection against transient overvoltages caused by direct lightning strikes or lightning strikes in the vicinity of buildings protected by a lightning protection system.

 **Note:** The requirements do not take into account surge protective components, which may be incorporated in the appliances connected to the installation. (See also Appendix 16 to BS 7671:2008 (incorporating Amendment No. 1).

- **Switchgear** – the requirements concerning the connection of cables to switchgear in relation to temperature rating have been modified.
- **Warning notices** – the regulation concerning the warning notice for voltage has been modified.
- **Appendices** – the following main changes have been made to the appendices:

    – Appendix 3 – minor changes to the time/current characteristics of overcurrent protective devices and residual current devices (RCDs), includes minor changes;
    – Appendix 4 – the current-carrying capacity and voltage drop for cables, now includes rating factors for triple harmonic currents in four-core and five-core cables and changes to the tables on rating factors and correction factors;
    – Appendix 5 – some minor changes have been introduced to the tables concerning the classification of external influences;
    – Appendix 6 – model forms for certification and reporting, now includes a new electrical installation condition report that replaces the periodic inspection report;
    – Appendix 11 – the effect of harmonic currents on balanced three-phase systems has been deleted and the content moved to Appendix 4;
    – Appendix 12 – the effect of a voltage drop in consumers' installations has been deleted and the content moved to Appendix 4;
    – Appendix 16 – advice concerning devices for protection against overvoltage has been added.

## Content of this book

To reflect the above changes, this 3rd edition of *Wiring Regulations in Brief* is structured as follows:

| | |
|---|---|
| Chapter 1 – Introduction | The background to BS 7671, what it contains and a description of its unique numbering system, objectives and legal status. The effect that the Wiring Regulations have on other statuary instruments and how BS 7671 can be implemented. |
| Chapter 2 – Domestic buildings | The requirements of the Building Act 1984 together with the Building Regulations 2010 and their Approved Documents (which provide guidance for conformance), and how these Building Control Regulations interrelate with the Wiring Regulations. A résumé of the responsibilities for electrical installations, the types of inspections and tests that have to be completed, and the requirements for records. The contents of Approved Document P for |

electrical safety and other relevant Approved Documents (such as those for Fire Safety, Access and Facilities for Disabled People, Conservation of Fuel and Power, Resistance to the Passage of Sound, etc.), together with a list of all the most important requirements of these Approved Documents that directly concern electrical installations.

**Note:** While the requirements from the Wiring Regulations are normally prefaced by the word 'shall' (meaning that this section **is** a mandatory requirement), you will notice that the Building Regulations use the words 'should' (i.e. recommended), 'may' (i.e. permitted) or 'can' (i.e. possible).

The reason for this is that Approved Documents reproduce the actual *requirements* contained in the Building Regulations relevant to a particular subject area. This is then followed by *Practical and Technical Guidance* (together with examples) showing how the requirements can be met in some of the more common building situations. There may, however, be alternative ways of complying with the requirements of the Building Regulations 2000 to those shown in the Approved Documents, and you are, therefore, under no obligation to adopt any particular solution in an Approved Document – if you prefer to meet the requirement(s) in some other way – but you **must** meet the requirement!

| | |
|---|---|
| Chapter 3 – Earthing | This chapter reminds the reader about the different types of earthing system and earthing arrangements. It then lists all the main requirements for earthing (particularly for special installations such as medical locations, which are new to Amendment No. 1 to BS 7671:2008) before briefly touching on the test requirements for earthing. |
| Chapter 4 – Safety protection | This chapter lists the mandatory and fundamental requirements for safety protection contained in BS 7671:2008 (Incorporating Amendment No. 1) as well as the Building Regulations 2010. It also includes information concerning the basic protection against electric shock, fault protection, protection against both direct and indirect contact, protective conductors and protective equipment, and then lists the test requirements for safety protection. |
| Chapter 5 – Electrical equipment, components, accessories and supplies | The amount of different types of equipment, components, accessories and supplies for electrical installations currently available is enormous, and any attempt to cover every type, model and/or manufacture would prove an impossible task for a book such as this. The intention of this chapter, therefore, is to provide a catalogue of all the different types |

identified and referred to in the Wiring Regulations (e.g. luminaires, RCDs, plugs and sockets) and then make a list of the specific requirements that are sprinkled throughout the Regulations. For your (hopeful!?) convenience this catalogue has been compiled in alphabetical order.

Chapter 6 – Cables and conductors

Within the Wiring Regulations there is frequent reference to different types of cables (e.g. single core, multi-core, fixed, flexible), conductors (e.g. live supply, protective, bonding) and conduits, cable ducting, cable trunking and so on. Unfortunately, as is the case for equipment and components, the requirements for these items are liberally sprinkled throughout the Regulations. The aim of this chapter, therefore, is to provide a catalogue of all the different types identified and referred to in the Wiring Regulations, under three main headings (namely cables, conductors and conduits/etc.), and then provide a list of their essential requirements.

Chapter 7 – Special installations and locations

While the Regulations apply to all electrical installations in buildings, there are also some indoor and outdoor special installations (e.g. floor and central heating systems) and locations that are subject to special requirements owing to the extra dangers they pose. This chapter considers the requirements for these special locations and installations, and includes the new regulations contained in BS 7671:2008 (Incorporating Amendment No. 1) for Medical Locations and Operating and Maintenance Gangways.

Chapter 8 – External influences

The new amendment to BS 7671:2008 now includes far more details of the regulations concerning external influences, and (i.e. as per Appendix 5 to BS 7671) provides a concise list of environmental influences. This chapter provides guidance on all forms of external influence, and includes extracts from the current Regulations that have an impact on the environment.

Chapter 9 – Inspection and testing

To meet the requirements for electrical safety, it is essential for any electrician engaged in the inspection, testing and certification of electrical installations to have a full working knowledge of the IET Wiring Regulations. The electrician must also have above average experience and knowledge of the type of installation under test in order to carry out any inspection and testing. Without this prerequisite, it could be quite dangerous. This chapter provides a consolidated list of how electrical installations shall be inspected and tested, as well as a brief insight into some of the test equipment that may be used.

Chapter 10 – Installation, maintenance and repair
This final chapter provides some guidance on the requirements for the installation, maintenance, inspection, certification and repair of electrical installations. It lists the Regulations' requirements for these activities with respect to electrical installations, and (in an appendix) provides an example Stage Audit Checklist for designers and engineers to use.

The ten chapters are then supported by the following annexes:

Annex A – Symbols used in electrical installations
Annex B – List of electrical and electromechanical symbols
Annex C – SI units for existing technology
Annex D – Acronyms and abbreviations
Annex E – British Standards currently used with the Wiring Regulations (by standard and by title)
Annex F – Books by the same author

plus a full Index.

The following symbols will help you get the most out of this book. In the margins you will find:

 An important requirement or point.

 A good idea or suggestion.

and within the text:

 **Notes:** are used to provide further amplification or information.

> ***"Bold italic text in double quotation marks is a quote from the relevant Act, Regulation or Approved Document."***

Shaded boxes are used in Chapter 6 to show either the full text of the *legal requirements* of the Building Regulation or a paraphrased version of these requirements. For example:

An RCD shall **not** be used in a TN-C system.  WR-411.4.4

For your convenience (and to save you having to look backwards and forwards through the book for the correct requirement), quite of a lot of these requirements have been shown more than once (i.e. in different chapters and/or sections of the book) as have a few of the figures and tables.

 **Note:** If any reader has any thoughts about the contents of this book (such as areas where perhaps they feel I have not given sufficient coverage, omissions and/or mistakes etc.), then please let know by e-mailing me at ray@herne.org. uk, and I will make suitable amendments in the next edition of this book.

Enjoy!!

Ray Tricker

# Foreword

BS 7671 *Requirements for Electrical Installations – IEE Wiring Regulations* has been an integral part of most of my professional life as an electrical engineer, and previously as an apprentice, only slipping out of sight during my time overseas. It is a document that has to guide many different people on safe and acceptable practice, including professional engineers, designers, technicians and electricians. It also has to remain up to date, taking cognisance of the evolution of legislation, technological developments, and changes in custom and practice. It is thus a difficult task to make the requirements readily accessible to such a diverse readership. Commentaries and further explanations are, therefore, often helpful, and are sometimes essential reading to gain sufficient understanding to carry out work associated with electrical installations.

I am, therefore, pleased to contribute in a small way to Ray's book, which sheds another useful light on the Wiring Regulations. Significantly for me, and perhaps many other older hands, it places these Regulations within their context in Part P of the Building Regulations and cross-references requirements. To have the Wiring Regulations virtually side by side with the legal requirements of the Building Regulations and other relevant background information in one handy volume should be particularly useful. Ray's book is a welcome contribution to working with the Wiring Regulations and Part P of the Building Regulations.

I like the simple approach, not delving too deeply into the fine detail and making information easy to find by grouping and laying it out logically. This is what most readers need much of the time. For me, this book is a must for work associated with electrical installations complying with the Building Regulations and the Wiring Regulations. This book can save the cost of buying parts of the Building Regulations, making it a cost-effective option. Some readers may find that they do not need to buy the Wiring Regulations as well.

We live every day with the multitude of benefits provided by electricity. Modern life would be impossible without it. Yet electricity is potentially dangerous to property, humans and many living creatures. Only the application of human ingenuity can reduce the risks to levels that are deemed acceptable. That we can live, work and play surrounded by live electrical equipment and installations is testimony to the dedication, understanding and skills of many people. It is also testimony to the success of the Wiring Regulations, in its many revisions/editions over the years, and books such as Ray's commentary, which helps to make the Regulations more accessible to every user.

Nigel Moore BSc (Hons), MBA, CEng, MIET, MCIM, HNC

# 1

# Introduction

The latest edition of the Institution of Engineering and Technology (IET) Wiring Regulations has now grown to a massive 464-page document that defines the way in which all electrical installation work must be carried out. It does not matter whether the work is carried out by a professional electrician or an unqualified DIY enthusiast, the installation **must** still comply with the Wiring Regulations.

The current edition of the Regulations is BS 7671:2010 (Incorporating Amendment No. 1:2011) and is entitled *Requirements for Electrical Installations – IET Wiring Regulations (Seventeenth Edition)*. Which is a bit of a mouthful to remember (!), and so it is normally referred to as *The Green Book* or *The 1st Amendment to the 17th Edition*.

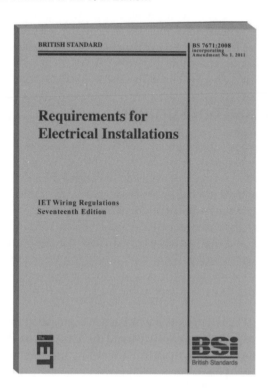

**Figure 1.1** Front cover of BS 7671:2010 incorporating Amendment No. 1:2011.

This British Standard is published with the full support of the BEC (i.e. the British Electrotechnical Committee – which is the UK national body responsible for formal standardization within the electrotechnical sector) in partnership with the BSI (i.e. the British Standards Institution – which has ultimate responsibility for all British Standards produced within this sector) and the Institution of Engineering and Technology (IET) – which, with more than 135,000 members, is Europe's largest grouping of professional engineers involved in power engineering, communications, electronics, computing, software, control, informatics and manufacturing.

The technical authority for this standard is the Joint IET/BSI Technical Committee (JPEL/64), which is responsible for all the work previously undertaken by the IET Wiring Regulations Committee and BSI Technical Committee PEL64.

Copyright is jointly held by BSI and the IET, and Amendment No. 1:2011 to BS 7671:2008 was issued on 1 July 2011 and came into effect on 1 January 2012.

 **All** installations that are designed after 31 December 2011 **must** comply with this edition, as amended and expanded.

 **Note:** All references made in this book to the 'Wiring Regulations' or the 'Regulation(s)' – where not otherwise specifically identified – refer to BS 7671:2010, Incorporating Amendment No. 1:2011 *Requirements for Electrical Installations*.

## 1.1 Historical background

The first public electricity supply in the UK was at Godalming in Surrey in November 1881 and mainly provided street lighting. At that time there were no existing rules and regulations available to control the installation, and so the electricity company just dug up the roads and laid the cables in the gutters. This particular electricity supply was discontinued in 1884.

On 12 January 1882, the steam-powered Holborn Viaduct Powerstation officially opened, and this facility supplied 110 V d.c. for both private consumption and street lighting. Once more, there was no one in authority to stipulate how the cables should be laid, and their positioning was, therefore, dependent on the electrician responsible for that particular section of the work.

Later in 1882, *The Electric Lighting Clauses Act* (modelled on the previous 1847 *Gas Act*) was passed by Parliament, and this enabled the Board of Trade to authorize the supply of electricity in any area by a local authority, company or person, and to grant powers to install this electrical supply (including breaking up the streets) through the use of the 1882 *Rules and Regulations for the Prevention of Fire Risks Arising from Electric Lighting*.

This document was the forerunner of today's IET Wiring Regulations.

Historically, since 1882 there has been a succession of amendments and new editions of the Regulations, as shown in Table 1.1.

**Table 1.1 BS 7671:2008 – publication details**

| Year | Cover colour | Edition | Comments |
| --- | --- | --- | --- |
| 1882 | | First edition | Entitled *Rules and Regulations for the Prevention of Fire Risks Arising from Electric Lighting* |
| 1888 | | Second edition | |
| 1897 | | Third edition | Entitled *General Rules recommended for Wiring for the Supply of Electrical Energy* |
| 1903 | | Fourth edition | Entitled *Wiring Rules* |
| 1907 | | Fifth edition | |
| 1911 | | Sixth edition | |
| 1916 | | Seventh edition | |
| 1924 | | Eighth edition | Entitled *Regulations for the Electrical Equipment of Buildings* |
| 1927 | | Ninth edition | |
| 1934 | | Tenth edition | |
| 1939 | | Eleventh edition | Revised issue (1943), reprinted with minor amendments (1945), supplement issued (1946), revised Section 8 (1948) |
| 1950 | | Twelfth edition | Supplement issued (1954) |
| 1955 | | Thirteenth edition | Reprinted, 1958, 1961, 1962 and 1964 |
| 1966 | | Fourteenth edition | Reprinted, 1968, 1969, 1970 (in metric units), 1972, 1973, 1974 and 1976 |

By now this continual updating was seen as a bit of a problem, particularly for designers and installers, who had to ensure that they were always working in compliance with the latest regulations. With the publication of the 15th edition, therefore, it was decided that in future all reprints of the same edition would be contained in one of five different coloured covers (i.e. red, green, yellow, blue and brown), and a new edition would be published when the brown-covered reprint required updating.

| Year | Cover colour | Edition | Comments |
| --- | --- | --- | --- |
| 1981 | Red | Fifteenth edition | Entitled *Regulations for Electrical Installations* |
| 1983 | Green | | Reprinted incorporating amendments |
| 1984 | Yellow | | Reprinted incorporating amendments |
| 1986 | Blue | | Reprinted incorporating amendments |
| 1987 | Brown | | Reprinted incorporating amendments |
| 1988 | Brown | | Reprinted with minor corrections |
| 1991 | Red | Sixteenth edition | Reprinted with minor corrections in 1992 Reprinted as BS 7671 in 1992 Amendment No. 1 issued in December 1994 |

**Table 1.1** (*continued*)

| Year | Cover colour | Edition | Comments |
|------|------|------|------|
| 1994 | Green | | Reprinted incorporating Amendment No. 1<br>Amendment No. 2 issued in December 1997 |
| 1997 | Yellow | | Reprinted incorporating Amendment No. 2<br>Amendment No. 3 issued in April 2000 |
| 2001 | Blue | | BS 7671:2001 issued in June 2001 (see Note below)<br>Amendment No. 1 issued in February 2002<br>Amendment No. 2 issued in March 2004 |
| 2004 | Brown | | Reprinted incorporating Amendments No. 1 and No 2 |
| 2008 | Red | Seventeenth Edition | New edition aligned with existing and new CENELEC, IEC and EN Harmonized Documents |
| 2010 | Green | | Reprinted incorporating Amendment No. 1 |

## 1.2  What does BS 7671:2008 contain?

This standard *"contains the rules for the design and erection of electrical installations so as to provide for safety and proper functioning for the intended use"* and is based on the plan agreed internationally (i.e. through the Comité Européen de Normalisation Electrotechnique (CENELEC)) for the *"arrangement of safety rules for electrical installations"*.

The structure of BS 7671:2008 is as shown in Table 1.2.

**Table 1.2  BS 7671:2008 – structure**

| Part | Description |
|------|------|
| Part 1 | Sets out the scope, object and fundamental principles |
| Part 2 | Defines certain terms that are used throughout the Regulations |
| Part 3 | Identifies the characteristics of an installation that will need to be taken into account in choosing and applying the requirements of the subsequent parts of the Regulations |
| | These characteristics may vary from one part of an installation to another and need to be assessed for each location to be served by the installation. |
| Part 4 | Describes the basic measures that are available for the protection of persons, property and livestock, and against the hazards that may arise from the use of electricity |
| Part 5 | Describes the precautions that need to be taken in the selection and erection of the equipment of an installation |
| Part 6 | Covers inspection and testing |
| Part 7 | Identifies particular requirements for special installations or locations |

Any intended departure from the requirements of Parts 1 to 6 requires special consideration by the installation designer – and these changes **must** be documented in the Electrical Installation Certificate specified in Part 6.

The seven parts of the standard are then supported by the appendices listed in Table 1.3.

**Table 1.3 BS 7671:2008 – appendices**

| Appendix | Title | Description and remarks |
|---|---|---|
| 1 | British Standards to which reference is made in the Regulations | Reproduced in the Reference section of this book. |
| 2 | Statutory regulations and associated memoranda | Details of all the statutory regulations, legislation and EU Harmonized Directives that electrical installations are required to comply with |
| 3 | Time/current characteristics of overcurrent protective devices and RCDs | Details of time/current characteristics for:<br>• fuses<br>• circuit breakers<br>• RCDs |
| 4 | Current-carrying capacity and voltage drop for cables | Schedules of:<br>• installation methods for conductors and cables (e.g. cleated, in conduits, on trays, in trenches)<br>• cable specifications and current rating tables (e.g. armoured cables, mineral-insulated cables, fire-resistant cables, screened cables)<br>• correction factors (for groups of cables, mineral-insulated cables, cables installed in trenches, cables direct in the ground, ambient temperature where protection is against short-circuits and overload)<br>• copper conductors<br>• aluminium conductors |
| 5 | Classification of external influences | Lists and schedules of external influences having an influence on electrical installations are detailed in Table 1.4) |
| 6 | Model forms for certification and reporting | Reproduced in Part 6 of this book |
| 7 | Harmonized cable-core colours | Current details of cable-core marking and colours that are to be used in all installations. (For details see inside front and back cover of this book.) |

**Table 1.3** (*continued*)

| Appendix | Title | Description and remarks |
|----------|-------|------------------------|
| 8 | Current-carrying capacity and voltage drop for busbar trunking and powertrack systems | Information concerning:<br><br>• the basis of current-carrying capacity<br>• rating factors for current-carrying capacity of busbar trunking systems<br>• effective current-carrying capacity<br>• protection against overload current<br>• voltage drop |
| 9 | Definitions – multiple source, d.c. and other systems | Examples of TN (TN-S and TN-C-S), TT and IT systems |
| 10 | Protection of conductors in parallel against overcurrent | Information concerning:<br><br>• overload protection of conductors in parallel<br>• short-circuit protection of conductors in parallel |
| 11 | Effect of harmonic currents on … | Now moved (and incorporated in) Appendix 4, Sections 5.5 and 5.6 |
| 12 | Voltage drop in consumers' installations | Now (and incorporated in) Appendix 4, Section 6.4 |
| 13 | Methods for measuring the insulation resistance/ impedance of floors and walls to earth or to the protective conductor system | Test methods for testing the impedance of floors and walls with a.c. voltage<br>Test electrode 1 |
| 14 | Measurement of Earth fault loop impedance: consideration of the increase in the resistance of conductors with increase in temperature | Informative |
| 15 | Ring and radial final circuit arrangements | Information concerning Section 433.1.5 |
| 16 | Devices for protection against overvoltage | Informative |

## 1.2.1  What about the standard's numbering system?

The numbering system used to identify specific requirements in BS 7671:2006 is as follows:

**Table 1.4 List of external influences relevant to electrical installations**

| Environment | Utilization | Buildings |
|---|---|---|
| Altitude (metres) | Capability | Materials |
| Ambient temperature (°C) | Contact with Earth | Structure |
| Corrosion | Evacuation | |
| Electromagnetic | Materials | |
| Fauna | Resistance | |
| Flora | | |
| Foreign bodies | | |
| Impact | | |
| Lightening | | |
| Movement of air | | |
| Other mechanical stresses | | |
| Seismic | | |
| Solar | | |
| Temperature and humidity | | |
| Vibration | | |
| Water | | |
| Wind | | |

- the first digit signifies a Part;
- the second digit a Chapter;
- the third digit a Section;
- and subsequent digits the Regulation number.

**Example**

Section number **413** is made up as follows:

- Part **4** – Protection for Safety
- Chapter **41** (first chapter of Part 4) – Protection against Electric Shock.
- Section **413** (third section of Chapter 41) – Protective Measure: Electrical Separation

## 1.3 What are the objectives of the IET Wiring Regulations?

Current legal requirements for employee competence in electrical work now call for everyone involved in certain electrical activities – even simply choosing the size of cable or fuse – to be aware of the regulative requirements associated with such work. BS 7671:2008 (i.e. The IET Wiring Regulations) is the traditionally approved Code of Practice for those who are involved in (or supervise) electrical work such as electrical maintenance, control and/or instrumentation.

The stated (IET) intention of wiring safety codes is to *"provide technical, performance and material standards that will allow sufficient distribution of electrical energy and communication signals, at the same time protecting persons in the building from electric shock and preventing fire and explosion*". In other words:

> **To ensure the protection of people and live stock from fire, shock or burns from any installation that complies with their requirements.**

The Regulations form the basis of safe working practice throughout the electrical industry.

## 1.4  What is the legal status of the IET Wiring Regulations?

Although the IET Wiring Regulations have always been held in high esteem throughout Europe, they had no legal status and did not require Continentals who were carrying out installation work in the UK to abide by them. This problem was overcome in October 1992, when the IET Wiring Regulations became British Standard 7671 – thus providing the Regulations with a national/international status.

 **Note:** Although the Regulations are *non-statutory regulations*, they may be used as evidence in a court of law to claim compliance with a statutory requirement.

## 1.5  What do the IET Wiring Regulations cover?

As shown below, the IET Wiring Regulations cover both electrical installations and electrical equipment.

### 1.5.1  Electrical installation

> **Definition:** For the purpose of the Regulations:
>
> *"Electrical installations (or Installation) means an assembly of associated electrical equipment having certain co-ordinated characteristics."*

The Regulations apply to the design, erection and verification of electrical installations such as those of:

- agricultural and horticultural premises;
- caravans, caravan parks and similar sites;
- commercial premises;

- construction sites, exhibitions, shows, fairgrounds and other installations for temporary purposes, including professional stage and broadcast applications;
- external lighting and similar installations;
- industrial premises;
- marinas;
- medical locations;
- mobile or transportable units;
- operating and maintenance gangways;
- photovoltaic systems;
- prefabricated buildings;
- public premises;
- low-voltage generating sets;
- highway equipment and street furniture;
- medical locations;
- operating and maintenance gangways;
- residential premises.

 **Note:** 'Premises' covers the land and all facilities (including buildings) belonging to it.

The Regulations include requirements for:

- the addition to (or alteration of) installations and parts of existing installations affected by an addition or alteration;
- circuits supplied at nominal voltages up to and including 1000 V a.c. or 1500 V d.c.;

 **Notes:**

1. The standard nominal supply voltage for domestic single-phase 50 Hz installations in the UK has been 230 V a.c. (rms) since 1 January 1995. Previously it was 240 V.
2. Although the preferred frequencies are 50, 60 and 400 Hz, the use of other frequencies for special purposes is not excluded.

- circuits (but not apparatus and/or equipment internal wiring) that is operating at voltages greater than 1000 V and derived from an installation having a voltage not exceeding 1000 V a.c. (e.g. discharge lighting, electrostatic precipitators);
- all consumer installations external to buildings;
- fixed wiring for communication and information technology, signalling, command and control etc. (but not apparatus and/or equipment internal wiring);
- wiring systems and cables not specifically covered by the standards for appliances.

Although the Regulations are intended as a standard for electrical installations, in certain cases they may need to be supplemented by the requirements and/or recommendations of other British Standards or by the requirements of the person ordering the work. Such cases could include (among others) the following:

- design and installation of temporary distribution systems delivering a.c. electrical supplies for lighting, technical services and other entertainment-related purposes – BS 7909;
- electrical equipment for explosive gas atmospheres – BS EN 60079;
- electrical equipment for use in the presence of combustible dust – BS EN 50281 and BS EN 61241;
- electrical installations for open-cast mines and quarries – BS 6907;
- electric signs and high-voltage luminous discharge-tube installations – BS 559 and BS EN 50107;
- electric surface-heating systems – BS 6351;
- emergency lighting – BS 5266;
- fire detection and alarm systems in buildings – BS 5839;
- life safety and fire-fighting applications – BS 8519;
- telecommunications systems – BS 6701;
- temporary electrical systems for entertainment and related purposes – BS 7909.

The Regulations do **not** apply to the following installations:

- aircraft equipment;
- electric fences covered by BS EN 50126 60335-2-76;
- electrical equipment or machines covered by BS EN 60204;
- equipment of mobile and fixed offshore installations;
- equipment on board ships covered by BS 8450;
- lightning protection systems for buildings and structures covered by BS EN 62305;
- motor vehicle equipment (except that to which the requirements of the Regulations concerning caravans or mobile units is applicable);
- radio interference suppression equipment (except insofar as it affects safety of the electrical installation);
- railway traction equipment, rolling stock and signalling equipment;
- systems for the distribution of electricity to the public;
- those aspects of lift installations covered by relevant parts of BS 5655 and BS EN 81-1;
- those aspects of mines and quarries specifically covered by statutory regulations.

 **Note:** For installations in premises that are subject to statutory control (e.g. via a licensing or other authority), the requirements of that authority will need to be confirmed, and these requirements then complied with in the design and implementation of that installation.

### 1.5.2 Electrical equipment

**Definition:** For the purpose of these Regulations:

*"Electrical equipment (or Equipment) means any item used for generation, conversion, transmission, distribution or utilisation of electrical energy, such as machines, transformers, apparatus, measuring instruments, protective devices, wiring systems, accessories, appliances and luminaires."*

The Regulations are only applicable to the actual selection and application of items of electrical equipment **in** an electrical installation.

The Regulations do **not** deal with requirements for the construction of assemblies of electrical equipment, which are required to comply with the appropriate standards.

## 1.6 What effect does using the Regulations have on other Statutory Instruments?

The requirements of the IET Wiring Regulations also have an effect on the implementation of other Statutory Instruments, such as:

- the Building Act 1984;
- the Disability and Equality Act 2010;
- the Electricity at Work Regulations 1989;
- the Fire Precautions Act 1971;
- the Health and Safety at Work Act 1974.

### 1.6.1 What is the Building Act 1984?

The Building Act 1984 (as implemented by the Building Regulations 2010) is the enabling Act under which all Building Regulations have been made. The Secretary of State (under the power given in the Building Act 1984) is required to:

- secure the health, safety, welfare and convenience of persons in or about buildings and of others who may be affected by buildings or matters connected with buildings;
- further the conservation of fuel and power;
- prevent waste, undue consumption, misuse or contamination of water;

and may make regulations with respect to the design and construction of buildings and the provision of services, fittings and equipment in (or in connection with buildings).

 **Note:** The current regulations governing the Building Regulations 2000 are SI 2000/2531 (as amended) – a copy of which can be downloaded from: http://www.odpm.gov.uk/stellent/groups/odpm_buildreg/documents/division-homepage/041014.hcsp.

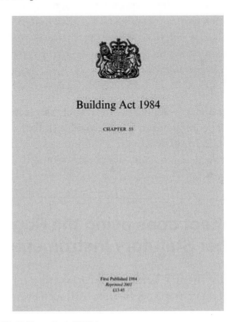

**Figure 1.2** The Building Act 1984.

For many years, the UK has managed to maintain a relatively high standard of electrical safety within buildings (domestic and non-domestic) based on voluntary controls centred around BS 7671. With the growing number of electrical accidents occurring in the 'home', the government has now been forced to implement a legal requirement for safety in **all** electrical installation work in dwellings

 As from 1 January 2005, therefore, **all** new electrical wiring or electrical components for domestic premises (or small commercial premise linked to domestic accommodation) must be designed and installed in accordance with the Building Regulations, Part P (which is based on the fundamental principles set out in Chapter 13 of BS 7671:2010).

In addition, **all** fixed electrical installations (i.e. wiring and appliances fixed to the building fabric – such as socket-outlets, switches, consumer units and ceiling fittings) **must** now be designed, installed, inspected, tested and certified to BS 7671:2010.

Part P of the Building Regulations also introduces the requirement for the cable-core colours of all a.c power circuits to align with BS  7671:2010.

 **Note:** Part P only applies to fixed electrical installations that are intended to operate at low voltage or extra-low voltage, which are not controlled by the Electricity Supply Regulations 1988 as amended, or the Electricity at Work Regulations 1989 as amended.

 **Competent Person Scheme**

Under Part P of the Building Regulations, all domestic installation work **must** now be inspected by Local Authority Building Control officers **unless** the work has been completed out by a *competent person*, who is able to self-certify the work. The IET supports the Part  P Competent Person Scheme.

 For more details about the Building Regulations, see: http://www.communitie\s.gov.uk/index.asp?id=1130474, or *Building Regulations in Brief*, 7th edition, by Ray Tricker and Sam Alford (Routledge).

### 1.6.2  What is the Disability and Equality Act 2010?

From 1 October 2010, the Equality Act replaced most of the Disability Discrimination Act (DDA). The Equality Act 2010 aims to protect disabled

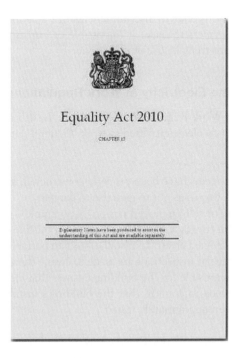

**Figure 1.3** The Equality Act 2010.

people and prevent disability discrimination. It provides legal rights for disabled people in the areas of:

- employment;
- education;
- access to goods, services and facilities, including larger private clubs and land-based transport services;
- buying and renting land or property;
- functions of public bodies, for example the issuing of licences.

The Equality Act also provides rights for people not to be directly discriminated against or harassed because they have an association with a disabled person. This can apply to a carer or parent of a disabled person. In addition, people must not be directly discriminated against or harassed because they are wrongly perceived to be disabled.

 **Note:** More information about the Equality Act, and how you can obtain copies of the Act, can be found on the Government Equalities Office website: http://homeoffice.gov.uk/equalities.

From the point of view of the BS 7671:2008, the Equality Act makes it unlawful:

- for a trade organization to discriminate against a disabled person;
- for a qualifications body to discriminate against a disabled person;
- for service providers to make it impossible or unreasonably difficult for disabled persons to make use of that service.

### 1.6.3 What is the Electricity at Work Regulations 1989?

The Electricity at Work Regulations 1989 impose health and safety requirements with respect to electricity used at work. General duties are imposed to ensure that:

- all electrical systems have been properly constructed, maintained and are used in such a way so as not to give rise to danger;
- the employer (or self-employed person) who employs one or more individuals under a contract of employment, is responsible for safety;
- maintenance of fixed electrical installations and portable appliances is carried out and regular inspections are made to ensure their continued safety;
- the person responsible for the building ensures that electrical test certificates are in place to confirm that the building's installations and appliances have been appropriately tested.

 **Note:** The Electricity at Work Regulations 1989 also state that, where an accident occurs and it is found that the systems are not covered by a valid

**Figure 1.4** The Electricity at Work Regulations 1989.

test certificate, the Health and Safety Executive (HSE) takes a keen interest in prosecutions resulting from electrocution or death within the workplace. Reducing the risk of such an accident *is* a legal requirement.

Overall, the Regulations require that:

* all electrical systems shall be constructed and maintained to prevent danger;
* all electrical equipment and installations are maintained in a safe condition;
* all people working with electricity are competent to do the job.

Complicated tasks (i.e. equipment repairs, alterations, installation work and testing) may require a suitably qualified electrician.

* all staff are aware of an organization's electrical safety arrangements;
* all work activities are carried out so as not to give rise to danger;
* equipment and procedures are safe and suitable for the working environment;
* equipment is switched off and/or unplugged before making adjustments – *live working* must be eliminated from work practices.

Electricity is recognized as a major hazard, for not only can it kill (research has shown that the majority of electric-shock fatalities occur at voltages up to

230 V), it can also cause fires and explosions. Even non-fatal shocks can cause severe and permanent injury. Most of the electrical risks can be controlled by using suitable equipment, following safe procedures when carrying out electrical work, and/or ensuring that all electrical equipment and installations are properly maintained.

Additional precautions will also be required for harsh and particular conditions (e.g. wet surroundings, cramped spaces, work out of doors or near live parts of equipment). For this reason, the Electricity at Work Regulations 1989 is used to impose health and safety requirements on all electricity used at work.

While the majority of the Regulations concern hardware requirements, others are more generalized. For example:

- *"installations shall be of proper construction*
- *conductors shall be insulated*
- *means of cutting off the power and for electrical isolation shall be available."*

In brief, the Regulations concern:

| | | |
|---|---|---|
| Systems, work activities and protective equipment | All systems shall at all times be so constructed to prevent, insofar as is reasonably practicable, danger. | Regulation 4 |
| Strength and capability of electrical equipment | No electrical equipment is to be used where its strength and capability may be exceeded so as to give rise to danger. | Regulation 5 |
| Adverse or hazardous environments | Electrical equipment sited in adverse or hazardous environments must be suitable for those conditions. | Regulation 6 |
| Insulation, protection and placing of conductors | Permanent safeguarding or suitable positioning of live conductors is required. | Regulation 7 |
| Earthing and other suitable precautions | Equipment must be earthed or other suitable precautions must be taken (e.g. the use of RCDs, double-insulated equipment, reduced-voltage equipment.). | Regulation 8 |
| Integrity of reference conductors | Nothing is to be placed in an earthed circuit conductor that might, without suitable precautions, give rise to danger | Regulation 9 |

|  |  |  |
|---|---|---|
|  | by breaking the electrical continuity or by introducing a high impedance. |  |
| Connections | All joints and connections in a system must be mechanically and electrical suitable for use. | Regulation 10 |
| Means for protecting from excess current | Suitable protective devices should be installed in each system to ensure all parts of the system **and** users of the system are safeguarded from the effects of fault conditions. | Regulation 11 |

 **Note:** Regulations 4 to 11, in effect, therefore, place a duty on the designer, installer **and** the end user to ensure the suitability and protection of all electrical equipment.

|  |  |  |
|---|---|---|
| Means of cutting off the supply and for isolation | Where necessary to prevent danger, suitable means shall be available for cutting off the electrical supply to any electrical equipment. | Regulation 12 |
| Precautions for work on equipment made dead | Adequate precautions must be taken to prevent electrical equipment, which has been made dead in order to prevent danger, from becoming live while any work is being carried out. | Regulation 13 |
| Work on or near live conductors | No work can be carried out on live electrical equipment unless this can be properly justified. | Regulation 14 |
|  | Which means that risk assessments are required. |  |
|  | If such work is to be carried out, suitable precautions must be taken to prevent injury. |  |
| Working space, access and lighting | Adequate working space, adequate means of access and adequate lighting shall be provided at all electrical equipment on or near | Regulation 15 |

| | which work is being done in circumstances that may give rise to danger. | |
|---|---|---|
| Competence to prevent danger and injury | No person shall engage in work that requires technical knowledge or experience to prevent danger or injury, unless he has that knowledge or experience, or is under appropriate supervision. | Regulation 16 |

For more information about the Electricity at Work Regulations 1989, contact the Health and Safety Executive Local Authorities Enforcement Liaison Committee (HELA):

> http://www.hse.gov.uk/lau/lacs/19-3.htm
> Tel.: 020 7717 6441
> Fax: 020 7717 6418
> HSE Infoline: 0845 345 0055 (a 'one-stop' shop, providing rapid access to expert advice and guidance)
> Email: LAU.enquiries@hse.gsi.gov.uk

Or download a copy of the Electricity at Work Regulations 1989 (Statutory Instrument 1989 No. 635) at: http://www.opsi.gov.uk/si/si1989/Uksi_19890635_en_1.htm.

### 1.6.4  What are the Fire Precautions (Workplace) Regulations 1997?

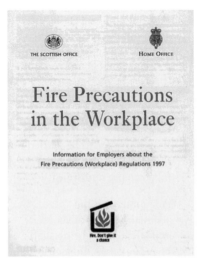

**Figure 1.5** The Fire Precautions (Workplace) Regulations 1997.

The Fire Precautions (Workplace) Regulations 1997 (as amended by the Fire Precautions (Workplace) (Amendment) Regulations 1999) stipulate that:

> *"All offices, shops, railway premises and factories which have more than 20 persons employed in the building (or more than 10 person employed anywhere other than on the ground floor) require a Fire Certificate."*

> *"Any hotel or boarding house provided sleeping accommodation for more than six persons (guests or staff) or where this sleeping accommodation is above the first floor or below the ground floor, require a Fire Certificate."*

When a Fire Certificate is issued the owner or occupier is required to provide and maintain:

- the means of escape;
- other means for ensuring that the means of escape can be safely and effectively used at all material times;
- means of fighting fire;
- means of providing warning in case of fire.

These requirements are reflected in the electrical installation.

 For further information about the Fire Precautions (Workplace) (Amendment) Regulations 1999 (Statutory Instrument 1999 No. 1877) visit: http://www.odpm.gov.uk/stellent/groups/odpm_fire/documents/sectionhomepage/odpm_fire_page.hcsp

Or for a copy of the Act, go to: http://www.legislation.gov.uk/uksi/1999/1877/contents/made.

### 1.6.5 What is the Health and Safety at Work Act 1974?

Any company with more than five employees is legally obliged to possess a comprehensive health and safety policy.

Over the years, the IET Wiring Regulations have been regularly used by the HSE in its guidance and installation notices, and installations that conform to BS 7671 (as amended) are regarded by the HSE as likely to achieve conformity with the relevant parts of the Electricity at Work Regulations 1989. In certain instances where the Regulations have been used, they may also be accompanied by Codes of Practice approved under Section 16 of the Health and Safety at Work Act 1974.

 Although some existing installations may have been designed and installed to conform to the standards set by earlier editions of the Wiring Regulations, this does **not** necessarily mean that they will fail to achieve conformity with the relevant parts of the Electricity at Work Regulations 1989.

**Figure 1.6** The Health and Safety at Work Act 1974.

 For further information about the Health and Safety at Work Act 1974 see: http://www.legislation.gov.uk/ukpga/1974/37/contents.

## 1.7  How are the IET Wiring Regulations implemented?

Although the IET Wiring Regulations rely (primarily) on British Standards for their implementation (see Annex E for details), they do, however, include the policy decisions made in a number of Statutory Instruments and by the Council of European Communities in the relative EU Harmonized Directives.

### 1.7.1  Statutory Instruments

In Great Britain the following classes of electrical installations are required to comply with the statutory regulations.

 The full text of **all** Statutory Instruments that have been published since 1987 are now available from the Office of Public Sector Information (OPSI) via their website: http://www.hmso.gov.uk/stat.htm.

With effect from July 1999, Statutory Instruments that have also been made by the National Assembly for Wales have been published via the Wales Legislation section: http://www.opsi.gov.uk/legislation/wales/w-stat.htm.

And the series of Scottish Statutory Instruments have been published via the Scottish Legislation: http://www.opsi.gov.uk/legislation/scotland/s-stat.htm.

**Table 1.5  Statutory Instruments affecting electrical installations**

| Type of electrical installation | Statutory Instrument |
|---|---|
| Building generally (subject to certain exemptions) | Building Regulations 2010 (as amended) (for England and Wales): <br><br> • SI 2010 No. 2214 (as amended by SI 2011/1515) <br><br> Building (Scotland) Amendment Regulations 2011(as amended): <br><br> • Scottish SI 2011 No. 120 <br><br> Building Regulations (Northern Ireland) 2000 (as amended) <br><br> • Statutory Rule 2000 No. 398 |
| Cinematograph installations | Cinematograph (Safety) Regulations 1955 (as amended under the Cinematograph Act, 1909, and/or Cinematograph Act, 1952): <br><br> • SI 1982 No. 1856 |
| Distributors' installations generally (subject to certain exemptions) | Electricity Safety, Quality and Continuity Regulations 2002: <br><br> • SI 2002 No. 2665 <br> • SI 2006 No. 1521 |
| High-voltage luminous tube | Conditions of licence under: <br><br> • in England and Wales – Local Government (Miscellaneous provisions) Act 1982 <br> • in Scotland – Civic Government (Scotland) Act 1982 |
| Machinery | The Supply of Machinery (Safety) Regulations 1992 as amended: <br><br> • SI 1992 No. 3073 <br> • SI 1994 No. 2063 |
| Theatres and other places licensed for public entertainment, music, dancing etc. | Conditions of licence under: <br><br> • in England and Wales – Local Government (Miscellaneous provisions) Act 1982 <br> • in Scotland – Civic Government (Scotland) Act 1982 |
| Work activity Places of work Non-domestic installations | The Electricity at Work Regulations 1989 as amended: <br><br> • SI 1989 No. 635 <br> • SI 1996 No. 192 <br> • SI 1997 No. 1993 <br> • SI 1999 No. 2024 <br><br> The Electricity at Work Regulations (Northern Ireland) 1991: <br><br> • Statutory Rule No. 13 |

## 1.7.2  British Standards, International Standards and Harmonized Documents

As well as British and International Standards, the Wiring Regulations also take account of the technical substance of agreements reached in CENELEC.

For your convenience, a list of the current (i.e. at the time of writing this book) Standards and Directives relevant to the IET Wiring Regulations are listed in Annex E to this book as a follows:

- British Standards currently used with the wiring regulations:
    - by standard; and
    - by subject;
- other standards to which reference is made in the regulations:
    - IEC and ISO;
    - CENELEC Harmonized Directives:
- by subject;
- by directive.

By the time you read this edition of *Wiring Regulations in Brief*, it is quite possible that some of the Standards and Directives listed in this book **and** the IET Wiring Regulations will have been reviewed and updated and, although the ones listed in Annex E are still relevant, the latest edition of these documents should always be taken into account. For this reason, the reader is recommended to always have a quick check via Google (or some other search engine) to make sure that they are using the most up-to-date regulation and/or recommendation.

# 2

# Domestic buildings

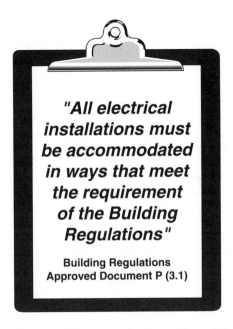

"All electrical
installations must
be accommodated
in ways that meet
the requirement
of the Building
Regulations"

**Building Regulations
Approved Document P (3.1)**

**Figure 2.1** Mandatory requirements for domestic buildings.

## 2.1 The Building Act 1984

The Building Act 1984 is the mechanism by which the Secretary of State ensures that the health, welfare and convenience of persons living in or working in or near buildings is secured.

The primary purpose of the Building Act 1984 is to assist in the conservation of fuel and power, and to prevent waste, undue consumption, misuse and contamination of water. The Building Act 1984 imposes a set of requirements on owners and occupiers of buildings, which cover the design and construction of buildings, and the provision of services, fittings and equipment used in (or in connection with) buildings. These involve and cover:

- a method of controlling (inspecting and reporting) buildings;
- how services, fittings and equipment may be used;
- the inspection and maintenance of any service, fitting or equipment used.

 The Building Act 1984 does **not** apply to Scotland or to Northern Ireland – see Section 2.6 for details.

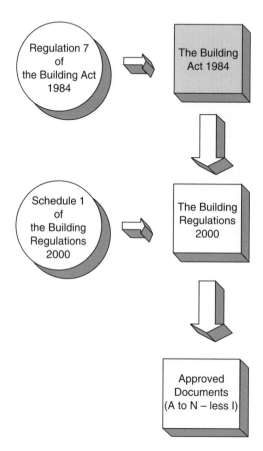

**Figure 2.2** Implementing the Building Act.

## 2.2  Who polices the Building Act?

Under the terms of the Building Act 1984, **local authorities** are responsible for ensuring that any building work that takes place in their area conforms to the requirements of the associated Building Regulations. They have the authority to:

- make you complete alterations so that your work complies with the Building Regulations;

- make you take down and remove or rebuild anything that contravenes a regulation;
- employ a third party (and then send you the bill!) to take down and rebuild non-conforming buildings or parts of buildings.

You can be prosecuted or ordered to carry out remedial work on a property whether you are the owner or merely the occupier.

They can, in certain circumstances, even take you to court and have you fined – especially if you fail to complete the removal or rebuilding of the non-conforming work.

### 2.2.1  What does the Building Act 1984 contain?

The Building Act 1984 is made up of five parts:

| | |
|---|---|
| Part 1 | The Building Regulations |
| Part 2 | Supervision of building work, etc., other than by a local authority |
| Part 3 | Other provisions about buildings |
| Part 4 | General |
| Part 5 | Supplementary |

**Note:** Regulation 7 of the Building Act 1984 covers materials and workmanship and states that building work shall be carried out

- with adequate and proper materials; which:

  - are appropriate for the circumstances in which they are used;
  - are adequately mixed or prepared; and
  - are applied, used or fixed so as adequately to perform the functions for which they are designed; and

- in a workmanlike manner.

## 2.3  What are the Supplementary Regulations?

The Supplementary Regulations make up Part 5 of the Building Act. They comprise seven schedules the function of which is to list the principal areas requiring regulation and to show how the Building Regulations are to be controlled by local authorities. These schedules are listed in Table 2.1.

**Note:** Schedule 1 is the most important Schedule (from the point of view of builders and electricians), as it shows, in general terms, how the Building Regulations are to be administered by local authorities, the approved methods of construction, and the approved types of materials that are to be used in (or in connection with) buildings.

**Table 2.1 The seven schedules of the Supplementary Regulations contained in Part 5 of the Building Act**

| Schedule | Title | Description |
|---|---|---|
| Schedule 1 | Building Regulations | Schedule 1 describes the mandatory requirements for completing **all** building work |
| Schedule 2 | Relaxation of Building Regulations for existing work | Schedule 2 provides guidance in connection with work that has been carried out prior to a local authority (under the Building Act 1984 Section 36) dispensing with or relaxing some of the requirements contained in the Building Regulations |
| | | Schedule 2 is quite difficult to understand and, if it affects you, then I would strongly advise that you discuss it with the local authority before proceeding any further! |
| Schedule 3 | Inner London | Schedule 3 applies to how Building Regulations are to be used in Inner London, and (as well as ruling which sections of the Act may be omitted) also details how by-laws concerning the relation to the demolition of buildings (in Inner London) may be made |
| Schedule 4 | Provisions consequential upon a public body's notice | Schedule 4 concerns the authority and ruling of public bodies' notices and certificates |
| Schedule 5 | Transitional provisions | Schedule 5 lists the transitional effect of the Building Act 1984 on existing Acts of Parliament |
| Schedule 6 | Consequential amendments | Schedule 6 lists the consequential amendments that will have to be made to existing Acts of Parliament owing to the acceptance of the Building Act 1984 |
| Schedule 7 | Repeals | Schedule 7 lists the cancellation (repeal) of some sections of existing Acts of Parliament, owing to acceptance of the Building Act 1984 |

# 2.4 The Building Regulations

Building Regulations 2010 (SI 2010/2214) form Schedule 1 of the Building Act and are a set of minimum requirements and basic performance standards designed to secure the health, safety and welfare of people in and around buildings, and to conserve fuel and energy in England and Wales.

They are legal requirements laid down by Parliament and based on the Building Act 1984. The Building Regulations:

- are approved by parliament;
- are designed to ensure structural stability;
- contribute to meeting the needs of disabled people;

- deal with the minimum standards of design and building work for the construction of domestic, commercial and industrial buildings;
- ensure the health and safety of people in and around buildings (by providing functional requirements for building design and construction);
- promote energy efficiency in buildings;
- promote the use of suitable materials to provide adequate durability, fire and weather resistance, and the prevention of damp;
- set out the procedure for ensuring that building work meets the standards laid down;
- stipulate the minimum amount of ventilation and natural light to be provided for habitable rooms.

The level of safety and standards acceptable is set out as guidance in **Approved Documents**, which are discussed below. Compliance with the detailed guidance of the Approved Documents is usually considered as evidence that the Building Regulations themselves have been fulfilled.

### 2.4.1 How is my building work evaluated for conformance with the Building Regulations?

Part of the local authority's duty is to make regular checks that all building work being completed is in conformance with the approved plan and the Building Regulations. These checks would normally be completed at certain stages of the work (e.g. the excavation of foundations), and tests will include:

- tests of any service, fitting or equipment that has been, is being or is proposed to be provided in or in connection with a building;
- tests of any material, component or combination of components that has been, is being or is proposed to be used in the construction of a building;
- tests of the soil or sub-soil of the site of the building;
- the cost of carrying out these tests will normally be charged to the owner or occupier of the building.

 The local authority has the power to ask the person responsible for the building work to complete some of these tests on their behalf.

## 2.5 Approved Documents

The Secretary of State makes available a series of documents (called **Approved Documents**) that contain practical and technical guidance on ways in which the requirements of Schedule 1 and Regulation 7 of the Building Act 1984 can be met.

 More details concerning Approved Documents are contained on the Planning Portal at: http://www.planningportal.gov.uk/buildingregulations/approved-documents.

Each Approved Document reproduces the actual *requirements* contained in the Building Regulations relevant to the subject area. This is then followed by *practical and technical guidance* (together with examples) showing how the requirements can be met in some of the more common building situations.

There may, however, be alternative ways of complying with the requirements to those shown in the Approved Documents, and you are, therefore, under no obligation to adopt any particular solution contained in an Approved Document if you prefer to meet the requirement(s) in some other way.

If you intend to carry out building work, you should **always** check with your local authority (or an approved inspector) that your proposals comply with Building Regulations.

The current sets of Approved Documents are in 14 Parts, A to P (less I and O), and consist of:

A   Structure
B   Fire safety
C   Site preparation and resistance to contaminants and moisture
D   Toxic substances
E   Resistance to the passage of sound
F   Ventilation
G   Sanitation, hot water safety and water efficiency
H   Drainage and waste disposal
J   Combustion appliances and fuel storage systems
K   Protection from falling collision and impact
L   Conservation of fuel and power
M   Access to and use of buildings
N   Glazing – safety in relation to impact, opening and cleaning
P   Electrical safety – Dwellings

For some reason, the Planning Portal actually shows these as different titles!

In accordance with Regulation 8 of the Building Regulations, the requirements in Parts A to D, F to K and N (except for paragraphs H2 and J6) of Schedule 1 to the Building Regulations do not require anything to be done except for the purpose of securing reasonable standards of health and safety for persons in or about buildings (and any others who may be affected by buildings or matters connected with buildings).

**Notes:**
Paragraphs H2 and J6 are excluded from Regulation 8 because they deal directly with the prevention of contamination of water.

Parts E and M (which deal, respectively, with resistance to the passage of sound, and access to and use of buildings) are excluded from Regulation 8 because they address the welfare and convenience of building users.

Part L is excluded from Regulation 8 because it addresses the conservation of fuel and power.

## 2.6  What about the rest of the UK?

The Building Act 1984 only applies to England and Wales. Separate Acts and Regulations apply in Scotland or Northern Ireland; these are shown in Table 2.2.

**Table 2.2  Building legislation within the UK**

|  | Act | Regulations | Implementation |
|---|---|---|---|
| England and Wales | Building Act 1984 | Building Regulations 2010 | Approved Documents |
| Scotland | Building (Scotland) Act 2004 | Building (Scotland) Regulations 2004 (as amended) | Technical Handbooks |
| Northern Ireland | Building Regulations (Northern Ireland) Order 1979 | Building Regulations (Northern Ireland) 2000 (as amended) | Technical Booklets |

### 2.6.1  Scotland

In Scotland, the requirements for buildings are controlled by the Building (Scotland) Regulations 2004 as amended by the Building (Scotland) Amendment Regulations 2010. The methods for implementing these requirements are similar to those in England and Wales, except that the guidance documents for achieving compliance are contained in two Technical Handbooks (one for domestic work and the other for non-domestic work) and other associated guidance.

The main procedural difference between the Scottish system and the others is that a building warrant is **still** required before work can start.

### 2.6.2  Northern Ireland

In Northern Ireland the primary legislation for buildings is the Building Regulations (Northern Ireland) Order 1979, under which Building Regulations (Northern Ireland) 2000 are made. These regulations revoked and replaced with amendments the previous 1994 Building Regulations, and they have since been amended by the Building (Amendment) Regulations (Northern Ireland) 2010. The Principal Regulations comprise 15 Parts and are supported by Technical Booklets, which are used to ensure that the requirements are implemented.

Table 2.3 provides a summary of the titles of the sections of Building Regulations that apply in the UK.

**Table 2.3 Building Regulations in the UK**

| England and Wales | Scotland | | Northern Ireland |
|---|---|---|---|
| Part A  Structure | Section 1 – Structure | Technical Handbooks 2010 | Part D – Structure |
| Part B  Fire safety | Section 2 – Fire | Technical Handbooks 2010 | Part E – Fire safety |
| Part C  Site preparation and resistance to contaminants and water | Section 3 – Environment | Technical Handbooks 2010 | Part C – Preparation of site and resistance to moisture |
| Part D  Toxic substances | Section 3 – Environment | Technical Handbooks 2010 | Part B – Materials and workmanship |
| Part E  Resistance to the passage of sound | Section 5 – Noise | Technical Handbooks 2010 | Part G – Sound insulation of dwellings |
| Part F  Ventilation | Section 3 – Environment | Technical Handbooks 2010 | Part K – Ventilation |
| Part G  Sanitation, hot water safety and water efficiency | Section 3 – Environment | Technical Handbooks 2010 | Part P – Sanitary appliances and unvented hot water storage systems |
| Part H  Drainage and waste disposal | Section 3 – Environment | Technical Handbooks 2010 | Part N – Drainage |
| Part J  Combustion appliances and fuel storage | Section 3 – Environment Section 4 – Safety | Technical Handbooks 2010 | Part L – Heat-producing appliances and liquefied petroleum gas installations |
| Part K  Protection from falling, collision and impact | Section 4 – Safety | Technical Handbooks 2010 | Part H – Stairs, ramps and protection from impact |
| Part L  Conservation of fuel and power | Section 6 – Energy | Technical Handbooks 2010 | Part F – Conservation of fuel and power |
| Part M  Access and facilities for disabled people | Section 4 – Safety | Technical Handbooks 2010 | Part R – Access and facilities for disabled people |
| Part N  Glazing | Section 6 – Energy | Technical Handbooks 2010 | Part V – Glazing |
| Part P  Electrical safety | Section 4 – Safety | Technical Handbooks 2010 | |

For full details of the Building Act 1984, the Building Regulations 2010 and all the associated Approved Documents, see the latest edition (i.e. 7th edition) of *Building Regulations in Brief* by Ray Tricker and Sam Alford (Routledge).

## 2.7 Electrical safety

For many years, the UK has managed to maintain relatively high electrical safety standards with the support of voluntary controls based on BS 7671. However, with a growing number of electrical accidents occurring in the 'home', the government has been forced to consider the legal requirement for safety in electrical installation work in dwellings.

As from 1 January 2005, therefore, all new electrical wiring or electrical components for domestic premises (or small commercial premises linked to domestic accommodation) have had to be designed and installed in accordance with the Building Regulations Part P, which is based on the fundamental principles set out in Chapter 13 of BS 7671:2008 (i.e. the IET Wiring Regulations). In addition, all fixed electrical installations (i.e. wiring and appliances fixed to the building fabric, such as socket-outlets, switches, consumer units and ceiling fittings) must also be designed, installed, inspected, tested and certified to BS 7671.

Part P also introduced new requirements for cable-core colours for a.c. power circuits and, with effect from 31 March 2006, **all** new installations or alterations to existing installations **must** use the new (harmonized) colour cables. (Further information on cable identification colours for extra-low voltage and d.c. power circuits is available from the Institution of Engineering and Technology (IET) website at: http://www.theiet.org.)

Table 2.4 Identification of conductors in a.c. power and lighting circuits

| Conductor | Colour |
|---|---|
| Protective conductor | Green and yellow |
| Neutral | Blue |
| Phase of single phase circuit | Brown |
| Phase 1 of 3-phase circuit | Brown |
| Phase 2 of 3-phase circuit | Black |
| Phase 3 of 3-phase circuit | Grey |

For single-phase installations in domestic premises, the new colours are the same as those for flexible cables to appliances (namely green-and-yellow, blue and brown for the protective, neutral and phase conductors, respectively).

**Note:** Part P only applies to fixed electrical installations that are intended to operate at low voltage or extra-low voltage, which are **not** controlled by the Electricity Safety, Quality and Continuity Regulations 2002 (as amended) or the Electricity at Work Regulations 1989 (as amended).

# 2.8  What is the aim of Approved Document P?

The aim of Part P is to increase the safety of householders by improving the design, installation, inspection and testing of electrical installations in dwellings when they (i.e. the installations) are being newly built, extended or altered.

 The government is currently in the process of introducing a mandatory scheme whereby domestic installations shall be checked at regular intervals (as well as when they are sold and/or purchased) to make sure that they comply. This will mean, of course, that if you had an installation that was not correctly certified, then your house insurance might well **not** be valid!

### 2.8.1  Who is responsible for electrical safety?

Basically, there are three people who are responsible for the electrical safety of (and within) buildings. These are:

- **The owner** – needs to determine whether the works carried out are minor or notifiable work. If the work is notifiable, then the owner needs to make sure that the person(s) carrying out the work is either registered under one of the self-certified schemes (see Figure 2.3) or is able to certify their work under the local authority building control approval route.
- **The designer** – needs to ensure that all electrical work is designed, constructed, inspected and tested in accordance with the BS 7671 (current issue) and either falls under a competent persons scheme or the local authority building control approval route.
- **The builder/developer** – needs to ensure that they have electricians who can self-certify their work or who are qualified/experienced enough to enable them to sign off under the Electrical Installation Certification form.

### 2.8.2  What are the statutory requirements?

**All** electrical installations need to:

- be designed and installed to protect against mechanical and thermal damage;
- be designed and installed so that they will **not** present an electrical shock and/or fire hazard;
- be tested and inspected to meet relevant equipment/installation standards;
- provide sufficient information so that persons wishing to operate, maintain or alter an electrical installation can do so with reasonable safety;
- comply with such requirements as placed by the Building Regulations.

| Authorized competent person self-certification schemes for installers who can do all electrical installation work | | Authorized competent person self-certification schemes for installers who can do electrical work only if it is necessary when they are carrying out other work | |
| --- | --- | --- | --- |
| | **BRE Certification Ltd**<br>Tel 01923 664000<br>http://www.bre.co.uk | | **Benchmark Certification Limited**<br>Tel: 0844 8794798<br>http://www.benchmark-certification.com |
| **BSI** | **British Standards Institution**<br>Tel 0840 765600<br>http://www.bsi-global.com/kitemark | | **ELECSA Ltd**<br>Phone: 0845 634 9043<br>http://www.elecsa.org.uk |
| | **ELECSA Ltd**<br>Tel: 0845 634 9043<br>http://www.elecsa.org.uk | | **NAPIT Certification Ltd**<br>Tel. 0870 444 1392<br>http://www.napit.org.uk |
| | **NAPIT Certification Ltd**<br>Tel. 0870 444 1392<br>http://www.napit.org.uk | | **NICEIC Group Ltd**<br>Tel. 0870 013 0382<br>http:// www.niceic.com |
| | **NICEIC Group Ltd**<br>Tel. 0870 013 0382<br>http:// www.niceic.com | | **OFTEC (Oil Firing Technical Association)**<br>Tel. 0845 658 5080<br>http:// www.oftec.co.uk |

**Figure 2.3** Authorised competent person self-certification schemes for installers.

## 2.8.2.1 What does all this mean?

With a few exceptions, **any** electrical work undertaken in a home that includes the addition of a new electrical circuit, or involves work in the

- kitchen;
- bathroom;
- garden area;

**must** be reported to the local authority building control for inspection. This includes any work undertaken professionally, or by you or another family member or by a friend.

The **ONLY** exception is when the installer has been approved by a competent persons organisation such as ELECSA (see Figure 2.3).

### 2.8.3  What types of building does Approved Document P cover?

Part P applies to **all** electrical installations in (**and around**) buildings or parts of buildings comprising:

- dwelling houses and flats;
- dwellings and business premises that have a common supply;
- land associated with domestic buildings;
- fixed lighting and pond pumps in gardens;
- shops and public houses with a flat above;
- common access areas in blocks of flats, such as corridors and stairways;
- shared amenities of blocks of flats, such as laundries and gymnasiums.

Table 2.5 provides the details of works that are notifiable to the local authority and/or must be completed by a company registered as a *competent firm*.

**Table 2.5  Notifiable work**

| Location where work is being completed | Extensions and modifications to circuits | New circuits |
| --- | --- | --- |
| Bathrooms | Yes | Yes |
| Bedrooms | Yes | Yes |
| Ceiling (overhead) heating | Yes | Yes |
| Communal area of flats | Yes | Yes |
| Computer cabling | | |
| Conservatories | | Yes |
| Dining rooms | | Yes |
| Garages (integral) | | Yes |
| Garages (detached) | | Yes* |
| Garden – lighting | Yes | Yes |
| Garden – power | Yes | Yes |
| Greenhouses | Yes | Yes |
| Halls | | Yes |
| Hot air saunas | Yes | Yes |
| Kitchen | Yes | Yes |

| | | |
|---|---|---|
| Kitchen diners | Yes | Yes |
| Landings | | Yes |
| Lounge | | Yes |
| Paddling pools | Yes | Yes |
| Remote buildings | Yes | Yes |
| Sheds | Yes | Yes |
| Shower rooms | Yes | Yes |
| Small-scale generators | Yes | Yes |
| Solar power systems | Yes | Yes |
| Stairways | | Yes |
| Studies | | Yes |
| Swimming pools | Yes | Yes |
| Telephone cabling | | Yes |
| TV rooms | | Yes |
| Underfloor heating | Yes | Yes |
| Workshops (remote) | Yes | Yes |

\* If the installation requires outdoor wiring.

### 2.8.4  What is a competent firm?

For the purposes of Part P, the government has defined *competent firms* as electrical contractors:

* who work in conformance with the requirements to BS 7671:2010;
* whose standard of electrical work has been assessed by a third party;
* who are registered under the NICEIC (National Inspection Council for Electrical Installation Counselling) Approved Contractor Scheme and the Electrotechnical Assessment Scheme.

### 2.8.5  What is a competent person responsible for?

When a competent person undertakes installation work, that person is responsible for:

* ensuring compliance with BS 7671:2010 and all relevant Building Regulations;
* providing the person ordering the work with a signed Building Regulations self-certification certificate;
* providing the relevant Building Control Body with an information copy of the certificate;
* providing the person ordering the work with a completed Electrical Installation Certificate.

### 2.8.6 Who is entitled to self-certify an installation?

Part P affects **every** electrical contractor carrying out fixed installation and/or alteration work in homes. **Only** registered installers are entitled to self-certify the electrical work, however, and they **must** be registered as a competent person under one of the schemes shown in Figure 2.3.

Working with industry and consumer organisations, the government has developed the Trustmark initiative for builders and specialist firms that work on (and in) the home. Schemes that are capable of delivering 'agreed competence and customer care standards' are approved to use the Trustmark brand (Figure 2.4) by a board consisting of industry and consumer representatives, with government observers. The brand is owned by the DTI, which licences the board.

**Figure 2.4** The TrustMark initiative. (Logo produced courtesy of TrustMark.)

The TrustMark replaced the Quality Mark scheme, which closed on 31 December 2004 because too few firms joined. For more information about TrustMark, see their website at: http://www.trustmark.org.uk.

### 2.8.7 When do I have to inform the local authority Building Control Body?

All proposals to carry out electrical installation work **must** be notified to the local authority's Building Control Body before work begins, **unless** the proposed installation work is undertaken by a person who is a competent person registered under a government-approved Part P self-certification scheme or the work is agreed as non-notifiable work, such as:

- connecting an electric gate or garage door to an existing isolator (but, be careful, the installation of the circuit up to the isolator **is** notifiable!);
- fitting and replacing cookers and electric showers (unless a new circuit is required);

- installing equipment (e.g. security lighting, air-conditioning equipment and radon fans) that is attached to the outside wall of a house (unless there are exposed outdoor connections and/or the installation is a new circuit, or an extension of a circuit in a kitchen, or special location, or is associated with a special installation);
- installing fixed equipment where the final connection is via a 13 A plug and socket (unless it involves fixed wiring and the installation of a new circuit or the extension of a circuit in a kitchen or special location);
- installing prefabricated, 'modular' systems such as kitchen lighting systems and armoured garden cabling that are linked by plug-and-socket connectors (provided that products are CE-marked and that any final connections in kitchens and special locations are made to existing connection units or points, e.g. a 13 A socket-outlet);
- installing or upgrading main or supplementary equipotential bonding (provided that the work complies with other applicable legislation, such as the Gas Safety (Installation and Use) Regulations);
- installing mechanical protection to existing fixed installations (provided that the circuit's protective measures and the current-carrying capacity of conductors are unaffected by increased thermal insulation);
- re-fixing or replacing the enclosures of existing installation components;
- replacement, repair and maintenance jobs;
- replacing fixed electrical equipment (e.g. socket-outlets, control switches and ceiling roses) that do not require the provision of any new fixed cabling;
- replacing the cable of a single circuit cable (where damaged, for example, by fire, rodent or impact – provided that the replacement cable has the same current-carrying capacity, follows the same route and does not serve more than one sub-circuit through a distribution board);
- work that is not in a kitchen or special location, which does not involve a special installation and which only consists of:

  - adding lighting points (light fittings and switches) to an existing circuit;
  - adding socket-outlets and fused spurs to an existing ring or radial circuit (provided that the existing circuit protective device is suitable and supplies adequate protection for the modified circuit);

- work that is not in a special location and only concerns:

  - adding a telephone, extra-low voltage wiring and equipment for communications, information technology, signalling, command, control and other similar purposes;
  - adding prefabricated equipment sets (and their associated flexible leads) with integral plug and socket connections.

All this work can be completed by a DIY enthusiast (family member or friend) but still needs to be installed in accordance with manufacturers' instructions and done in such a way that it does not present a safety hazard. This work does not need to be notified to a local authority Building Control Body (unless it is installed in an area of high risk, such as a kitchen or a bathroom), but all DIY electrical work (unless completed by a qualified professional – who is responsible for issuing a Minor Electrical Installation Certificate) will still need to be checked, certified and tested by a competent electrician.

Any work that involves adding a new circuit to a dwelling will need to be either notified to the Building Control Body (who will then inspect the work) or needs to be carried out by a competent person who is registered under a government-approved Part P self-certification scheme.

Work involving any of the following will also **have** to be notified:

- consumer unit replacements;
- electric floor or ceiling heating systems;
- extra-low voltage lighting installations (other than pre-assembled, CE-marked lighting sets);
- garden lighting and/or power installations;
- installation of a socket-outlet on an external wall;
- installation of outdoor lighting and/or power installations in the garden or that involves crossing the garden;
- installation of new central heating control wiring;
- solar photovoltaic (PV) power supply systems;
- small-scale generators (such as micro combined heat and power (CHP) units).

 **Note:** Where a person who is **not** registered to self-certify intends to carry out the electrical installation, a Building Regulation (i.e. a Building Notice or Full Plans) application will need to be submitted, together with the appropriate fee, based on the estimated cost of the electrical installation. The Building Control Body will then arrange to have the electrical installation inspected at first-fix stage and tested upon completion.

In any event, the electrical work will still need to be certified under BS 7671:2010 by a suitably competent person, who will be responsible for the design, installation, inspection and testing of the system (on completion), and have the confidence of completing a certificate to say that the work is satisfactory and complies with current codes of practice.

The main things to remember are:

- Is the work notifiable or non notifiable?
- Does the person undertaking the work need to be registered as a competent person?
- What records (if any) need to be kept of the installation?

### 2.8.8 What if the work is completed by a friend, a relative or me?

You do **not** need to tell your local authority's building control department about non-notifiable work such as:

• repairs, replacements and maintenance work;
• extra power points or lighting points or other alterations to existing circuits (unless they are in a kitchen or bathroom, or are outdoors).

You **do**, however, need to tell them about most other work.

If you are not sure about this, or you have any questions, ask your local authority's building control department.

### 2.8.9 What if the work is completed by a contractor or an installer?

If the work is of a notifiable nature, then the installer(s) must be registered with one of the schemes shown in Figure 2.3.
   Figure 2.5 provides a quick guide to the requirements.

## 2.9 What inspections and tests will have to be completed and recorded?

As shown in Table 2.6, there are four types of electrical installation certificate and one Building Regulation compliance certificate that have to be completed.

Copies of these various certificates and forms are contained in Chapter 6 of this book.

### 2.9.1 What should be included in the records of the installation?

All *original* certificate should be retained in a safe place and be shown to any person inspecting or undertaking further work on the electrical installation in the future. If you later vacate the property, this certificate will demonstrate to the new owner that the electrical installation complied with the requirements of BS 7671 at the time the certificate was issued. The Construction (Design and Management) Regulations require that, for a project covered by those Regulations, a copy of this certificate (together with schedules and test results) is included in the project health and safety documentation.
   Figure 2.6 indicates how to choose what type of inspection is required.

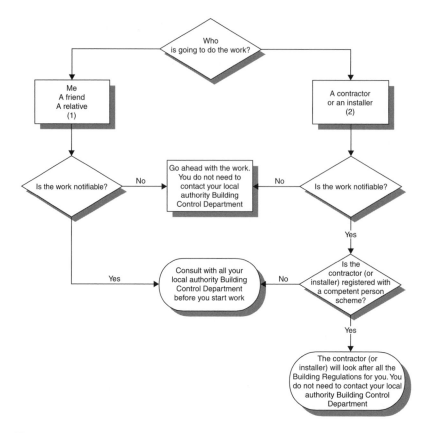

**Figure 2.5** How to meet the new rules.

**Table 2.6  Types of installation**

| Type of inspection | When is it used? | What should it contain? | Remarks |
|---|---|---|---|
| Electrical Installation Certificate | For the initial certification of a new installation or for the alteration of or addition to an existing installation where new circuits have been introduced | A schedule of inspections and test results as required by Part 6 (of BS 7671) A certificate, including guidance for recipients (standard form from Appendix 6 of BS 7671) | The original certificate shall be given to the person ordering the work and a duplicate retained by the contractor |
| Minor Electrical Installation Works Certificate | For additions and alterations to an installation such as an extra socket-outlet or lighting | Relevant provisions of Part 6 of BS 7671 | This certificate may also be used for the replacement of equipment such as accessories or luminaires, but *not* for the replacement |

| | | | |
|---|---|---|---|
| | point to an existing circuit, the relocation of a light switch, etc. | | of distribution boards (or similar items) or the provision of a new circuit |
| Electrical Installation Certificate | For the inspection of an existing electrical installation | A schedule of inspections and a schedule of test results as required by Part 6 (of BS 7671) | For safety reasons, the electrical installation will need to be inspected at appropriate intervals by a competent person |
| Building Regulations Compliance Certificate | Confirmation that the work carried out complies with the Building Regulations | The basic details of the installation, the location, completion date and the name of the installer | A purchaser's solicitor may request this document when you come to sell your property. Looking further ahead, it may be required as one of the documents that will make up your 'Home Information Pack' |

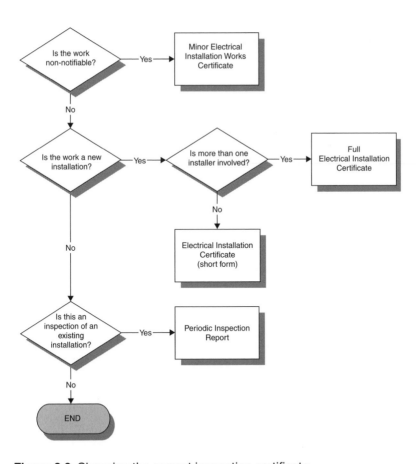

**Figure 2.6** Choosing the correct inspection certificate.

### 2.9.2 Where can I get more information about the requirements of Part P?

Further guidance concerning the requirements of Part P (Electrical Safety) is available from:

- the IET (Institution of Engineering Technology) at http://electrical. theiet.org/building-regulations/part-p/index.cfm;
- the NICEIC (National Inspection Council for Electrical Installation Contracting) at http://www.niceic.org.uk;
- the ECA (Electrical Contractors' Association) at http://www.eca. co.uk.

For details of fixed wire colour changes see: http://www.niceic.org.uk/downloads/WiringSupp.pdf

## 2.10 Requirements from the Approved Documents

Although:

- Part E (Resistance to the passage of sound);
- Part J (Combustion appliances and fuel storage systems); and
- Part K (Protection from falling, collision and impact).

have a number of requirements concerning electrical safety and electrical installations (see below for details), the main requirements are contained in:

- Part P (Electrical safety), together with:
  - Part M (Access and use of buildings);
  - Part L (Conservation of fuel and power);
  - Part B (Fire safety).

**Figure 2.7** Building Regulations.

**Note:** Part L consists of four separate sub-parts: L1A, L1B (for domestic buildings) and L2A and L2B (for non-domestic buildings).

### 2.10.1  Part P – Electrical safety

*"Reasonable provision shall be made in the design, installation, inspection and testing of electrical installations in order to protect persons from fire or injury."*

(Approved Document P1)

*"Sufficient information shall be provided so that persons wishing to operate, maintain or alter an electrical installation can do so with reasonable safety."*

(Approved Document P2)

**Note:** While Part P lays out the requirements for the safety of fixed electrical installations, it does not cover system functionality (such as electrically powered fire alarm systems, fans and pumps), which is covered by other parts of the Building Regulations and other legislation.

### 2.10.2  Part M – Access and facilities for disabled people

*"In addition to the requirements of the Disability and the Equality Act 2010, precautions need to be taken to ensure that:*

- *new non-domestic buildings and/or dwellings (e.g. houses and flats used for student living accommodation etc);*
- *extensions to existing non-domestic buildings;*
- *non-domestic buildings that have been subject to a material change of use (e.g. so that they become a hotel, boarding house, institution, public building or shop).*

*are capable of allowing people, regardless of their disability, age or gender to:*

- *gain access to buildings;*
- *gain access within buildings;*
- *use sanitary conveniences in the principal storey of any new dwelling;*
- *be able to use the facilities of the buildings (both as visitors and as people who live or work in them)."*

(Approved Document M)

### 2.10.3  Part L – Conservation of fuel and power

*"Energy efficiency measures shall be provided which:*

- *provide lighting systems that utilise energy-efficient lamps with manual switching controls or, in the case of external lighting fixed to the building, automatic switching, or both manual and automatic switching controls as appropriate, such that the lighting systems can be operated effectively as regards the conservation of fuel and power;*

- *provide information, in a suitably concise and understandable form (including results of performance tests carried out during the works) that shows building occupiers how the heating and hot water services can be operated and maintained."*

(Approved Document L1)

Responsibility for achieving compliance with the requirements of Part L rests with the person carrying out the work. That *person* may be, for example, a developer, a main (or sub-) contractor, or a specialist firm directly engaged by a private client.

**Note:** The person responsible for achieving compliance should either themselves provide a certificate, or obtain a certificate from the sub-contractor, that commissioning has been successfully carried out. The certificate should be made available to the client and the Building Control Body.

### 2.10.4 Part B – Fire safety

*"The building shall be designed and constructed so that there are appropriate provisions for the early warning of fire, and appropriate means of escape in case of fire from the building to a place of safety outside the building capable of being safely and effectively used at all material times."*

(Approved Document B1)

Requirement B1 does not apply to any prison covered by Section 33 of the Prison Act 1952 (power to provide prisons, etc.).

*"To inhibit the spread of fire within the building, the internal linings shall:*

- *adequately resist the spread of flame over their surfaces; and*
- *have, if ignited, a rate of heat release or a rate of fire growth which is reasonable in the circumstances."*

(Approved Document B2)

'Internal linings' means the materials or products used in lining any partition, wall, ceiling or other internal structure.

(1) *"The building shall be designed and constructed so that, in the event of fire, its stability will be maintained for a reasonable period.*

(2) *A wall common to two or more buildings shall be designed and constructed so that it adequately resists the spread of fire between those buildings. For the purposes of this sub-paragraph a house in a terrace and a semi-detached house are each to be treated as a separate building.*

(3) *Where reasonably necessary to inhibit the spread of fire within the building, measures shall be taken, to an extent appropriate to the size and intended use of the building, comprising either or both of the following:*

- *Sub-division of the building with fire-resisting construction;*
- *Installation of suitable automatic fire suppression systems.*

(4)  *The building shall be designed and constructed so that the unseen spread of fire and smoke within concealed spaces in its structure and fabric is inhibited."*

(Approved Document B3)

Requirement B3(3) does not apply to material alterations to any prison covered by Section 33 of the Prison Act 1952.

### 2.10.5 Design

Electrical installations must be designed and installed (suitably enclosed and appropriately separated) so that they:

| | |
|---|---|
| • *are safe to use, maintain and alter;* | P1.7* |
| • *comply with the requirements of BS 7671;* | P1.4 |
| • *comply with Part P (and any other relevant parts) of the Building Regulations);* | P1.7 and 3.1 |
| • *comply with the relevant equipment and installation standards;* | P0.1b |
| • *do not present an electric shock or fire hazard to people;* | P0.1a |
| • *provide adequate protection against mechanical and thermal damage;* | P0.1a |
| • *provide adequate protection for persons against the risks of electric shock, burn or fire injuries.* | P0.1a |

**Note:** See Appendix A of Part P to the Building Regulations for details of the types of electrical services normally found in dwellings, some of the ways that they can be connected, and the complexity of wiring and protective systems that can be used to supply them.

* Refers to relevant Building Regulations requirement

### 2.10.6 Extensions, material alterations and material changes of use

In accordance with Regulation 4(2), the whole of the existing installation does not have to be upgraded to current standards, but only to the extent necessary for the new work to meet the current standards, except where upgrading is required by the energy-efficiency requirements of the Building Regulations.

Where any electrical installation work is classified as an extension, a material alteration or a material change of use, the work must consider and include:

| | |
|---|---|
| • confirmation that the mains supply equipment is suitable (and can) carry the additional loads envisaged; | P2.1b– P2.2 |
| • the amount of additions and alterations that will be required to the existing fixed electrical installation in the building; | P2.1a |
| • the earthing and bonding systems required are satisfactory and meet the requirements; | P2.1a– P2.2c |
| • the necessary additions and alterations to the circuits which feed them; | P2.1a |
| • the rating and the condition of existing equipment (belonging to both the consumer and the electricity distributor) is sufficient; | P2.2a |
| • the protective measures required to meet the requirements. | P2.1a– P2.2b |

See Figure 2.8 for details of some of the types of electrical services normally found in dwellings, some of the ways they can be connected, and the complexity of wiring and protective systems that can be used to supply them.

**Note:** Appendix C to Part P of the Building Regulations offers guidance on some of the older types of installations that might be encountered during alteration work, while Appendix D provides guidance on the application of the now harmonized European cable identification system.

### 2.10.7 Electricity distributors' responsibilities

The electricity distributor is responsible for:

| | |
|---|---|
| • ensuring that the installation is mechanically protected and can be safely maintained; | P1.5 |
| • evaluating and agreeing proposals for new installations or significant alterations to existing ones; | P1.2 |
| • installing the cut-out and meter in a safe location; | P1.5 |
| • taking into consideration the possible risk of flooding. | P1.5 |

**Note:** See the Planning Portal's publication *Preparing for Floods* at: http://www.planningportal.gov.uk/uploads/odpm/4000000009282.pdf.

The Environment Agency also has some interesting information about preparing for flooding, at: http://www.environment-agency.gov.uk/subjects/flood/826674/830330/?lang=_e.

Distributors are required to:

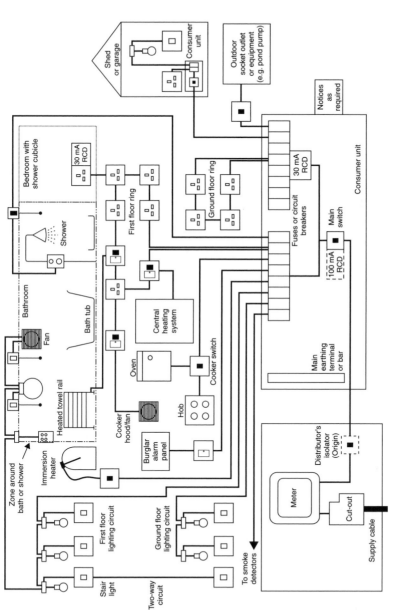

**Figure 2.8** Typical fixed installations that might be encountered in new or upgraded existing dwellings.

- maintain the supply within defined tolerance limits;      P3.8
- provide an earthing facility for new connections;         P3.8
- provide certain technical and safety information to       P3.8
  consumers to enable them to design their installations.

Distributors and meter operators must ensure that their equipment on consumers' premises:

- clearly shows the polarity of the conductors;     P3.9
- is safe in its particular environment;            P3.9
- is suitable for its purpose.                      P3.9

Distributors:

- are prevented by the Regulations from connecting         P3.12
  installations to their networks that do not comply with BS
  7671;
- may disconnect consumers' installations that are a source  P3.12
  of danger or cause interference with their networks or other
  installations.

**Note:** Detailed guidance on these Regulations is available at: http://www.dti.gov.uk/electricity-regulations

### 2.10.8 Earthing

**Note:** The most usual type of earthing is an electricity distributor's earthing terminal, which is provided for this purpose, near the electricity meter.

| | |
|---|---|
| Distributors are required to provide an earthing facility for all new connections. | P3.8 |
| All electrical installations shall be properly earthed. | P AppC1 |
| All lighting circuits shall include a circuit-protective conductor. | P AppC |
| All socket-outlets that have a rating of 32 A or less and that may be used to supply portable equipment for use outdoors **shall** be protected by a residual current device (RCD). | P AppC |
| It is **not** permitted to use a gas, water or other metal service pipe as a means of earthing for an electrical installation (this does not rule out, however, | |

| | |
|---|---|
| equipotential bonding conductors being connected to these pipes). | P AppC |
| New or replacement, non-metallic light fittings, switches or other components do not require earthing (e.g. non-metallic varieties) unless new circuit-protective (earthing) conductors are provided. | P AppC |
| Socket-outlets that will accept unearthed (2-pin) plugs must **not** use supply equipment that needs to be earthed. | P AppC |
| Where electrical installation work is classified as an extension, a material alteration or a material change of use, the work must consider and include that the earthing and bonding systems are satisfactory and meet the requirements. | P2.1a–P2.2c |

 See Figure 2.9 for details of some Earth and bonding conductors that might be part of an electrical installation.

| | |
|---|---|
| Accessible consumer units should be fitted with a child-proof cover or installed in a lockable cupboard. | P1.6 |

### 2.10.9 Electrical installations

All electrical installations shall provide adequate protection for persons against the risks of electric shock, burn or fire injuries, and should be designed and installed (suitably enclosed and appropriately separated) to provide mechanical and thermal protection.

Electrical installations must be inspected and tested during installation, at the end of installation, and before they are taken into service, in order to verify that they:

| | |
|---|---|
| • are safe to use, maintain and alter; | P1.7 |
| • comply with Part P (and any other relevant parts) of the Building Regulations; | P1.7 |
| • meet the relevant equipment and installation standards; | P0.1b |
| • meet the requirements of the Building Regulations. | P3.1 |

Any proposal for a new mains supply installation (or where significant alterations are going to be made to an existing mains supply) **must** be agreed with the electricity distributor.

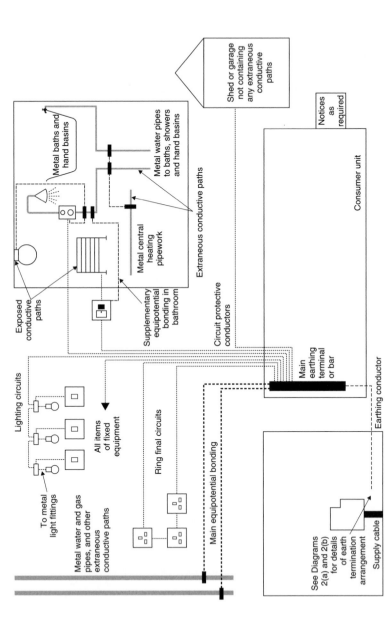

**Figure 2.9** Typical Earth and bonding conductors that might be part of the electrical installation of consumer units.

## 2.10.10 Electrical installation work

| | |
|---|---|
| • Shall be carried out professionally. | P1.1 |
| • Shall comply with the Electricity at Work Regulations 1989 (as amended). | P1.1 |
| • May only be carried out by persons who are competent to prevent danger and injury while doing it, or who are appropriately supervised. | P3.4a |

**Note:** Persons installing domestic combined heat and power equipment must advise the local distributor of their intentions before (or at the time of) commissioning the source.

### 2.10.10.1 Types of wiring or wiring system

| | |
|---|---|
| Cables concealed in floors and (in certain circumstances) walls are required to have an earthed metal covering, be enclosed in steel conduit, or have additional mechanical protection (see BS 7671 for more information). | P AppA 2d |
| Cables to an outside building (e.g. garage or shed), if run underground, should be routed and positioned so as to give protection against electric shock and fire as a result of mechanical damage to a cable. | P AppA 2d |
| Heat-resisting flexible cables are required for the final connections to certain equipment (see makers' instructions). | P AppA 2d |

PVC-insulated and sheathed cables are likely to be suitable for much of the wiring in a typical dwelling.

### 2.10.10.2 Equipotential bonding conductors

| | |
|---|---|
| Main equipotential bonding conductors are required to water service pipes, gas installation pipes, oil supply pipes and certain other 'earthy' metalwork that may be present on the premises. | P AppC |
| The installation of supplementary equipotential bonding conductors is required for installations and locations where there is an increased risk of electric shock (e.g. bathrooms and shower rooms). | P AppC |
| The minimum size of supplementary equipotential bonding conductors (without mechanical protection) is 4 mm$^2$. | P AppC |

## 2.10.11 Inspection and test

Electrical installations must be inspected and tested during, at the end of installation and before they are taken into service, to verify that they:

> - are reasonably safe and that they comply with BS 7671:2008; P1.7
> - meet the relevant equipment and installation standards. P0.1b

All electrical work should be inspected (during installation as well as on completion) to verify that the components have:

> - been selected and installed in accordance with BS 7671; P1.11a ii)
> - been made in compliance with appropriate British P1.11a i) Standards or harmonized European Standards;
> - been evaluated against external influences (e.g. the P1.11a ii) presence of moisture);
> - not been visibly damaged (or are defective) so as to be P1.11a iii) unsafe;
> - been tested to check its satisfactory performance P1.1b with respect to continuity of conductors, insulation resistance, separation of circuits, polarity, earthing and bonding arrangements, Earth fault loop impedance and functionality of all protective devices, including residual current devices;
> - been inspected and tested for conformance with the P1.9 requirements of BS 7671:2008;
> - been tested using appropriate and accurate P1.10 instruments.

 **Note**: Inspections and testing of DIY work should **also** meet the above requirements.

## 2.10.12 Additional requirements and facilities for disabled people

During 2002/3, Approved Document M was thoroughly overhauled and restructured in order to meet the changed requirements of the Disability Discrimination Act 1995 (DDA).

 **Note:** From 1 October 2010, the Equality Act replaced most of the DDA. However, the Disability Equality duty in the DDA continues to apply.

This major rewrite of Part M became effective on 1 May 2004, and was as a result of amendments made to Section 6 of DDA, which previously stated that: *"reasonable adjustments to physical features of premises"* shall be made *"in certain circumstances"* and *"shall apply to all employers with 15 or more employees"*. As such, an employer with only a few employees (as well as occupations such as police, fire-fighters and prison officers) was not required to alter the physical characteristics of a building that met existing requirements at the time the building works were carried out and **still** continued to meet those particular requirements.

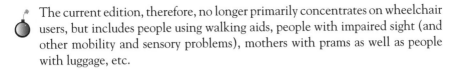

From 1 October 2004, however, under the DDA (Amendment) Regulations 2003 (SI 2003/1673), this exemption ceased and *"all those who provide services to the public, irrespective of their size"* were then required *"to take reasonable steps to remove, alter or provide a reasonable means of avoiding a physical feature of their premises, which makes it unreasonably difficult or impossible for disabled people to make use of their services"*.

These amendments have naturally resulted in Part M being completely overhauled, and it now covers:

- the conversion of a building for use as a shop being redefined as a *material change of use*;
- amendments to omit specific references to (and a definition of) disabled people;
- expansion of the terms to include parents with children, elderly people and people with all types of disabilities (e.g. mobility, sight and hearing);
- the use of a building by disabled people as residents, visitors, spectators, customers or employees, or participants in sports events, performances and conferences (which resulted in amendments being made to M1 (accessibility), M2 (sanitary accommodation) and (M3) audience and/or spectator seating).

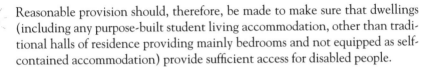

The current edition, therefore, no longer primarily concentrates on wheelchair users, but includes people using walking aids, people with impaired sight (and other mobility and sensory problems), mothers with prams as well as people with luggage, etc.

Reasonable provision should, therefore, be made to make sure that dwellings (including any purpose-built student living accommodation, other than traditional halls of residence providing mainly bedrooms and not equipped as self-contained accommodation) provide sufficient access for disabled people.

### 2.10.13 Electrical components and installations

New or replacement, non-metallic light fittings, switches or other components do **not** require earthing (e.g. non-metallic varieties) **unless** new circuit-protective (earthing) conductors are provided.

### 2.10.13.1 Controls and switches

The aim should be to ensure that all controls and switches should be easy to operate, visible and free from obstruction, and:    M (5.4i)

- they should be located between 750 mm and 1200 mm above the floor;
- they should not require the simultaneous use of both hands (unless necessary for safety reasons) to operate;
- switched socket-outlets should indicate whether they are ON;
- mains and circuit isolator switches should clearly indicate whether they are ON or OFF;
- individual switches on panels and on multiple socket-outlets should be well separated;
- front plates should contrast visually with their backgrounds.

Controls that need close vision (e.g. thermostats) should be located between 1200 mm and 1400 mm above the floor.    M (4.30f)

The operation of all switches, outlets and controls should not require the simultaneous use of both hands (unless necessary for safety reasons).    M (4.30j)

Where possible, light switches with large push pads should be used in preference to pull cords.    M (5.3)

The colours red and green should **not** be used in combination as indicators of ON and OFF for switches and controls.    M (5.3)

### 2.10.13.2 Lighting circuits

All lighting circuits shall include a circuit-protective conductor.

### 2.10.13.2.1 Fixed lighting

In locations where lighting can be expected to have most use, fixed lighting (e.g. fluorescent tubes and compact fluorescent lamps – but not GLS tungsten lamps with bayonet cap or Edison screw bases) with a luminous efficacy greater than 40 lumens per circuit-watt should be available.    L

**Note:** The following is an indication of recommended number of locations (excluding garages, lofts and outhouses) that need to be equipped with efficient lighting:

| Number of rooms created (hall, stairs and landing(s) count as one room as does a conservatory) | Recommended minimum number of locations |
|---|---|
| 1–3 | 1 |
| 4–6 | 2 |
| 7–9 | 3 |
| 10–12 | 4 |

Hall, stairs and landing(s) count as one room, as does a conservatory.

### 2.10.13.2.2 External lighting fixed to the building

External lighting (including lighting in porches, but not lighting in garages and carports) should:

| | |
|---|---|
| Automatically extinguish when there is enough daylight, and when not required at night. | L |
| Have sockets that can only be used with lamps having an efficacy greater than 40 lumens per circuit-watt (such as fluorescent or compact fluorescent lamp types, and **not** GLS tungsten lamps with bayonet cap or Edison screw bases). | L |

### 2.10.13.3 Emergency alarms

| | |
|---|---|
| Emergency alarm pull cords should be: <ul><li>coloured **red**;</li><li>located as close to a wall as possible;</li><li>have two **red** 50 mm diameter bangles.</li></ul> | M (4.30e) |
| Front plates should contrast visually with their backgrounds. | M (4.30m) |
| The colours red and green should **not** be used in combination as indicators of ON and OFF for switches and controls. | M (4.28) |

### 2.10.13.4 Fire alarms

| | |
|---|---|
| Fire alarms should emit an audio and visual signal to warn occupants with hearing or visual impairments. | M (5.4g) |
| Emergency assistance alarm systems should have: <ul><li>visual and audible indicators to confirm that an emergency call has been received;</li></ul> | M (5.4h) |

- a reset control reachable from a wheelchair, WC or a shower/changing seat;
- a signal that is distinguishable visually and audibly from the fire alarm.

### 2.10.13.5 Heat emitters

| | |
|---|---|
| Heat emitters should either be screened or have their exposed surfaces kept at a temperature below 43°C. | M (5.4j) |

In toilets and bathrooms, heat emitters (if located) should not restrict:  M (5.10p)

- the minimum clear wheelchair manoeuvring space;
- the space beside a WC used to transfer from the wheelchair to the WC.

### 2.10.13.6 Power-operated doors

Doors to accessible entrances shall be provided with a power-operated door opening and closing system if a force greater than 20 N is required to open or shut a door.

Once open, all doors to accessible entrances should be wide enough to allow unrestricted passage for a variety of users, including wheelchair users, people carrying luggage, people with assistance dogs, and parents with pushchairs and small children.

The effective clear width through a single-leaf door (or one leaf of a double-leaf door) should be in accordance with Table 2.7.  M (2.13b)

**Table 2.7 Minimum effective clear widths of doors**

| Direction and width of approach | New buildings (mm) | Existing buildings (mm) |
|---|---|---|
| Straight on (without a turn or oblique approach) | 800 | 750 |
| At right angles to an access route at least 1500 mm wide | 800 | 750 |
| At right angles to an access route at least 1200 mm wide | 825 | 775 |
| External doors to buildings used by the general public | 1000 | 775 |

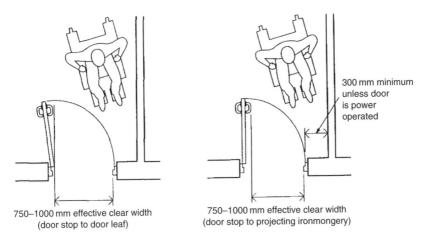

750–1000 mm effective clear width
(door stop to door leaf)

300 mm minimum
unless door
is power
operated

750–1000 mm effective clear width
(door stop to projecting ironmongery)

**Figure 2.10** Effective clear width and visibility requirements for doors.

Power-operated entrance doors should have a sliding, swinging or folding action, controlled manually (by a push pad, card swipe, coded entry or remote control) or automatically by a motion sensor or proximity sensor such as contact mat.

Power-operated entrance doors should:

| | |
|---|---|
| • be provided with a manual or automatic opening device in the event of a power failure where and when necessary for health or safety; | K5 (5.2d) |
| • open towards people approaching the doors; | M (2.21a) |
| • provide visual and audible warnings that they are operating (or are about to operate); | M (2.21c) |
| • incorporate automatic sensors to ensure that they open early enough (and stay open long enough) to permit safe entry and exit; | M (2.21c) |
| • incorporate a safety stop that is activated if the doors begin to close when a person is passing through; | M (2.21b) |
| • have a readily identifiable and accessible stop switch; | K5 (5.2d) |
| • have safety features to prevent injury to people who are struck or trapped (e.g. a pressure-sensitive door edge that operates the power switch); | K5 (5.2d) |
| • revert to manual control (or fail-safe) in the open position in the event of a power failure; | M (2.21d) |
| • when open, should not project into any adjacent access route; | M (2.21e) |

| | |
|---|---|
| • ensure that its manual controls: | M (2.21f) |
|    ◦ are located between 750 mm and 1000 mm above floor level; | |
|    ◦ are operable with a closed fist; | |
| • be set back 1400 mm from the leading edge of the door when fully open; | M (2.21g) |
| • be clearly distinguishable against the background; | M (2.21g) |
| • contrast visually with the background. | M (2.19 and 2.21g) |

**Note:** Revolving doors are **not** considered 'accessible', as they create particular difficulties (and possible injury) for people who are visually impaired, people with assistance dogs or mobility problems, and for parents with children and/or pushchairs.

### 2.10.13.7 Switches and socket-outlets

| | |
|---|---|
| Switches and socket-outlets for lighting and other equipment should be located so that they are easily reachable. | M2 (8.2) |
| Switches and socket-outlets (for lighting) should be installed between 450 mm and 1200 mm from the finished floor level (see Figure 2.11). | M2 (8.3) |

The aim is to help people with limited reach (e.g. seated in a wheelchair) to access a dwelling's wall-mounted switches and socket-outlets.

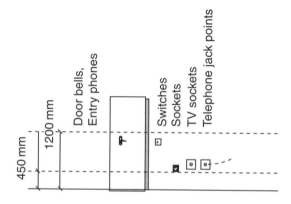

**Figure 2.11** Heights of switches and sockets, etc.

### 2.10.13.7.1 Socket-outlets

| | |
|---|---|
| Older types of socket-outlet designed for non-fused plugs must not be connected to a ring circuit. | P AppC |
| Socket-outlets that will accept unearthed (2-pin) plugs must not be used to supply equipment that needs to be earthed. | P AppC |
| Sensitive RCD protection is required for all socket-outlets that have a rating of 32 A or less and that may be used to supply portable equipment for use outdoors. | P AppC |
| Socket-outlets should comply with the requirements of Part M (see Section 2.13.24). | P1.5 |

### 2.10.13.7.2 Portable equipment for use outdoors

| | |
|---|---|
| Sensitive RCD protection is required for all socket-outlets that have a rating of 32 A or less and that may be used to supply portable equipment for use outdoors. | P AppC |

### 2.10.13.7.3 Switched socket-outlets

| | |
|---|---|
| Switched socket-outlets should indicate whether they are ON. | M (4.30k) |
| Mains and circuit isolator switches should clearly indicate whether they ON or OFF. | M (4.30l) |
| Individual switches on panels and on multiple socket-outlets should be well separated. | M (4.29) |
| All socket-outlets should be wall-mounted. | M (4.30a and b) |
| Socket-outlets should be located no nearer than 350 mm from room corners. | M (4.30g) |
| Front plates should contrast visually with their backgrounds. | M (4.30m) |
| The colours red and green should **not** be used in combination as indicators of ON and OFF for switches and controls. | M (4.28) |

### 2.10.13.7.4 Wall sockets

Wall sockets shall meet the requirements listed in Table 2.8.

**Table 2.8 Building Regulations requirements for wall sockets**

| Type of wall | Requirement | Section |
|---|---|---|
| Timber framed | Power points may be set in the linings, provided there is a similar thickness of cladding behind the socket box | E (p. 14) |
| | Power points should not be placed back to back across the wall | E (p. 14) |
| Solid masonry | Deep sockets and chases should **not** be used in separating walls | E 2.32 |
| | Stagger the position of sockets on opposite sides of the separating wall | E 2.32f |
| Cavity masonry | Stagger the position of sockets on opposite sides of the separating wall | E 2.65e |
| | Deep sockets and chases should **not** be used in a separating wall | E 2.65d2 |
| | Deep sockets and chases in a separating wall should **not** be placed back to back | E 2.65d2 |
| Framed walls with absorbent material | Sockets should: | |
| | • be positioned on opposite sides of a separating wall | E 2.146b |
| | • not be connected back to back | E 2.146b2 |
| | • be staggered a minimum of 150 mm edge to edge | E 2.146b2 |

## 2.10.13.7.5  Light switches

| | |
|---|---|
| Light switches should:<br><br>• have large push pads;<br>• align horizontally with door handles;<br>• be within 900 mm and 1100 mm from the entrance door opening. | M (4.30h and i) |
| Switches and controls should be located between 750 mm and 1200 mm above the floor. | M (4.30c and d) |
| Where possible, light switches with large push pads should be used in preference to pull cords. | M (5.3) |
| The colours red and green should **not** be used in combination as indicators of ON and OFF for switches and controls. | M (5.3) |

### 2.10.13.7.6 Telephone points and TV sockets

| | |
|---|---|
| All telephone points and TV sockets should be located between 400 mm and 1000 mm above the floor (or 400 mm and 1200 mm above the floor for permanently wired appliances). | M (4.30a and b) |

## *2.10.13.8 Other considerations*

### 2.10.13.8.1 Lecture/conference facilities

Artificial lighting should be designed to:

| | |
|---|---|
| • give good colour rendering of all surfaces;<br>• be compatible with other electronic and radio-frequency installations; | M (4.9) |
| • be compatible with other electronic and radio-frequency installations;<br>• give good colour rendering of all surfaces; | M (4.12.1) |
| • be compatible with other electronic and radio-frequency installations; | M (4.36f) |
| • be compatible with other electronic and radio-frequency installations;<br>• give good colour rendering of all surfaces. | M (4.34) |

### 2.10.13.8.2 Swimming pools and saunas

| | |
|---|---|
| Swimming pools and saunas are subject to special requirements specified in Part 7 of BS 7671:2008. | P AppA 1 |

### 2.10.13.8.3 Cellars or basements

| | |
|---|---|
| Liquid petroleum gas (LPG) storage vessels and LPG-fired appliances fitted with automatic ignition devices or pilot lights must not be installed in cellars or basements. | J (3.5i) |

# 3

# Earthing

From an electrical point of view, the world is effectively a huge conductor at zero potential and is used as a reference point which is called *Earth* (in the UK) or *ground* (in the USA). People and animals are normally in contact with the Earth, and so, if another part, which is open to touch, becomes charged at a different voltage from Earth, a shock hazard will exist.

**Figure 3.1** Typical lightning strike. (Courtesy StingRay.)

 One lightning bolt has enough electricity to service 200,000 homes!!!

This chapter reminds the reader about the different types of earthing systems and earthing arrangements. It then lists the main requirements from the various Approved Documents and Regulations for safety protection (direct and indirect contact), protective conductors and protective equipment, before briefly touching on the test requirements for earthing.

 Similar to other chapters, it should be noted that these lists of requirements are **only** the author's impression of the most important aspects of the Wiring Regulations, and electricians should **always** consult the latest BS 7671 to satisfy compliance.

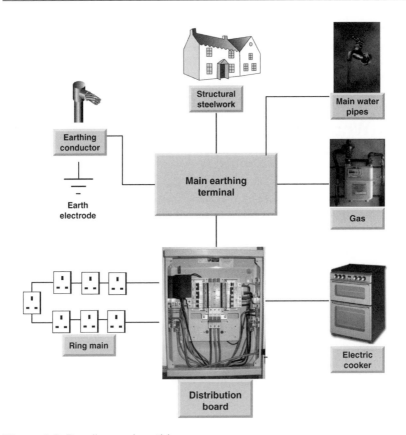

**Figure 3.2** Bonding and earthing.

 Readers are reminded that BS 7671:2008 included some new requirements for bonding and earthing in order to maintain technical alignment with CENELEC harmonization documents. Earthing requirements, therefore, now include the following:

- Protection of low-voltage installations against temporary overvoltages due to Earth faults in high- and low-voltage systems.

New appendices concerning earthing were also included in BS 7671:2008. These included:

- Measurement of Earth fault loop impedance (consideration of the increase in the resistance of conductors with increase of temperature);
- Methods for measuring the insulation resistance/impedance of floors and walls to Earth or to the protective conductor system.

It has now been agreed that the use of a gas, water or other metal service pipe as a means of earthing for an electrical installation is **not** permitted – although this does not rule out, however, equipotential bonding conductors being connected to these pipes.

## 3.1  What is Earth?

In electrical terms, 'Earth' is defined as:

> *"The conductive mass of the Earth, whose electric potential at any point is conventionally taken as zero."*

From an astrological and geophysical point of view, on the other hand:

> *"Earth (also known as The Earth, Terra, and – mostly in the 19th century – Tellus) is the third planet outward from the Sun. It is the largest of the solar system's terrestrial planets and the only planetary body that modern science confirms as harbouring life. The planet formed around 4.57 billion $(4.57 \times 10^9)$ years ago and 'shortly' thereafter (i.e. about 4.533 billion years ago – to be precise!) acquired its single natural satellite, the Moon. Its astronomical symbol consists of a circled cross, representing a meridian and the equator."*

## 3.2  What is meant by 'earthing' and how is it used?

Definition:

Earthing. *Connection of the exposed-conductive-parts of an installation to the main earthing terminal of that installation.*

*Earthing* is a process that is used to connect all of the parts that could become charged, to the general mass of Earth, and in so doing provide a path for fault currents that will hold these parts as close as possible to Earth (i.e. zero) potential. In doing so, this will prevent a potential difference occurring between Earth and earthed parts, as well as letting the flow of fault current to operate the protective systems.

An *earthing system*, on the other hand, defines the electrical potential of the conductors relative to that of the Earth's conductive surface. It should be noted that the choice of earthing system has implications for the safety and electromagnetic compatibility of the power supply.

An *Earth electrode* is that part of the system that is directly in contact with the Earth, and this can be just a metal (usually copper) rod or stake driven into the Earth or a connection to a buried metal service, pipe or a complex system of buried rods and wires.

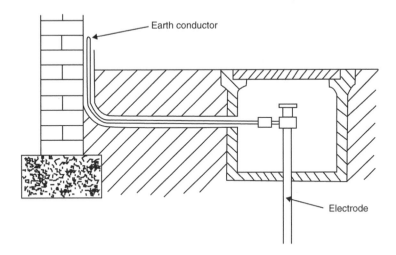

**Figure 3.3** Earth conductor and electrode.

The resistance of the electrode-to-Earth connection will determine its quality and this can be improved by:

- increasing the surface area of the electrode that is in contact with the Earth;
- increasing the depth to which the electrode is driven;
- using several connected ground rods;
- increasing the moisture of the soil;
- improving the conductive mineral content of the soil; and
- increasing the land area covered by the ground system.

A protective Earth (PE) connection ensures that all exposed-conductive surfaces are at the same electrical potential as the surface of the Earth, and thus avoids the risk of an electric shock if a person, or an animal, touches a piece of equipment (or device) in which an insulation fault has occurred. PE also ensures that, if an insulation fault occurs, a high fault current will flow, which will trigger an overcurrent protection device (e.g. a fuse), which will disconnect the power supply.

A functional Earth (FE) connection, as well as providing protection against electric shock, can also carry a current during the normal operation of a device – a facility that is often required by devices such as surge suppression and electromagnetic-compatibility filters, some types of antennas as well as a number of measuring instruments.

In a mains (i.e. a.c. power) wiring installation, the 'ground' wire is (directly or indirectly) connected to one or more Earth electrodes and carries currents away under fault conditions. These Earth electrodes may be located locally or in the supplier's network some distance away, and the ground wire is also usually bonded to pipework so as to keep it at the same potential as the electrical ground during a fault.

## 3.3 Conductor arrangement and system earthing

Within BS 7671:2008 (Incorporating Amendment No. 1) the following current-carrying conductors are considered (Figure 3.4).

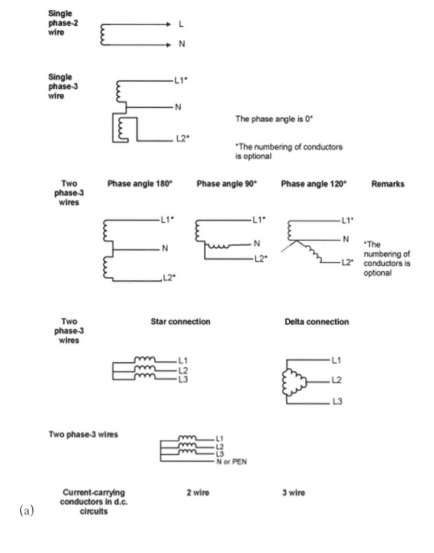

(a)

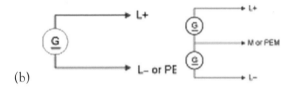

(b)

**Figure 3.4** Conductor arrangements and system earthing.

## 3.4 Current-carrying conductors

The standard method of attaching the electrical supply system to Earth is to make a direct connection between the two at the supply transformer so that the neutral conductor (often the star point of a three-phase supply – see Figure 3.5) is connected to Earth using an Earth electrode or the metal sheath and/or armouring of a buried cable.

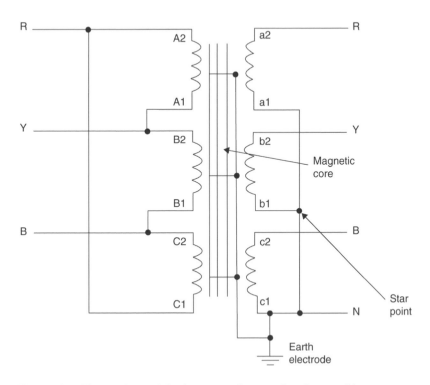

**Figure 3.5** Three-phase delta/star transformer showing earthing arrangements.

**Note:** Lightning conductor systems must be bonded to the installation Earth with a conductor that is not larger (i.e. in cross-sectional area) than that of the earthing conductor.

## 3.5 Advantages of earthing

The main advantage to earthing is that the whole electrical system is tied to the potential of the general mass of Earth and cannot *float* at another potential. By connecting Earth to metalwork (that is not intended to carry current), a path is provided for fault current, which can be detected by a protective conductor and, if necessary, broken. The path for this fault current is shown in Figure 3.6.

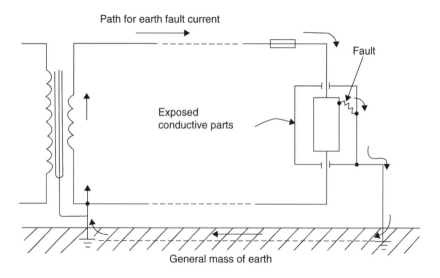

**Figure 3.6** Path for Earth fault current (shown by arrows).

On the other hand, the main disadvantage of earthing is primarily the cost of having to provide protective conductors and Earth electrodes, etc.!

 **Note:** Until the mid-1900s, power outlets generally lacked protective Earth terminals.

Earthing arrangements may be used jointly or separately for protective and functional purposes, according to the requirements of the installation. They should, however, ensure that:

• they are sufficiently robust (or have additional mechanical protection) to external influences;

- the impedance from the consumer's main earthing terminal to the earthed point of the supply meets the protective and functional requirements of the installation;
- Earth fault currents and protective conductor currents that may occur are carried without danger.

**Note:** This particularly applies with respect to thermal, thermomechanical and electromechanical stresses.

If a number of installations have separate earthing arrangements, then any protective conductor that is common to one of these installations:

- shall either be capable of carrying the maximum fault current likely to flow through them; or
- Earth one installation and be insulated from the earthing arrangements of other installation(s).

Precautions should be taken against possible damage to other metallic parts through electrolysis, and, if the protective conductor forms part of a cable, then this shall only be earthed in the installation containing the associated protective device.

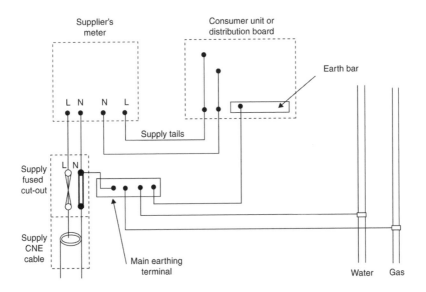

**Figure 3.7** Domestic earthing arrangement Earth electrodes.

## 3.6 What types of earthing system are there?

In the Regulations, an electrical system is defined as consisting *"of a single source of electrical energy and an installation"*, and the type of system depends on the link between the source and the exposed-conductive-parts of the installation, to Earth.

 **Note:** In this context, an *exposed-conductive-part* means a conductive-part of an equipment that can be touched and which is not (i.e. currently) a live part, but which *may* become live under fault conditions.

BS 7671:2010 lists two types of main earthing systems. The first concerns single-sourced three-phase systems, while the second deals with IT, multiple-source, d.c. and other systems. The three most common three-phase systems (i.e. TN-S, TN-C-S and TT) are shown in Figure 3.8.

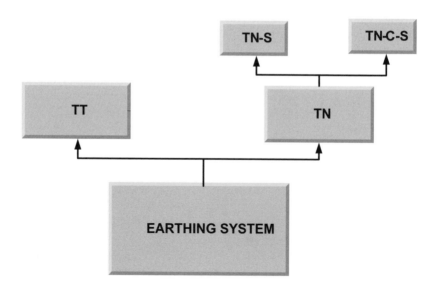

**Figure 3.8** Earthing systems.

 **Note:** T = Earth (from the French word *terre*), N = neutral, S = separate, C = combined.

### 3.6.1 System classification

In order to identify the different systems, a unique four-letter code is used, whereby:

The **first letter** indicates the type of supply earthing, so that:

T  indicates that one or more points of the supply are directly earthed (e.g. the earthed neutral at the transformer);
I  indicates either that the supply system is not earthed (at all) or that the earthing includes a deliberately inserted impedance, the purpose of which is to limit fault current.

The **second letter** indicates the earthing arrangement in the installation, so that:

T  indicates that all exposed-conductive metalwork is connected directly to Earth;
N  indicates that all exposed-conductive metalwork is connected directly to an earthed supply conductor provided by the electricity supply company.

The **third and fourth** letters indicate the arrangement of the earthed supply conductor system, so that:

S  ensures that neutral and Earth conductor systems are quite separate; and
C  ensures that neutral and Earth are combined into a single conductor.

### 3.6.2 TN systems

In a TN system:

- the integrity of the earthing of the installation depends on a reliable and effective connection of the PEN (combined protective and neutral conductors) or the PE conductors to Earth;
- one or more points in the generator or transformer is connected to Earth (usually the star point in a three-phase system); the body of the electrical device is then connected to Earth via this Earth connection at the transformer, and exposed-conductive-parts of the installation are then connected to that point by protective conductors.

The conductor that connects together the exposed metallic parts of the consumer installation is called the *protective Earth* (PE), while the conductor that connects to the start point in a three-phase system (or which carries the return current in a single-phase system) is called the *neutral* (N).

### 3.6.3 TN-S system

TN-S systems have separate protective Earth (PE) and neutral (N) conductors from the transformer to the consuming device, and these conductors remain separated throughout the system (Figure 3.10).

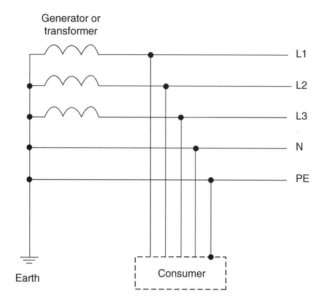

**Figure 3.9** TN system.

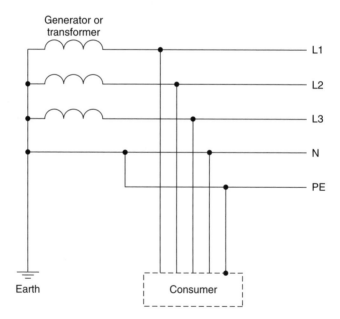

**Figure 3.10** TN-S system.

The TN-S is the most common earthing system in the UK, and one where the electricity supply company provides an Earth terminal at the incoming mains position.

This Earth terminal is then connected by the supply PE back to the star point (neutral) of the secondary winding of the supply transformer, which is also connected, at that point, to an Earth electrode.

The Earth conductor is usually the armour and sheath (if applicable) of the underground supply cable.

In TN-S systems a residual current device (RCD) can be used as an additional protection.

In the TN-S system (Figure 3.11) the neutral of the source of energy is connected with Earth at one point only, at (or as near as is reasonably practicable) the source, and the consumer's earthing terminal is connected to the metallic sheath or armour of the distributor's service cable into the premises.

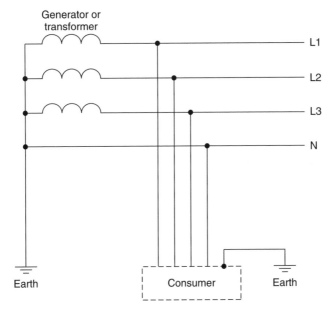

**Figure 3.11** Typical supplier's installation of a TN-S System. (Based on a diagram by Pathos Electrician – Cyprus.)

### 3.6.3.1 Electromagnetic compatibility

One of the advantages of a TN-S system concerns electromagnetic compatibility, whereby the consumer has a low-noise connection to Earth and does not suffer from the voltage that appears on the neutral conductor as a result of the return currents and the impedance of that conductor. This is of particular importance with some types of telecommunication and measurement equipment.

TN-S networks also save costs by having a fairly low-impedance Earth connection near each consumer.

### 3.6.4 TN-C-S system

A TN-C-S system is one that uses a combined PEN conductor between the transformer and the building distribution point substation and the entry point into the building, and then splits into separate PE and N lines within the building (Figure 3.12) to fixed indoor wiring and flexible power cords.

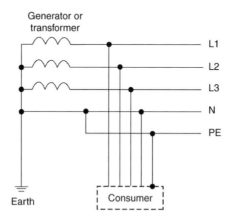

**Figure 3.12** TN-C-S system.

**Note:** In the UK, this system is also known as *protective multiple earthing* (PME), as it connects the combined neutral and Earth to real Earth at many locations – and thereby reduces the risk of broken neutrals.

Although TN-C networks save the cost of an additional conductor to separate N and PE connections, special cable types and lots of connections to Earth are required.

In the TN-C-S system, shown in Figures 3.12 and 3.13, the supply neutral conductor of a distribution main is connected with Earth at source and at intervals along its run. This is usually referred to as 'protective multiple earthing' (PME). With this arrangement the distributor's neutral conductor is also used to return Earth fault currents arising in the consumer's installation safely to the source. To achieve this, the distributor will provide a consumer's earthing terminal, which is linked to the incoming neutral conductor.

Any connection between the combined neutral and Earth core and the body of the Earth could end up carrying significant current under normal conditions.

The use of a TN-C-S system is **not** recommended for locations such as petrol stations etc. where there is a combination of lots of buried metalwork and explosive gasses.

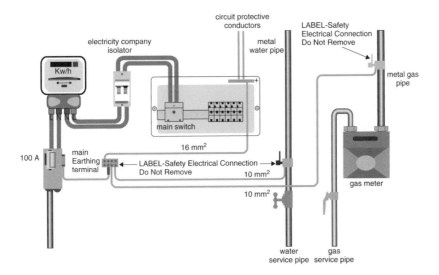

**Figure 3.13** Typical supplier's installation of a TN-C-S system. (Based on a diagram by Pathos Electrician – Cyprus.)

 Owing to the possibility of a lost neutral, the use of TN-C-S supplies is **banned** for caravans and boats in the UK, and it is often recommended to make outdoor wiring TT with a separate rod.

### 3.6.5 TT system

A TT system is one that has one point of the energy source directly earthed, and the exposed-conductive-parts of the consumer's installation are provided with a local connection to Earth, independent of any Earth connection at the generator (Figure 3.14).

This type of installation is usually found in rural locations where the system is not provided with an Earth terminal by the electricity supply company and the installation is fed from an overhead supply. Neutral and Earth (protective) conductors must be kept quite separate throughout the installation, and the final Earth terminal must be connected to an Earth electrode – via an earthing conductor.

TT systems (similar to TN-S systems) have a low-noise connection to Earth, which is particularly important with some types of telecommunication and measurement equipment.

The TT system, shown in Figures 3.14 and 3.15, has the neutral of the source of energy connected as for TN-S, but **no** facility is provided by the distributor

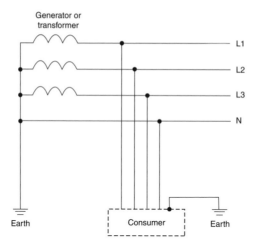

**Figure 3.14** TT system.

for the consumer's earthing. This usually occurs when the distributor cannot guarantee the Earth connection back to the source (for example, a low voltage overhead supply, where there is the likelihood of the Earth wire either somehow becoming disconnected or even stolen).

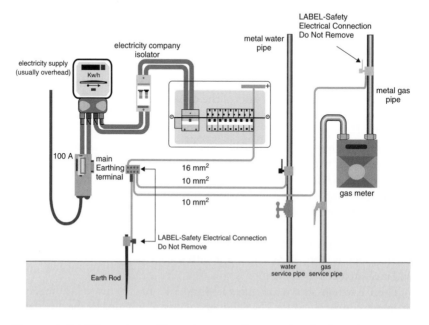

**Figure 3.15** TT system. (Based on a diagram by Pathos Electrician – Cyprus.)

With a TT system the consumer must provide his own connection to Earth, i.e. by installing a suitable Earth electrode local to the installation.

### 3.6.6 IT system

An IT system is one that has no direct connection between live parts and Earth, and where the exposed-conductive-parts of the electrical installation are earthed (Figure 3.16).

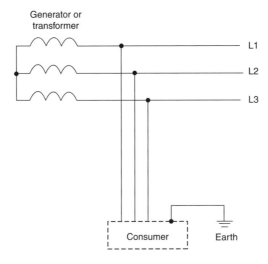

**Figure 3.16** IT system.

An IT system is similar to a TT system, except that the supply earthing in an IT system can either be from an unearthed supply or one which (although not totally earthed) is connected to Earth through a current-limiting impedance.

This lack of Earth will usually mean that normal protective methods cannot be used, and for this reason IT systems are not normally allowed in the UK public supply system – except for hospitals and other medical locations where such systems are recommended for use with circuits supplying medical equipment intended for life-support of patients.

## 3.7 Earthing points

It has been proved that the resistance area around an Earth electrode depends on the size of the electrode and the type of soil – and that this electrode resistance is particularly important with regard to the voltage at the surface of the ground. For example, as shown in Figure 3.17, for a 2 m rod with its

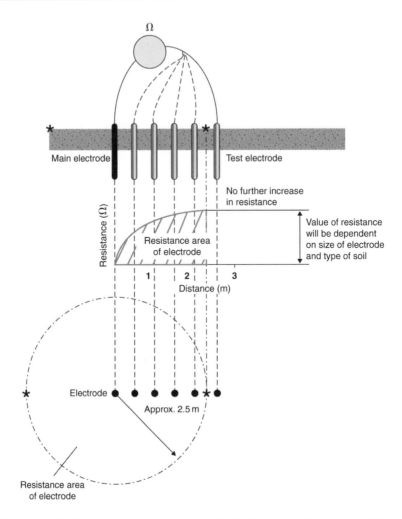

**Figure 3.17** The resistance area of an Earth electrode. (Courtesy of Brian Scaddan.)

top at ground level, approximately 80–90% of the voltage appearing at the electrode under fault conditions will be dropped in the first 2.5–3 m from the electrode.

This can be particularly dangerous where livestock is concerned. For example, in some circumstances, a grazing cow might have its forelegs inside the resistive area, while its hind legs are outside of the area. Bearing in mind that a potential difference of 25 V can be lethal, measures have to be taken to reduce this risk. One method is to house the Earth electrode in a pit that is below ground level (Figure 3.18).

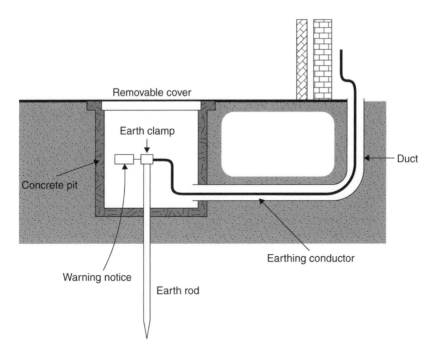

**Figure 3.18** An Earth electrode protected by a pit below ground level. (Courtesy of Brian Scaddan.)

# 3.8 Main earthing terminals

The main earthing terminal acts as the single reference point and can be composed of a bar, a plate or even a copper internal 'ring' conductor. This is often connected directly to an effective Earth electrode, and this connection must be of copper because of the corrosion risk if aluminium or copper-clad aluminium were used.

The main earthing terminal shall be connected to Earth as shown below and in Figure 3.19.

| System | Main earthing terminal |
| --- | --- |
| TT or IT | Connected via an earthing conductor to an Earth electrode |
| TN-S | Connected to the earthed point of the energy source |
| TN-C-S | Connected (by the distributor) to the neutral of the energy source |

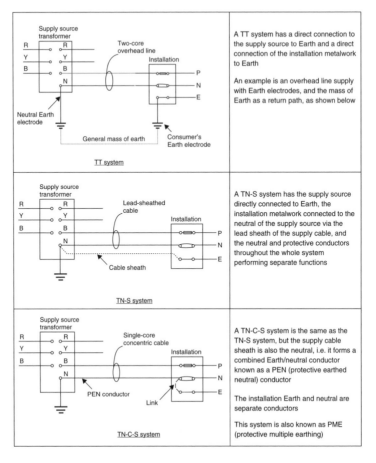

**Figure 3.19** Earth provision.

 The Earth electrode should be positioned as close as possible to the main earthing terminal.

## 3.9 Earth electrodes

An Earth electrode is a conductor, or group of conductors, that connects the main earthing terminal of an installation to an Earth electrode or to other means of earthing. Earth electrodes shall be designed and constructed so that they can:

- withstand damage;
- take into account a possible increase in resistance due to corrosion.

The following types of Earth electrode may be used for electrical installations:

- Earth rods or pipes;
- Earth tapes or wires;
- Earth plates;
- lead sheaths and other metal cable covers;
- other suitable underground metalwork;
- structural metalwork embedded in foundations;
- welded metal reinforced concrete (except pre-stressed concrete) embedded in the Earth.

## 3.10 Earthing conductors

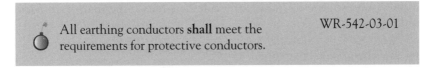

All earthing conductors **shall** meet the requirements for protective conductors.                WR-542-03-01

The earthing conductor (Figure 3.20) is an important part of the Earth fault loop impedance as it is a protective conductor that connects the main

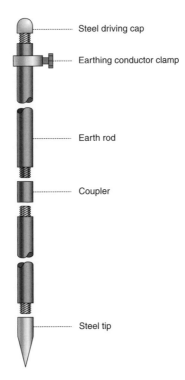

Steel driving cap

Earthing conductor clamp

Earth rod

Coupler

Steel tip

**Figure 3.20** An example of an earthing conductor. (Courtesy of Brian Scaddan.)

earthing terminal of an installation to an Earth electrode, or to some other means of earthing.

Rods should be driven into virgin (as opposed to backfilled or previously disturbed) ground so as to make an effective contact with the surrounding material.

## 3.11 Earth fault loop impedance

Earth fault loop impedance ($Z_s$) is the impedance of the intended path of an Earth fault current (i.e. the Earth fault loop) starting and ending at the point of the fault to Earth.

As shown in Figure 3.21, the Earth fault loop starts at the point of the fault, and comprises:

- the circuit protective conductor (CPC);
- the consumer's main earthing terminal (MET) and earthing conductor;
- (for TT and IT systems) the Earth return path, or (for TN systems) the metallic return path;
- the path through the earthed neutral point of the transformer;
- the transformer winding;
- the line (phase) conductor from the transformer to the point of fault.

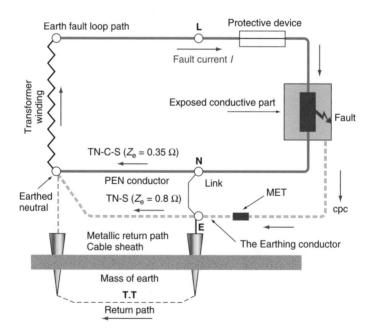

**Figure 3.21** Example of a TT Earth fault loop impendence path. (Courtesy of Pathos Electrician – Cyprus.)

It is recommended that radial wiring patterns are used to avoid 'Earth loops' that may cause electromagnetic interference – particularly in medical locations.

## 3.12 Requirements from the Regulations

### 3.12.1 Additions and alterations to an installation

No addition or alteration, temporary or permanent, shall      WR-132.6
be made to an existing installation unless it has been
ascertained that the earthing and bonding arrangements
used as a protective measure for the safety of the addition
or alteration are adequate.

### 3.12.2 Automatic disconnection of supply

Automatic disconnection of supply is a protective measure, in which fault protection is provided by protective earthing, protective equipotential bonding and automatic disconnection in the case of a fault.

### 3.12.3 Autotransformers and step-up transformers

A step-up autotransformer **shall not** be connected
to an IT system.                                         WR-555.1.2

#### 3.12.3.1 Transformers in medical locations

Transformers **shall** be installed in close proximity to a medical location and with the following additional requirements:

- The leakage current of the output winding to      WR-710.512.1.1
  Earth and the leakage current of the enclosure
  shall not exceed 0.5 mA.
- At least one single-phase transformer per room
  or functional group of rooms **shall** be used to
  form the IT systems for mobile and fixed
  equipment, and the rated output **shall** be not
  less than 0.5 kVA and **shall not** exceed
  10 kVA.

Where several transformers are needed to supply equipment in one room, they **shall not** be connected in parallel.

- If the supply of three-phase loads via an IT system is also required, a separate three-phase transformer **shall** be provided for this purpose.

Capacitors **shall not** be used in transformers for medical IT systems.

### 3.12.4  Cables

A cable passing through a joist within a floor or ceiling construction or through a ceiling support (e.g. under floorboards), shall:    WR-522.6.100

- incorporate an earthed metallic covering; or
- be enclosed in an earthed conduit; or
- be enclosed in earthed trunking or ducting; or
- be mechanically protected against damage sufficient to prevent penetration of the cable by nails, screws etc.; or
- be at least 50 mm measured vertically from the top, or bottom as appropriate, of the joist or batten.

A cable concealed in a wall or partition at a depth of less than 50 mm from a surface of the wall or partition shall:    WR-522.6.101

- incorporate an earthed metallic covering; or
- be enclosed in earthed conduit; or
- be enclosed in earthed trunking or ducting complying; or
- be mechanically protected against damage sufficient to prevent penetration of the cable by nails, screws etc.; or
- be installed in a zone within 150 mm from the top of the wall or partition, or within 150 mm of an angle formed by two adjoining walls or partitions.

Where the installation is not intended to be under the supervision of a skilled or instructed person, consideration shall be given to providing additional protection by means of an RCD.

If the cables of an installation not intended to be under the supervision of a skilled or instructed person are concealed in a wall or partition (the internal construction of which includes metallic parts, other than metallic fixings such as nails, screws and the like) then it shall:                                        WR-522.6.103

- incorporate an earthed metallic covering; or
- be enclosed in an earthed conduit; or
- be enclosed in earthed trunking or ducting; or
- be mechanically protected sufficiently to avoid damage to the cable during construction of the wall or partition and during installation of the cable; or
- be provided with additional protection by means of an RCD.

A cable buried in the ground (that is not installed in a conduit or duct) shall incorporate an earthed armour or metal sheath or both, suitable for use as a protective conductor.                             WR-522.8.10

**Note:** This degree of protection for electrical equipment shall be maintained after installation of the cables and conductors.

### 3.12.5 Earth electrodes

The following types of Earth electrode are recognized as being suitable for the purposes of these Regulations:

- Earth rods or pipes;
- Earth tapes or wires;
- Earth plates;
- underground structural metalwork embedded in foundations;
- welded metal reinforcement of concrete (except pre-stressed concrete) embedded in the ground;
- lead sheaths and other metal coverings of cables;
- other suitable underground metalwork.

**Note:** Further information on Earth electrodes can be found in BS 7430.

The type and embedded depth of an Earth electrode shall ensure that soil drying and freezing will not increase its resistance above the required level.                WR-542.4

The use, as an Earth electrode, of the lead sheath or other metal covering of a cable shall be subject to all of the following conditions:      WR-542.2.5

- adequate precautions have been taken to prevent excessive deterioration by corrosion;
- the sheath or covering shall have an effective contact with Earth;
- the consent of the owner of the cable has been obtained;
- arrangements exist for the owner of the electrical installation to be warned of any proposed change to the cable that might affect its suitability as an Earth electrode.

### 3.12.6  Earth fault current

The choice and type of wiring system and the method of installation shall include consideration of the electromechanical stresses likely to occur due to short-circuit and Earth fault currents.      WR-132.7

The characteristics of protective equipment shall be determined with respect to their function, including protection against the effects of Earth fault current.      WR-132.8

### 3.12.7  Earth faults

This particular section provides requirements for the safety of a low-voltage installation in the event of:

- a fault between the high-voltage system and Earth in the transformer substation that supplies the low-voltage installation;
- the loss of the supply neutral in the low-voltage system;
- a short-circuit between a line conductor and neutral in the low-voltage installation;
- an accidental earthing of a line conductor of a low-voltage IT system.

An earthing arrangement may be considered electrically independent of another earthing arrangement if a rise in potential with respect to Earth in one      WR-442.1.2

earthing arrangement does not cause an unacceptable rise in potential with respect to Earth in the
other earthing arrangement.

The magnitude and duration of the power frequency        WR-442.2.2
stress voltages on the low-voltage equipment in the
low-voltage installation due to an Earth fault in the
high-voltage system shall not exceed the requirements given in Table 3.1.

In the event of a first fault to Earth, a total loss        WR-710.512.1.2
of supply in medical Group 2 locations shall be
prevented.

**Table 3.1** Permissible power frequency stress voltage

| Duration of the Earth fault in the high-voltage system, $t$ | Permissible power frequency stress voltage on equipment in low-voltage installations, $U$ |
|---|---|
| >5 s | $U_o + 250\,V$ |
| <5 s | $U_o + 1200\,V$ |

## 3.12.8 Earthing arrangements

For a TN-S system, means shall be provided for        WR-542.1.2.1
the main earthing terminal of the installation to
be connected to the earthed point of the source of
energy.

> Part of the connection may be formed by
> the distributor's lines and equipment.

For a TN-C-S system, where protective multiple        WR-542.1.2.2
earthing is provided, the main earthing terminal of
the installation shall be connected by the distributor to the neutral of the source of energy.

For a TT or IT system, the main earthing terminal        WR-542.1.2.3
shall be connected via an earthing conductor to an
Earth electrode.

Where the supply to an installation is at high volt-        WR-542.1.2.4
age, protection against faults between the high-
voltage supply and Earth shall be provided.

The earthing arrangements shall be such that:          WR-542.1.3.1

- the value of impedance from the consumer's main earthing terminal to the earthed point of the supply for TN systems, or to Earth for TT and IT systems, is in accordance with the protective and functional requirements of the installation, and considered to be continuously effective;
- Earth fault currents and protective conductor currents that may occur are carried without danger, particularly from thermal, thermomechanical and electromechanical stresses; and
- they are adequately robust or have additional mechanical protection appropriate to the assessed conditions of external influence.

Precautions shall be taken against the risk of damage to other metallic parts through electrolysis.          WR-542.1.3.2

Where a number of installations have separate earthing arrangements, any protective conductors common to any of these installations shall either:          WR-542.1.3.3

- be capable of carrying the maximum fault current likely to flow through them; or
- be earthed within one installation only and insulated from the earthing arrangements of any other installation.

Where earthing for combined protective and functional purposes is required, the requirements for protective measures shall take precedence.          WR-543.5.1

Equipment having a protective conductor current exceeding 10 mA shall be connected to the supply:          WR-543.7.1.102

- permanently via the wiring of the installation;
- via a flexible cable with a plug and socket-outlet; or
- via a protective conductor with an Earth monitoring system.

When this is a *permanent* connection, it may be by means of a flexible cable.

The wiring of every final and distribution circuit     WR-543.7.1.103
intended to supply one or more items of equipment
with a total protective conductor current likely to
exceed 10 mA shall be equipped with a high-integrity protective connection complying with one or
more of the following:

- a single protective conductor with a cross-
  sectional area greater than 10 mm$^2$;
- a single copper protective conductor having
  a cross-sectional area of not less than
  4 mm$^2$;
- two individual protective conductors;
- an Earth monitoring system which, in the
  event of a continuity fault occurring in the
  protective conductor, automatically disconnects the supply to the equipment;
- connection of the equipment to the supply
  by means of a double-wound transformer
  or equivalent unit, such as a motor-
  alternator set.

## 3.12.9  Earthing conductors

The connection of an earthing conductor to the Earth     WR-542.3.2
electrode shall be:

- soundly made;
- electrically and mechanically satisfactory;
- suitably labelled;
- suitably protected against corrosion.

Where buried in the ground, the earthing conduc-     WR-542.3.1
tor shall have a cross-sectional area not less than that
stated in Table 3.2.

Where PME (protective multiple earthing) exists, the     WR-542.3.1
requirements for main equipotential bonding
conductors (i.e. with respect to the cross-sectional
area of a main equipotential bonding conductor) shall
be met.

**Table 3.2 Minimum cross-sectional area of a buried earthing conductor**

| | Protected against mechanical damage | Not protected against mechanical damage |
|---|---|---|
| Protected against corrosion by a sheath | 2.5 mm² copper<br>10 mm² steel | 16 mm² copper<br>16 mm² coated steel |
| Not protected against corrosion | 25 mm² copper<br>50 mm² steel | |

The thickness of tape or strip conductors shall be capable of withstanding mechanical damage and corrosion (see BS 7430).

### 3.12.10 Electrical separation

Electrical separation is a protective measure in which:

*   basic protection is provided by basic insulation of live parts or by barriers or enclosures; and
*   fault protection is provided by simple separation of the separated circuit from other circuits and from Earth.

| | |
|---|---|
| Electrical separation may only supply one item of current-using equipment from one unearthed source with simple separation. | WR-413.1.2 |

### 3.12.11 Electrical services

| | |
|---|---|
| A voltage Band I circuit shall **not** be contained in the same wiring system as a Band II circuit unless (for a multi-core cable) the cores of the Band I circuit are separated from the cores of the Band II circuit by an earthed metal screen of equivalent current-carrying capacity to that of the largest core of a Band II circuit. | WR-528.1 |

### 3.12.12 Emergency switching

Emergency switching may be emergency switching **ON** or emergency switching **OFF**.

| | |
|---|---|
| Unless the neutral conductor can be regarded as being reliably connected to Earth in a TN-S or TN-C-S system, the neutral conductor need **not** be isolated or switched. | WR-537.4.1.2 |

### 3.12.13 Earthing terminals or bars

| | |
|---|---|
| In every installation, a main earthing terminal shall be provided to connect the following to that earthing conductor: | WR-542.4.1 |

- the circuit protective conductors;
- the protective bonding conductors;
- functional earthing conductors (if required);
- lightning protection system bonding conductor, if any.

| | |
|---|---|
| To enable the resistance of the earthing arrangements to be measured, the earthing conductor needs to be capable of being easily disconnected. | WR-542.4.2 |
| Joints in an earthing conductor shall be capable of being disconnected **only** by means of a tool. | WR-542.4.2 |
| The neutral (star) point of the secondary windings of three-phase transformers and generators (or the midpoint of the secondary windings of a single-phase transformer or generator) shall be connected to Earth. | WR 411.8.4.2 |
| The main protective bonding conductors of each installation shall connect extraneous-conductive-parts to the main earthing terminal, including the following: | WR-411.3.1.2 |

- central heating and air-conditioning systems;
- exposed metallic structural parts of the building;
- gas installation pipes;
- water installation pipes;
- other installation pipework and ducting.

 Where an installation serves more than one building, the above requirement shall be applied to **each** building.

| | |
|---|---|
| In a TN system, exposed-conductive-parts of the installation shall be connected by a protective conductor to the main earthing terminal of the installation (which shall, in turn, be connected to the earthed point of the power supply system). | WR-411.4.2 |
| In a TT system, exposed-conductive-parts of the installation shall be protected by a single protective device connected (via the main earthing terminal) to a common Earth electrode. | WR-411.5.1 |

> For a TN-S system, the main earthing terminal of the installation shall be connected to the earthed point of the source of energy.
>
> WR-542.1.2.1

 Part of the connection may be formed by the distributor's lines and equipment.

> For a TN-C-S system, where protective multiple earthing is provided, the main earthing terminal of the installation shall be connected, by the distributor, to the neutral of the source of energy.
>
> WR-542.1.2.2
>
> For a TT or IT system, the main earthing terminal shall be connected via an earthing conductor to an Earth electrode.
>
> WR-542.1.2.3

### 3.12.14 Fault protection

> All exposed-conductive-parts of the reduced low-voltage system shall be connected to Earth.
>
> WR-411.8.3
>
> The Earth fault loop impedance at every point of utilization, including socket-outlets, shall be such that the disconnection time does not exceed 5 s.
>
> WR-411.8.3
>
> Where a circuit-breaker is used, the maximum value of the Earth fault loop impedance ($Z_s$) shall be determined by the formula $Z_s \times I_a\ U_o$.
>
> WR-411.8.3

Where:

$Z_s$ is the impedance in ohms ($\Omega$) of the fault loop BS;

$I_a$ is the current in amperes (A) causing the automatic operation of the disconnecting device within the time specified in Table 3.3;

$U_o$ is the nominal a.c. rms or d.c. line voltage to Earth in volts (V).

**Table 3.3 Maximum disconnection times**

| | Maximum disconnection time (s) | | | | | | | |
|---|---|---|---|---|---|---|---|---|
| System | $50\,V < U_o\ 120\,V$ | | $120\,V < U_o\ 230\,V$ | | $230\,V < U_o\ 5.400\,V$ | | $U_o > 400\,V$ | |
| | a.c. | d.c. | a.c. | d.c. | a.c. | d.c. | a.c. | d.c. |
| TN | 0.8 | Not required for protection against electric shock | 0.4 | 5 | 0.2 | 0.4 | 0.1 | 0.1 |
| TT | 0.3 | | 0.2 | 0.4 | 0.07 | 0.2 | 0.04 | 0.1 |

Where a fuse is used, the maximum values of Earth fault loop impedance ($Z_s$) for a 5 s disconnection time and $U_o$ of 55 V (single phase) and 63.5 V (three phase) shall be in accordance with the values stated in Table 41.6 (page 61 of BS 7671:2008).

| | |
|---|---|
| Where fault protection is provided by an RCD, the product of the rated residual operating current ($I_a$) in amperes and the Earth fault loop impedance in ohms ($\Omega$) shall not exceed 50. | WR-411.8.3 |
| Fault protection may be omitted for unearthed street furniture supplied from an overhead line and inaccessible in normal use. | WR-410.3.9 |
| Live parts of the separated circuit **shall not** be connected at any point to another circuit or to Earth or to a protective conductor. | WR-413.3.3 |
| No exposed-conductive-part of the separated circuit shall be connected either to the protective conductor or to the exposed-conductive-parts of other circuits, or to Earth. | WR-413.3.6 |

### 3.12.14.1 Medical locations

In Group 1 and Group 2 medical locations, the following shall apply:

| | |
|---|---|
| • For TN, TT and IT systems, the voltage presented between simultaneously accessible exposed-conductive-parts and/or extraneous-conductive-parts shall not exceed 25 V a.c. or 60 V d.c.<br>• For TN and TT systems, the requirements listed in Table 3.4 **shall** apply. | WR-710.411.3.2.5 |

**Table 3.4 Maximum disconnection times**

| System | Maximum disconnection time (s) | | | | | | | | | |
|---|---|---|---|---|---|---|---|---|---|---|
| | $25V < U_o$ $50V$ | | $50V < U_o$ $120V$ | | $120V < U_o$ $5230V$ | | $230V < U_o$ $400V$ | | $1.10 > 400V$ | |
| | a.c. | d.c. | a.c. | d.c. | a.c. | d.c. | a.c. | d.c. | a.c. | d.c. |
| TN | 5 | 5 | 0.3 | 2 | 0.3 | 0.5 | 0.05 | 0.06 | 0.02 | 0.02 |
| TT | 5 | 5 | 0.15 | 0.2 | 0.05 | 0.1 | 0.02 | 0.06 | 0.02 | 0.02 |

**Note:** In TN systems, 25 V a.c. or 60 V d.c. may be met with protective equipotential bonding, by complying with the disconnection time in accordance with Table 3.4.

### 3.12.15 High leakage current

| | |
|---|---|
| Any insulation or insulating arrangement of extraneous-conductive-parts shall not pass a leakage current exceeding 1 mA in normal conditions of use. | WR-612.5.2 |
| In supply systems, a residual current monitor (RCM) may be installed to reduce the risk of the protective device operating in the event of excessive leakage current of the installation, or the connected appliances. | WR-538.4.1 |

 **Note:** An RCM permanently monitors any leakage current in the downstream installation or part of it.

| | |
|---|---|
| Where an RCM is used in an a.c. IT system, the use of a directionally discriminating RCM is recommended so as to avoid undesirable signalling of leakage current where high leakage capacitances are liable to exist downstream from the point of installation of the RCM. | WR-538.4 |

### 3.12.16 Isolation

Isolation is intended, for reasons of safety, to make a circuit dead by separating an installation or section from every source of electric energy.

| | |
|---|---|
| Every circuit shall be capable of being isolated from each of the live supply conductors. | WR-537.2.1.1 |

 **Note:** In a TN-S or TN-C-S system, it is not necessary to isolate or switch the neutral conductor where it is regarded as being reliably connected to Earth by suitably low impedance.

| | |
|---|---|
| Where an installation, item of equipment or enclosure contains live parts that are connected to more than one supply, a warning notice shall be placed so that any person who might inadvertently gain access to these live parts, **must** isolate those parts from the various supplies – unless an interlocking arrangement is provided to ensure that all the circuits concerned are isolated. | WR-537.2.1.3 |

**Note:** Where necessary, suitable means shall be provided for the discharge of stored electrical energy.

| | |
|---|---|
| Where an isolating device for a particular circuit is remote from the equipment to be isolated, the means of isolation should always be secured in the open position. | WR-537.2.1.5 |

Notes:
1.  If this means of isolation is a lock or removable handle, then the key or handle shall be non-interchangeable with any other used for a similar purpose within the premises.
2.  If a switch is provided for this purpose:

    •   it shall be capable of cutting off the full load current of the relevant part of the installation, but
    •   if used as a device for switching off for mechanical maintenance, the switch need not necessarily interrupt the neutral conductor.

### 3.12.17 Protective and neutral (PEN) conductors

PEN conductors shall not be used in medical locations and medical buildings downstream of the main distribution board.

| | |
|---|---|
| PEN conductors may only be used within an installation where the installation is supplied by a privately owned transformer or converter in such a way that there is no metallic connection (except for the earthing connection) with the distributor's network. | WR-543.4.1 and 543.4.2 |

The PEN conductor shall be connected to     WR-543.4.2
the terminals or bar intended for the protective earthing conductor and the neutral
conductor.

 **Note:** In Great Britain, Regulation 8(4) of the *Electricity Safety, Quality and Continuity Regulations 2002* prohibits the use of PEN conductors in consumers' installations.

### 3.12.18 Protective bonding conductors (PME)

Except where protective multiple earthing (PME)     WR-544.1.1
conditions apply, a main protective bonding conductor
shall have a cross-sectional area not less than half the
cross-sectional area required for the earthing conductor
of the installation and not less than 6 mm².

Where supplementary bonding is to be applied to a fixed     WR-544.2.5
appliance (which is supplied via a short length of flexible
cord from an adjacent connection unit or other accessory,
incorporating a flex outlet) the circuit-protective conductor within the flexible cord shall be deemed to provide
the supplementary bonding connection to the exposed-
conductive-parts of the appliance, from the earthing terminal in the connection unit or other accessory.

### 3.12.19 Supplementary equipotential bonding

For Group 1 and Group 2 medical locations,     WR-710.415.2.1
supplementary equipotential bonding shall be
installed for the following parts which are
located in (or that may be moved into) the
*patient environment*:

• protective conductors;
• extraneous-conductive-parts;
• screening against electrical interference fields,
  if installed;
• connection to conductive floor grids, if
  installed;
• metal screens of isolating transformers, via the
  shortest route to the earthing conductor.

Supplementary equipotential bonding connection points for medical electrical equipment shall be provided in each medical location as follows:

WR-710.415.2.1

- Group 1: one per patient location;
- Group 2: one per medical IT socket-outlet.

Unless they are intended to be isolated from Earth, fixed conductive non-electrical patient supports, such as operating-theatre tables, physiotherapy couches and dental chairs, should be connected to the equipotential bonding conductor.

WR-710.415.2.1

The equipotential bonding busbar shall be located in or near the medical location, and all connections shall be accessible, labelled, clearly visible and be capable of being easily disconnected individually.

WR-710.415.2.3

 **Note:** It is recommended that radial wiring patterns are used to avoid 'Earth loops', which may cause electromagnetic interference.

In locations intended for livestock, supplementary bonding shall connect all exposed-conductive-parts and extraneous-conductive-parts that can be touched by livestock.

WR-705.415.2.1

### 3.12.20 Protective measures

A protective measure shall consist of:

- a combination of basic protection and an independent provision for fault protection, or
- an enhanced protective provision that provides both basic **and** fault protection.

 **Note:** An example of an enhanced protective measure is reinforced insulation.

Protective measures such as Earth-free local equipotential bonding shall **only** be applied where the installation is under the supervision of skilled or instructed persons, so that unauthorized changes cannot be made.

WR-410.3.6

Protection by extra-low voltage is a protective measure that consists of either separated extra-low voltage (SELV) or protective extra-low voltage (PELV), and these are considered to be protective measures in all situations.

| | |
|---|---|
| Protection by extra-low voltage provided by SELV requires basic insulation between the SELV system and Earth. | WR-414.1.1 |

### 3.12.21 Protective conductors

 A gas pipe, an oil pipe, flexible or pliable conduit, support wires or other flexible metallic parts, or constructional parts that are subject to mechanical stress in normal service, shall NOT be selected as a protective conductor.

| | |
|---|---|
| Where the protective conductor is formed by metal conduit, trunking, ducting or the metal sheath and/or armour of a cable, the earthing terminal of each accessory shall be connected by a separate protective conductor to an earthing terminal incorporated in the associated box or other enclosure. | WR-543.2.7 |
| Except where the circuit-protective conductor is formed by a metal covering or enclosure containing all of the conductors of the ring, the circuit-protective conductor of every ring final circuit shall also be run in the form of a ring – with both ends connected to the earthing terminal at the **origin** of the circuit. | WR-543.2.9 |
| A switching device **shall not** be inserted in a protective conductor unless:<br><br>• the switch has been inserted in the connection between the neutral point and the means of earthing, and<br>• that the switch is a linked switch arranged to disconnect and connect the earthing conductor for the appropriate source, at substantially the same time as the related live conductors. | WR-543.3.4 |
| Where electrical monitoring of earthing is used, no dedicated devices (e.g. operating sensors, coils) shall be connected in series with the protective conductor. | WR-543.3.5 |

## 3.12.22 Protective devices

The rated breaking capacity of a protective device **shall not** be less than the maximum prospective short-circuit or Earth fault current at the point at which the device is installed unless back-up protection is provided.

WR-536.1

In a TN-S or TN-C-S system, the neutral conductor need not be isolated or switched where it can be regarded as being reliably connected to Earth by suitably low impedance.

WR-537.1.2

**Note:** A lower breaking capacity is permitted if another protective device such as a back-up protective device) is installed on the supply side, provided that the characteristics of the device ensures that the energy let-through of the upstream device does not exceed that which can be withstood by the downstream device.

Where an installation is supplied from more than one source of energy (one of which requires a means of earthing independent of the means of earthing of other sources, and there is a need to ensure that no more than one means of earthing is applied at any time), a switch may be inserted in the connection between the neutral point and the means of earthing, **provided** that the switch is a linked switch arranged to disconnect and connect the earthing conductor for the appropriate source, at substantially the same time as the related live conductors.

WR-537.1.5

## 3.12.23 Protective switches

Switches, circuit-breakers (except where linked) or fuses shall be inserted in an earthed neutral conductor.

WR-132.14.2

Any linked switch or linked circuit-breaker that is used with an earthed neutral conductor shall be capable of breaking **all** of the related line conductors.

WR-132.14.2

### 3.12.24 Protective earthing

Exposed-conductive-parts shall be connected to a pro-     WR-411.3.1.1
tective conductor.

Simultaneously accessible exposed-conductive-parts     WR-411.3.1.1
shall be connected to the same earthing system, either
individually, in groups or collectively.

A circuit-protective conductor shall be run to (and     WR-411.3.1.1
terminated at) each point in the wiring and at each
accessory (except a lampholder having no exposed-
conductive-parts and suspended from such a point).

#### 3.12.24.1 Protective equipotential bonding – mobile and transportable units

An IT system can be provided by either:     WR-717.411.6.2

- an isolating transformer or a low-voltage generating set (provided that it is equipped with
  an insulation monitoring device); or
- an installation fault location system; or
- a transformer providing simple separation via:

  o   an RCD; and
  o   an Earth electrode installed that provides
      automatic disconnection of the supply
      in case of failure in the transformer
      (Figure 3.23).

The protective measures of:     WR-717.417
                                WR-717.418
- obstacles and placing out of reach, are **not**
  permitted;
- non-conducting location, is **not** permitted;
- Earth-free local equipotential bonding, is not
  recommended.

Where an alternative system for the automatic disconnection of supply is
available, an IT system **shall not** be used.

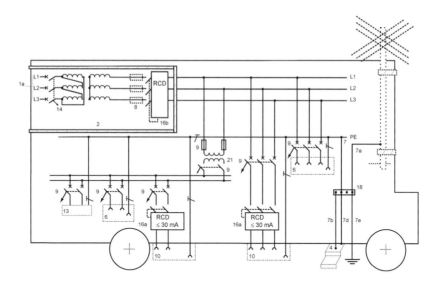

**Figure 3.22** Example of connection to a fixed installation with any type of earthing system by using an IT system without automatic disconnection in the event of a fault. (Courtesy of BSI.)

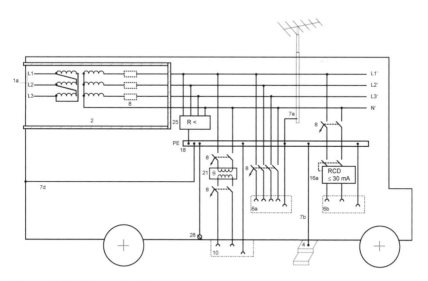

**Figure 3.23** Example of connection to a fixed installation with any type of earthing system by using a simple separation transformer and an internal IT system, with an Earth electrode. (Courtesy of BSI.)

### 3.12.25 Residual current devices (RCDs)

The rated residual operating current of the protective device shall comply with the requirements of the type of system earthing used. — WR-531.2.3

In a TN system, where the characteristics of a protective device (for a particular equipment in a certain part of the installation) does not satisfy the requirements, that part may be protected by an RCD. — WR-531.3.1

 **Note:** In these sorts of circumstances, the exposed-conductive-parts of that part of the installation shall be connected to the TN earthing system's protective conductor or to a separate Earth electrode that affords impedance appropriate to the operating current of the RCD.

 In this latter case the circuit shall be treated as if it were a TT system.

For a TN-S system where the neutral is not isolated, RCDs shall be positioned so as to avoid incorrect operation due to the existence of any parallel neutral–Earth path. — WR-551.6.2

In an IT system (where protection is provided by an RCD and disconnection following a first fault is not envisaged) the non-operating residual current of the device shall be at least equal to the current that circulates on the first fault to Earth. — WR2'-531.5.1

 In an IT system, an RCD may not operate unless one of the Earth faults is on a part of the system that is on the supply side of the device. If a medical IT system is used, additional protection by means of an RCD **shall not** be used.

### 3.12.25.1 Medical locations

Care shall be taken to ensure that simultaneous use of a number of items of equipment connected to the same circuit cannot cause unwanted tripping of the RCD. — WR-710.411.3.2.1

In Group 1 and Group 2 medical locations, where RCDs are required, **only** type A (according to BS EN 61008 and BS EN 61009) or type B (according to IEC 62423) shall be selected, depending on the possible fault current arising.

WR-710.411.3.2.1

 Type AC RCDs **shall not** be used.

In Group 1 final circuits that are rated up to 63 A, RCDs shall be used, and in TN-S systems and the insulation level of all live conductors shall be monitored.

WR-710.411.4

In Group 2 medical locations (except for a medical IT system), protection by automatic disconnection of supply by means of RCDs **shall only** be used on circuits for:

WR-710.411.4

• the supply of movements of fixed operating tables; or
• X-ray units; or
• large equipment with a rated power greater than 5 kVA.

In Group 1 and Group 2 medical locations using a TT system, RCDs shall be used.

WR-710.411.5

For each circuit that is protected by an RCD, the possibility of the RCD's unwanted tripping due to excessive protective conductor currents produced by equipment in normal operation shall be considered.

WR-710.531.2.4

### 3.12.25.2 Warning notices: periodic inspection and testing

Where an installation incorporates an RCD a notice shall be fixed in a prominent position at or near the origin of the installation and shall read as shown in Figure 3.24.

WR-514.12.2

> This installation, or part of it, is protected by a device which automatically switches off the supply if an Earth fault develops. Test quarterly by pressing the button marked 'T' or 'Test'. The device should switch off the supply and should then be switched on to restore the supply. If the device does not switch off the supply when the button is pressed, seek expert advice.

**Figure 3.24** Warning notice – RCD protection.

### 3.12.26 Functional extra-low voltage (FELV)

In medical locations, functional extra-low voltage (FELV) is **not** permitted as a method of protection against electric shock.                    710.411.7

### 3.12.27 Requirements for separated extra-low voltage (SELV) and protective extra-low voltage (PELV) circuits

SELV circuits shall have basic insulation between live parts and Earth.                    WR-414.4.1

The PELV circuits and/or exposed-conductive-parts of equipment supplied by the PELV circuits may be earthed.                    WR-414.4.1

The earthing of PELV circuits may be achieved by a connection to Earth or to an earthed protective conductor within the source itself.                    WR-414.4.1

**Note:** In normally dry conditions, basic protection is generally unnecessary for PELV circuits, where the nominal voltage does not exceed 25 V a.c. or 60 V d.c. and exposed-conductive-parts and/or the live parts are connected by a protective conductor to the main earthing terminal.

#### 3.12.27.1 Medical locations

When using SELV and/or PELV circuits in Group 1 and Group 2 medical locations, the nominal voltage applied to current-using equipment shall not exceed 25 V a.c. rms or 60 V ripple-free d.c.                    WR-710.414

Protection by basic insulation of live parts or by barriers or enclosures **shall** be provided.                    WR-710.414

Where PELV is used in Group 2 medical loca-    WR-710.414.4.1
tions, exposed-conductive-parts of equipment
(such as an operating theatre luminaire) shall be
connected to the circuit-protective conductor.

### 3.12.28 Sources

The neutral (star) point of the secondary windings of    WR-411.8.4.2
three-phase transformers and generators (or the midpoint of the secondary windings of single-phase transformers and generators) shall be connected to Earth.

### *3.12.28.1 Low-voltage generating sets*

The prospective short-circuit current and prospec-    WR-551.2.2
tive Earth fault current shall be assessed for each
source of supply (or combination of sources) that
can operate independently of other sources (or other
combinations).

Protection by automatic disconnection of supply    WR-551.4.3.2.1
shall not rely upon the connection to the earthed
point of the public electricity distribution system
when the generator is operating as a switched alternative to a TN system.

A suitable means of earthing **shall** be provided.    WR-551.4.3.2.1

 The connection of live parts of the generator with Earth may affect the fault
protection that is provided by the equipment.

### *3.12.28.2 Warning notices: earthing and bonding connections*

A permanent label to BS 951 will be available for all    WR-514.13.1
earthing and bonding connections with the words
shown in Figure 3.25.

**Safety Electrical Connection – Do Not Remove**

**Figure 3.25** Warning notice – earthing and bonding.

This notice shall be permanently fixed in a visible position at or near:                                    WR-514.13.1

- the point of connection of every earthing conductor to an Earth electrode; and
- the point of connection of every bonding conductor to an extraneous-conductive-part; and
- the main Earth terminal, where this is separate from the main switchgear.

Where electrical separation to the supply to more than one current-using equipment is used (Regulations 418.2.5 or 418.3), the warning notice shall read as shown in Figure 3.26.                                    WR-514.13.2

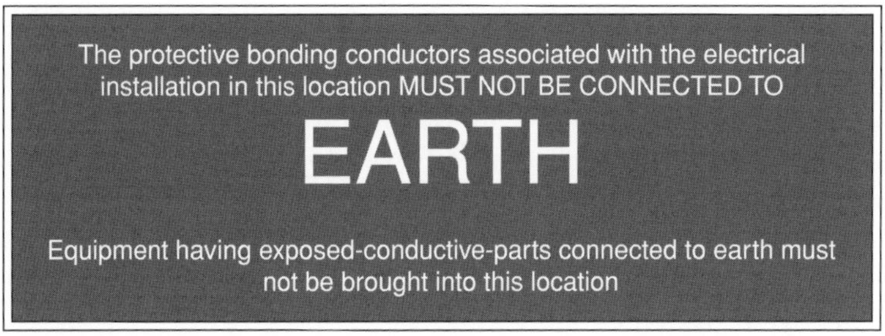

> The protective bonding conductors associated with the electrical installation in this location MUST NOT BE CONNECTED TO
>
> # EARTH
>
> Equipment having exposed-conductive-parts connected to earth must not be brought into this location

**Figure 3.26** Warning notice – protective bonding conductors.

### 3.12.29  Testing

#### 3.12.29.1  Insulation resistance

The insulation resistance shall be measured:                                    WR-612.3.1

- between live conductors; and
- between live conductors and the protective conductor connected to the earthing arrangement.

The insulation resistance measured with the test voltages shown in Table 3.5 shall be considered satisfactory if the main switchboard and each distribution circuit tested separately (with all of its final circuits connected but with current-using equipment disconnected) has an insulation resistance not less than the appropriate value given in Table 3.5.                                    WR-612.3.2

More stringent requirements are applicable for the wiring of fire alarm systems in buildings (see BS 5839-1).

**Table 3.5** Minimum values of insulation resistance

| Circuit nominal voltage (V) | Test voltage d.c. (V) | Minimum insulation resistance (MΩ) |
|---|---|---|
| SELV and PELV | 250 | 0.5 |
| Up to and including 500V, with the exception of the above systems above systems | 500 | 1.0 |
| Above 500V | 1000 | 1.0 |

**Note:** Where the circuit includes electronic devices that are likely to influence the results or be damaged, only a measurement between the live conductors connected together and the earthing arrangement shall be made.

### 3.12.29.2 Insulation resistance/impedance of floors and walls

In a non-conducting location, at least three measurements shall be made in the same location. One of these measurements being approximately 1 m from any accessible extraneous-conductive-part in the location. The other two measurements shall be made at greater distances.

WR-612.5.1

 These measurements shall be repeated for each relevant surface of the location.

**Note:** Further information on the measurement of the insulation resistance/impedance of floors and walls can be found in Appendix 13 of BS 7671:2008 Incorporating Amendment No. 1.

### 3.12.29.3 Medical locations

In addition to the requirements of Chapter 61 and HTM 06-01 (Part A), the following tests **shall** be carried out, prior to commissioning, after alteration or repairs and before re-commissioning:

- complete functional tests of all insulation monitoring devices (IMDs) associated with the medical IT system (including insulation failure, trans

former high temperature, overload, discontinuity
and the acoustic/visual alarms linked to them);
WR-710.6.1

- measurements of leakage current from the IT transformers of the output circuit and enclosure in no-
load condition;
- measurements to verify that the resistance of the
supplementary equipotential bonding meets the
stipulated requirements (see Section 710.415.2.2
of the Wiring Regulations).

- **Annually** – complete functional tests of all IMDs
associated with the medical IT system (including
insulation failure, transformer high temperature,
overload, discontinuity and the acoustic/visual
alarms linked to them).
WR-710.6.2
- **Annually** – perform measurements to verify that
the resistance of the supplementary equipotential
bonding is within the stipulated limits.
- **Every 3 years** – carry out measurements of leakage
current of the output circuit, and of the enclosure of the medical IT transformers in no-load
condition.

 The dates and results of each verification shall be recorded.

### 3.12.29.4 *Protection by electrical separation*

The separation of the live parts from those of other
circuits and from Earth shall be confirmed by measuring the insulation resistance. The resistance values
obtained shall be in accordance with those given in
Table 3.5.
WR-612.4.3

### 3.12.29.5 *Polarity*

A test of polarity shall be made (except for E14 and E27
lampholders to BS EN 60238) to verify that:
WR-612.6

- every fuse, single pole control and protective device
is connected to the line conductor – only; and
- wiring has been correctly connected to the socket-
outlets and similar accessories.

### 3.12.29.6 Earth electrode resistance

| | |
|---|---|
| Where the earthing system incorporates an Earth electrode as part of the installation, the electrode resistance to Earth shall be measured. | WR-612.7 |
| If the installation incorporates an Earth electrode, the electrode resistance to Earth shall be measured before the installation is energized. | WR-612.1 |
| If any test indicates a failure to comply, then, after the fault has been rectified, that test and any preceding test (the results of which may have been influenced by the fault indicated) shall be repeated. | WR-612.1 |

### 3.12.29.7 Earth fault loop impedance

| | |
|---|---|
| Where protective measures are used that require knowledge of Earth fault loop impedance, the relevant impedances shall be measured. | WR-612.9 |

 **Note:** Further information on the measurement of Earth fault loop impedance can be found in Appendix 14 to BS 7671:2008 Incorporating Amendment No. 1.

### 3.12.29.8 Prospective fault current

| | |
|---|---|
| The prospective short-circuit current and prospective Earth fault current shall be measured at the origin, as well as at other relevant points in the installation. | WR-612.11 |

## 3.12.30 Special locations and installations

### 3.12.30.1 Agricultural and horticultural premises

 In agricultural and horticultural premises, a TN-C system **shall not** be used in the installation of a protective device that is going to be used for the automatic disconnection of supply.

| | |
|---|---|
| The protective measures of non-conducting location and Earth-free local equipotential bonding) are **not** permitted. | WR-705.410.3.6 |

| | |
|---|---|
| In locations intended for livestock, supplementary bonding shall connect all exposed-conductive- and extraneous-conductive-parts that can be touched by livestock. | WR-705.415.2.1 |
| In circuits (whatever the type of earthing system used) the following type of disconnection device shall be provided: | WR-705.411.1 |

- in final circuits supplying socket-outlets with rated current not exceeding 32 A, an RCD with an operating time not exceeding 40 ms;
- In final circuits supplying socket-outlets with rated current more than 32 A, an RCD with a rated residual operating current not exceeding 100 mA;
- in all other circuits, RCDs with a rated residual operating current not exceeding 300 mA.

### 3.12.30.2 Conducting locations with restricted movement

| | |
|---|---|
| If a functional Earth is required for certain equipment (e.g. measuring and control equipment), equipotential bonding shall be provided between all exposed-conductive- and extraneous-conductive-parts inside a conducting location that has restricted movement, and the functional Earth. | WR-706.411.1.2 |
| The unearthed source shall have simple separation and shall be situated outside any conducting location with restricted movement, unless the source is part of the fixed installation within that particular conducting location. | WR-706.413.1.2 |

### 3.12.30.3 Electrical installations at construction and demolition sites

A PME facility **shall not** be used for earthing an installation at a construction or demolition site unless all extraneous-conductive-parts are reliably connected to the main earthing terminal.

### 3.12.30.4 Electrical installations in caravans and motor caravans

The use of a PME facility for earthing a caravan is **prohibited** by the Electricity Safety, Quality and Continuity Regulations 2002.

The protective measures of non-conducting loca-     WR-721.410.3.6
tion and Earth-free local equipotential bonding
are **not** permitted.

Structural metallic parts that are accessible from     WR-721.411.3.1.2
within the caravan shall be connected through
main protective bonding conductors to the main
earthing terminal within the caravan.

Every low-voltage socket-outlet (other than those     WR-721.55.2.1
supplied by an individual winding of an isolating
transformer) shall incorporate an Earth contact.

The protective measures of non-conducting loca-     WR-708.410.3.6
tion and Earth-free local equipotential bonding
are not permitted.

### 3.12.30.5 *Temporary electrical installations for amusement parks and circuses etc.*

 A PME facility **shall not** be used.

### 3.12.30.6 *Electrode water heaters and boilers*

If an electrode water heater or electrode boiler is con-     WR-554.1.5
nected to a three-phase low-voltage supply, the shell of
the electrode water heater or electrode boiler shall be
connected to the neutral of the supply as well as to the
earthing conductor.

If the supply to an electrode water heater or electrode     WR-554.1.6
boiler is single-phase and one electrode is connected to
a neutral conductor earthed by the distributor, then the
shell of the electrode water heater or electrode boiler
shall be connected to the neutral of the supply as well
as to the earthing conductor.

If the electrode water heater or electrode boiler is not     WR-554.1.7
piped to a water supply or is in physical contact with any
earthed metal (and where the electrodes and the water
in contact with the electrodes are sufficiently shielded
in insulating material that they cannot be touched
while the electrodes are live) a fuse in the line conductor may be substituted for the circuit-breaker and the
shell of the electrode water heater or electrode boiler
need not be connected to the neutral of the supply.

### 3.12.30.7 Medical locations

Section 710 of the Wiring Regulations is a new section specifically aimed at electrical installations in medical locations, and is specifically aimed at ensuring the safety of both patients and medical staff. Although the requirements of Section 710 mainly refer to hospitals, private clinics, medical and dental practices, healthcare centres and dedicated medical rooms in the workplace, they also equally apply to electrical installations in locations designed for medical research and (where applicable) to veterinary clinics.

 The requirements of Section 710 do not, however, apply to **medical** electrical equipment itself!

#### 3.12.30.7.1 General

- PEN conductors **shall not** be used in medical locations and medical buildings downstream of the main distribution board.
- It is recommended that radial wiring patterns are used so as to avoid 'Earth loops', which may cause electromagnetic interference.
- In the event of a first fault to Earth, a total loss of supply in medical Group 2 locations shall be prevented.

#### 3.12.30.7.2 Operational conditions and external influences

##### 3.12.30.7.2.1 Transformers for IT systems

In medical locations, IT transformers shall be installed in close proximity to the medical location and shall ensure that:

| | |
|---|---|
| • the leakage current of the output winding to Earth and the leakage current of the enclosure (when measured in no-load condition) and the transformer do not exceed 0.5 mA | WR-710.512.1.1 |
| • at least one single-phase transformer per room or functional group of rooms is used to form the IT systems for mobile and fixed equipment and the rated output shall be no less than 0.5 kVA and shall not exceed 10 kVA. | |

> Where several transformers are needed to supply equipment in one room, they shall **not** be connected in parallel.

- if the supply of three-phase loads via an IT system is also required, a separate three-phase transformer is provided for this purpose.

 Capacitors **shall not** be used in transformers for medical IT systems.

### 3.12.30.7.3  Protection against electric shock

#### 3.12.30.7.3.1  Residual current devices (RCDs)

> Care shall be taken to ensure that simultaneous use of large numbers of equipment connected to the same circuit cannot cause unwanted tripping of the RCD.

WR-710.411.3.2.1

In Group 1 and Group 2 medical locations, the following shall apply:

- where RCDs are required, **only** type A (according to BS EN 61008 and BS EN 61009) or type B (according to IEC 62423) shall be selected, depending on the possible fault current arising;

  WR-710.411.3.2.1

   Type AC RCDs **shall not** be used.

- for TN, TT and/or IT systems, the voltage presented between simultaneously accessible exposed-conductive-parts and/or extraneous-conductive-parts shall not exceed 25 V a.c. or 60 V d.c.;

  WR-710.411.3.2.5

- for TN and TT systems, the requirements of Table 3.6 shall apply.

**Table 3.6** Maximum disconnection times

| System | Maximum disconnection time (s) | | | | | | | | | |
|---|---|---|---|---|---|---|---|---|---|---|
| | 25 V $<U_o$ 50 V | | 50 V $<U_o$ 120 V | | 120 V $<U_o$ 5230 V | | 230 V $<U_o$ 400 V | | 1.10 > 400 V | |
| | a.c. | d.c. | a.c. | d.c. | a.c. | d.c. | a.c. | d.c. | a.c. | d.c. |
| TN | 5 | 5 | 0.3 | 2 | 0.3 | 0.5 | 0.05 | 0.06 | 0.02 | 0.02 |
| TT | 5 | 5 | 0.15 | 0.2 | 0.05 | 0.1 | 0.02 | 0.06 | 0.02 | 0.02 |

 **Note:** In TN systems, 25 V a.c. or 60 V d.c. may be met with protective equipotential bonding, by complying with the disconnection time in accordance with Table 3.6.

### 3.12.30.7.3.1.1 TN systems

| | |
|---|---|
| In Group 1 final circuits that are rated up to 63 A, RCDs **shall** be used. | WR-710.411.4 |
| In TN-S systems, the insulation level of all live conductors shall be monitored. | WR-710.411.4 |
| In Group 2 medical locations (except for a medical IT system), protection by automatic disconnection of supply by means of RCDs **shall only** be used on the following circuits:<br><br>• circuits for the supply of movements of fixed operating tables; or<br>• circuits for X-ray units; or<br>• circuits for large equipment with a rated power greater than 5 kVA. | WR-710.411.4 |

### 3.12.30.7.3.1.2 TT Systems

| | |
|---|---|
| In Group 1 and Group 2 medical locations using a TT system, RCDs shall be used. | WR-710.411.5 |

### 3.12.30.7.3.1.3 IT systems

 Where a medical IT system is used, additional protection by means of an RCD **shall not** be used.

In Group 2 medical locations, an IT system shall always be used for all:

| | |
|---|---|
| • final circuits supplying medical electrical equipment and systems intended for life support;<br>• surgical applications; and<br>• other electrical equipment located or that may be moved into the *patient environment*. | WR-710.411.6.3.1 |

For each group of rooms serving the same function, at least one IT system is necessary, and this shall be equipped with an IMD that has:

| | |
|---|---|
| • an acoustic and visual alarm system that is located in a suitable place so that it can be permanently monitored (i.e. audible and visual signals) by the medical staff and the technical staff; | WR-710.411.6.3.1 |

This alarm system shall have:

- a **green** signal lamp to indicate normal operation;                                    WR-710.411.6.3.1
- a **yellow** signal lamp that lights when the minimum value set for the insulation resistance is reached;

 It shall **not** be possible for this light to be cancelled or disconnected – but the **yellow** signal should automatically go out when the normal condition is restored.

- an audible alarm that sounds when the minimum value set for the insulation resistance is reached.

 **Note:** This audible alarm may be capable of being be silenced.

Documentation of all faults occurring in a medical location shall be maintained, and shall include:

- the meaning of each type of signal; and
- the procedure to be followed in case of an alarm at first fault.

 **Note:** See Figure 3.27 for an illustration of a typical theatre layout.

### 3.12.30.7.3.1.4 Socket-outlets

For each circuit that is protected by an RCD, the possibility of the RCD's unwanted tripping due to excessive protective conductor currents produced by equipment in normal operation shall be considered.                                    WR-710.531.2.4

It is a mandatory requirement that all IT systems in Group 2 medical locations shall observe the following:

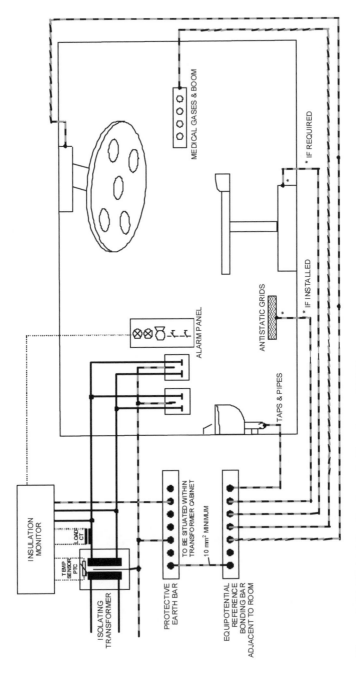

**Figure 3.27** Typical theatre layout. (Courtesy of BSI.)

Socket-outlets intended to supply medical elec-   WR-710.553.1
trical equipment **shall** be unswitched.

 Socket-outlets used on medical IT systems, **shall**
be coloured **blue** and clearly and permanently
marked '*Medical equipment only*'.

In addition,

At each patient's place of treatment (e.g. bedheads):   WR-710.553.1

- each socket-outlet shall be supplied by an individually protected circuit; or
- several socket-outlets shall be separately supplied
  by a minimum of two circuits.

### 3.12.30.7.3.2  Separated extra-low voltage (SELV) and protective extra-low voltage (PELV)

When using SELV and/or PELV circuits in   WR-710.414
Group 1 and Group 2 medical locations, the
nominal voltage applied to current-using equipment shall not exceed 25 V a.c. rms or 60 V ripple-free d.c.

Protection by basic insulation of live parts or by   WR-710.414
barriers or enclosures shall be provided.

In Group 2 medical locations, where PELV is   WR-710.414.4.1
used, exposed-conductive-parts of equipment
(e.g. operating theatre luminaires) shall be connected to the circuit-protective conductor.

### 3.12.30.7.4  Inspection and testing

### 3.12.30.7.4.1  Initial verification

In addition to the requirements of Chapter 61 of the Wiring Regulations and
HTM 06-01 (Part A), the following tests **shall** be carried out, both prior to commissioning, after alteration or repairs, and before re-commissioning:

- complete functional tests of all IMDs associated with   WR-710.6.1
  the medical IT system, including insulation failure,
  transformer high temperature, overload, discontinuity and the acoustic/visual alarms linked to them;
- measurements of leakage current from the IT transformers of the output circuit and enclosure in no-
  load condition;
- measurements to verify that the resistance of the
  supplementary equipotential bonding is within
  stipulated limits.

### 3.12.30.7.4.2 Periodic inspection and testing

In addition to the requirements of Chapter 62, periodic inspection and testing should be carried out in accordance with Health Technical Memorandum
(HTM) 06-01 (Part B) and local health authority requirements as follows, and
at the given intervals:

- **Annually** – complete functional tests of all IMDs   WR-710.6.2
  associated with the medical IT system, including
  insulation failure, transformer high temperature,
  overload, discontinuity and the acoustic/visual
  alarms linked to them.
- **Annually** – perform measurements to verify that
  the resistance of the supplementary equipotential
  bonding is within the stipulated limits.
- **Every 3 years** – complete measurements of leakage
  current of the output circuit and of the enclosure of
  the medical IT transformers in no-load condition.

 The dates and results of each verification **shall** be recorded.

### 3.12.30.8 Exhibitions, shows and stands

The protective measures of non-conducting loca-   WR-711.410.3.6
tion and Earth-free local equipotential bonding
are **not** permitted.

Structural metallic parts that are accessible from   WR-711.411.3.1.2
within the stand, vehicle, wagon, caravan or
container shall be connected through the main
protective bonding conductors to the main earthing terminal within the unit.

### 3.12.30.9 Marinas and similar locations

| | |
|---|---|
| The protective measures of non-conducting location and Earth-free local equipotential bonding are **not** permitted. | WR-709.410.3.6 |
| For marinas, particular attention shall be given to the likelihood of corrosive elements, movement of structures, mechanical damage, presence of flammable fuel and the increased risk of electric shock due to: | WR-709.512.2 |

- presence of water;
- reduction in body resistance;
- contact of the body with Earth potential.

 The Electricity Safety, Quality and Continuity Regulations 2002 prohibit the connection of a PME facility to the metalwork in a boat.    WR-709.411.4

### 3.12.30.10 Mobile or transportable units

| | |
|---|---|
| Accessible conductive-parts of the unit, such as the chassis, shall be connected through the main protective bonding conductors to the main earthing terminal within the unit. | WR-717.411.3.1.2 |
| A PME system **shall not** be used as a means of earthing, except: | WR-717.411.4 |

- where the installation is continuously under the supervision of a skilled or instructed person; and
- the suitability and effectiveness of the means of earthing has been confirmed before the connection is made.

A permanent notice shall be fixed to the unit in a prominent position (preferably adjacent to the supply inlet connector), and shall state in clear and unambiguous terms the following:    WR-717.514

- the type of supplies that may be connected to the unit;
- the voltage rating of the unit;
- the number of supplies, phases and their configuration;

- the on-board earthing arrangement;
- the maximum power requirement of the unit.

In each installation, main protective bonding               WR-411.3.1.2
conductors shall connect extraneous-conductive-
parts to the main earthing terminal, including the
following:

- central heating and air-conditioning systems;
- exposed metallic structural parts of the
  building;
- gas installation pipes;
- water installation pipes;
- other installation pipework and ducting.

 Where an installation serves more than one building, the above requirement
shall be applied to each building.

### 3.12.30.11 Outdoor lighting

The earthing conductor of a street electrical fixture       WR-559.10.3.4
shall have a minimum copper equivalent cross-sec-
tional area not less than that of the supply neutral
conductor at that point or not less than 6 mm$^2$,
whichever is the smaller.


For an outdoor lighting installation, where the             WR-559.10.4
protective measure for the whole installation is by
double or reinforced insulation, no protective conductor need be provided and the conductive-parts of
the lighting column need not be intentionally connected to the earthing system.

### 3.12.30.12 Solar photovoltaic (PV) power supply systems

Earthing of one of the live conductors of the d.c.          WR-712.312.2
side is permitted if there is at least simple separation
between the a.c. side and the d.c. side.

 **Note:** Any connections with Earth on the d.c. side should be electrically connected so as to avoid corrosion (see BS 7361-1:1991).

> PV string cables, PV array cables and PV d.c.     WR-712.522.8.1
> main cables shall be selected and erected so
> as to minimize the risk of Earth faults and
> short-circuits.

The protective measures of non-conducting location and Earth-free local equipotential bonding are **not** permitted on the d.c. side.

### 3.12.30.13 Swimming pools and other basins

**Note:** The following requirements apply to the basins of swimming pools, the basins of fountains and the basins of paddling pools, as well as to the surrounding zones of these basins. In these areas, in normal use, the risk of electric shock is increased by a reduction in body resistance and contact of the body with Earth potential.

> It is permitted to install an electric heating unit     WR-702.55.1
> embedded in the floor, provided that it:
>
> - is protected by SELV; or
> - incorporates an earthed metallic sheath con-
>   nected to the supplementary equipotential bond-
>   ing, and its supply circuit is additionally protected
>   by an RCD; or
> - is covered by an embedded earthed metallic grid
>   connected to the supplementary equipotential
>   bonding, and its supply circuit is additionally pro-
>   tected by an RCD.
>
>
>
> The protective measures of non-conducting
> location and Earth-free local equipotential bond-
> ing are **not** permitted.

Special requirements may be necessary for swimming pools used for medical purposes.

### 3.12.30.14 Rooms and cabins containing sauna heaters

>
>
> The protective measures of non-conducting
> location Earth-free local equipotential bond-     WR-703.410.3.6
> ing are **not** permitted.

### 3.12.30.15 Temporary electrical installations for structures, amusement devices and booths at fairgrounds, amusement parks and circuses

 The protective measures of non-conducting location and Earth-free local equipotential bonding are **not** permitted.

| | |
|---|---|
| The Electricity Safety, Quality and Continuity Regulations 2002 prohibit the use of a PME facility for the supply to caravans or similar constructions. | WR-740.411.4 |
| Where a generator supplies a temporary installation, forming part of a TN, TT or IT system, care shall be taken to ensure that the earthing arrangements are in accordance with the Regulations. | WR-740.551.8 |

### 3.12.30.16 Water heaters having immersed and uninsulated heating elements

| | |
|---|---|
| All metal parts of the heater or boiler that are in contact with the water (other than current-carrying parts) shall be solidly and metallically connected to a metal water pipe through which the water supply to the heater or boiler is provided, and that water pipe shall be connected to the main earthing terminal independently of the circuit-protective conductor. | WR-554.3.2 |

### 3.12.31 TN systems

| | |
|---|---|
| In a TN installation, all exposed-conductive-parts **shall** be connected (by a protective conductor) to the main earthing terminal of the installation, and that terminal shall be connected to the earthed point of the supply source. | WR-411.4.2 |
| The neutral or midpoint of the power supply system **shall** be earthed. | WR-411.4.2 |
| The maximum disconnection time for final circuits not exceeding 32 A shall be in accordance with Table 3.7. | WR-411.3.2.2 |

**Table 3.7** Maximum disconnection times

| System | Maximum disconnection time (s) | | | | | | | | |
|---|---|---|---|---|---|---|---|---|---|
| | 50V <$U_o$ 120V | | 120V <$U_o$ 230V | | 230V <$U_o$ 5400V | | $U_o$ >400V | |
| | a.c. | d.c. | a.c. | d.c. | a.c. | d.c. | a.c. | d.c. |
| TN | 0.8 | Not required for protection against electric shock | 0.4 | 5 | 0.2 | 0.4 | 0.1 | 0.1 |
| TT | 0.3 | | 0.2 | 0.4 | 0.07 | 0.2 | 0.04 | 0.1 |

Where a fuse is used to satisfy the requirements of Regulation 411.3.2.2 (see above), maximum values of Earth fault loop impedance ($Z_s$) corresponding to a disconnection time of 0.4 s shall meet the requirements of Table 41.2 (page 55 of BS 7671:2008 Incorporating Amendment No. 1) for a nominal voltage ($U_o$) of 230V.

In a TN system, a disconnection time not exceeding 5 s is permitted for a distribution circuit and/or a circuit **not** covered by Table 3.7 (page 55 of BS 7671:2008 Incorporating Amendment No. 1).      WR-411.3.2.3

**Note:** Where a circuit-breaker is used to satisfy the requirements of Regulation 411.3.2.3, the maximum value of the Earth fault loop impedance ($Z_s$) shall be determined by the following formula:

$$Z_s \times I_a \leq U_o$$

where:

$Z_s$  is the impedance in ohms ($\Omega$) of the fault loop;
$I_a$  is the current in amperes (A) causing the automatic operation of the disconnecting device within the time specified in Table 3.7;
$U_o$  is the nominal a.c. rms or d.c. line voltage to Earth in volts (V).

This maximum disconnection time of 5 s shall apply to **all** circuits feeding fixed equipment used in highway power supplies.      WR-559.10.3.3

Alternatively, for a nominal voltage ($U_o$) of 230V and a disconnection time of 5 s (in accordance with Regulation WR-411.3.2.3), the values specified in Table 41.3 (page 56 of BS 7671:2008 (Incorporating Amendment No. 1)).      WR-411.4.7

Where a fuse is used for a distribution circuit or        WR-411.4.8
a final circuit in accordance with Regulation
411.3.2.3, maximum values of Earth fault loop
impedance ($Z_s$) corresponding to a disconnection
time of 5 s are stated in Table 41.4 (page 56 of BS
7671:2008 (Incorporating Amendment No. 1)) for a
nominal voltage ($U_o$) of 230 V.

### 3.12.31.1 Protection of line conductors in a TN system

Normally, detection of overcurrent shall be provided for all line conductors,
and shall cause the disconnection of the conductor in which the overcurrent
is detected.

In a TN system, however, a circuit supplied between line conductors (and
in which the neutral conductor is not distributed), overcurrent detection need
not be provided for one of the line conductors, provided that:

- there exists, in the same circuit or on the supply        WR-431.1.2
  side, differential protection intended to detect
  unbalanced loads and cause disconnection of all
  the line conductors; **and**
- the neutral conductor is not distributed from an
  artificial neutral point of the circuits situated on
  the load side of this differential protective device.

In a TN system, exposed-conductive-parts of the instal-     WR-411.4.2
lation shall be connected by a protective conductor to
the main earthing terminal of the installation (which
shall, in turn, be connected to the earthed point of the
power supply system).

The following types of protective device may be used for    WR-411.4.4
fault protection in a TN system:

- an overcurrent protective device;
- an RCD (in which case the circuit should also
  incorporate an overcurrent protective device).

 **Note:** Compliance with the above Regulation may be verified by:

- measurement of the Earth fault loop impedance;
- verification of the characteristics and/or the effectiveness of the associated protective device.

### 3.12.31.2 Protective and neutral (PEN) conductors in a TN system

| | |
|---|---|
| In a fixed installation, a single conductor may serve both as a protective conductor **and** as a neutral conductor (i.e. a PEN conductor). | WR-411.4.3 |

### 3.12.31.3 Protective multiple earthing (PME) facilities in a TN system

 The Electricity Safety, Quality and Continuity Regulations 2002 prohibit the connection of a PME facility of a caravan or similar construction.   WR-740.411.4

### 3.12.31.4 Neutral conductors in a TN system

In a TN system:   WR-431.2.1

- the neutral conductor shall be protected against short-circuit current;
- overcurrent protective devices that are used as fault   WR-531.1.1 protection devices shall be selected and erected in compliance with the requirements of Chapter 41 of the Wiring Regulations.

### 3.12.31.5 Residual current devices (RCDs) in a TN system

In a TN system, where a protective device fails to   WR-531.3.1 satisfy the requirements of the Regulations, then that part of the system may be protected by an RCD. Where this happens, the exposed-conductive-parts of that part of the installation shall be connected to the TN earthing system protective conductor, or to a separate Earth electrode that provides impedance appropriate to the operating current of the RCD.

 In this latter case, the circuit shall be treated as a TT system.

### 3.12.31.6 Luminaires in a TN system

> In a TN system, the outer contact of every Edison    WR-559.6.1.8
> screw or single-centre bayonet-cap type lampholder
> (other than E14 and E27 lampholders) shall be con-
> nected to the neutral conductor.

 This regulation also applies to track-mounted systems!

### 3.12.31.7 Medical locations

> In medical locations, RCDs shall be used for final    WR-710.411.4
> circuits of Group 1 rated up to 63 A.
>
> In Group 2 medical locations (except for the medical    WR-710.411.4
> IT system), protection by automatic disconnection of
> supply by means of RCDs shall only be used on the
> following circuits:
>
> • circuits for the supply of a movement capability
>   for fixed operating tables; or
> • circuits for X-ray units; or
> • circuits for large equipment with a rated power
>   greater than 5 kVA.

## 3.12.32 TN-C systems

>  An RCD **shall not** be used in a TN-C system.    WR-411.4.4
>
>  In agricultural and horticultural premises, a    WR-705.411.4
> TN-C system **shall not** be used.

## 3.12.33 TN-S systems

> For a TN-S system, the main earthing terminal of the    WR-542.1.2
> installation **shall** be capable of being connected to the
> earthed point of the source of energy. Part of the connec-
> tion may be formed by the distributor's lines and equipment.

**Note:** This particularly applies to installations where the generating set provides a supply as a switched alternative to the system for distribution of electricity to the public (standby systems).

 In TN-S systems at medical locations, the insulation level of all live conductors **shall** be monitored.

### 3.12.33.1 Neutral conductors in a TN-S system

| | |
|---|---|
| For a TN-S system where the neutral is **not** isolated, any RCD shall be positioned to avoid incorrect operation due to the existence of any parallel neutral–Earth path. | WR-551.6.2 |
| In a TN-S system the neutral conductor need not be isolated or switched where it can be regarded as being reliably connected to Earth by suitably low impedance. | WR-537.1.2 WR-537.2.1.1 |

 **Note:** It may be desirable to disconnect the neutral of the installation from the neutral or PEN of the system for distribution of electricity to the public in order to avoid disturbances such as induced voltage surges caused by lightning.

### 3.12.33.2 Special installations and locations

| | |
|---|---|
| Except for a part of an installation in a building, a PME facility **shall not** be used to supply any metalwork in leisure accommodation vehicles (including a caravan) or mobile or transportable unit e (e.g. at exhibitions, shows and stands) except: | WR-708.411.4 WR-711.411.4 |

- where the installation is continuously under the supervision of a skilled or instructed person; and
- the suitability and effectiveness of the means of earthing has been confirmed before the connection is made.

### 3.12.34 TN-C-S systems

| | |
|---|---|
| In a TN-C-S system the neutral conductor need not be isolated or switched where it can be regarded as being reliably connected to Earth by suitably low impedance. | WR-537.1.2 |

Where an RCD is used in a TN-C-S system, a PEN conductor **shall not** be used on the load side.

### 3.12.34.1 Protective multiple earthing in a TN-C-S system

Part of the TN-C-S system uses a combined PEN conductor, which is at some point is split up into separate PE and N lines (see Figure 3.12). The combined PEN conductor typically occurs between the substation and the entry point into the building, whereas within the building, separate PE and N conductors are used. This type of system is known as 'protective multiple earthing' (PME), because of the practice of connecting the combined neutral-and-Earth conductor to real Earth at many locations so as to reduce the risk of broken neutrals.

| | |
|---|---|
| For a TN-C-S system where protective multiple earthing is provided, the main earthing terminal of the installation shall be connected by the distributor to the neutral of the source of energy. | WR-542.1.2.2 |
| Except where PME conditions apply, a main protective bonding conductor **shall** have a cross-sectional area not less than half that of the earthing conductor and not less than 6 mm². | WR-544.1.1 |
| Except for highway power supplies and street furniture, where PME conditions apply, the main protective bonding conductor shall be selected in accordance with the neutral conductor of the supply and Table 3.8. | WR-544.1.1 |

**Table 3.8 Minimum cross-sectional area of the main protective bonding conductor in relation to the neutral of the supply**

| Copper equivalent cross-sectional area of the supply neutral conductor | Minimum copper equivalent* cross-sectional area of the main protective bonding conductor |
|---|---|
| 35 mm² or less | 10mm² |
| Over 35 mm² up to 50 mm² | 16mm² |
| Over 50 mm² up to 95 mm² | 2.25mm² |
| Over 95 mm² up to 150 mm² | 2.35mm² |
| Over 150 mm² | 2.50mm² |

## 3.12.34.2 *Special installations and locations*

| | |
|---|---|
| A PME facility **shall not** be used for earthing an installation at a construction or demolition site unless all extraneous-conductive-parts are reliably connected to the main earthing terminal. | WR-704.411.3.1 |
| In locations intended for livestock, supplementary bonding shall connect all exposed-conductive-parts and extraneous-conductive-parts that can be touched by livestock. | WR-705.415.2.1 |
| Where a metal grid is laid on the floor, it shall be included within the supplementary bonding of the location as shown in Figure 705 on page 220 of BS 7671:2008 (Incorporating Amendment No. 1). | WR-705.415.2.1 |

 **Note:** Unless a metal grid is laid on the floor, the use of a PME facility as the means of earthing for the electrical installation is **not** recommended.

| | |
|---|---|
| Except for a part of an installation in a building, a PME facility **shall not** be used to supply a mobile or transportable unit except:<br><br>• where the installation is continuously under the supervision of a skilled or instructed person; and<br>• the suitability and effectiveness of the means of earthing has been confirmed before the connection is made. | WR-717.411.4 |

### 3.12.35 TT systems

 In Group 1 and Group 2 medical TT locations, RCDs shall be used as disconnection devices.

| | |
|---|---|
| In a TT system, all exposed-conductive-parts that are protected by a single protective device **shall** be connected, via the main earthing terminal, to a common Earth electrode. | WR-411.5.1 |

It cannot be overemphasised that consideration should be given to the fact that, if the neutral conductor in a three-phase TT system is interrupted, basic, double and reinforced insulation, as well as components rated for the voltage between line and neutral conductors, can be temporarily stressed with the line-to-line voltage.

| | |
|---|---|
| For a TT system, the main earthing terminal shall be connected via an earthing conductor to an Earth electrode. | WR-542.1.2.3 |

The earthing arrangements shall be such that:   WR-542.1.3.1

- the value of impedance from the consumer's main earthing terminal to the earthed point of the supply for TN systems (or to Earth for TT systems) is in accordance with the protective and functional requirements of the installation and are continuously effective;
- Earth fault currents and protective conductor currents that may occur are carried without danger, particularly from thermal, thermomechanical and electromechanical stresses; and
- they are adequately robust or have additional mechanical protection appropriate to the assessed conditions of external influence.

### 3.12.35.1 Protective devices in a TT system

One or more of the following types of protective device shall be used:   WR-411.5.2

- an RCD (the preferred option);
- an overcurrent protective device.

### 3.12.35.2 Overcurrent protective devices in a TT system

In a TT system, if overcurrent protective devices are used for fault protection, they shall be selected and erected in compliance with the requirements of Chapter 41 of the Wiring Regulations.   WR-531.1.2

### 3.12.35.3 Residual current devices (RCDs) in a TT system

If an installation that is part of a TT system is protected    WR-531.4.1
by a single RCD, it shall be placed at the origin of the
installation (unless that part of the installation between
the origin and the device complies with the requirements for protection by the use of Class II equipment or
equivalent insulation).

**Note:** Where there is more than one origin, this requirement applies to **each**
origin.

Where an RCD is used for Earth fault protection:    WR-411.5.3

- the circuit should also incorporate an overcurrent protective device;
- the disconnection time shall be:    WR-411.5.3

$$R_A \times I_{\Delta n} \le 50\,V$$

where:

$R_A$    is the sum of the resistances (in ohms)of the Earth electrode and the
protective conductor connecting it to the exposed-conductive-parts;

$I_{\Delta n}$    is the rated residual operating current of the RCD.

**Note:** The requirements of this regulation are met if the Earth fault loop impedance of the circuit protected by the RCD meets the requirements of Table 3.9.

**Table 3.9** Maximum Earth fault loop impedance ($Z_s$) for non-delayed RCDs to BS EN
61008–1 and BS EN 61009–1 for $U_o$ of 240 V

| Rated residual operating current (mA) | Maximum Earth fault loop impedance, $Z_s$ ($\Omega$) |
|---|---|
| 30 | 16667 |
| 100 | 500 |
| 300 | 167 |
| 500 | 100 |

Where RCDs are also used for protection against fire,    WR-612.8
the conditions for protection by automatic disconnection of the supply shall be verified by:

- measurement of the resistance of the Earth electrode for exposed-conductive-parts of the installation;
- verification of the characteristics and/or effectiveness of the associated protective device.

In a TT system, protection by means of one or more RCDs (with a rated residual operating current of not more than 30 mA) shall be installed to protect every circuit.                                                WR-551.4.4.2

In locations where there is a risk of fire due to the nature of processed or stored materials (other than mineral insulated cables, busbar trunking systems or powertrack systems) a wiring system shall be protected against insulation faults by an RCD having a rated residual operating current not exceeding 300 mA.                                                WR-422.3.9

### 3.12.35.4 Protection of line conductors in a TT system

 **Note:** The neutral conductor shall be protected against short-circuit current.

In a TT system, where a circuit is supplied between line conductors and where the neutral conductor is not distributed, overcurrent detection need not be provided for one of the line conductors, provided that:                                                WR-431.1.2

- there exists, in the same circuit (or on the supply side) differential protection intended to detect unbalanced loads and that will cause disconnection of all the line conductors; **and**
- the neutral conductor is not distributed from an artificial neutral point of the circuits situated on the load side of this differential protective device.

### 3.12.35.5 Luminaires in a TT system

In a TT system, the outer contact of every Edison screw or single-centre bayonet-cap type lampholder (other than E14 and E27 lampholders) shall be connected to the neutral conductor.                                                WR-559.6.1.8

This regulation also applies to track-mounted systems.

## 3.12.36 IT systems

 Where an IT system has been selected for continuity of service, it is recommended that the insulation monitoring device (IMD) is combined with other devices to enable the fault to be located while the circuit is operating.

Earthing arrangements may be used jointly or separately for protective and functional purposes, according to the requirements of the installation, such that:

WR-542.1.1
WR-542.1.3.1

- the value of impedance from the consumer's main earthing terminal to the earthed point of the supply for systems is in accordance with the protective and functional requirements of the installation, and considered to be continuously effective; and
- Earth fault currents and protective conductor currents that may occur are carried without danger (particularly from thermal, thermomechanical and electromechanical stresses); and
- they are adequately robust or have additional mechanical protection appropriate to the assessed conditions of external influence.

**Note:** The earthing arrangement may be considered electrically independent of another earthing arrangement provided that a rise of potential (with respect to Earth) in one earthing arrangement does not cause an unacceptable rise of potential (with respect to Earth) in the other earthing arrangement.

In an IT system:

WR-411.6.1

- all live parts shall be insulated from Earth or connected to Earth through a sufficiently high impedance connector, either at the neutral point or midpoint of the system – or at an artificial neutral point;
- precautions shall be taken to avoid the risk of a person being in contact with simultaneously accessible exposed-conductive-parts in the event of two faults occurring at the same time;

- exposed-conductive-parts shall be earthed indi-    WR-411.6.2
  vidually, in groups, or collectively;
- the main earthing terminal shall be connected    WR-542.1.2.3
  via an earthing conductor to an Earth electrode.

It is strongly recommended that IT systems with distributed neutrals should **not** be employed. Where no neutral point or midpoint exists, a line conductor may be connected to Earth through a high impedance.

### 3.12.36.1 Protective equipotential bonding in an IT system

In each installation, main protective bonding    WR-411.3.1.2
conductors shall connect extraneous-conductive-
parts to the main earthing terminal including the
following:

- central heating and air-conditioning systems;
- exposed metallic structural parts of the building;
- gas installation pipes;
- water installation pipes;
- other installation pipework and ducting.

Where an installation serves more than one building, the above requirement shall be applied to **each** building.

Protective equipotential bonding shall be applied to    WR-411.3.1.2
the metallic sheath of telecommunication cables.

Earth-free local equipotential bonding is intended to prevent the appearance of a dangerous touch voltage and shall **only** be used in special circumstances.

### 3.12.36.2 Protection of the neutral conductor in an IT system

In an IT system, the neutral conductor **shall not** be dis-    WR-431.2.2
tributed unless:

- overcurrent detection is provided for the neutral
  conductor of every circuit; or
- the neutral conductor is effectively protected
  against short-circuit by a protective device
  installed on the supply side; or

- the circuit is protected by an RCD with a rated
  residual operating current not exceeding 0.2 times
  the current-carrying capacity of the corresponding
  neutral conductor.

If an IT system does not have a neutral conductor, it is permitted to omit the overload protective device in one of the line conductors **provided** that an RCD is installed in each circuit.

### 3.12.36.3  Overload protection in an IT system

The omission of devices for protection against overload is permitted for circuits supplying current-using equipment where unexpected disconnection of the circuit could cause danger or damage.

Examples of such circuits are:

- the exciter circuit of a rotating machine;
- the supply circuit of a lifting magnet;
- the secondary circuit of a current transformer;
- a circuit supplying a fire-extinguishing device;
- a circuit supplying a safety service, such as a fire alarm or a gas alarm;
- a circuit supplying medical equipment used for life support in specific medical locations where an IT system is incorporated.

### 3.12.36.4  Supplies to an IT system

In an IT system:

- equipment **shall** be suitable for the nominal voltage ($U_o$);
- equipment **shall** be insulated for the nominal voltage between lines.

Where a generator supplies a temporary installation,    WR-740.551.8
forming part of an IT system:

- care shall be taken to ensure that the earthing arrangements are in accordance with the
  Regulations;
- the neutral conductor of the star point
  of the generator shall **not** be connected
  to the exposed-conductive-parts of the
  generator.

### 3.12.36.5 Monitoring devices in an IT system

| | |
|---|---|
| In IT systems, continuous insulation monitoring devices shall be provided that give an auditable and visual indication in the event of a first fault. | WR-560.5.3<br>WR-411.6.3.1 |

 **Note:** This device shall initiate an audible and/or visual signal which shall continue as long as the fault persists.

| | |
|---|---|
| The following monitoring devices and protective devices may be used: | WR-411.6.3 |

- residual current devices (RCDs);
- residual current monitoring devices (RCMs);
- insulation monitoring devices (IMDs);
- insulation fault location systems;
- overcurrent protective devices.

| | |
|---|---|
| Consideration should be given to the possibility that, if a line conductor of an IT system is earthed accidentally, then the insulation (or components) that are normally rated for the voltage between line and neutral conductors can be temporarily stressed with the line-to-line voltage. | WR-442.4 |

 The object of this particular regulation is to provide requirements for the safety of a low-voltage installation in the event of:

- a fault between the high-voltage system and Earth in the transformer sub-station that supplies the low-voltage installation;
- loss of the supply neutral in the low-voltage system;
- short-circuit between a line conductor and neutral in the low-voltage installation;
- accidental earthing of a line conductor of a low-voltage IT system.

| | |
|---|---|
| In locations where there is a strong risk of fire due to the nature of processed or stored materials, the wiring system of an IT system (except for mineral-insulated cables, busbar trunking systems or powertrack systems) **shall** be protected against insulation faults, by an IMD with audible and visual signals. | WR-422.3.9 |

### 3.12.36.5.1 RCDs in an IT system

| | |
|---|---|
| In an IT system, protection by means of one or more RCDs (with a rated residual operating current of not more than 30 mA) **shall** be installed to protect every circuit. | WR-551.4.4.2 |
| In addition to: | WR-717.411.6.2 |

- an isolating transformer or a low-voltage generating set; or
- an installation fault location system;

an IT system in a mobile or transportable unit can be provided by a transformer providing simple separation via an RCD.

| | |
|---|---|
| In mobile (or transportable) units, additional protection by an RCD **shall** be provided for every socket-outlet intended to supply current-using equipment outside the unit, with the exception of socket-outlets that are supplied from circuits with protection by: | WR-717.415 |

- SELV; or
- PELV; or
- electrical separation.

| | |
|---|---|
| Where protection is provided by an RCD (and disconnection following a first fault is not envisaged), the non-operating residual current of the device **shall** be at least equal to the current that circulates on the first fault to Earth of negligible impedance affecting a line conductor. | WR-531.5.1 |

 In an IT system, an RCD should **not** operate unless one of the Earth faults is on a part of the system that is on the supply side of the device.

### 3.12.36.5.2 Residual current monitoring devices (RCMs) in an IT system

| | |
|---|---|
| Except where a protective device is installed to interrupt the supply in the event of the first Earth fault, an RCM **shall** be provided to indicate the occurrence of a first fault from a live part to an exposed-conductive-part or to Earth. | WR-411.6.3.2 |

**Note:** This device shall initiate an audible and/or visual signal which shall continue as long as the fault persists.

| | |
|---|---|
| After the occurrence of a first fault and in the event of a second fault occurring on a different live conductor, automatic disconnection of supply shall occur. | WR-411.6.4 |
| Where an RCM is used in an a.c. IT system, it is recommended that a directionally discriminating RCM is used so as to avoid undue signalling of leakage current where high leakage capacitances are liable to exist downstream from the point of installation of the RCM. | WR-538.4 |
| In an IT system where interruption of the supply in case of a first insulation fault to Earth is not required or not permitted, an RCM may be installed to assist in the location of a fault. | WR-538.4.2 |

 An RCM is **not** intended to provide protection against electric shock.

### 3.12.36.5.3 IMDs for an IT system

An IMD is designed to indicate any important reduction in the insulation level of the system in order to find the cause before a second insulation fault occurs, thus avoiding any power-supply interruption. An IMD is intended to be permanently connected to an IT system and to continuously monitor the insulation resistance of the complete system (secondary side of the power supply and the complete installation supplied by this power supply) to which it is connected.

 An IMD is **not** intended to provide protection against electric shock.

| | |
|---|---|
| The IMD shall be connected between Earth and a live conductor of the monitored equipment. | WR-538.3 |
| An IMD **shall** indicate when an fault Earth fault is detected, to enable that the fault shall be located and eliminated, as soon as possible, in order to restore normal operating conditions. | WR-538.1 |
| An IMD shall be used in a circuit comprising safety equipment that is normally de-energized by a switching device which disconnects all live poles and which is only energized in the event of an emergency. . | WR-538.3 |

**Note:** In these circumstances the reduction in the insulation level shall be indicated locally by a visual or an audible signal, with the choice of remote indication.

IMDs, installed in locations where persons other than instructed persons or skilled persons have access to their use, shall be designed or installed in such a way that it shall be impossible to modify the settings, except by the use of a key, a tool or a password.   WR-538.1.3

 The Earth or functional Earth terminal of the IMD shall be connected to the main Earth terminal of the installation.

The supply circuit of the IMD shall be connected either to the installation on the same circuit of the connecting point of the 'line' terminal and as close as possible to the origin of the system, or to an auxiliary supply.   WR-538.1.2

Where the installation is supplied from more than one power supply, connected in parallel, one IMD per supply shall be used, provided they are interlocked in such a way that only one IMD remains connected to the system.   WR-538.1.2

For d.c. installations, the line terminal(s) of the IMD shall be connected either directly to the midpoint, if any, or to one or all of the supply conductors.   WR-538.1.2

In some d.c. IT two-conductor installations, a passive IMD that does not inject current into the system may be used, **provided** that:

- the insulation of all live distributed conductors is monitored; and
- all exposed-conductive-parts of the installation are interconnected; and
- circuit conductors are selected and installed so as to reduce the risk of an Earth fault to a minimum.

The connecting point to the installation shall be selected so that the IMD is able to monitor the insulation of the installation under all operating conditions.   WR-538.1.2

The line terminal(s) of the IMD shall be connected as close as practicable to the origin of the system to either:   WR-538.1.2

- the neutral point of the power supply; or
- an artificial neutral point with impedances connected to the line conductors; or
- a line conductor or two or more line conductors.

Where the IMD is connected between one line and Earth, it shall be suitable to withstand at least the line-to-line voltage between its 'line' terminal and its Earth terminal.   WR-538.1.2

The IMD shall be set to a lower value corresponding to the normal insulation of the system when operating normally with the maximum of loads connected.   WR-538.1.3

### 3.12.36.5.4  Overcurrent protective devices for an IT system

Overcurrent protective devices that are used in IT systems for fault protection, in the event of a second fault shall:   WR-531.1.3

- be suitable for line-to-line voltage applications for operation in case of a second insulation fault;
- (in the event of a second fault) disconnect all corresponding live conductors, including the neutral conductor, if any.

### 3.12.36.5.5  IT systems in medical locations

 Where a medical IT system is used, additional protection by means of an RCD shall **not** be used.

In Group 2 medical locations, an IT system shall always be used for:

- final circuits supplying medical electrical equipment and systems intended for life support;   WR-710.411.6.3.1
- surgical applications; and
- 'other' electrical equipment located in, or that may be moved into, the 'patient environment'.

For each group of rooms serving the same function, at least one IT system is necessary, and this shall be equipped with an IMD that has:

an acoustic and visual alarm system that is located    WR-710.411.6.3.1
in a suitable place so that it can be permanently
monitored (audible and visual signals) by the
medical staff and the technical staff.

This alarm system shall have:

- a **green** signal lamp to indicate normal operation;
- a **yellow** signal lamp that lights when the minimum value set for the insulation resistance is reached;

  It shall **not** be possible for this light to be cancelled or disconnected – but the **yellow** signal should automatically go out when the normal condition is restored.

- an audible alarm that sounds when the minimum value set for the insulation resistance is reached.

  Note: This audible alarm may be capable of being silenced.

- Documentation of all faults occurring in a medical location shall be maintained, and shall include:

  o   the meaning of each type of signal; and
  o   the procedure to be followed in case of an alarm at first fault.

  Note: See Figure 3.28 for an illustration of a typical theatre layout.

It is also a mandatory for socket-outlet circuits in a medical IT system Group 2 that:

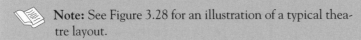

                                                    WR-710.553.1
Socket-outlets intended to supply medical electrical equipment **shall** be unswitched.

Socket-outlets used on medical IT systems, **shall** be coloured **blue** and clearly and permanently marked 'Medical equipment only'.

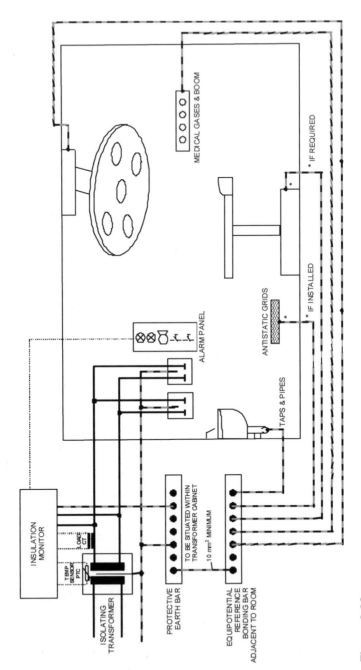

INSULATION MONITOR

LOAD CT

T EMP SENSOR PTC

ISOLATING TRANSFORMER

PROTECTIVE EARTH BAR

TO BE SITUATED WITHIN TRANSFORMER CABINET

10 mm² MINIMUM

EQUIPOTENTIAL REFERENCE BONDING BAR ADJACENT TO ROOM

ALARM PANEL

TAPS & PIPES

MEDICAL GASES & BOOM

ANTISTATIC GRIDS

* IF INSTALLED

* IF REQUIRED

**Figure 3.28**

In addition

At each patient's place of treatment (e.g. bedheads):        710.553.1

- each socket-outlet shall be supplied by an individually protected circuit; or
- several socket-outlets shall be separately supplied by a minimum of two circuits.

# 4

# Safety protection

Around 1000 electrical accidents at work are reported to the Health and Safety Executive (HSE) every year, and about 30 people die of their injuries. Indeed, electrocution is one of the top five causes of workplace deaths!

Many of these deaths and injuries arise from:

- use of poorly maintained electrical equipment;
- working near overhead power lines;
- contact with underground power cables during excavation work;
- work on or near 230 V domestic electricity supplies;
- use of unsuitable electrical equipment in explosive areas such as car paint spraying booths etc.

In addition, fires started by poor electrical installation and faulty electrical appliances cause many additional deaths and injuries.

For this reason, protection against electric shock and safety protection methods are an essential part of the Wiring Regulations (i.e. BS 7671:2008 (Incorporating Amendment No. 1)) and the following mandatory requirement, therefore, needs to be observed:

 **All installations shall comply with the requirements for safety protection in respect of:**

- electric shock;
- thermal effects;
- overcurrent;
- undervoltage;
- isolation and switching.

 **Note:** 'Installation' in this context is taken to mean installation either as a whole or in its several parts.

In electrical installations, risk of injury may result from:

- arcing or burning (likely to cause blinding effects, excessive pressure and/or toxic gases);
- excessive temperatures (likely to cause burns, fires and other injurious effects);
- ignition of a potentially explosive atmosphere;
- mechanical movement of electrically actuated equipment;
- power-supply interruptions and/or interruption of safety services;
- shock currents;
- undervoltages, overvoltages and electromagnetic influences likely to cause or result in injury or damage.

# 4.1 Basic safety requirements

The fundamental safety requirements of the Wiring Regulations are as follows.

## 4.1.1 Mandatory requirements

- Protective safety measures shall be applied in every installation, part installation and/or equipment.
- Installations shall comply with the requirements for safety protection in respect of:

  ○ electric shock;
  ○ thermal effects;
  ○ overcurrent;
  ○ fault current;
  ○ undervoltage;
  ○ isolation and switching.

- There shall be no detrimental influence between various protective measures used in the same installation, part installation or equipment.

## 4.1.2 Fundamental safety requirements

The following are summarized details of the most important elements of the Institution of Engineering and Technology (IET) Wiring Regulations that meet these fundamental design requirements.

### 4.1.2.1 Design

Electrical installations shall be designed for:

- the protection of persons, livestock and property;
- the proper functioning of the electrical installation;

- protection against mechanical and thermal damage; and
- protection of people from an electric shock or fire hazard.

### 4.1.2.2 Characteristics of available supply or supplies

Detailed design characteristics shall be available for all supplies. These shall include:

- nature of current (a.c. and/or d.c.);
- purpose and number of conductors;

| For a.c. | For d.c. |
|---|---|
| • Phase conductor(s) | • Outer conductor |
| • Neutral conductor | • Middle conductor |
| • Protective conductor | • Earthed conductor |
| • PEN conductor | • Live conductor |
| | • Protective conductor |
| | • PEN conductor |

- values and tolerances of:
  - Earth fault loop impedance;
  - nominal voltage and voltage tolerances;
  - nominal frequency and frequency tolerances;
  - maximum current allowable;
  - particular requirements of the distributor;
  - prospective short-circuit current;
  - protective measures inherent in the supply (e.g. Earth, neutral or mid-wire).

### 4.1.2.3 Electricity distributor's responsibilities

The electricity distributor is responsible for:

- evaluating and agreeing proposals for new installations or significant alterations to existing ones;
- ensuring that their equipment on consumers' premises:
  - is suitable for its purpose;
  - is safe in its particular environment;
  - clearly shows the polarity of the conductors.
- installing the cut-out and meter in a safe location;
- ensuring that the cut-out and meter are mechanically protected and can be safely maintained;
- providing an earthing facility for all new connections;
- maintaining the supply within defined tolerance limits;

- providing certain technical and safety information to the consumer to enable him to design his installations.

### 4.1.2.4 Installation and erection

- All electrical joints and connections shall meet stipulated requirements concerning conductance, insulation, mechanical strength and protection.
- Conductors shall be identified by colour, lettering and/or numbering.
- Connections and joints shall be accessible for inspection, testing and maintenance, unless:

  ○ they are in a compound-filled or encapsulated joint;
  ○ the connection is between a cold tail and a heating element;
  ○ the joint is made by welding, soldering, brazing or compression tool.

- Design temperatures shall not be exceeded by the installation of electrical equipment.
- Electrical equipment shall be arranged such that it is fully accessible (i.e. for operation, inspection, testing, maintenance and repair) and that there is sufficient space for later replacement.
- Equipment used for the supply of safety services shall be arranged to allow easy access for periodic inspection, testing and maintenance.
- Exposed parts of electrical equipment shall be located (or guarded) so as to prevent accidental contact and/or injury to persons or livestock.
- Good workmanship and proper materials shall be used.
- Installed electrical equipment shall minimize the risk of igniting flammable materials.
- Installed equipment must be accessible for operational, inspection and maintenance purposes.
- Installations shall be divided into circuits in order to:

  ○ avoid danger and minimize inconvenience in the event of a fault;
  ○ facilitate safe operation, inspection, testing and maintenance.

- The process of erection shall not impair the characteristics of electrical equipment.

### 4.1.2.5 Identification and notices

Wiring shall be marked and/or arranged so that it can be quickly identified for inspection, testing, repair or alteration of the installation.

### 4.1.2.6 Inspection and testing

Precautions shall be taken to avoid danger to persons and livestock, and to avoid damage to property and installed equipment, during inspection and testing.

- Every electrical installation must be inspected and tested during erection and on completion **before** being put into service.

- Details of the general design characteristics of the electrical installation must be made available. These shall include the result of the assessment of general characteristics.
- Information (e.g. diagrams, charts, tables and/or schedules) must be made available to the person carrying out the inspection and testing, and these (as a minimum) shall indicate:

  o the type and composition of each circuit (points of utilization served, number and size of conductors, type of wiring);
  o the method used;
  o the identification (and location) of all protection, isolation and switching devices;
  o circuits or equipments that are susceptible to a particular test.

- If the inspection and tests are satisfactory, a signed Electrical Installation Certificate together with a Schedule of Inspections and a Schedule of Test Results (see Chapter 9 in this book) are to be given to the person responsible for ordering the work.
- Precautions shall be taken to avoid danger to persons, and to avoid damage to property and installed equipment, during inspection and testing.

### 4.1.2.7 Maintenance

An assessment shall be made of the frequency and type of maintenance (e.g. periodic inspection, testing, maintenance and repair) that an installation can reasonably be expected to receive during its intended life.

## 4.1.3 Building Regulations requirements

The following are summarized details of the most important elements of the Building Regulations Approved Documents and Standards concerning safety protection. They include:

- design, installation, inspection and testing of electrical installations;
- conservation of fuel and power;
- access and facilities for disabled people;
- extensions, material alterations and material changes of use.

### 4.1.3.1 Design, installation, inspection and testing of electrical installations

- All proposals to carry out electrical installation work **must** be notified to the local authority's Building Control Body before work begins, **unless** the proposed installation work:

  o is undertaken by a person who is a competent person registered with an electrical self-certification scheme; and
  o does not include the provision of a new circuit.

- Any work that involves adding a new circuit to a dwelling needs to be either notified to the Building Control Body (who will then inspect the work) or needs to be carried out by a competent person who is registered under a government Approved Part P Self-Certification Scheme.

**Note:** Where a person who is **not** registered to self-certify, intends to carry out the electrical installation, a Building Regulation (i.e. a Building Notice or Full Plans) application will need to be submitted together with the appropriate fee, based on the estimated cost of the electrical installation. The Building Control Body will then arrange to have the electrical installation inspected at first-fix stage and tested upon completion.

- Reasonable provision shall be made in the design, installation, inspection and testing of electrical installations in order to protect persons from fire or injury.
- Sufficient information shall be provided so that persons wishing to operate, maintain or alter an electrical installation can do so with reasonable safety.

Work involving any of the following will also have to be notified to the Building Control Body:

- locations containing a bathtub or shower basin;
- swimming pools or paddling pools;
- hot-air saunas;
- electric floor or ceiling heating systems;
- garden lighting or power installations;
- solar photovoltaic (PV) power supply systems;
- small-scale generators such as micro combined heat and power (micro-CHP) units;
- extra-low-voltage lighting installations (other than pre-assembled, CE-marked lighting sets).

**Note:** While Part P of the Building Regulations makes requirements for the safety of fixed electrical installations, it does not cover system functionality (such as electrically powered fire alarm systems, fans and pumps), which are covered in other parts of the Building Regulations and other legislation.

### 4.1.3.2 Conservation of fuel and power

Energy-efficiency measures shall be provided which:

- provide lighting systems that utilize energy-efficient lamps with manual switching controls (in the case of external lighting fixed to the building) or automatic switching, or both manual and automatic switching controls, as appropriate, such that the lighting systems can be operated effectively with regard to the conservation of fuel and power;

- provide information, in a suitably concise and understandable form (including results of performance tests carried out during the works), which shows building occupiers how the heating and hot water services can be operated and maintained.

The person responsible for achieving compliance should either himself provide a certificate or obtain a certificate from the sub-contractor that commissioning has been successfully carried out. The certificate should be made available to the client and the Building Control Body.

Responsibility for achieving compliance with these requirements rests with the person carrying out the work. That *person* may be, for example, a developer, a main (or sub-) contractor, or a specialist firm directly engaged by a private client.

### 4.1.3.3 Access and facilities for disabled people

In addition to the requirements of the Disability and the Equality Act 2010 (which replaced the previous Disability Discrimination Act 1995 on 1 October 2011), precautions need to be taken to ensure that:

- new non-domestic buildings and/or dwellings (e.g. houses and flats used for student living accommodation);
- extensions to existing non-domestic buildings; and
- non-domestic buildings that have been subject to a material change of use (e.g. so that they become a hotel, boarding house, institution, public building or shop);

are capable of allowing people, regardless of their disability, age or gender, to be able to safely use the facilities of the buildings (both as visitors and as people who live or work in them).

### 4.1.3.4 Extensions, material alterations and material changes of use

Where any electrical installation work is classified as an extension, a material alteration or a material change of use, the work must consider and include:

- confirmation that the mains supply equipment is suitable and can carry the additional loads envisaged;
- the amount of additions and alterations that will be required to the existing fixed electrical installation in the building;
- the earthing and bonding systems being satisfactory and meeting the requirements;
- the necessary additions and alterations to the circuits that feed them;
- the protective measures required to meet the requirements;

- the rating and the condition of existing equipment (belonging to both the consumer and the electricity distributor) is sufficient.

**Note:** Appendix C to Part P of the Building Regulations offers guidance on some of the older types of installations that might be encountered during alteration work, and Appendix D provides guidance on the application of the now harmonized European cable identification system

### 4.1.4 Protection from electric shock

- Protection against electric shock shall be provided.
- Protection against both direct contact (i.e. basic protection) and against indirect contact (i.e. fault protection) shall be provided.
- Persons and livestock shall be protected against dangers that may arise from contact with exposed-conductive-parts during a fault.
- Persons and livestock shall be protected against dangers that may arise from contact with live parts of the installation.
- Live parts shall be completely covered with insulation which:

  o can only be removed by destruction;
  o is capable of durably withstanding electrical, mechanical, thermal and chemical stresses normally encountered during service.

- Live parts shall be inside enclosures (or behind barriers) protected to at least IP2X or IPXXB.
- Bare (or insulated) overhead lines being used for distribution between buildings and structures shall be installed in accordance with the Electricity Safety, Quality and Continuity Regulations 2002.
- Bare live parts (other than overhead lines) shall not be within arm's reach.
- Bare live parts (other than overhead lines) shall not be within 2.5 m of:

  o an exposed-conductive-part;
  o an extraneous-conductive-part;
  o a bare live part of any other circuit.

- Simultaneously accessible exposed-conductive-parts shall be connected to the same earthing system, either individually, in groups or collectively.
- Exposed parts of electrical equipment shall be located (or guarded) so as to prevent accidental contact and/or injury to persons or livestock.

The following methods are used for protection against direct contact (i.e. basic protection) and for protection against indirect contact (fault protection).

| 1–2 mA | Barely perceptible, no harmful effects |
| 5–10 mA | Throw off, painful sensation |
| 10–15 mA | Muscular contraction, can't let go! |
| 20–30 mA | Impaired breathing |
| 50 mA and above | Ventricular fibrillation and death |

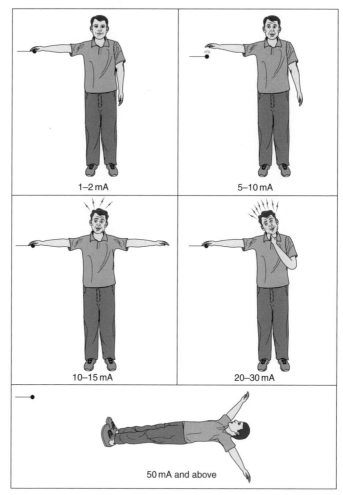

1–2 mA

5–10 mA

10–15 mA

20–30 mA

50 mA and above

**Figure 4.1** The effects of electric shock. (Courtesy Brian Scaddan.)

## 4.2 Basic protection against electric shock

A person may perform work involving direct contact with electrical parts **only** if the electrical part:

- is isolated from all sources of electricity;
- is tested to ensure its isolation from all sources of electricity; and
- is earthed if it is of high voltage.

Work may be performed by a person operating plant or vehicle coming within the exclusion zone for an electrical part only if the electrical part:

- is isolated from all sources of electricity;
- is tested to ensure its isolation from all sources of electricity; and
- is earthed if it is of high voltage.

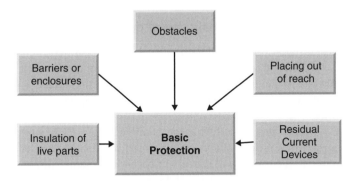

**Figure 4.2** Basic protection against electric shock.

To meet these requirements, the Regulations state that one of the following, basic, measures shall be used for protection against indirect contact:

- insulating live parts;
- using a barrier or an enclosure;
- using obstacles;
- placing equipment out of reach;
- use of a residual current device (RCD).

### 4.2.1 Protection by insulation of live parts

As the title suggests, this is a basic form of insulation protection, and is intended to prevent direct contact with a live parts of an electrical installation. Paint, lacquers and varnishes do **not** provide adequate protection.

### 4.2.2 Protection by barriers or enclosures

This intention of this form of protection is to prevent or deter any contact with a live part. While, generally speaking, this method is for protection against direct contact, it also provides a degree of protection against indirect contact.

### 4.2.3 Protection by obstacles and placing out of reach

Obstacles and placing out of reach will only provide basic protection, and the intention of this form of protection is to prevent unintentional contact with a live part, but **not** an intentional contact caused by deliberately circumnavigating the obstacle. Protection by obstacles and placing out of reach is primarily intended for installations that are controlled or supervised by skilled persons.

**Note:** Protection by placing out of reach is intended only to prevent unintentional contact with live parts.

### 4.2.4 Protection by residual current devices (RCDs)

In electrical installations, an RCD or a residual current operated circuit breaker (RCCB) with integral overcurrent protection is a circuit breaker that operates to disconnect a particular circuit whenever it detects that current leaking out of that circuit (e.g. current leaking to Earth through a ground fault) exceeds safety limits.

Figure 4.3 illustrates the construction of an RCD. An RCD works on the principle that, in a normal (i.e. healthy) circuit, the magnetic effects of the phase and neutral currents will cancel out because the same current will pass through the phase coil, the load and then back through the neutral coil. In a faulty circuit, where the phase or the neutral are to Earth, the currents will no longer be equal and the out-of-balance current will produce some residual magnetism in the core. As the magnetism will be alternating, it will link with the turns of the search coil and induce an electromotive force (EMF), which will drive a current through the trip coil and cause the tripping mechanism to operate.

**Note:** The use of RCDs is not recognized as a sole means of protection and does not obviate the need to apply one of the other protective measures (such as automatic disconnection of supply, double or reinforced insulation, separated extra-low voltage (SELV) or protective extra-low voltage (PELV)).

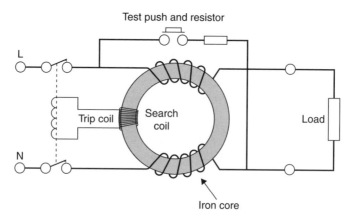

**Figure 4.3** The construction of a residual current device (RCD). (Courtesy Brian Scaddan.)

The use of RCDs with a rated residual operating current not exceeding 30 mA and an operating time not exceeding 40 ms is recognized in a.c. systems as providing additional protection in the event of failure of:

- one of the other methods of basic protection against electric shock; and/or
- the provision for fault protection; or
- the carelessness by users.

 Although RCDs reduce the risk of electric shock, they should **not** be used as the sole means of protection against direct contact.

### 4.2.5 Requirements from the Regulations – basic protection against electric shock

| | |
|---|---|
| All electrical equipment shall comply with one of the provisions for basic protection (e.g. basic insulation; barriers, enclosures, obstacles or placing out of reach) where appropriate. | WR-411<br>WR-419 and WR 417 |
| Basic protection is deemed to be provided where:<br><br>• the nominal voltage cannot exceed the upper limit of voltage Band I; and<br><br>• the SELV and/or PELV supply is from a recognized source such as a safety isolating transformer, battery, diesel-driven generator, insulation testing equipment, monitoring device or motor-generator (see Section 414.3 of BS 7671:2008 (Incorporating Amendment No. 1)) | WR-414.2 |

Exposed-conductive-parts of an SELV circuit shall not be connected:

- to Earth; or
- to protective conductors; or
- to exposed-conductive-parts of another circuit.

| | |
|---|---|
| Basic protection is generally unnecessary in normal dry conditions for:<br><br>• SELV circuits where the nominal voltage does not exceed 25 V a.c. or 60 V d.c.<br>• PELV circuits where the nominal voltage does not exceed 25 V a.c. or 60 V d.c., and exposed- | WR-414.4.5 |

conductive-parts and/or the live parts are connected by a protective conductor to the main earthing terminal.

In **all** other cases, basic protection is not required if the nominal voltage of the SELV or PELV system does not exceed 12 V a.c. or 30 V d.c.

### 4.2.5.1 Protection by insulation of live parts

 Live parts **shall** be completely covered with insulation that can **only** be removed by destruction.

WR 416.1

### 4.2.5.2 Protection by obstacles and placing out of reach

Obstacles shall prevent:

WR-417.2.1

- unintentional bodily approach to live parts; and
- unintentional contact with live parts during the operation of live equipment in normal service.

Obstacles may be removable without using a key or tool.   WR-417.2.2

Obstacles shall be secured to prevent unintentional removal.

WR-417.2.2

 Protection by placing out of reach is intended only to prevent unintentional contact with live parts.

Simultaneously accessible parts at different potentials **shall not** be within arm's reach.

WR-417.3.1

A bare live part (other than an overhead line) **shall not** be within arm's reach or within 2.5 m of:

WR-417.3.1

- an exposed-conductive-part;
- an extraneous-conductive-part;
- a bare live part of any other circuit.

**Note:** Bare (or insulated) overhead lines used for distribution between buildings and structures shall be installed in accordance with the Electricity Safety, Quality and Continuity Regulations 2002 (as amended).

| | |
|---|---|
| The protective measures of placing out of reach and obstacles **shall not** be used **except where** the maintenance of equipment is restricted to skilled persons who are specially trained. | WR-559.10.1 |

 Items of street furniture that are within 1.5 m of a low-voltage overhead line must be protected by something other than placing out of reach!

| | |
|---|---|
| The protective measures of obstacles and placing out of reach are **not** permitted in the following: | |
| • agricultural and horticultural premises; | WR-705.410.3.5 |
| • construction and demolition site installations; | WR-704.410.3.5 |
| • conducting locations with restricted movement; | WR-706.410.3.5 |
| | WR-708.410.3.5 |
| • electrical installations in caravan/camping parks and similar locations; | WR-721.410.3.5 |
| • electrical installations in caravans and motor caravans; | WR-711.410.3.5 |
| • exhibitions, shows and stands; | |
| • floor and ceiling heating systems; | |
| • locations containing a bath or shower; | WR-710.410.3.5 |
| • medical locations; | WR-717.410.3.5 |
| • mobile or transportable units; | WR-703.410.3.5 |
| • rooms and cabins containing sauna heaters; | WR-702.410.3.5 |
| • swimming pools and other basins. | |
| Live parts shall be inside enclosures or behind barriers providing at least the degree of protection IPXXB or IP2X. | WR- 416.2.1 |
| A barrier or enclosure shall be: | WR-416.2.3 |
| • firmly secured in place; | |
| • have sufficient stability and durability to maintain the required degree of protection and appropriate separation from live parts. | |

If it is necessary to remove a barrier or open an    WR-416.2.4
enclosure or remove parts of enclosures, then this
shall **only** be possible:

- by the use of a key or tool; or
- after disconnection of the supply to live
  parts; or
- where an intermediate barrier (with a degree
  of protection of at least IPXXB or IP2X)
  prevents contact with live parts.

This Regulation does not apply to:

- a ceiling rose complying with BS 67;
- a cord-operated switch complying with BS 3676;
- a bayonet lampholder complying with BS EN 61184;
- an Edison-screw lampholder complying with BS EN 60238.

A horizontal top surface of a barrier or enclosure    WR-416.2.2
that is readily accessible shall provide a degree of
protection of at least IPXXD or IP4X.

If an item of equipment (such as a capacitor) is    WR-416.2.5
installed behind a barrier or in an enclosure (and
that equipment could retain a dangerous electrical
charge after it has been switched off) a warning label
shall be provided.

Where the protective measure 'automatic    WR-559.10.3.1
disconnection of supply' is used:

- all live parts of electrical equipment shall be
  protected by insulation; or
- by barriers or enclosures providing basic
  protection.

All conductive-parts of operational electrical    WR-412.2.2.1
equipment that are only separated from live parts by
basic insulation shall be contained in an insulating
enclosure affording at least the degree of protection
IPXXB or IP2X.

**Note:** The insulating enclosure shall not:

- be traversed by conductive-parts likely to transmit a potential; or
- contain any screws or other fixing means that might need to be removed (e.g. during installation and maintenance) and that 'could' be replaced by metallic screws or some other type of fixing which could affect the enclosure's insulation.

| | |
|---|---|
| If the insulating enclosure must be traversed by mechanical joints or connections (e.g. for operating handles of built-in equipment), then these should be arranged so that protection against shock in case of a fault is not impaired. | WR-412.2.2.2 |
| Where a lid or door in an insulating enclosure can be opened without the use of a tool or key, **all** conductive-parts that are accessible (if the lid or door is open) shall be behind an insulating barrier (providing a degree of protection not less than IPXXB or IP2X) which prevents persons from coming unintentionally into contact with those conductive-parts. | WR-412.2.2.3 |

 This insulating barrier shall be removable **only** by the use of a tool or key.

| | |
|---|---|
| No conductive-part contained in the insulating enclosure shall be connected to a protective conductor. | WR-412.2.2.4 |
| No exposed-conductive-part or intermediate part shall be connected to any protective conductor unless explicit provision for this is made in the specification for the equipment concerned. | WR-412.2.2.4 |
| The enclosure shall not affect the operation of the equipment that it is protecting. | WR-412.2.2.5 |

### 4.2.5.3 *Protection by residual current devices (RCDs)*

Every installation shall be divided into circuits, as necessary, to reduce the possibility of unwanted tripping of RCDs due to excessive protective conductor (PE) currents produced by equipment.

WR-314.1

In a.c. systems, additional protection by means of an RCD shall be provided for:

WR-411.3.3

- socket-outlets with a rated current not exceeding 20 A that are used by ordinary persons, unless:

  o they are used under the supervision of skilled or instructed persons e.g. in some commercial or industrial locations; or
  o a suitably labelled and/or identified socket-outlet is provided for connection of a particular item of equipment.

- mobile equipment with a current rating not exceeding 32 A for use outdoors.

If a generating set is connected, protection by RCDs shall remain effective for every intended combination of sources of supply.

WR-551.4.2

# 4.3 Fault protection (protection against indirect contact)

Persons and livestock shall be protected against dangers that may arise from contact with exposed-conductive-parts during a fault.

WR-131.2.2

**Note:** Indirect contact (i.e. when part of the body touches or is in dangerous proximity to any object that is in contact with energized electrical equipment or exposed-conductive-parts that might become live under fault conditions) has always been a potential problem to the unwary when installing, maintaining or inspecting electrical installations. It should also be remembered that, as voltages increase, the potential for arcing increases, and injuries and/or fatalities through arcing, will often occur **even** if no actual bodily contact with high-voltage lines and/or equipment is made!

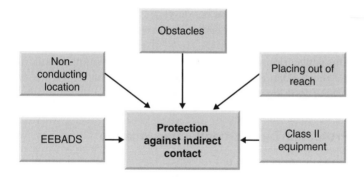

**Figure 4.4** Protection against indirect contact. EEBADS, equipotential bonding and automatic disconnection of supply.

To meet these requirements, the Regulations state that one of the following, basic, measures shall be used for protection against indirect contact:

- earthed equipotential bonding and automatic disconnection of supply (EEBADS);
- non-conducting location;
- protection by obstacles and placing out of reach;
- Class II equipment or equivalent insulation.

### 4.3.1 Protection by earthed equipotential bonding and automatic disconnection of supply (EEBADS)

An earthed equipotential zone is a zone within which exposed-conductive-parts and extraneous-conductive-parts are maintained at substantially the same potential by bonding, such that, under fault conditions, the differences in potential between simultaneously accessible exposed- and extraneous-conductive-parts will not cause electric shock.

Earthed equipotential bonding provides a very good form of protection against indirect contact, by joining together (i.e. bonding) all the metallic parts and then connecting them to Earth. This ensures that all metalwork is at (or near) zero volts, and so, under fault conditions, all metalwork will rise to a similar potential, and simultaneous contact with two metal parts will not result an electric shock as there is no significant potential difference between them.

For installations and locations with an increased risk of shock (e.g. bathrooms and saunas) additional measures may be required, such as:

- automatic disconnection of supply by means of an RCD with a rated residual operating current ($I_{\Delta n}$) not exceeding 30 mA;

- supplementary equipotential bonding;
- reduction of the maximum fault clearance time.

 **Note:** The application of protection by earthed equipotential bonding (and automatic disconnection of supply) will depend on the requirements of the type of system earthing in use (e.g. TN, TT or IT).

### 4.3.2 Protection by non-conducting location

 This method of protection is **not** recognized for general application, and may not be used in installations such as agricultural and horticultural buildings (plus saunas, caravans etc.) that are subject to an increased risk of shock.

This method of protection is intended to prevent simultaneous contact with parts that may be at different potentials (i.e. through the failure of the basic insulation of live parts). A *non-conducting location* is a location where there is no earthing or protective system because:

- there is nothing that needs to be earthed;
- exposed-conductive-parts are arranged so that it is impossible to touch two of them (or one exposed-conducting-part and one extraneous-conductive-part) at the same time.

### 4.3.3 Protection by obstacles and placing out of reach

Obstacles and placing out of reach will only provide basic protection, and the prime intention is to prevent unintentional contact with a live part, but **not** an intentional contact caused by deliberately circumnavigating the obstacle. Protection by obstacles and placing out of reach is primarily intended for installations that are controlled or supervised by skilled persons.

### 4.3.4 Protection by Class II equipment or equivalent insulation

Class II equipment is unique in that, as well as providing the basic insulation for live parts, it also has a second layer of insulation, which can be used to either prevent contact with exposed-conductive-parts or to make sure that there can never be any contact between such exposed-conductive-parts and live parts.

Class II protection is provided by one or more of the following:

- electrical equipment having double or reinforced insulation;
- low-voltage switchgear;
- low-voltage controlgear assemblies;
- supplementary insulation;
- reinforced insulation applied to uninsulated live parts.

### 4.3.5  Requirements from the Regulations – fault protection against electric shock

*4.3.5.1  Protection by earthed equipotential bonding and automatic disconnection of supplies (EEBADS)*

> In each installation, main protective bonding      WR-411.3.1.2
> conductors shall connect extraneous-conductive-
> parts to the main earthing terminal, including the
> following:
>
> • central heating and air-conditioning systems;
> • exposed metallic structural parts of the building;
> • gas installation pipes;
> • water installation pipes;
> • other installation pipework and ducting.

 Where an installation serves more than one building, the above requirement shall be applied to each building.

> Protective equipotential bonding shall be applied to      WR-411.3.1.2
> any metallic sheath of a telecommunication cable.

 Connection of a lightning protection system to the protective equipotential bonding shall be made in accordance with BS EN 62305.

*4.3.5.2  Protection by obstacles and placing out of reach*

> Obstacles shall prevent:      WR-417.2.1
>
> • unintentional bodily approach to live parts; and
> • unintentional contact with live parts during the
>   operation of live equipment during normal service.
>
> Obstacles may be removable without using a key or      WR-417.2.2
> tool.
>
> Obstacles shall be secured to prevent unintentional      WR-417.2.2
> removal.

 Protection by placing out of reach is only intended to prevent unintentional contact with live parts.

> Simultaneously accessible parts at different potentials **shall not** be within arm's reach.     WR-417.3.1
>
> A bare live part (other than an overhead line) **shall not** be within arm's reach or within 2.5 m of:     WR-417.3.1
>
> * an exposed-conductive-part;
> * an extraneous-conductive-part;
> * a bare live part of any other circuit.

**Note:** Bare (or insulated) overhead lines used for distribution between buildings and structures shall be installed in accordance with the Electricity Safety, Quality and Continuity Regulations 2002.

> The protective measures of placing out of reach and obstacles **shall not** be used **except where** the maintenance of equipment is restricted to skilled persons who are specially trained.     WR-559.10.1

Items of street furniture that are within 1.5 m of a low-voltage overhead line must be protected by something other than placing out of reach!

## 4.4 Protection against both direct and indirect contact

The Regulations state that one of the following, basic, measures shall be used for protection against both direct contact and indirect contact:

* SELV; or
* limitation of discharge of energy.

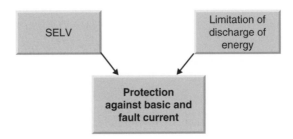

**Figure 4.5** Protection against basic and fault contact.

## 4.4.1 Protection by SELV or PELV

SELV (separated extra-low voltage) is an extra-low-voltage system that is electrically separated from Earth and from other systems so that a single fault cannot give rise to the risk of electric shock. PELV (protective extra-low voltage) is an extra-low-voltage system that is not electrically separated from Earth, but which otherwise satisfies all the requirements for SELV.

SELV is a term used to describe the highest voltage level that can be contacted by a person without causing injury. It is usually defined as 60 V d.c.

This type of protection may also be used for electric fences supplied from electric fence controllers complying with BS EN 61011 or BS EN 61011–1.

### 4.4.1.1 Requirements from the regulations for protection by SELV

| | |
|---|---|
| The separation of the live parts from those of other circuits and from Earth shall be confirmed by a measurement of the insulation resistance. The resistance values obtained shall be in accordance with Table 4.1. | WR-612.4.1 |

More stringent requirements are applicable for the wiring of fire alarm systems in buildings – see BS 5839–1.

**Table 4.1  Minimum values of insulation resistance**

| Circuit nominal voltage (V) | Test voltage d.c. (V) | Minimum insulation resistance (M.(2) |
|---|---|---|
| SELV and PELV | 250 | 0.5 |
| Up to and including 500 V, with the exception of the above systems | 500 | 1.0 |
| Above 500 V | 1000 | 1.0 |

| | |
|---|---|
| When using SELV and/or PELV circuits in Group 1 and/or Group 2 medical locations, protection by basic insulation of live parts or by barriers or enclosures shall be provided. | WR-710.414.1 |
| In Group 2 medical locations, where PELV is used, exposed-conductive-parts of equipment (e.g. operating theatre luminaires) shall be connected to the circuit-protective conductor. | WR-710.414.4.1 |

In medical locations, functional extra-low        WR-710.411.7
voltage (FELV) is not permitted as a method
of protection against electric shock.

## 4.5  Additional requirements

The following are additional requirements for installations and locations where the risk of electric shock is increased by a reduction in body resistance and/or by contact with Earth potential.

### 4.5.1  Protective bonding conductors

Equipotential bonding ensures that protective devices will operate, and it removes dangerous potential differences before a hazardous shock can be delivered. This is achieved by making sure that all the installation's earthed metalwork (i.e. exposed-conductive-parts) is connected to other metalwork (i.e. extraneous-conductive-parts) via the Earth conductor, to provide an Earth fault current path which ensures that dangerous potential differences cannot occur.

**Main equipotential bonding conductors** connect together the installation earthing system and the metalwork of other services, such as gas, electricity and water, as close as possible to their point of entry to the building.

**Supplementary bonding conductors** connect together extraneous-conductive-parts – that is, metalwork which is not associated with the electrical installation but which may provide a conducting path that could give rise to shock.

### 4.5.2  Main equipotential bonding conductors

The Regulations require that the main equipotential bonding conductors for every electrical installation are connected to the main earthing terminal of that particular installation, and that these shall include the following:

- water service pipes (but see requirements for domestic buildings in Chapter 2);
- gas installation pipes;
- other service pipes and ducting;
- central heating and air-conditioning systems;
- exposed metallic structural parts of the building;
- the lightning protective system.

 **Note:** where an installation serves more than one building, the above requirement shall be applied to each building.

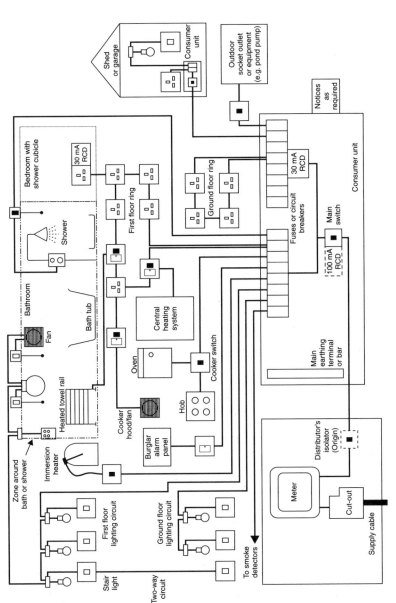

**Figure 4.6** Typical fixed installations that might be encountered in new (or upgraded) existing dwellings.

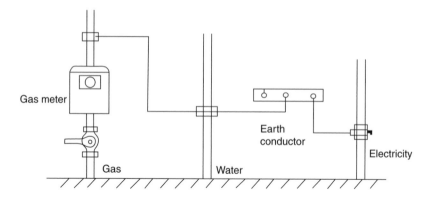

**Figure 4.7**  Main equipotential bonding.

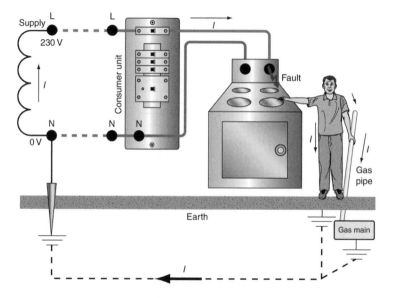

**Figure 4.8**  Earthed equipotential bonding. (Courtesy Brian Scaddan.)

### 4.5.2.1 Protective earthing

Automatic disconnection of supply is a protective measure in which fault protection is provided by protective earthing

> **Safety Electrical Connection – Do Not Remove**

**Figure 4.9**  Earthing and bonding notice.

### 4.5.3 Supplementary bonding conductors

For installations and locations where there is an increased risk of shock (such as agricultural and horticultural premises, building sites etc.) additional measures may be required, such as reduction in maximum fault-clearance time and the use of supplementary equipotential bonding.

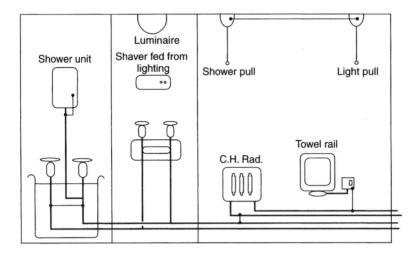

**Figure 4.10** Supplementary equipotential bonding. (Courtesy Brian Scaddan.)

Locations that contain a bath or shower (and where body resistance is lowered as a result of water) are potentially very hazardous environments, and it is important to ensure that no dangerous potentials exist between exposed and extraneous-conductive-parts. For this reason, local supplementary equipotential bonding needs to be provided to connect together the terminals of the protective conductors of each circuit supplying Class I and Class II equipment with extraneous-conductive-parts in those zones, such as:

- metallic pipes supplying services and metallic waste pipes (e.g. water, gas);
- metallic central heating pipes;
- air-conditioning systems;
- accessible metallic structural parts of the building;
- metallic baths and shower basins.

### 4.5.4 Protective conductors

A protective conductor is a conductor that provides a measure of protection against electric shock, and it is used to connect together any of the following parts:

- exposed-conductive-parts;
- extraneous-conductive-parts;
- the main earthing terminal;
- Earth electrode(s);
- the earthed point of the source.

A circuit-protective conductor, on the other hand, is an arrangement of conductors that join all the exposed-conductive-parts together and connect them to the main earthing terminal. There are many types of circuit-protective conductor, such as.

- a separate conductor;
- a conductor included in a sheathed cable with other conductors;
- the metal sheath and/or armouring of a cable;
- a conducting-cable enclosure (e.g. conduit or trunking);
- exposed-conductive-parts (e.g. the conducting cases of equipment).

A gas pipe, an oil pipe, flexible or pliable conduit, support wires or other flexible metallic parts, or constructional parts that are subject to mechanical stress in normal service, **shall NOT** be selected as a protective conductor.

### 4.5.5 Protective equipment (devices and switches)

The type of protective equipment chosen will depend on the type of protection that is required (e.g. whether overcurrent, Earth fault current, overvoltage or undervoltage).

### 4.5.6 Protection against overvoltage

Overvoltage is the hazardous condition that occurs when the voltage in a circuit (or part of a circuit) is suddenly raised over its upper limit. An overvoltage incident can be permanent or transient, and it is often referred to as a 'voltage spike'. A typical example of a naturally occurring transient overvoltage is by lightening. Man-made sources are usually due to electromagnetic induction when switching on or off inductive loads (e.g. electric motors or electromagnets). Transient overvoltage might last microseconds and reach hundreds– sometimes thousands – of volts in amplitude.

In accordance with the Regulations, additional protection against overvoltages of atmospheric origin is not necessary for:

- installations that are supplied by low-voltage systems that do not contain overhead lines;
- installations that are supplied by low-voltage networks that contain overhead lines and their location is subject to less than 25 thunderstorm days per year;

**Figure 4.11** Lightning.

- installations that contain overhead lines and their location is subject to less than 25 thunderstorm days per year;

**provided** that they meet the required minimum equipment impulse to withstand the voltages shown in Table 4.2.

Table 4.2 Required minimum impulse to withstand voltage

| Nominal voltage of the installation (V) | Required minimum impulse (kV) | | | |
| | Category IV (equipment with very high impulse voltage) | Category III (equipment with high impulse voltage) | Category II (equipment with normal impulse voltage) | Category I (equipment with reduced impulse voltage) |
|---|---|---|---|---|
| 230/240 277/480 | 6 | 4 | 2.5 | 1.5 |
| 400/690 | 8 | 6 | 4 | 2.5 |
| 1000 | Values to be determined by system engineer or, in the absence of information, the values for 400/690 can be chosen | | | |

Suspended cables with insulated conductors that have earthed metallic coverings are considered to be an 'underground cable'.

## 4.5.7 Requirements from the Regulations

### 4.5.7.1 Accessibility of electrical equipment

Electrical equipment shall be arranged so that:                    WR-132.12

- there is sufficient space for the initial installation and later replacement of individual items of electrical equipment;
- the equipment is accessible for operation, inspection, testing, fault detection, maintenance and repair.

### 4.5.7.2 Additions and alterations to an installation

No addition or alteration, temporary or permanent,                WR-132.16
shall be made to an existing installation:

- unless it has been ascertained that the rating and the condition of any existing equipment (including that of the distributor) will be adequate for the altered circumstances;
- the earthing and bonding arrangements used as a protective measure for the safety of the addition or alteration are adequate.

Additions and/or alterations to an existing installation          WR-610.4
shall be verified to ensure that the addition or alteration complies with the Regulations and does not impair the safety of the existing installation.

### 4.5.7.3 Automatic supply

Safety services may be required to operate at all certain (and significant) times where people or livestock are at risk – including during mains and local supply failure and through fire conditions. To meet this requirement, specific sources, equipment, circuits and wiring are necessary.

For safety services required to operate in fire                   WR-560.5.2
conditions:

- a safety source of supply shall be selected that will maintain a supply of adequate duration;

- equipment shall be provided, either by construction or by erection, with protection ensuring fire-resistance of adequate duration.

**Note:** The safety source is generally additional to the normal source. The normal source is, for example, the public supply network.

A failure in the control or bus system of a normal installation shall not adversely affect the function of safety services.

### 4.5.7.4 Circuits for safety services

| Circuits of safety services: | WR-560.7.1 |
|---|---|

- shall be independent of other circuits;

- shall not pass through zones exposed to explosion risk (BE3);   WR-560.7.2

- shall not pass through locations exposed to fire risk (BE2);   WR-560.7.3

**unless** they are fire-resistant.

Protection against overload may be omitted where the loss of supply may cause a greater hazard, provided that the occurrence of an overload is indicated.

| Overcurrent protective devices shall be selected and erected so as to avoid an overcurrent in one circuit weakening the correct operation of other circuits of safety services. | WR-560.7.4 |
|---|---|
| Switchgear and controlgear shall be clearly identified and grouped in locations accessible only to skilled or instructed persons (BA5 or BA4). | WR-560.7.5 |
| If equipment is supplied by two different circuits, then a fault occurring in one circuit **shall not** affect the protection against electric shock – or the correct operation of – the other circuit. | WR-560.7.6 |
| Safety circuit cables, other than metallic-screened, fire-resistant cables, shall be separated by distance or by barriers from other circuit cables and other safety circuit cables. | WR-560.7.7 |

For battery cables, special requirements may apply.

> With the exception of wiring for fire and rescue
> service lift supply cables and wiring for lifts with
> special requirements, circuits for safety services
> **shall not** be installed in lift shafts or other flue-like
> openings.
>
> WR-560.7.8

 **Note:** While fire-resistant cables will survive most fires, if they are located in an unstopped vertical shaft the upward air draught of a fire can generate excessive temperatures that can damage the (otherwise) fire-resistant cable.

> As well as a general schematic diagram, full details of
> all electrical safety sources shall be provided adjacent
> to the distribution board.
>
> WR-560.7.9

 **Note:** In these cases, a single-line diagram is sufficient.

> Drawing(s) of the electrical safety installations shall
> be available showing the exact location of all:
>
> WR-560.7.10
>
> • electrical equipment and distribution
>   boards (together with their equipment
>   designations);
> • safety equipment (together with its final circuit
>   designation and purpose);
> • special switching and monitoring equipment for
>   the safety of power supply (e.g. area switches,
>   visual or acoustic warning equipment).
>
> A list of all the current-using equipment permanently
> connected to the safety power supply (indicating
> the nominal electrical power, rated nominal voltage,
> current and starting current, together with its
> duration) shall be available.
>
> WR-560.7.11
>
>  **Note:** This information may be included in
>   the circuit diagrams.

 Operating instructions for all safety equipment and electrical safety services shall be available.

### 4.5.7.5  *Combined protective and neutral (PEN) conductors*

| | |
|---|---|
| PEN conductors shall **only** be used within an installation: | WR-543.4.1<br>WR-543.4.2 |

- where any necessary authorization for use of a PEN conductor has been obtained and where the installation complies with the conditions for that authorization; or
- where the installation is supplied by a privately owned transformer or converter in such a way that there is no metallic connection (except for the earthing connection) with the distributor's network; or
- where the supply is obtained from a private generating plant.

| | |
|---|---|
| A PEN conductor shall be connected to the terminals of the bar intended for the protective earthing conductor and the neutral conductor. | WR-543.4.3 |
| The outer conductor of a concentric cable shall not be common to more than one circuit. | WR-543.4.4 |
| The conductance of the outer conductor of a concentric cable (measured at a temperature of 20°C) shall: | WR-543.4.5 |

- for a single-core cable, be not less than that of the internal conductor;
- for a multi-core cable serving a number of points (contained within one final circuit or having the internal conductors connected in parallel), be not less than that of the internal conductors connected in parallel.

| | |
|---|---|
| At every joint in the outer conductor of a concentric cable and at a termination, the continuity of that joint shall be supplemented by a conductor additional to any means used for sealing and clamping the outer conductor. | WR-543.4.6 |

 No means of isolation or switching shall be inserted in the outer conductor of a concentric cable.

| | |
|---|---|
| The PEN conductors of every cable shall be insulated or have an insulating covering suitable for the highest voltage to which it may be subjected. | WR-543.4.8 |

| | |
|---|---|
| If, from any point of the installation, the neutral and protective functions are provided by separate conductors, those conductors shall not then be reconnected together beyond that point. | WR-543.4.2 |
| At the point of separation, separate terminals or bars shall be provided for the protective and neutral conductors. | WR-543.4.2 |

**Note:** In Great Britain, Regulation 8(4) of the Electricity Safety, Quality and Continuity Regulations 2002 **prohibits** the use of PEN conductors in consumers' installations.

### 4.5.7.6 Conditions of installation

| | |
|---|---|
| Electrical equipment shall be selected so as to withstand, safely, the stresses, the environmental conditions and the characteristics of its location. | WR-133.3 |

### 4.5.7.7 Cross-sectional area of conductors

| | |
|---|---|
| The cross-sectional area of conductors shall be determined for both normal operating conditions and, where appropriate, for fault conditions according to: | WR-132.6 |

- the admissible maximum temperature;
- the voltage drop limit;
- the electromechanical stresses likely to occur due to short-circuit and Earth fault currents;
- other mechanical stresses to which the conductors are likely to be exposed;
- the maximum impedance for operation of short-circuit and Earth fault protection;
- the method of installation;
- harmonics;
- thermal insulation.

### 4.5.7.8 Design

| | |
|---|---|
| The electrical installation shall be designed to provide for: | WR-132.1 |

- the protection of persons, livestock and property;
- the proper functioning of the electrical installation for the intended use.

### 4.5.7.9 Disconnecting devices

| | |
|---|---|
| Disconnecting devices shall be provided so as to allow electrical installations, circuits or individual items of equipment to be switched off or isolated for the purposes of operation, inspection, fault detection, testing, maintenance and repair. | WR-132.10 |

### 4.5.7.10 Earthing arrangements and protective conductors

| | |
|---|---|
| Where protective bonding conductors are installed (especially in photovoltaic (PV) power supply systems) they shall be parallel to and in as close contact as possible with d.c. cables and a.c. cables and accessories. | WR-712.54 |

### 4.5.7.11 Earthing arrangements for protective purposes

| | |
|---|---|
| Where overcurrent protective devices are used for fault protection, the protective conductor shall be incorporated in the same wiring system as the live conductors or in their immediate proximity. | WR-543.6.1 |

### 4.5.7.12 Earthing requirements for the installation of equipment having high protective-conductor currents

| | |
|---|---|
| Equipment having a protective conductor current exceeding 3.5 mA but not exceeding 10 mA shall be either permanently connected to the fixed wiring of the installation without the use of a plug and socket-outlet or connected by means of a plug and socket-outlet complying with BS EN 60309–2. | WR-543.7.1.101 |
| Equipment having a protective conductor current exceeding 10 mA shall be connected to the supply:<br><br>• permanently via the wiring of the installation;<br>• via a flexible cable with a plug and socket-outlet; or<br>• via a protective conductor with an Earth monitoring system. | WR-543.7.1.102 |

The wiring of every final circuit and distribution circuit intended to supply one or more items of equipment (such that the total protective conductor current is likely to exceed 10 mA) shall have a high-integrity protective connection complying with one or more of the following:   WR-543.7.1.103

- a single protective conductor with a cross-sectional area greater than 10 mm²;
- a single copper protective conductor having a cross-sectional area of not less than 4 mm²;
- two individual protective conductors;
- an Earth monitoring system that, in the event of a continuity fault occurring in the protective conductor, automatically disconnects the supply to the equipment;
- connection of the equipment to the supply by means of a double-wound transformer or equivalent unit, such as a motor-alternator set.

Where two protective conductors are used, the ends of the protective conductors shall be terminated independently of each other at all connection points throughout the circuit (e.g. the distribution board, junction boxes and socket-outlets).   WR-543.7.1.104

At the distribution board, information shall be provided indicating those circuits having a high protective conductor current.   WR-543.7.1.105

 **Note:** This information shall be positioned so as to be visible to a person who is modifying or extending the circuit.

In agricultural and horticultural premises, protective bonding conductors shall be protected against mechanical damage and corrosion, and shall be selected to avoid electrolytic effects.   WR-705.544.2

The socket-outlet protective conductors used for electrical installations in caravan/camping parks (or similar locations) **shall not** be connected to any PEN conductor of the electricity supply.   WR-708.553.1.14

## 4.5.7.13 Electrical safety service supply

An electrical safety service supply is either:

- a non-automatic supply, the starting of which is initiated by an operator; or
- an automatic supply, the starting of which is independent of an operator.

## 4.5.7.14 Emergency control

| | |
|---|---|
| An interrupting device shall be installed in such a way that it can be easily recognized and effectively (and rapidly) operated in the case of danger. | WR-132.9 |

## 4.5.7.15 Environmental conditions

| | |
|---|---|
| The design of the electrical installation shall take into account the environmental conditions to which it will be subjected. | WR-132.5.1 |
| Equipment in surroundings susceptible to risk of fire or explosion shall be so constructed or protected (or utilize some other special precautionary method) to prevent danger. | WR-132.5.2 |

## 4.5.7.16 Erection of electrical installations

Electrical equipment shall be installed in accordance with the instructions provided by the manufacturer of the equipment.

| | |
|---|---|
| The characteristics of the electrical equipment shall not be impaired by the process of erection. | WR-134.1.2 |
| Electrical joints and connections shall be properly constructed with regard to conductance, insulation, mechanical strength and protection. | WR-134.1.4 |
| Electrical equipment shall be installed so that design temperatures are not exceeded. | WR-134.1.5 |
| Electrical equipment that is likely to cause high temperatures or electric arcs shall be placed (or guarded) so as to minimize the risk of igniting flammable materials. | WR-134.1.6 |

> Where the temperature of an exposed part          WR-134.1.6
> of electrical equipment is likely to cause injury
> to persons or livestock, that part shall be so
> located or guarded in order to prevent accidental
> contact.

**Note:** Where necessary, suitable safety warning signs and/or notices shall be provided.

### 4.5.7.17 External influences

> The selection of equipment according to external          WR-512.2.4
> influences is necessary for proper functionality and to
> ensure the reliability of the measures of protection for
> safety.

### 4.5.7.18 Initial verification

> The person or persons responsible for the          WR-632.3
> design, construction, inspection and testing of the
> installation shall provide the person ordering the
> work with a certificate which takes account of their
> respective responsibilities for the safety of that
> installation.

### 4.5.7.19 Inspection

**Note:** Inspection shall precede testing and shall normally be done with that part of the installation under inspection being disconnected from the supply.

> The inspection shall be made to verify that the installed          WR-611.2
> electrical equipment is:
>
> - in compliance;
> - correctly selected and erected; and
> - not visibly damaged or defective so as to impair
>   safety.

## 4.5.7.20 *Installation of equipment for insulation fault location in an IT system*

An insulation monitoring device (IMD) shall be used on a circuit comprising safety equipment that is normally de-energized by a switching means disconnecting all live poles and which is only energized in the event of an emergency (provided that the IMD is automatically deactivated whenever the safety equipment is activated).   WR-538.3

## 4.5.7.21 *Isolation and switching*

Effective means shall be provided so that all voltage can be cut off (when required) from every installation, circuit and equipment, so as to prevent or remove danger.   WR-132.15.1

Fixed electric motors shall be provided with an efficient means of switching off, and be readily accessible, easily operated and located so as to prevent danger.   WR-132.15.2

## 4.5.7.22 *Luminaires*

A luminaire with a lamp that could eject flammable materials in case of failure should be equipped with a safety protective shield.   WR-422.3.1
WR-422.4.2

Any flexible cable between the fixing means and the luminaire shall be installed so that any expected stresses in the conductors, terminals and terminations do not interfere with the safety of the installation.   WR-559.6.1.5

## 4.5.7.23 *New materials and inventions*

Where the use of a new material or invention leads to departures from the Regulations, the resulting degree of safety of the installation **shall not** be less than that obtained by compliance with the Regulations.   WR-133.5

### 4.5.7.24 Omission of devices for protection against overload for safety reasons

The omission of devices for protection against overload is permitted for circuits supplying current-using equipment where unexpected disconnection of the circuit could cause danger or damage.

Examples of such circuits are:

• the exciter circuit of a rotating machine;
• the supply circuit of a lifting magnet;
• the secondary circuit of a current transformer;
• a circuit supplying a fire-extinguishing device;
• a circuit supplying a safety service, such as a fire alarm or a gas alarm;
• a circuit supplying medical equipment used for life support in specific medical locations where an IT system is incorporated.

 In such situations, consideration should be given to the provision of an overload alarm.

### 4.5.7.25 Periodic inspection and testing

Periodic inspection comprising a detailed examination of an installation shall be carried out by appropriate tests to show that the requirements for disconnection times for protective devices, are complied with, to provide for:   WR-621.2

• safety of persons and livestock against the effects of electric shock and burns;
• protection against damage to property by fire and heat arising from an installation defect;
• confirmation that the installation is not damaged or deteriorated so as to impair safety;
• the identification of installation defects and departures from the requirements of the Regulations that may give rise to danger.

### 4.5.7.26 Precautions within a fire-segregated compartment

The risk of spread of fire shall be minimized by the selection of appropriate materials and erection.   WR-527.1.1

A wiring system shall be installed so that the general building structural performance and fire safety are not reduced.   WR-527.1.2

Where safety depends on the direction of rotation of      WR-537.5.4.3
a motor, provision shall be made for the prevention of
reverse operation due to, for example, a phase reversal.

### 4.5.7.27 Preservation of electrical continuity of protective conductors

A protective conductor shall be suitably protected       WR-543.3.1
against mechanical and chemical deterioration and
electrodynamic effects.

Every connection and joint shall be accessible for        WR-543.3.2
inspection, testing and maintenance.

 See Regulation 526.3 of BS 7671:2008 (Incorporating Amendment No. 1) for exceptions to this rule.

Where the sheath of a cable incorporating an              WR-543.3.2
uninsulated protective conductor of cross-sectional
area up to and including 6 mm$^2$ is removed adjacent to
joints and terminations, the protective conductor shall
be protected by insulating sleeving complying with the
BS EN 60684 series.

Every connection and joint shall be accessible for        WR-543.3.3
inspection, testing and maintenance.

Where electrical monitoring of earthing is used, no       WR-543.3.4
dedicated devices (e.g. operating sensor or coils) shall
be connected in series with the protective conductor.

Any exposed-conductive-part(s) of equipment shall         WR-543.3.5
not be used to form a protective conductor for other
equipment.

 See Regulations 543.2.1, 543.2.2 and 543.2.5 of BS 7671:2008 (Incorporating Amendment No. 1) for exceptions to this rule.

Every joint in metallic conduit shall be mechanically     WR-543.3.6
and electrically continuous.

### 4.5.7.28 Prevention of harmful effects

| | |
|---|---|
| Electrical equipment shall not cause harmful effects on other equipment or interfere with the supply during normal service, including switching operations. | WR-133.4 |

### 4.5.7.29 Prevention of mutual detrimental influence

| | |
|---|---|
| An electrical installation shall be arranged in such a way that no mutual detrimental influence will occur between electrical installations and non-electrical installations. | WR-132.11 |

 Electromagnetic interference shall be taken into account.

### 4.5.7.30 Protection against fault current

| | |
|---|---|
| Conductors other than live conductors, and any other parts intended to carry a fault current, shall be capable of carrying that current without attaining any excessive temperature. Electrical equipment, including conductors, shall be provided with mechanical protection against electromechanical stresses caused by fault currents in order to prevent injury or damage to persons, livestock or property. | WR-131.5 |

### 4.5.7.31 Protection against overcurrent

| | |
|---|---|
| Persons and livestock shall be protected against injury, and property shall be protected against damage, due to excessive temperatures or electromechanical stresses caused by any overcurrents likely to arise in live conductors. | WR-131.4 |

### 4.5.7.32 Protection against power supply interruption

| | |
|---|---|
| Where danger or damage is expected to arise due to an interruption of supply, suitable provisions shall be made in the installation or installed equipment. | WR-131.7 |

## 4.5.7.33 Protection against thermal effects

Electrical installation shall be so arranged that:    WR-131.3.1

- the risk of ignition of flammable materials due to high temperature or electric arc is minimized;
- during normal operation of the electrical equipment, there shall be minimal risk of burns to persons or livestock.

Persons, livestock, fixed equipment and fixed materials    WR-131.3.2
adjacent to electrical equipment shall be protected
against harmful effects of heat or thermal radiation
emitted by electrical equipment.

## 4.5.7.34 Protection against voltage disturbances and measures against electromagnetic influences

Persons and livestock shall be protected against    WR-131.6.1
injury, and property shall be protected against any
harmful effects, as a consequence of a fault
between live parts of circuits supplied at different
voltages.

Persons and livestock shall be protected against    WR-131.6.2
injury, and property shall be protected against
damage:

- as a consequence of overvoltages such as those originating from atmospheric events or from switching;
- as a consequence of undervoltage and any    WR-131.6.3 subsequent voltage recovery.

The installation shall have an adequate level of    WR-131.6.4
immunity against electromagnetic disturbances
so as to function correctly in the specified
environment.

The installation design shall take into consideration    WR-131.6.4
the anticipated electromagnetic emissions,
generated by the installation or the installed
equipment.

## 4.5.7.35 Protective conductors

Where a number of installations have separate earthing arrangements, any protective conductors common to any of these installations shall either:  WR-542.1.3.3

- be capable of carrying the maximum fault current likely to flow through them; or
- be earthed within one installation only and insulated from the earthing arrangements of any other installation.

If the protective conductor:  WR-543.1.1

- is not an integral part of a cable;
- is not formed by conduit, ducting or trunking; or
- is not contained in an enclosure formed by a wiring system;

then the cross-sectional area shall be not less than 2.5 mm$^2$ copper equivalent if protection against mechanical damage is provided, and 4 mm$^2$ copper equivalent if mechanical protection is not provided.

Where a protective conductor is common to two or more circuits, its cross-sectional area shall be:  WR-543.1.2

- calculated for the most onerous of the values of fault current and operating time encountered in each of the various circuits; or
- selected so as to correspond to the cross-sectional area of the largest line conductor of the circuits.

## 4.5.7.36 Protective bonding conductors

In each installation, main protective bonding conductors shall connect extraneous-conductive-parts to the main earthing terminal, including the following:  WR-411.3.1.2

- central heating and air-conditioning systems;
- exposed metallic structural parts of the building;
- gas installation pipes;
- water installation pipes;
- other installation pipework and ducting.

> Protective bonding conductors shall interconnect
> every simultaneously accessible exposed-conductive-
> part and extraneous-conductive-part.          WR-418.2.2

 Unless protection by automatic disconnection of supply can be applied, local protective bonding conductors should not:

- be in electrical contact with Earth directly;
- through exposed-conductive-parts; nor
- through extraneous-conductive-parts.

> The exposed-conductive-parts of the separated circuit
> shall be connected together by insulated, non-earthed
> protective bonding conductors.          WR-418.3.4

 These conductors **shall not** be connected to the protective conductor or exposed-conductive-parts of any other circuit or to any extraneous-conductive-parts.

> Where electrical separation to the supply to more
> than one current using equipment is used, the
> warning notice shall read as follows:          WR-514.13.2

> The protective bonding conductors associated with the electrical
> installation in this location MUST NOT BE CONNECTED TO
>
> # EARTH
>
> Equipment having exposed-conductive-parts connected to earth must
> not be brought into this location

## 4.5.7.37 Main equipotential bonding conductors

> Except where protective multiple earthing (PME)
> conditions apply, a main protective bonding conductor
> shall have a cross-sectional area not less than half the
> cross-sectional area required for the earthing conductor
> of the installation, and not less than $6\,mm^2$.          WR-544.1.1

| | |
|---|---|
| Where an installation has more than one source of supply to which PME conditions apply, a main protective bonding conductor shall be selected according to the largest neutral conductor of the supply. | WR-544.1.1 |
| The main equipotential bonding connection to any gas, water or other service shall be made as near as is practicable to the point of entry of that service into the premises. | WR-544.1.2 |
| Where there is an insulating section or insert at that point or there is a meter, the connection shall be made to the consumer's hard metal pipework and before any branch pipework. | WR-544.1.2 |
| Where practicable, the connection shall be made within 600 mm of the meter outlet union, or at the point of entry to the building if the meter is external. | WR-544.1.2 |
| In agricultural and horticultural premises, protective bonding conductors shall be protected against mechanical damage and corrosion, and shall be selected to avoid electrolytic effects. | WR-705.544.2 |
| In mobile units, accessible conductive-parts of the unit, such as the chassis, shall be connected through the main protective bonding conductors to the main earthing terminal within the unit. | WR-717.411.3.1.2 |

### 4.5.7.38  Types of protective conductor

Where a metal enclosure or frame of a low-voltage switchgear or controlgear assembly or busbar trunking system is used as a protective conductor:

- its electrical continuity should be ensured, either by construction or by suitable connection, in such a way as to be protected against mechanical, chemical or electrochemical deterioration;
- its cross-sectional area shall be in accordance with BS EN 60439–1;
- it shall permit the connection of other protective conductors at every predetermined tap-off point.

| | |
|---|---|
| A gas pipe, an oil pipe, flexible or pliable conduit, support wires or other flexible metallic parts, or constructional parts subject to mechanical stress in normal service, **shall not** be selected as a protective conductor. | WR-543.2.1 |

A protective conductor may consist of one or more of the following:    WR-543.2.2

- a single-core cable;
- a conductor in a cable;
- an insulated or bare conductor in a common enclosure with insulated live conductors;
- a fixed bare or insulated conductor;
- a metal covering (e.g. the sheath, screen or armouring of a cable);
- a metal conduit, metallic cable management system or other enclosure or electrically continuous support system for conductors;
- an extraneous-conductive-part.

 This protective conductor shall be of copper with a cross-sectional area of no more than 10 mm$^2$.

The metal covering (including the sheath – bare or insulated) of a mineral insulated cable, trunking, ducting and metal conduit, may be used as a protective conductor for the associated circuit.    WR-543.2.5

An extraneous-conductive-part may be used as a protective conductor if:    WR-543.2.6

- electrical continuity can be ensured and the part is either constructed or connected so that it is protected against mechanical, chemical or electrochemical deterioration;
- precautions have been taken against its removal;
- it has been considered for such a use and, if necessary, suitably adapted.

Where the protective conductor is formed by a conduit, trunking, ducting or the metal sheath and/or armour of a cable, the earthing terminal of each accessory shall be connected by a separate protective conductor to an earthing terminal that is part of the associated box or other enclosure.    WR-543.2.7

An exposed-conductive-part of equipment shall not be used to form a protective conductor for other equipment.    WR-543.2.8

See Regulations 543.2.1, 543.2.2 and 543.2.5 of the IET Wiring Regulations for exceptions to this rule.

| | |
|---|---|
| Except where the circuit-protective conductor is formed by a metal covering or enclosure containing all the conductors of the ring, the circuit-protective conductor of every ring final circuit shall also be run in the form of a ring having both ends connected to the earthing terminal at the origin of the circuit. | WR-543.2.9 |
| A separate metal enclosure for cables shall not be used as a PEN conductor. | WR-543.2.10 |

### 4.5.7.39 Protective devices and switches

| | |
|---|---|
| Single-pole fuses, switches or circuit-breakers shall only be inserted in the line conductor. | WR-132.14.1 |
| No switches, circuit-breaker, (except where linked) or fuses shall be inserted in an earthed neutral conductor. | WR-132.14.2 |
| Any linked switch or linked circuit-breaker inserted in an earthed neutral conductor shall be capable of breaking all the related line conductors. | WR-132.14.2 |

### 4.5.7.40 Protective earthing

The type of earthing system that is going to be used for an installation shall be determined, taking into consideration the characteristics of the source of energy and (in particular) any earthing facilities.

| | |
|---|---|
| If an overcurrent protective device is used for protection against electric shock, the protective conductor shall be incorporated in the same wiring system as the live conductors, or be located nearby. | WR-543.6.1 |

 Where earthing is required for protective as well as functional purposes, then the requirements for protective measures shall take precedence.

| | |
|---|---|
| Exposed-conductive-parts shall be connected to a protective conductor. | WR-411.3.1.1 |

| | |
|---|---|
| Simultaneously accessible exposed-conductive-parts shall be connected to the same earthing system individually, in groups or collectively. | WR-411.3.1.1 |
| A circuit-protective conductor shall be run to and terminated at each point in the wiring and at each accessory (except a lampholder having no exposed-conductive-parts and suspended from such a point). | WR-411.3.1.1 |
| The characteristics of protective equipment shall be determined with respect to their function, including protection against the effects of Earth fault current. | WR-132.8 |
| No switches, circuit-breakers (except where linked) or fuses shall be inserted in an earthed neutral conductor. | WR-132.14.2 |
| Any linked switch or linked circuit-breaker inserted in an earthed neutral conductor shall be capable of breaking all the related line conductors. | WR-132.14.2 |
| All exposed-conductive-parts of the reduced low-voltage system shall be connected to Earth. | WR-411.8.3 |
| The Earth fault loop impedance at every point of utilization, including socket-outlets, shall be such that the disconnection time does not exceed 5 s. | WR-411.8.3 |
| Live parts of the separated circuit shall not be connected at any point to another circuit, or to Earth or to a protective conductor. | WR-413.3.3 |
| No exposed-conductive-part of the separated circuit shall be connected either to the protective conductor or exposed-conductive-parts of other circuits, or to Earth. | WR-413.3.6 |

## 4.5.7.41 Protective equipment (devices and switches)

| | |
|---|---|
| The characteristics of protective equipment shall be determined with respect to their function, including protection against the effects of: | WR-132.8 |

- overcurrent (overload, short-circuit);
- Earth fault current;
- overvoltage;
- undervoltage and no-voltage.

| The protective devices shall operate at values of current, voltage and time that are suitably related to the characteristics of the circuits and to the possibilities of danger. | WR-132.8 |

### 4.5.7.42 Protective measures

| If a protective measure does not satisfy the requirements, then supplementary provisions shall be applied so that together the protective provisions achieve the same degree of safety. | WR-410.3.7 |

### 4.5.7.43 Protective measure: extra-low voltage provided by SELV

| In swimming pools and other basins where SELV is used, whatever the nominal voltage, protection shall be provided by:<br><br>• basic insulation; or<br>• barriers or enclosures affording a degree of protection of at least IPXXB or IP2X. | WR-702.414.4.5 |

### 4.5.7.44 Safety isolating transformers and electronic convertors

In temporary installations in fairgrounds, amusement parks and circuses:

| Safety isolating transformers shall comply with BS EN 61558–2–6 or provide an equivalent degree of safety. | WR-740.55.5 |
| A manually reset protective device shall protect the secondary circuit of each transformer or electronic convertor. | WR-740.55.5 |
| Safety isolating transformers shall be mounted out of arm's reach or be mounted in a location that provides equal protection and they shall have adequate ventilation. | WR-740.55.5 |
| Access by competent persons for testing (or by a skilled person competent in such work for protective device maintenance) shall be provided. | WR-740.55.5 |

Electronic convertors shall conform to BS EN
61347–2–2.                                                    WR-740.55.5

Enclosures containing rectifiers and transformers shall     WR-740.55.5
be adequately ventilated, and the vents shall not be
obstructed when in use.

### 4.7.5.45 Electrical sources for safety services

Safety services need to regulated.                          WR-351

**Note:** Examples of safety services include:

- emergency lighting;
- fire pumps;
- fire rescue service lifts;
- fire detection and alarm systems;
- carbon monoxide (CO) detection and alarm systems;
- fire evacuation systems;
- smoke ventilation systems;
- fire services communication systems;
- essential medical systems;
- industrial safety systems.

### 4.7.5.46 Safety sources

Electrical sources for safety services shall be installed as fixed equipment, in
such a manner that they cannot be adversely affected by failure of the normal
source.

Safety sources shall be placed in a suitable location         WR-560.6.3
and be accessible only to skilled persons or instructed
persons (BA4 or BA5).

The location of a safety source shall be properly and         WR-560.6.4
adequately ventilated so that exhaust gases, smoke or
fumes from the safety source cannot penetrate areas
occupied by persons.

Separated independent feeders from a supply network           WR-560.6.5
shall not serve as electrical safety sources unless
assurance can be obtained that the two supplies are
unlikely to fail concurrently.

| | |
|---|---|
| The safety source shall have sufficient capability to supply its related services. | WR-560.6.6 |
| A safety source may, in addition, be used for purposes other than safety services, **provided** that a fault occurring in a circuit for purposes other than safety services does not cause the interruption of any circuit for safety services. | WR-560.6.7 |
| Protection against fault current and against electric shock in case of a fault shall be ensured whether the installation is supplied separately by either of the two sources or by both in parallel. | WR-560.6.8.1 |

 **Note:** The following sources for safety services are recognized:

- storage batteries;
- primary cells;
- generator sets independent of the normal supply;
- a separate feeder of the supply network effectively independent of the normal feeder.

### 4.7.5.47  Seismic effects

Wiring systems shall be selected and erected with due regard to the seismic hazards of the physical location of the installation.

| | |
|---|---|
| Where the seismic hazards experienced are of severity (i.e. AP2) or higher, particular attention shall be paid to:<br><br>- the fixing of wiring systems to the building structure;<br>- the connections between the fixed wiring and all items of essential equipment (e.g. safety services) shall be selected for their flexible quality. | WR-522.12.2 |

### 4.5.7.48  Sources for SELV and PELV

| | |
|---|---|
| The following sources may be used for SELV and PELV systems:<br><br>- an electrochemical source (e.g. a battery) or another source independent of a higher voltage circuit (e.g. a diesel-driven generator); | WR-414.3 |

- an electronic device such as insulation testing equipment or monitoring device;
- a motor-generator with windings providing equivalent isolation to a safety isolating transformer;
- a safety isolating transformer.

### 4.5.7.49 *Supplementary bonding conductors*

A supplementary bonding conductor connecting two exposed-conductive-parts shall have a conductance (if sheathed or otherwise provided with mechanical protection) not less than that of the smaller protective conductor connected to the exposed-conductive-parts.  WR-544.2.1

 If mechanical protection is not provided, its cross-sectional area shall be not less than 4 mm².

A supplementary bonding conductor connecting an exposed-conductive-part to an extraneous-conductive-part shall have a conductance (if sheathed or otherwise provided with mechanical protection) not less than half that of the protective conductor connected to the exposed-conductive-part.  WR-544.2.2

A supplementary bonding conductor connecting two extraneous-conductive-parts shall have a cross-sectional area not less than 2.5 mm² if sheathed or otherwise provided with mechanical protection, or 4 mm² if mechanical protection is not provided.  WR-544.2.3

Supplementary bonding shall be provided by a supplementary conductor, a conductive-part of a permanent and reliable nature, or by a combination of these.  WR-544.2.4

Where supplementary bonding is to be applied to a fixed appliance (which is supplied via a short length of flexible cable from an adjacent connection unit or other accessory incorporating a flex outlet), the circuit-protective conductor within the flexible cable shall be deemed to provide the supplementary bonding connection to the exposed-conductive-parts of the appliance.  WR-544.2.5

In agricultural and horticultural premises, supplementary bonding conductors shall be protected against mechanical damage and corrosion, and shall be selected to avoid electrolytic effects.

WR-705.544.2

### 4.5.7.50 Testing

When undertaking testing in a potentially explosive atmosphere, appropriate safety precautions in accordance with BS EN 60079–17 and BS EN 61241–17 are necessary.

WR-612.1

### 4.5.7.51 Transformers and converters

A safety isolating transformer for an extra-low-voltage lighting installation shall comply with BS EN 61558–2–6 and:

WR-559.11.3.1

- either the transformer shall be protected on the primary side by a protective device; or
- the transformer shall be short-circuit proof (both inherently and non-inherently).

### 4.5.7.52 Type of wiring and method of installation

The choice of the type of wiring system and the method of installation shall include consideration of the following:

WR-132.7

- accessibility of wiring to persons and livestock;
- electromagnetic interference;
- the electromechanical stresses likely to occur due to short-circuit and Earth fault currents;
- the nature of the location;
- the nature of the structure supporting the wiring;
- voltage;
- other external influences (e.g. mechanical, thermal and those associated with fire) to which the wiring is likely to be exposed during the erection of the electrical installation or in service.

## 4.5.7.53  Uninterruptible power supply sources

A static-type uninterruptible power supply (UPS)    WR-560.6.12
source shall be able to:

- operate distribution circuit-protective devices; and
- start the safety devices (when operating in the emergency condition from the convertor supplied by the battery).

## 4.5.7.54  Agricultural and horticultural premises

For high-density livestock rearing, systems operating    WR-705.560.6
for the life support of livestock shall meet the
following requirements:

- where the supply of food, water, air and/or lighting to livestock is not ensured in the event of power supply failure, a secure source of supply shall be provided (e.g. an alternative or back-up supply) and separate final circuits for ventilation and lighting units shall also be provided;
- where electrically powered ventilation is necessary in an installation, one of the following shall be provided:

  o  a standby electrical source ensuring sufficient supply for ventilation equipment, or
  o  temperature and supply voltage monitoring.

 **Note:** A notice should be placed adjacent to the standby electrical source, indicating that it should be tested periodically according to the manufacturer's instructions.

## 4.5.7.55  Construction and demolition site installations

Safety and standby supplies shall be connected    WR-704.537.2.2
by means of devices arranged to prevent
interconnection of the different supplies.

### 4.5.7.56 Electrical installations in caravan/camping parks and similar locations

| | |
|---|---|
| The protective measures of obstacles and placing out of reach are **not** permitted. | WR-708.410.3.5 |
| The protective measures of non-conducting location and Earth-free local equipotential bonding are **not** permitted. | WR-708.410.3.6 |

 In the UK, the Electricity Safety, Quality and Continuity Regulations 2002 prohibit the use of a TN-C-S system for the supply to a caravan or similar construction.

### 4.5.7.57 Locations containing a bath or shower

| | |
|---|---|
| The protective measures of obstacles and placing out of reach are **not** permitted. | WR-701.410.3.5 |
| The protective measures of non-conducting location and Earth-free local equipotential bonding are **not** permitted. | WR-701.410.3.6 |

### 4.5.7.58 Swimming pools and other basins

### 4.5.7.58.1 Zones 0 and 1 of fountains

| | |
|---|---|
| In zones 0 and 1, one or more of the following protective measures shall be employed:<br><br>• SELV;<br>• automatic disconnection of supply using an RCD;<br>• electrical separation. | WR-702.410.3.4.2 |

### 4.5.7.58.2 Zone 2 (swimming pools and other basins)

| | |
|---|---|
| One or more of the following protective measures shall be employed:<br><br>• SELV;<br>• automatic disconnection of supply;<br>• electrical separation. | WR-702.410.3.4.3 |

The protective measures of obstacles and placing out of reach are **not** permitted.  WR-702.410.3.5

The protective measures of non-conducting location and Earth-free local equipotential bonding are **not** permitted.  WR-702.410.3.6

### 4.5.7.58.3 Supplementary equipotential bonding

All extraneous-conductive-parts in zones 0, 1 and 2 shall be connected by supplementary protective bonding conductors to the protective conductors of exposed-conductive-parts of equipment situated in these zones.  WR-702.411.3.3

The socket-outlet of a circuit supplying such equipment and the control device of such equipment shall have a notice in order to warn the user that this equipment shall be used only when the swimming pool is not occupied by persons.

### 4.5.7.59 Medical locations

The main requirement of an electrical installation in a medical location is to provide safety for the patient and medical personnel, and this is achieved by providing an isolated power supply, equipotential Earth bonding and anti-static discharge. Leakage currents must always be controlled within strict limits. There should also be a monitoring device with an alarm for disconnection, insulation failure, overload and high temperature.

To achieve this aim, and ensure the protection of patients from possible electrical hazards, additional protective measures need to be applied in medical locations.

### 4.5.7.59.1 Protection against electric shock

The protective measures:  WR-710.410.3.5

- of obstacles and placing out of reach;
- non-conducting location;
- Earth-free local equipotential bonding; and
- electrical separation for the supply of more than one item of current-using equipment;

are **not** permitted in medical locations.

### 4.5.7.59.2 Protective measure: extra-low voltage provided by SELV or PELV

| | |
|---|---|
| When using SELV and/or PELV circuits in medical locations of Group 1 and Group 2, protection by basic insulation of live parts (e.g. covering live parts with insulation), or barriers or enclosures shall be provided. | WR-710.414.1 |
| In Group 2 medical locations, where PELV is used, exposed-conductive-parts of equipment (e.g. operating theatre luminaires) shall be connected to the circuit-protective conductor. | WR-710.414.4.1 |

### 4.5.7.59.3 Functional extra-low voltage (FELV)

| | |
|---|---|
| In medical locations, functional extra-low voltage (FELV) is not permitted as a method of protection against electric shock. | WR-710.411.7 |

### 4.5.7.59.4 Additional protection: supplementary equipotential bonding

| | |
|---|---|
| In each Group 1 or Group 2 medical location, supplementary equipotential bonding shall be installed to equalize the potential differences between: | WR-710.415.2.1 |

- extraneous-conductive-parts;
- screening against electrical interference fields, if installed;
- connection to conductive floor grids, if installed;
- metal screens of isolating transformers, via the shortest route to the earthing conductor;

which are located in, or that may be moved into, the *patient environment*.

### 4.5.7.59.5 Risk of explosion

| | |
|---|---|
| All electrical devices (e.g. socket-outlets and/or switches) shall be installed at least 0.2 m below any medical-gas outlet, so as to minimize the risk of ignition of flammable gases. | WR-710.512.2.1 |

### 4.5.7.59.6 Safety services

Medical locations shall have a standby power supply available that will energize the installations required for continuous operation in case of failure of the general power system.

 Safety services that have been provided for a number of locations with different classifications should meet that classification that gives the highest security of supply.

| | |
|---|---|
| The safety power supply system shall be capable of automatically taking over if the main power supply has dropped for more than 0.5 s and by more than 10%, and shall be capable of providing power for a period of at least 3 h for: | WR-710.56 |

- operating theatre table luminaires;
- medical electrical equipment containing light sources (endoscopes and monitors etc.);
- life-supporting electrical medical equipment.

### 4.5.7.59.7 General requirements for safety power supply sources for Group 1 and Group 2

| | |
|---|---|
| Primary cells are not allowed as safety power sources. | WR-710.560.5.5 |

An additional main incoming power supply, from the general power supply, is not regarded as a source of the safety power supply.

The availability (readiness for service) of safety power sources shall be monitored.

| | |
|---|---|
| Where socket-outlets are supplied from the safety power supply source, they shall be identified according to their safety services classification. | WR-710.560.5.7 |

### 4.5.7.59.8 Power supply sources with a changeover period greater than 15 s

| | |
|---|---|
| Hospital maintenance services equipment such as: | WR-710.560.6.1.3 |

- sterilization equipment;

- technical building installations, in services and waste disposal systems;
- cooling equipment;
- catering equipment;
- storage battery chargers;

shall be connected either automatically or manually to a safety power supply source capable of maintaining it for a minimum period of 24 h.

### 4.5.7.59.9 Emergency lighting systems

In the event of mains power failure, the changeover period to the standby safety services sources such as:

WR-710.560.9.1

- emergency lighting and exit signs;
- switchgear and control gear for emergency generating sets;
- normal power supply and the standby safety services power supply distribution boards;
- essential service rooms;
- locations of central fire alarm and monitoring systems;

shall not exceed 15 s.

- Group 1 medical location rooms shall be supplied with at least one luminaire in case of emergency.
- Group 2 medical locations rooms shall have at least 90% of their normal lighting requirements supplied from the standby safety service.

 The luminaires of the escape routes shall be arranged on alternate circuits.

### 4.5.7.59.10 Other services

Other services such as

WR-710.560.11

- fire-fighter lifts;
- smoke-extraction ventilation systems;
- paging systems;

- vitally important medical electrical equipment used in Group 2 medical locations;
- electrical equipment of medical gas supply including (e.g. compressed air, vacuum supply and narcosis (anaesthetics) – exhaustion as well as their monitoring devices;

shall be provided with a changeover system to a standby safety service within 15 s.

# 5

# Electrical equipment, components, accessories and supplies

The amount of different types of equipment, components, accessories and supplies for electrical installations currently available is enormous, and any attempt to cover every type, model and/or manufacture would prove an impossible task for a book such as this. The intention of this chapter, therefore, is to provide a catalogue of all the different types identified and referred to in the Wiring Regulations (e.g. luminaires, residual current devices (RCDs), plugs and sockets), and then make a list of the specific requirements that are sprinkled throughout the Regulations. For your (hopeful!?) convenience, this catalogue has been compiled in alphabetical order.

 Similar to other chapters, please remember that these lists of requirements are **only** the author's impression of the most important aspects of the Wiring Regulations, and electricians should **always** consult the latest edition of BS 7671 to satisfy compliance!

## 5.1 Installation

| | |
|---|---|
| Except where specifically designed for direct connection to flexible wiring, equipment should be fixed so that connections between wiring and equipment **shall not** be subject to undue stress or strain resulting from the normal use of the equipment. | WR-530.4.1 |
| Unenclosed equipment shall be mounted in a suitable mounting box or enclosure. | WR-530.4.2 |
| Socket-outlets, connection units, plate switches and similar accessories shall be fitted to a mounting box. | WR-530.4.2 |

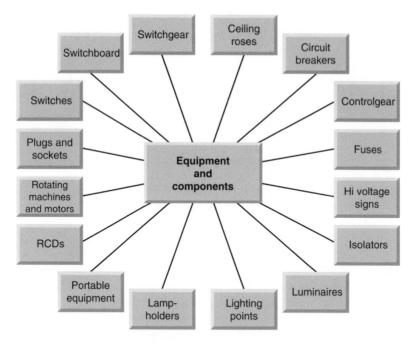

**Figure 5.1** Electrical equipment and components.

| | |
|---|---|
| Wherever equipment is fixed on or in cable trunking, skirting trunking or in mouldings, it **shall not** be fixed on covers that can be removed inadvertently. | WR-530.4.3 |
| Any wiring system within Group 2 medical locations shall be exclusive for the use of equipment and accessories within those locations. | WR-710.52 |

### 5.1.1  Circuit-breakers

Every circuit shall be provided with a means of isolation from all live supply conductors by a linked switch or a linked circuit-breaker.

| | |
|---|---|
| A circuit-breaker providing protection against both overload and fault current shall be capable of 'making' any overcurrent up to and including the maximum prospective fault current at the point where the device is installed. | WR-432.1 |

| | |
|---|---|
| Where a circuit-breaker is used, the maximum value of the Earth fault loop impedance ($Z_s$) shall be determined using the formula $Z_s \times I_a \ U_o$. | WR-411.8.3 |
| Circuit-breakers shall be inserted in the line conductor only. | WR-132.14.1 |
| Circuit-breakers (except where linked) shall be inserted in an earthed neutral conductor. | WR-132.14.2 |

 Any linked circuit-breaker inserted in an earthed neutral conductor shall be capable of breaking all the related line conductors.

| | |
|---|---|
| Where a circuit-breaker can be operated by a person other than a skilled person or instructed person, it shall be designed or installed so that it is not possible to modify the setting or the calibration of its overcurrent release without a deliberate act involving the use of either a key or a tool and resulting in a visible indication of its setting or calibration. | WR-533.1.2 |
| A main linked switch or linked circuit-breaker shall be provided as near as is practicable to the origin of every installation as a means of switching the supply on load and as a means of isolation. | WR-537.1.4 |
| The supply to an electrode water heater or electrode boiler shall be controlled by a linked circuit-breaker. | WR-554.1.2 |

 **Note:** If the electrode water heater or electrode boiler is not piped to a water supply or in physical contact with any earthed metal (and where the electrodes and the water in contact with the electrodes are so shielded in insulating material that they cannot be touched while the electrodes are live), a fuse in the line conductor may be substituted for the circuit-breaker, and the shell of the electrode water heater or electrode boiler need not be connected to the neutral of the supply.

| | |
|---|---|
| Single-phase water heaters and boilers with an uninsulated heating element immersed in the water shall not have a non-linked circuit-breaker fitted in the neutral conductor, in any part of the circuit between the heater or boiler or the origin of the installation. | WR-554.3.1 WR-554.3.4 |

### 5.1.1.1 Protection against fault current only

A device shall be capable of breaking (and, for a circuit-breaker, making) the fault current up to and including the prospective fault current.

A circuit-breaker with a short-circuit release providing protection against fault current shall only be installed where overload protection is achieved by other means.

## 5.1.2 Isolators

| | |
|---|---|
| The location of each disconnector (isolator) shall be indicated with a suitable notice, unless there is no possibility of confusion. | WR-514.11.1 |
| Off-load isolators (disconnectors) **shall not** be used for functional switching. | WR-537.5.2.3 |

# 5.2 Fuses

| | |
|---|---|
| Single-pole fuses shall only be inserted in the line conductor. | WR-132.14.1 |
| Fuses shall be inserted in an earthed neutral conductor. | WR-132.14.2 |
| For every fuse and circuit-breaker there shall be provided on, or adjacent to it, an indication of its intended rated current as appropriate to the circuit it protects. | WR-533.1 |

## 5.2.1 Devices for isolation and switching

| | |
|---|---|
| For TN systems where it is intended that isolation and switching is carried out **only** by instructed persons, the means of switching the supply on load and the means of isolation can be provided by a suitably rated fuse carrier. | WR-559.10.6.1 |

**Note:** Table 5.1 provides an indication of the maximum values of Earth fault loop impedance corresponding to a disconnection time of 5 s for a nominal voltage to Earth ($U_o$) of 230 V.

**Table 5.1** General-purpose fuses – maximum earth fault loop impedance (Zs) for circuit-breakers of Uo = 230 V

| Rating (A) | 6 | 10 | 16 | 20 | 25 | 32 | 40 | 50 |
|---|---|---|---|---|---|---|---|---|
| $Z_s$ (Ω) | 12.8 | 7.19 | 4.18 | 2.95 | 2.30 | 1.84 | 1.35 | 1/04 |

**Note:** The circuit loop impedances given in Table 5.1 should not be exceeded when the conductors are at their normal operating temperature. If the conductors are at a different temperature when tested, the reading should be adjusted accordingly.

Where the device is a general-purpose type fuse to BS 88–2.1, BS 88–6 or BS 1361 (or a semi-enclosed fuse to BS 3036), the conditions shown in Table 5.2 shall apply.

All low-voltage fused plug and socket-outlets shall conform to the applicable British Standard listed in Table 5.3.

**Table 5.2** Coordination between conductor and protective device (Courtesy of Stingray)

| | Design current ($I_B$) of the circuit | Lowest current-carrying capacity ($I_Z$) of any conductor in a circuit | Operating current of any protective device ($I_2$) |
|---|---|---|---|
| **Requirements** | | | |
| | Shall be greater than the nominal current or current setting ($I_n$) of a protective device | Shall be greater than the nominal current or current setting ($I_n$) of a protective device | Shall not exceed 1.45 times the lowest of the current-carrying capacities ($I_Z$) of any of the conductors in the circuit |
| BS 88–2.1 fuses | | Yes | Yes |
| BS 88–6 fuses | | Yes | Yes |
| BS 1361 fuses | | Yes | Yes |
| BS 3036 (semi-enclosed) fuses | | | Yes, provided ($I_n$) does not exceed 0.725 ($I_Z$) |

**Table 5.3** Plugs and socket-outlets for low-voltage circuits

| Type of plug and socket-outlet | Rating (A) | Applicable British Standard |
|---|---|---|
| Fused plugs and shuttered socket-outlets, two-pole and Earth, for a.c. | 13 | BS 1363 (fuses to BS 1362) |
| Plugs, fused or non-fused, and socket-outlets, two-pole and Earth | 2, 5, 15, 30 | BS 546 (fuses, if any, to BS 646) |
| Plugs, fused or non-fused, and socket-outlets, protected type, two-pole and Earth | 5, 15, 30 | BS 196 |
| Plugs and socket-outlets (industrial type) | 16, 32, 63, 125 | BS EN 60309–2 |

## 5.2.2 Devices for protection against overcurrent

Every fuse shall be provided with an indication of its intended rated current as appropriate to the circuit it protects.

| | |
|---|---|
| Fuses shall preferably be of the cartridge type. | WR-533.1.1.3 |
| Fuse bases shall be arranged so as to exclude the possibility of the fuse carrier making contact between conductive parts belonging to two adjacent fuse bases. | WR-533.1.1.1 |
| A fuse base using screw-in fuses shall be connected so that the centre contact is connected to the conductor from the supply and the shell contact is connected to the conductor to the load. | WR-533.1.1.1 |
| A fuse with fuse links that is likely to be removed or replaced by persons other than instructed persons or skilled persons shall either: | WR-533.1.1.2 |

- have marked on or adjacent to it an indication of the type of fuse link that should be used; or
- be of a type such that there is **no possibility** of inadvertently replacing the fuse with one that has a higher rated current or a higher fusing factor than that intended.

 **Note:** Fuses or combination units that have fuse links which are likely to be removed and replaced only by skilled or instructed persons shall be installed in such a manner that it is ensured that the fuse links can **only** be removed or replaced without unintentional contact with live parts.

| | |
|---|---|
| Where a semi-enclosed fuse is selected, it shall be fitted with an element in accordance with the manufacturer's instructions, or (in the absence of such instructions) it shall be fitted with a single element of tinned copper wire of the appropriate diameter specified in Table 5.4. | WR-533.1.1.3 |

## 5.2.3 Devices for protection against overvoltages

The following requirements (paraphrased from Chapter 544 of BS 7671:2008 (Incorporating Amendment No. 1, 2011)) concern the selection and erection of surge protective devices (SPDs) that are intended:

**Table 5.4** Sizes of tinned copper wire for use in semi-enclosed fuses

| Nominal current (A) | Nominal diameter of wire (mm) |
| --- | --- |
| 3 | 0.15 |
| 5 | 0.2 |
| 10 | 0.35 |
| 15 | 0.5 |
| 20 | 0.6 |
| 25 | 0.75 |
| 30 | 0.85 |
| 45 | 1.25 |
| 60 | 1.53 |
| 80 | 1.8 |
| 100 | 2.0 |

- to limit transient overvoltages of atmospheric origin that are transmitted via the supply distribution system and against switching overvoltages to electrical installations of buildings;
- to protect against transient overvoltages caused by direct lightning strikes (or lightning strikes in the near vicinity) of buildings protected by a lightning protection system.

### 5.2.3.1 Selection and erection of SPDs

Where required or otherwise specified, SPDs shall be installed:

- near the origin of an installation; or
- in the main distribution assembly nearest the origin of an installation.

If additional SPDs are required to protect sensitive and critical equipment (e.g. hospital equipment and fire/security alarm systems), these SPDs shall be coordinated with the SPDs installed upstream and be installed as close as is practicable to the equipment to be protected.

### 5.2.3.2 Connection of SPDs

SPDs at or near the origin of the installation shall be connected between specific conductors according to Table 5.5.

**Notes:**

1. If more than one SPD is connected on the same conductor, it is necessary to ensure co-ordination between them.
2. The SPD assembly shall be installed as close as possible to the origin of the installation, and shall have sufficient surge-withstand capability for this location.

**Table 5.5** Types of protection for various LV systems

| SPDs connected between: | TN-C-S, TN-S or TT: installation in accordance with: | | IT without distributed neutral |
| --- | --- | --- | --- |
| | Connection Type 1 | Connection Type 2 | |
| Each line conductor and neutral conductor | Optional | SPD required | Not applicable |
| Each line conductor and PE conductor | SPD required | Not applicable | SPD required |
| Neutral conductor and PE conductor | SPD required | SPD required | Not applicable |
| Each line conductor and PEN conductor | Not applicable | Not applicable | Not applicable |
| Line conductors | Optional | Optional | Optional |

3.  Where the equipment to be protected is located close to the main distribution board, one SPD assembly may be sufficient.

Sensitive and critical equipment will require protection in both common and differential modes to ensure further protection against switching transients.

### 5.2.3.3 Selection of SPDs

SPDs shall be selected in accordance with the following regulations.

### 5.2.3.3.1 Selection with regard to voltage protection level (Up)

SPDs shall be selected in accordance with the impulse-withstand voltage of the equipment (or impulse immunity of critical equipment) to be protected, and the nominal voltage of the system shall be considered in selecting the preferred voltage protection level ($U_p$) value of the SPD.

SPDs shall be selected to provide a voltage protection level that is lower than the impulse-withstand capability of the equipment or (where the continuous operation of the equipment is critical) is lower than the impulse immunity of the equipment.

SPDs shall be selected that will protect equipment from failure, remain operational during surge activity and withstand most temporary overvoltage conditions.

**Note:** If the distance between the SPD and the equipment to be protected ('protective distance') is greater than 10 m, oscillations could lead to a voltage at the equipment terminals of up to twice the voltage protection level of the SPD. Consideration should, therefore, be given to providing additional co-ordinated SPDs closer to the equipment, or the selecting SPDs with a lower voltage protection level.

Where the voltage protection level required cannot be obtained with a single SPD, additional, co-ordinated SPDs should be installed to ensure the required voltage protection level.

### 5.2.3.3.2 Selection with regard to continuous operating voltage

The maximum continuous operating voltage $U_c$ of SPDs shall be equal to or higher than that required by Table 5.6.

**Table 5.6** Minimum required $U_c$ of the SPD, depending on the supply system configuration

| SPDs connected between: | TN-C-S, TN-S or TT | IT without distributed neutral |
|---|---|---|
| Line conductor and neutral conductor | $1.1\,U_{aspd}$ | Not applicable |
| Each line conductor and PE conductor | $1.1\,U_{aspd}$ | Line to line voltage (Note 2) |
| Neutral conductor and PE conductor | $U_{aspd}$ (Note 2) | Not applicable |
| Each line conductor and PEN conductor | Not applicable | Not applicable |

 Notes:

1. $U_{aspd}$ is the nominal a.c. rms line voltage of the low-voltage system to Earth.
2. These values are related to worst-case fault conditions; therefore, the tolerance of 10% is not taken into account.
3. In extended IT systems, higher values of $U_c$ may be necessary.

### 5.2.3.3.3 Selection with regard to temporary overvoltages

SPDs shall be selected and erected in accordance with manufacturers' instructions.

SPDs are expected to fail safely.

### 5.2.3.3.4 Selection with regard to nominal discharge current ($I_{nspd}$) and impulse current ($I_{imp}$)

SPDs shall be selected according to their withstand capability, as classified in BS EN 61643–11 (for power systems) and in BS EN 61643–21 for telecommunication systems.

The nominal discharge current $I_{nspd}$ of the SPD shall be not less than 5 kA, with a waveform characteristic of 8/20 for each mode of protection.

Type 1 SPDs shall be installed where a structural lightning protection system is fitted, or where the installation is supplied by an overhead line that is at risk of a direct lightning strike.

Where Type 1 SPDs are required, the value of $I_{imp}$ shall be not less than 12.5 kA for each mode of protection, if $I_{imp}$ cannot be calculated.

Where Connection Type 2 (CT 2) installations are required, the lightning impulse current $I_{imp}$ for the SPD connected between the neutral conductor and the protective conductor shall be calculated in accordance with BS EN 62305–4.

**Note:** If the current value cannot be established, the value of $I_{imp}$ shall be not less than 50 kA for three-phase systems and 25 kA for single-phase systems.

### 5.2.3.3.5 Selection with regard to the prospective fault current and the follow current interrupt rating

The short-circuit withstand of the combination SPD and overcurrent protective device (OCPD), as stated by the SPD manufacturer shall be equal to or higher than the maximum prospective fault current expected at the point of installation.

**Note:** The OCPD may be either internal or external to the SPD.

When a follow current interrupt rating is declared by the manufacturer, it shall be equal to, or higher than, the prospective line to neutral fault current at the point of installation.

SPDs connected between the neutral conductor and the protective conductor in TT or TN systems, which allow a power frequency follow current after operation (e.g. spark gaps), shall have a follow current interrupt rating greater than or equal to 100 A.

In IT systems, the follow current interrupt rating for SPDs connected between the neutral connector and the protective conductor shall be the same as for SPDs connected between line and neutral.

### 5.2.3.3.6 Co-ordination of SPDs

SPDs shall be selected and erected such as to ensure operational coordination.

### 5.2.3.3.7 Protection against overcurrent and consequences of the end of the life of the SPD

Protection against SPD short-circuits is provided by OCPDs.

### 5.2.3.4 *Fault-protection integrity*

Fault protection shall remain effective in the protected installation, even in case of failures of SPDs.

In TN systems, automatic disconnection of supply shall be obtained by correct operation of the OCPD on the supply side of the SPD.

In TT systems this shall be obtained by the installation of SPDs either on the load side or on the supply side of an RCD.

**Notes:**

1.   If SPDs are installed on the load side of an RCD, they should have an immunity to surge currents of at least 3 kA 8/20.
2.   SPDs shall be provided with a status indicator (e.g. electrical, visual or audible alarm system) to indicate when they no longer provide overvoltage protection.

### 5.2.3.4.1  Critical length of connecting conductors

To gain maximum protection:

- supply conductors shall be kept as short as possible so as to minimize additive inductive voltage drops across the conductors;
- current loops shall be avoided;
- the total lead length of supply conductors should not exceed 1.0 m (preferably no longer than 0.5 m);
- if an SPD is fitted in-line, the protective conductor shall not exceed 1.0 m (preferably not longer than 0.5 m).

To ensure that SPD connections are as short and their inductance as low as possible, SPDs may be connected to the main earthing terminal (or to the protective conductor) via the metallic enclosures of the assembly that are connected to the protective conductor.

### 5.2.3.4.2  Cross-sectional area of connecting conductors

The connecting conductors of SPDs shall either:

- have a cross-sectional area of not less than 4 mm$^2$ copper (or equivalent), **if** the cross-sectional area of the line conductors is greater than or equal to 4 mm$^2$; or
- have a cross-sectional area not less than that of the line conductors, where the line conductors have a cross-sectional area less than 4 mm$^2$.

**Note:** The minimum cross-sectional area for Type 1 SPDs shall be 16 mm$^2$ copper, or equivalent, where there is a structural lightning protection system.

### 5.2.4 Electrode water heaters and boilers

| | |
|---|---|
| If an electrode water heater or electrode boiler is not piped to a water supply or in physical contact with any earthed metal (and where the electrodes and the water in contact with the electrodes are so shielded in insulating material that they cannot be touched while the electrodes are live), a fuse in the line conductor may be substituted for the circuit-breaker and the shell of the electrode water heater or electrode boiler need not be connected to the neutral of the supply. | WR-554.1.7 |
| Single-phase water heaters and boilers with an uninsulated heating element immersed in the water shall not have a fuse fitted in the neutral conductor, in any part of the circuit between the heater or boiler or the origin of the installation. | WR-554.3.1 WR-554.3.4 |

### 5.2.5 Functional switching devices

| | |
|---|---|
| Functional switching devices:<br><br>• shall be suitable for the most onerous duty they are intended to perform;<br>• may control the current without necessarily opening the corresponding poles. | WR-537.5.2.1 WR-537.5.2.2 |

 Fuses and links **shall not** be used for functional switching.

### 5.2.6 Protection against fault current only

| | |
|---|---|
| Fuses that provide protection against fault current shall only be installed where overload protection is achieved by other means. | WR-432.3 |

## 5.3 Heaters

| | |
|---|---|
| Measures shall be taken to prevent an enclosure of electrical equipment (such as a heater) from exceeding the following temperatures:<br><br>• 90°C under normal conditions; and<br>• 115°C under fault conditions. | WR-422.3.2 |

## 5.3.1 Electrode water heaters and boilers

| | |
|---|---|
| Electrode water heaters and boilers shall **only** be connected to an a.c. system. | WR-554.1.1 |
| The supply to electrode water heaters or boilers shall be controlled by a linked circuit-breaker. | WR-554.1.2 |
| The shell of an electrode water heater or boiler shall be bonded to the metallic sheath and armour, if any, of the incoming supply cable. | WR-554.1.3 |
| If an electrode water heater or boiler is directly connected to a supply at a voltage exceeding low voltage, the installation shall include an RCD. | WR-554.1.4 |
| If an electrode water heater or boiler is connected to a three-phase low-voltage supply, the shell of the electrode water heater or boiler shall be connected to the neutral of the supply as well as to the earthing conductor. | WR-554.1.5 |
| If the supply to an electrode water heater or boiler is single phase and one electrode is connected to a neutral conductor earthed by the distributor, the shell of the electrode water heater or boiler shall be connected to the neutral of the supply as well as to the earthing conductor. | WR-554.1.6 |
| If the electrode water heater or boiler is not piped to a water supply or in physical contact with any earthed metal (and where the electrodes and the water in contact with the electrodes are so shielded in insulating material that they cannot be touched while the electrodes are live), a fuse in the line conductor may be substituted for the circuit-breaker and the shell of the electrode water heater or electrode boiler need not be connected to the neutral of the supply. | WR-554.1.7 |

## 5.3.2 Electric floor heating systems

In locations containing a bath or shower:

| | |
|---|---|
| For electric floor heating systems, only heating cables or thin sheet flexible heating elements shall be erected, **provided** that they have either a metal sheath or a metal enclosure or a fine mesh metallic grid. | WR-701.753 |

> The fine mesh metallic grid, metal sheath or metal enclosure shall be connected to the protective conductor of the supply circuit.          WR-701.753

 Compliance with the latter requirement is not required if the protective measure of separated extra-low voltage (SELV) is provided for the floor heating system.

> For electric floor heating systems, the protective measure 'protection by electrical separation' is **not** permitted.          WR-701.753

### 5.3.3 Floor and ceiling heating systems

This section applies to the installation of electric floor and ceiling heating systems that are erected as either thermal storage heating systems or direct heating systems. It does not apply to the installation of wall heating systems.

#### 5.3.3.1 Additional protection – RCDs

> A circuit supplying heating equipment of Class II construction or equivalent insulation shall be provided with additional protection by the use of an RCD.          WR-753.415.1

#### 5.3.3.2 Automatic disconnection of supply

> RCDs with a rated residual operating current not exceeding 30 mA shall be used as disconnecting devices.          WR-753.411.3.2

#### 5.3.3.3 Compliance with standards

> Flexible sheet heating elements shall comply with the requirements of BS EN 60335–2–96.          WR-753.511
>
> Heating cables shall comply with the BS 6351 series.          WR-753.511

### 5.3.3.4 External influences

| | |
|---|---|
| Heating units for installation in ceilings shall have a degree of protection of not less than IPX1. | WR-753.512.2.5 |
| Heating units for installation in a floor of concrete or similar material shall have a degree of ingress protection not less than IPX7 and shall have the appropriate mechanical properties. | WR-753.512.2.5 |
| For cold tails (circuit wiring) and control leads installed in the zone of heated surfaces, the increase of ambient temperature shall be taken into account. | WR-753.522.1.3 |

### 5.3.3.5 Heating-free areas

| | |
|---|---|
| For the necessary attachment of room fittings, heating-free areas shall be provided in such a way that the heat emission is not prevented by such fittings. | WR-753.520.4 |

### 5.3.3.6 Heating units

To avoid the overheating of floor or ceiling heating systems in buildings, one or more of the following measures shall be applied within the zone where heating units are installed in order to limit the temperature to a maximum of 80°C:    WR-753.424.3.1

- appropriate design of the heating system;
- appropriate installation of the heating system in accordance with the manufacturer's instructions;
- use of protective devices.

Heating units shall be:    WR-753.424.1.1

- connected to the electrical installation via cold tails or suitable terminals;
- inseparably connected to cold tails (e.g. by a crimped connection).

As the heating unit may cause higher temperatures   WR-753.424.3.2
or arcs under fault conditions, special measures
should be taken when the heating unit is installed
close to easily ignitable building structures, such as
placing on a metal sheet, in metal conduit or at a
distance of at least 10 mm in air from the ignitable
structure.

### 5.3.3.7 Identification and notices

The designer of the installation/heating system or the   WR-753.514
installer shall provide a plan for each heating system,
containing the following details:

- manufacturer and type of heating units;
- number of heating units installed;
- length/area of heating units;
- rated power;
- surface power density;
- layout of the heating units in the form of a sketch,
  a drawing or a picture;
- position/depth of heating units;
- position of junction boxes;
- conductors, shields and the like;
- heated area;
- rated voltage;
- rated resistance (cold) of heating units;
- rated current of the OCPD;
- rated residual operating current of the RCD;
- insulation resistance of the heating installation
  and the test voltage used;
- leakage capacitance.

This plan shall be fixed to, or adjacent to, the distribution board of the heating
system.

### 5.3.3.8 Operational conditions

Precautions shall be taken not to stress the heating   WR-753.512.1.6
unit mechanically; for example, the material by
which it is to be protected in the finished installation shall cover the heating unit as soon as possible.

### 5.3.3.9 Presence of solid foreign bodies

| | |
|---|---|
| Where heating units are installed there shall be heating-free areas where drilling and fixing by screws, nails and the like are permitted. | WR-753.522.4.3 |
| The installer shall inform other contractors that no penetrating means (such as screws for door stoppers) shall be used in the area where floor or ceiling heating units are installed. | WR-753.522.4.3 |

### 5.3.3.10 Prevention of mutual detrimental influences

| | |
|---|---|
| Heating units shall not cross expansion joints of the building or structure. | WR-753.515.4 |

### 5.3.3.11 Protection against burns

| | |
|---|---|
| In floor areas where contact with skin or footwear is possible, the surface temperature of the floor shall be limited (e.g. to 35°C). | WR-753.423 |

### 5.3.3.12 Protection against electric shock

| | |
|---|---|
| The protective measures of obstacles and placing out of reach are **not** permitted. | WR-753.410.3.5 |
| The protective measures of non-conducting location and Earth-free local equipotential bonding are **not** permitted. | WR-753.410.3.6 |
| The protective measure of electrical separation is **not** permitted. | WR-753.410.3.10 |

## 5.3.4 Heaters for liquids or other substances having immersed heating elements

| | |
|---|---|
| Heaters for liquid and/or other substances shall have an automatic device to prevent a dangerous rise in temperature. | WR-554.2.1 |

### 5.3.5 Water heaters having immersed and uninsulated heating elements

All metal parts of the heater or boiler that are in con-     WR-554.3.2
tact with the water (other than current-carrying parts)
shall be solidly and metallically connected to a metal
water pipe through which the water supply to the
heater or boiler is provided, and that water pipe shall
be connected to the main earthing terminal by means
independent of the circuit-protective conductor.

Water heaters and boilers shall be permanently con-     WR-554.3.3
nected to the electricity supply via a double-pole
linked switch that is either:

- separate from and within easy reach of the heater/
  boiler; or
- part of the boiler/heater – provided that the
  wiring from the heater or boiler is directly connected to the switch without use of a plug and
  socket-outlet.

 If the heater or boiler is installed in a room containing a fixed bath, the switch
shall comply with Section 701 of BS 7671:2008.

Single-phase water heaters and boilers with an uninsu-     WR-554.3.1
lated heating element immersed in the water shall not     WR-554.3.4
have a single-pole switch, non-linked circuit-breaker
or fuse fitted in the neutral conductor, in any part of
the circuit between the heater or boiler or the origin of
the installation.

## 5.4 Luminaires

A luminaire is equipment that distributes, filters or transforms the light transmitted from one or more lamps. It includes all the parts necessary for supporting, fixing and protecting the lamps (but not the lamps themselves) and,
where necessary, circuit auxiliaries, together with the means of connecting
them to the supply.

 In locations where there may be fire hazards due to dust or fibres, luminaires
shall be installed so that dust or fibres cannot accumulate in dangerous
amounts.

### 5.4.1 General requirements for installations

| | |
|---|---|
| A track system for luminaires shall comply with the requirements of BS EN 60570. | WR-559.4.4 |
| Every luminaire shall comply with the relevant standard for manufacture and test of that luminaire, and shall also be selected and erected in accordance with the manufacturer's instructions. | WR-559.4.1 |
| Luminaires without transformers or converters (but which are fitted with extra-low-voltage lamps connected in series) shall be considered as low-voltage equipment as opposed to extra-low-voltage equipment. | WR-559.4.2 |
| Where a luminaire is installed in a pelmet, there shall be no adverse effects due to the presence or operation of curtains or blinds. | WR-559.4.3 |

Every plug, socket-outlet, luminaire supporting coupler (LSC) device:    WR-411.7.5

- for connecting a luminaire and cable coupler in a FELV system:

  o   shall have a protective conductor contact, and
  o   **shall not** be dimensionally compatible with those used for any other system in use in the same premises;

- for connecting a luminaire (i.e. a device for connecting a luminaire (DCL)) and cable coupler of a reduced low-voltage system:

  o   shall have a protective conductor contact, and
  o   **shall not** be dimensionally compatible with those used for any other system in use in the same premises.

 **Note:** This protective measure should **not** be applied to any circuit that includes a socket-outlet or cable coupler, or where a user may change items of equipment without authorization.

| | |
|---|---|
| Every socket-outlet and luminaire supporting coupler in an separated extra-low voltage (SELV) or protective extra-low voltage (PELV) system shall require the use of a plug that is incompatible dimensionally with those used for any other system in use in the same premises. | WR-414.4.3 |

Luminaires shall be:

- kept at an adequate distance from combustible materials;
- protected from foreseeable mechanical stresses.

WR-422.3.1
WR-422.4.2

A luminaire with a lamp that could eject flammable materials in case of failure should be equipped with a safety protective shield.

WR-422.3.1
WR-422.4.2

Every electric discharge lighting installation having an open-circuit voltage exceeding low voltage, and every circuit supplying luminaires at a voltage exceeding low voltage, shall be provided with:

WR-537.2.1.6

- an interlock on a self-contained luminaire;
- an effective local means for the isolation of the circuit from the supply;
- a switch with a lock or removable handle, or a distribution board which can be locked.

Every luminaire shall:

WR-422.3.8

- be appropriate for the location;
- be provided with an enclosure providing a degree of protection of at least IP5X;
- have a limited surface temperature in accordance with BS EN 60598–2–24; and
- be of a type that prevents lamp components from falling from the luminaire.

 **Note:** Luminaires marked ▽D̷ are designed to provide limited surface temperature.

Luminaires should only be installed and used at a reasonable (i.e. sufficient) distance from combustible materials.

WR-422.4.2

 All luminaire components (e.g. lamps and other components) shall be protected against all foreseeable mechanical stresses. Such protective means shall not be fixed to lampholders unless they form an integral part of the luminaire or are fitted in accordance with the manufacturer's instructions.

Parts of a cable or flexible cord within a luminaire shall be suitable for the temperatures likely to be encountered, or shall be provided with additional insulation suitable for those temperatures.

WR-522.2.100

The connection of fixed, suspended, current-using equipment (such as a luminaire for a fixed installation) shall be made by cable with flexible cores.

WR-522.7.2

At each fixed lighting point, one of the following shall be used:

WR-559.6.1.1

- a ceiling rose to BS 67;
- a luminaire-supporting coupler to BS 6972 or BS 7001;
- a batten lampholder or a pendant set to BS EN 60598;
- a luminaire to BS EN 60598;
- a suitable socket-outlet to BS 1363–2, BS 546 or BS EN 60309–2;
- a plug-in lighting distribution unit to BS 5733;
- a connection unit to BS 1363–4;
- appropriate terminals enclosed in a box complying with the relevant part of the BS EN 60670 series or BS 4662;
- a device for connecting a luminaire (DCL) outlet according to IEC 61995–1.

In suspended ceilings, one plug-in lighting distribution unit may be used for a number of luminaires.

A ceiling rose or lampholder for a filament lamp **shall not** be installed in any circuit operating at a voltage normally exceeding 250 V.

WR-559.6.1.2

A ceiling rose **shall not** be used for the attachment of more than one outgoing flexible cable unless it is specially designed for multiple pendants.

WR-559.6.1.3

**Note:** Luminaire-supporting couplers are designed specifically for the mechanical support and electrical connection of luminaires, and **shall not** be used for the connection of any other equipment.

## 5.4.2 Protection against fire

In the selection and erection of a luminaire, the ther-     WR-559.5.1
mal effects of radiant and convected energy on the surroundings shall be taken into account, including:

- the maximum permissible power dissipated by the
  lamps;
- the fire resistance of adjacent material:
  - o   at the point of installation, and
  - o   in the thermally affected areas;

- the minimum distance to combustible materials,
  including material in the path of a spotlight beam.

## 5.4.3  Fixing of the luminaire

Electrical installations shall be so arranged that:

- the risk of ignition of flammable materials due to high temperature or
  electric arc is minimized;
- during normal operation of the electrical equipment, there shall be minimal risk of burns to persons or livestock.

A lighting installation shall be appropriately          WR-559.6.1.9
controlled.

For example, by a switch, a combination of switches or by an automatic control system, which is suitable for discharge lighting circuits.

Adequate means to fix the luminaire shall be pro-    WR-559.6.1.5
vided, such as mechanical accessories (e.g. hooks
or screws), boxes or enclosures, which are able
to support luminaires or associated supporting
devices for connecting a luminaire.

Electronic switching devices should,          WR-559.6.1.100
where possible, include a neutral
conductor.

Bayonet lampholders B15 and B22 shall comply     WR-559.6.1.7
with BS EN 61184 and shall have the tempera-
ture rating T2 described in that standard.

> In places where the fixing means is intended to support a pendant luminaire, the fixing means shall be capable of carrying a mass of not less than 5 kg.   WR-559.6.1.5

**Note:** If the mass of the luminaire is greater than 5 kg, the installer shall ensure that the fixing means is capable of supporting the mass of the pendant luminaire.

> The weight of luminaires and their eventual accessories shall be compatible with the mechanical capability of the ceiling or suspended ceiling, or supporting structure where installed.   WR-559.6.1.5
>
> Lighting circuits incorporating B15, B22, E14, E27 or E40 lampholders shall be protected by an OCPD of maximum rating 16 A.   WR-559.6.1.6
>
> In circuits of a TN or TT system (except for E14 and E27 lampholders complying with BS EN 60238), the outer contact of every Edison screw, or single-centre bayonet- cap type lampholder, shall be connected to the neutral conductor. This Regulation also applies to track-mounted systems.   WR-559.6.1.8

### 5.4.3.1 Through wiring

The installation of through wiring in a luminaire is **only** permitted if the luminaire is designed for such wiring.

> A cable for through wiring shall be selected in accordance with the temperature information on the luminaire or on the manufacturer's instruction sheet.   WR-559.6.2.2
>
> Groups of luminaires divided between the three line conductors of a three-phase system with only one common neutral conductor shall be provided with at least one device that simultaneously disconnects all three line conductors.   WR-559.6.2.3

### 5.4.4 Ceiling roses

A ceiling rose or lampholder for a filament lamp shall **not** be installed in any circuit operating at a voltage that normally exceeds 250 V.    WR-559.6.1.2

A ceiling rose **shall not** be used for the attachment of more than one outgoing flexible cable unless it is specially designed for multiple pendants.    WR-559.6.1.3

### 5.4.5 Lampholders

A lampholder for a filament lamp **shall not** be installed in any circuit operating at a voltage normally exceeding 250 V.

Bayonet lampholders B15 and B22 shall comply with BS EN 61184 and shall have the temperature rating T2 described in that standard.    WR-559.6.1.7

In circuits of a TN or TT system (except for E14 and E27 lampholders complying with BS EN 60238) the outer contact of every Edison-screw or single-centre bayonet-cap type lampholder shall be connected to the neutral conductor. This Regulation also applies to track mounted systems.    WR-559.6.1.8

Insulation-piercing lampholders **shall not** be used unless the cables and lampholders are compatible, and unless the lampholders are non-removable once fitted to the cable.    WR-711.559.4.3

In exhibitions, shows and stands, lighting circuits incorporating B15, B22, E14, E27 or E40 lampholders shall be protected by an OCPD of maximum rating 16 A.    WR-559.6.1.6

Lampholders with ignitability characteristic P as specified in BS 476: Part 5 (or where separate overcurrent protection is provided) shall not be connected to any circuit where the rated current of the OCPD exceeds the appropriate value stated in Table 5.7.

 The above applies unless the wiring is enclosed in earthed metal or insulating material.

**Table 5.7** Overcurrent protection of lampholders

| Type of lampholder | | | Maximum rating of the overcurrent device protecting the circuit (A) |
|---|---|---|---|
| Bayonet lampholder (BS EN 61184): | B15 | SBC | 6 |
| | B22 | BC | 16 |
| Edison screw lampholder (BS EN 60238 | E14 | SES | 6 |
| | E27 | ES | 16 |
| | E40 | GES | 16 |

## 5.4.6 Small spotlights and projectors

Unless otherwise recommended by the manufacturer, small spotlights or projectors should be installed at the following minimum distance from combustible materials:

- rating up to 100 W: 0.5 m;
- over 100 and up to 300 W: 0.8 m;
- over 300 and up to 500 W: 1.0 m.

## 5.4.7 Independent lamp controlgear (e.g. ballasts)

Only independent lamp controlgear marked as suitable for independent use (according to the relevant standard) shall be used external to a luminaire.    WR-559.7

Only the following are permitted to be mounted on flammable surfaces:    WR-559.8

- a class P thermally protected ballast(s)/transformer(s), marked with the appropriate symbol;
- a temperature declared thermally protected ballast(s)/transformer(s), with a marked value equal to or below 130°C.

**Note:** For an explanation of symbols used in luminaires, see Table 5.8.

**Table 5.8** Explanation of symbols used in luminaires, in controlgear for luminaires and in the installation of luminaires

| BS EN 60598-1: 2004 | | BS EN 60598-1: 2008 | |
|---|---|---|---|
| $\overline{\vee}$ F | Luminaire suitable for direct mounting on normally flammable surfaces | (symbol) | Recessed luminaire not suitable for direct mounting on normally flammable surfaces |
| (symbol) F | Luminaire suitable for direct mounting on non-combustible surfaces only | (symbol) | Surface mounted luminaire not suitable for direct mounting on normally flammable surfaces |
| $\overline{\vee}$ F | Luminaire suitable for direct mounting in/on normally flammable surfaces when thermally insulating material may cover the luminaire | (symbol) | Luminaire not suitable for covering with thermally insulating material |

NOTE  Luminaire suitable for direct on normally flammable surfaces may be marked with the symbol shown according to BS EN 60598-1:2004

With the publication of BS EN 60598-1 Ed. 7. luminaires suitable for direct mounting on normally flammable surfaces have no special marking and only luminaires not suitable for mounting on normally flammable surfaces are marked with a symbol (see annex N.4 of BS EN 60598-1:2008 for further explanations).

| | | | |
|---|---|---|---|
| $\bigcirc$ t ..... °C (symbol) | Use of heat-resistant supply cables, interconnecting cables, or external wiring (number of conductors of cable is optional) (BS EN 60598 series) | $\triangle$ E | Luminaire for use with high pressure sodium lamps that require an external ignitor (BS EN 60598 series) |
| (symbol) | Luminaire designed for use with bowl mirror lamps (BS EN 60598 series) | $\triangle$ I | Luminaire for use with high pressure sodium lamps having an internal starting device (BS EN 60598 series) |
| $t_a$.... °C | Rated maximum ambient temperature (BS EN 60598 series) | $\triangledown$ D | Luminaire with limited surface temperature (BS EN 60598-2-24) |
| COOL BEAM (symbol) | Warning against the use of cool-beam lamps (BS EN 60598 series) | (symbol) | Short-circuit proof (inherently or non-inherently) safety isolating transformer (BS EN 61558-2-6) |
| (symbol) - - - m | Minimum distance from lighted objects (metres) (BS EN 60598 series) | $\triangledown$ ... | Temperature declared thermally protected lamp controlgear (... replaced by temperature) (BS EN 61347-1) |
| $\top$ | Rough service luminaire (BS EN 60598 series) | $\triangledown$ 110 | Electronic convertor for an extra-low voltage lighting installation |
| (symbol) | Replace any cracked protective screen (BS EN 60598 series) | $\triangledown$ P | Thermally protected lamp controlgear (class P) (BS EN 61347-1) |
| (symbol) | Luminaire designed for use with self-shielded tungsten halogen lamps or self-shielded metal halide lamps only (BS EN 60598 series) | (symbol) | Independent ballast EN 60417 sheet No. 5138 |

This requirement does not apply to capacitors forming part of the equipment.

### 5.4.7.1 Polarity

A test of polarity shall be made, and it shall be verified    WR-612.6
that (except for E14 and E27 lampholders to BS EN
60238) a circuit having an earthed neutral conductor,
centre-contact bayonet and Edison-screw lampholder
shall have:

- the outer or screwed contacts connected to the
  neutral conductor; and
- wiring correctly connected to socket-outlets and
  similar accessories.

## 5.4.8 Medical locations

In Group 2 medical locations, where PELV is    WR-710.414.4.1
used, exposed-conductive-parts of equipment,
(e.g. operating theatre luminaires) shall be
connected to the circuit protective conductor.

### 5.4.8.1 Safety lighting

In the event of mains power failure, the changeover period to the safety services source shall not exceed 15 s. The necessary minimum illuminance shall be provided for the following:

- emergency lighting and exit signs;
- locations for switchgear and controlgear for emergency generating sets, for main distribution boards of the normal power supply, and for power supply for safety services;
- rooms in which essential services are intended (in these rooms at least one luminaire shall be supplied from the power source, for safety services);
- locations of central fire alarm and monitoring systems;
- rooms of Group 1 medical locations (in these rooms at least one luminaire shall be supplied from the power source, for safety services);
- rooms of Group 2 medical locations (in theses rooms a minimum of 90% of the lighting shall be supplied from the power source, for safety services).

 Escape route luminaires shall be arranged on alternate circuits.

## 5.4.9 Outdoor lighting installation

An outdoor lighting installation comprises one or more luminaires, a wiring system and accessories, and includes lighting installations for:

- roads, parks, car parks, gardens, places open to the public, sporting areas, illumination of monuments and floodlighting;
- places such as telephone kiosks, bus shelters, advertising panels and town plans;
- road signs;
- temporary festoon lighting.

The following are excluded:

- temporary festoon lighting;
- owner-operated equipment;
- luminaires fitted to the outside of a building and supplied directly from the internal wiring of the building;
- road traffic signal systems.

## 5.4.10  Protection against fire

When selecting and erecting a luminaire, the thermal effects of radiant and convected energy on the surroundings shall be taken into account, including:

- the maximum permissible power dissipated by the lamps;
- the fire resistance of adjacent material:
  - ○ at the point of installation, and
  - ○ in the thermally affected areas;
- the minimum distance to combustible materials, including material in the path of a spotlight beam.

Electrical installations shall be so arranged that:

- the risk of ignition of flammable materials due to high temperature or electric arc is minimized;
- during normal operation of the electrical equipment there shall be minimal risk of burns to persons or livestock.

## 5.4.11  Stroboscopic effect

Lighting for premises where machines with moving parts   WR-559.9
are in operation should consider the stroboscopic effects,
which can give a misleading impression of moving parts
being stationary. Such effects may be avoided by selecting
luminaires with a suitable lamp controlgear (such as
high-frequency controlgear) or by distributing lighting
loads across all phases of a three-phase supply.

## 5.4.12 Requirements for outdoor lighting installations, highway power supplies and street furniture

The protective measures of placing out of reach and obstacles **shall not** be used except where:    WR-559.10.1

- the maintenance of equipment is restricted to skilled persons who are specially trained;
- items of street furniture are within 1.5 m of a low-voltage overhead line.

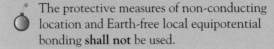 The protective measures of non-conducting location and Earth-free local equipotential bonding **shall not** be used.    WR-559.10.2

Where the protective measure of automatic disconnection of supply is used, all live parts of electrical equipment shall be protected by insulation or by barriers or enclosures providing basic protection.    WR-559.10.3.1

 A door in street furniture, used for access to electrical lighting equipment, shall not be used as a barrier or an enclosure.

Enclosures for live parts shall only be accessible with a key or a tool (unless the enclosure is in a location to which only skilled or instructed persons have access).    WR-559.10.3.1

A door giving access to electrical equipment and located less than 2.50 m above ground level shall:
- be locked with a key or shall require the use of a tool for access;
- be provided with basic protection when the door is open, either by the use of equipment having at least a degree of protection of IP2X or IPXXB by construction or by installation, or by installing a barrier or an enclosure giving the same degree of protection.

Access to the light source of a luminaire, which is at a height of less than 2.80 m above ground level, shall only be possible after removing a barrier or an enclosure requiring the use of a tool.

For an outdoor lighting installation, a metallic structure (e.g. a fence, grid), which is in the proximity of, but not part of, the outdoor lighting installation, need not be connected to the main earthing terminal.

| | |
|---|---|
| Lighting arrangements in places such as telephone kiosks, bus shelters and town plans shall be provided with additional protection by an RCD. | WR-559.10.3.2 |
| A maximum disconnection time of 5 s shall apply to all circuits feeding fixed equipment used in highway power supplies. | WR-559.10.3.3 |
| The earthing conductor of a street electrical fixture shall have a minimum copper equivalent cross-sectional area not less than that of the supply neutral conductor at that point or not less than 6 mm², whichever is the smaller. | WR-559.10.3.4 |

### 5.4.13  Double or reinforced insulation

| | |
|---|---|
| For an outdoor lighting installation, where the protective measure for the whole installation is double or reinforced insulation:<br><br>• no protective conductor shall be provided; and<br>• the conductive parts of the lighting column shall not be intentionally connected to the earthing system. | WR-559.10.4 |
| A device providing protection against the risk of fire shall meet the following requirements:<br><br>• the device shall continuously monitor the power demand of the luminaires;<br>• the device shall automatically disconnect the supply circuit within 0.3 s in the case of a short-circuit (or failure) which causes a power increase of more than 60 W;<br>• the device shall provide automatic disconnection while the supply circuit is operating with reduced power or if there is a failure that causes a power increase of more than 60 W;<br>• the device shall provide automatic disconnection from the supply circuit if there is a failure that causes a power increase of more than 60 W;<br>• the device shall be fail-safe. | WR-559.11.4.2 |

Suspension devices for extra-low-voltage luminaires, WR-559.11.6
including supporting conductors, shall be capable
of carrying five times the mass of the luminaires
(including their lamps) intended to be supported. In
all cases, the amount of support provided shall not
be less than 5 kg.

### 5.4.14 Underwater luminaires for swimming pools

A luminaire for use in the water or in contact with      WR-702.55.2
the water shall be fixed and shall comply with BS EN
60598–2–18.

Underwater lighting located behind watertight           WR-702.55.2
portholes, and serviced from behind, shall comply
with the appropriate part of BS EN 60598, and
be installed in such a way that no intentional or
unintentional conductive connection between any
exposed-conductive-part of the underwater luminaires
and any conductive parts of the portholes can occur.

### 5.4.15 Luminaires in fountains

A luminaire installed in zones 0 or 1 shall be fixed and   WR-702.55.3
shall comply with BS EN 60598–2–18.

Electrical equipment in zones 0 or 1 shall be provided    WR-702.55.3
with mechanical protection to medium severity
(AG2), e.g. by use of mesh glass or by grids which can
only be removed by the use of a tool.

An electric pump shall comply with the requirements      WR-702.55.3
of BS EN 60335–2–41.

### 5.4.16 Luminaires and lighting installations in exhibitions shows and stands

Stand installations containing a concentration          WR-711.422.4.2
of electrical equipment, luminaires or lamps that
are liable to generate excessive heat shall not be
installed unless adequate ventilation is available
(e.g. a well-ventilated ceiling constructed of
incombustible material).

Extra-low-voltage (ELV) lighting systems for filament lamps shall comply with BS EN 60598-2-23.

WR-711.559.4.2

Insulation-piercing lampholders **shall not** be used unless the cables and lampholders are compatible, and providing the lampholders are non-removable once they have been fitted to the cable.

WR-711.559.4.3

Luminaires mounted below 2.5 m (arm's reach) from floor level – or otherwise accessible to accidental contact – shall be firmly and adequately fixed, and sited or guarded so as to prevent possible risk of injuring persons or igniting of materials.

WR-711.559.5

## 5.4.17 Electric discharge lamp installations

Installations of any luminous tube, sign or lamp as an illuminated unit on a stand, or as an exhibit, which has a nominal power supply voltage higher than 230/400 V a.c., shall comply with the following:

WR-711.559.4.4

- location – the sign or lamp shall be installed out of arm's reach or shall be adequately protected to reduce the risk of injury to persons;

WR-711.559.4.4.1

- installation – the facia or stand fitting material behind luminous tubes, signs or lamps shall be non-ignitable.

WR-711.559.4.4.2

Emergency switching devices – a separate circuit shall be used to supply signs, lamps or exhibits, which shall be controlled by an emergency switch.

WR-711.559.4.4.3

The switch shall be easily visible, accessible and clearly marked.

## 5.4.18 Luminaires in caravans and motor caravans

Each luminaire in a caravan shall preferably be fixed directly to the structure or lining of the caravan.

WR-721.55.2.4

| | |
|---|---|
| Where a pendant luminaire is installed in a caravan, provision shall be made for securing the luminaire to prevent damage when the caravan is in motion. | WR-721.55.2.4 |
| Accessories for the suspension of pendant luminaires shall be suitable for the mass suspended and the forces associated with vehicle movement. | WR-721.55.2.4 |
| A luminaire intended for dual-voltage operation shall comply with the appropriate standard. | WR-721.55.2.5 |

### 5.4.19 *Luminaires in temporary installations*

| | |
|---|---|
| Every luminaire and decorative lighting chain shall:<br><br>• have a suitable IP rating;<br>• be installed so as not to impair its ingress protection; and<br>• be securely attached to the structure or support intended to carry it. | WR-740.55.1.1 |
| Its weight shall not be carried by the supply cable, unless it has been selected and erected for this purpose. | WR-740.55.1.1 |
| Luminaires and decorative lighting chains mounted less than 2.5 m (arm's reach) above floor level, or otherwise accessible to accidental contact, shall be firmly fixed and so sited or guarded as to prevent risk of injury to persons or ignition of materials. | WR-740.55.1.1 |
| Access to the fixed light source shall only be possible after removing a barrier or an enclosure which shall require the use of a tool. | WR-740.55.1.1 |
| Lighting chains shall use HO5RN-F (BS 7919) cable or equivalent. | WR-740.55.1.1 |
| Insulation-piercing lampholders shall not be used unless the cables and lampholders are compatible and the lampholders are non-removable once fitted to the cable. | WR-740.55.1.2 |
| All lamps in shooting galleries and other sideshows where projectiles are used shall be suitably protected against accidental damage. | WR-740.55.1.3 |

| | |
|---|---|
| Where transportable floodlights are used, they shall be mounted so that the luminaire is inaccessible. | WR-740.55.1.4 |
| Supply cables shall be flexible and have adequate protection against mechanical damage. | WR-740.55.1.4 |

Luminaires and floodlights shall be fixed and protected so that any focusing or concentration of heat is unlikely to cause ignition of any material.

### 5.4.19.1 Electric discharge lamp installations

Where the use of mobile equipment is envisaged, protection (i.e. by means of a non-conducting location):

- shall be ensured;
- shall be permanent; and
- it shall not be possible to make it ineffective.

| | |
|---|---|
| The location of a luminous tube, sign or lamp shall be installed out of arm's reach or shall be adequately protected to reduce the risk of injury to persons. | WR-740.55.3.1 |

## 5.5 Plug and socket-outlets

A plug and socket-outlet **shall not** be selected as a device for emergency switching.

| | |
|---|---|
| A plug and socket-outlet may be inserted in the main supply circuit to enable it to be switched off for mechanical maintenance. | WR-537.3.2.1 |
| If the rating of the plug and socket-outlet does not exceed 16 A, it shall be capable of cutting off the full load current of the relevant part of the installation. | WR-537.3.2.6 |
| A socket-outlet on a wall or similar structure shall be mounted at a height above the floor or any working surface that minimizes the risk of mechanical damage to the socket-outlet (or to an associated plug and its flexible cord) that might be caused during insertion, use or withdrawal of the plug. | WR-553.1.6 |

Equipment that has a protective conductor current exceeding 3.5 mA, but not exceeding 10 mA, shall either be permanently connected to the fixed wiring of the installation without the use of a plug and socket-outlet or be connected by means of a plug and socket complying with BS EN 60309–2.

WR-543.7.1.101

Equipment that has a protective conductor current exceeding 10 mA shall be connected to the supply:

WR-543.7.1.102

- via a flexible cable with a plug and socket-outlet;
- permanently via the wiring of the installation;
- via a flexible cable with a plug and socket-outlet complying with BSEN 60309–2; or
- via a protective conductor with an Earth monitoring system.

It **shall not** be possible for any pin of a plug to make contact with any live contact of any socket-outlet within the same installation other than the type of socket-outlet for which the plug is designed.

Every plug and socket-outlet (except for SELV) shall be of the non-reversible type, with provision for the connection of a protective conductor.

WR-553.1.2

Every socket-outlet for household and similar use shall be of the shuttered type, and for an a.c. installation shall preferably be of a type complying with BS 1363.

WR-553.1.100

A plug and socket-outlet not complying with BS 1363, BS 546, BS 196 or BS EN 60309–2 may be used in single-phase a.c. or two-wire d.c. circuits operating at a nominal voltage not exceeding 250 V for:

WR-553.1.5

- the connection of an electric clock;
- the connection of an electric shaver;
- a circuit having special characteristics (e.g. where danger would otherwise arise, or where it is necessary to distinguish the function of the circuit).

Except for the plug of a plug and socket-outlet identified in Table 5.9 as suitable for isolation, equipment of overvoltage categories I and II should **not** be used for isolation.

WR-537.2.2.1

**Table 5.9** Guidance on the election of protective, isolation and switching devices

| Device | Standard | Isolation | Emergency switching[2] | Functional switching |
|---|---|---|---|---|
| Switching device | BS 3676: Pt 1 1989 | Yes[4] | Yes | Yes |
| | BS EN 60669-1 | No | Yes | Yes |
| | BS EN 60669-2-1 | No | No | Yes |
| | BS EN 60669-2-2 | No | Yes | Yes |
| | BS EN 60669-2-3 | No | Yes | Yes |
| | BS EN 60669-2-4 | Yes | Yes | Yes |
| | BS EN 60947-3 | Yes[1] | Yes | Yes |
| | BS EN 60947-5-1 | No | Yes | Yes |
| Contactor | BS EN 60947-4-1 | Yes[1] | Yes | Yes |
| | BS EN 61095 | No | No | Yes |
| Circuit-breaker | BS EN 60898 | Yes | Yes | Yes |
| | BS EN 60947-2 | Yes[1] | Yes | Yes |
| | BS EN 61009-1 | Yes | Yes | Yes |
| RCD | BS EN 60947-2 | Yes[1] | Yes | Yes |
| | BS EN 61008-1 | Yes | Yes | Yes |
| | BS EN 61009-1 | Yes | Yes | Yes |
| Isolating switch | BS EN 60669-2-4 | Yes | Yes | Yes |
| | BS EN 60947-3 | Yes | Yes | Yes |
| Plug and socket-outlet (≤ 32 A) | BS EN 60309 | Yes | No | Yes |
| | IEC 60884 | Yes | No | Yes |
| | IEC 60906 | Yes | No | Yes |
| Plug and socket-outlet (> 32 A) | BS EN 60309 | Yes | No | No |
| Device for the connection of luminaire | BS IEC 61995-1 | Yes[3] | No | No |
| Control and protective switching device for equipment (CPS) | BS EN 60947-6-1 | Yes | Yes | Yes |
| | BS EN 60947-6-2 | Yes[1] | Yes | Yes |
| Fuse | BS 88 | Yes | No | No |
| Device with semiconductors | BS EN 60669-2-1 | No | No | Yes |
| Luminaire Supporting Coupler | BS 6972 | Yes[3] | No | No |
| Plug and unswitched socket-outlet | BS 1363-1 | Yes[3] | No | Yes |
| | BS 1363-2 | Yes[3] | No | Yes |
| Plug and switched socket-outlet | BS 1363-1 | Yes[3] | No | Yes |
| | BS 1363-2 | Yes[3] | No | Yes |
| Plug and socket-outlet | BS 5733 | Yes[3] | No | Yes |
| Switched fused connection unit | BS 1363-4 | Yes[3] | Yes | Yes |
| Unswitched fused connection unit | BS 1363-4 | Yes[3] (Removal of fuse link) | No | No |
| Fuse | BS 1362 | Yes | No | No |
| Cooker Control Unit switch | BS 4177 | Yes | Yes | Yes |

Yes      Function provided
No      Function not provided

[1] Function provided if the device is suitable and marked with the symbol for isolation (see BS EN 60617 identity number S00288) ─╱├

[2] See Regulation 5.7.4.2.5

[3] Device is suitable–or on-load isolation, i.e. disconnection whilst carrying load current.

[4] Function provided if the device is suitable and marked with ◯.

NOTE: In the above table, the functions provided by the devices for isolation and switching are summarised, together with the indication of the relevant product standards.

All low-voltage plug and socket-outlets shall conform with the applicable British Standard listed in Table 5.10.      WR-553.1.3

**Table 5.10** Plugs and socket-outlets for low-voltage circuits

| Type of plug and socket-outlet | Rating (A) | Applicable British Standard |
|---|---|---|
| Fused plugs and shuttered socket-outlets; two-pole and Earth | 13 | BS 1363 (fuses to BS 1362) |
| Plugs, fused and non-fused, and socket-outlets, two-pole and Earth | 2, 5, 15, 30 | BS 546 (fuses, if any, to BS 646) |
| Plugs and sockets (industrial type) | 16, 32, 63, 125 | BS EN 60309-2 |

 A plug and socket-outlet **shall not** be used as a device for connecting a water heater and/or boiler to the supply.

| | |
|---|---|
| Where mobile equipment is likely to be used, provision shall be made so that the equipment can be fed from an adjacent and conveniently accessible socket-outlet – taking into account the length of flexible cable normally fitted to portable appliances and luminaires. | WR-553.1.7 |
| For a final circuit with a number of socket-outlets (or connection units intended to supply two or more items of equipment, where it is known that the total protective conductor current in normal service will exceed 10 mA) the circuit shall be provided with a high-integrity protective conductor connection. | WR-543.7.2.101 |
| The following arrangements of the final circuit are acceptable: <br>• a radial final circuit with a ring protective conductor; <br>• a radial final circuit with a single protective conductor. | WR-543.7.2.1 |

 Plugs and socket-outlets in an SELV system shall **not** have a protective conductor contact.

### 5.5.1 Caravan and camping parks

| | |
|---|---|
| At least one socket-outlet shall be provided for each caravan pitch. | WR-708.553.1.11 |

Each socket-outlet and its enclosure forming part of the caravan pitch electrical supply equipment shall comply with BS EN 60309–2, and meet the degree of protection of at least IP44 in accordance with BS EN 60529.   WR-708.553.1.8

Each socket-outlet shall:   WR-708.553.1.12

• be provided with individual overcurrent protection;

• be protected individually by an RCD.   WR-708.553.1.13

The current rating of socket-outlets shall be not less than 16 A.   WR-708.553.1.10

The socket-outlets shall be placed at a height of 0.5 m to 1.5 m from the ground to the lowest part of the socket-outlet. In special cases (due to environmental conditions such as risk of flooding or heavy snowfall) the maximum height is permitted to exceed 1.5 m.   WR-708.553.1.9

 Socket-outlet protective conductors **shall not** be connected to any protective and neutral (PEN) conductor of the electricity supply.

## 5.5.2 Exhibitions, shows and stands

An adequate number of socket-outlets shall be installed to allow user requirements to be met safely.   WR-711.55.7

Where a floor-mounted socket-outlet is installed, it shall be adequately protected from accidental ingress of water and have sufficient strength to be able to withstand the expected traffic load.   WR-711.55.7

## 5.5.3 Marinas and similar locations

A maximum of four socket-outlets shall be grouped together in one enclosure.   WR-709.553.1.10

Every socket-outlet shall:   WR-709.553.1.8

• meet the degree of protection of at least IP44 or be protected by an enclosure;

- be located as close as is practicable to the     WR-709.553.1.9
  berth to be supplied;
- be installed in the distribution board or in a
  separate enclosure.

In general, 200 V to 250 V, 16 single-phase socket-     WR-709.553.1.12
outlets shall be provided.

One socket-outlet shall supply only one pleasure     WR-709.553.1.11
craft or houseboat.

Socket-outlets shall:     WR-709.553.1.8

o   comply with BS EN 60309–1 above 63 A, and
    BS EN 60309–2 up to 63 A;

o   be placed at a height of not less than 1 m     WR-709.553.1.13
    above the highest water level.

> **Note:** In the case of floating pontoons
> or walkways only, this height may be
> reduced to 300 mm above the highest
> water level, provided that appropriate
> additional measures are taken to protect
> against the effects of splashing.

## 5.5.4 Medical locations socket-outlets

The resistance of the protective conductors     WR-710.415.2.2
(including the resistance of the connections
between the terminals of the protective
conductor's socket-outlets and that of fixed
equipment or any extraneous-conductive-parts and
the equipotential bonding busbar) shall not exceed
the following:

- for Group 1 medical locations, $0.7\,\Omega$;
- in Group 2 medical locations, $0.2\,\Omega$.

Electrical devices (e.g. socket-outlets and switches)     WR-710.512.2.1
installed below any medical-gas outlets (such as
those used for oxidizing or flammable gases) shall
be located at a distance of at least 0.2 m from the
outlet (centre to centre), so as to minimize the risk
of ignition of the flammable gases.

For each socket-outlet in a medical location that is protected by an RCD, consideration shall be given to reduce the possibility of unwanted tripping of the RCD due to excessive protective conductor currents produced by equipment in normal operation.

WR-710.531.2.4

 In medical locations, all socket-outlets intended to supply medical electrical equipment shall be unswitched.

At each patient's place of treatment in a medical location (e.g. bedheads) the configuration of socket-outlets shall be as follows:

WR-710.553.1

- each socket-outlet supplied by an individually protected circuit; or
- several socket-outlets separately supplied by a minimum of two circuits.

Socket-outlets used on medical IT systems shall be coloured **blue** and clearly and permanently marked 'Medical equipment only'.

WR-710.553.1

### 5.5.5 Mobile and transportable units

Where mobile equipment is likely to be used, the equipment can be fed from an adjacent and conveniently accessible socket-outlet, taking account of the length of flexible cord normally fitted to portable appliances and luminaires.

Plugs and connectors (i.e. connecting devices) used to connect the unit to the supply shall comply with BS EN 60309–2 and shall also meet the following requirements:

WR-717.55.1

- connectors shall be within an enclosure of insulating material;
- connectors shall afford a degree of protection of not less than IP44, if located outside;
- enclosures containing the connector shall provide a degree of protection of at least IP55;
- connectors located outside the unit shall be provided with an enclosure affording a degree of protection not less than IP44.

WR-717.55.2

In a.c. systems, additional protection by means of an   WR-411.3.3
RCD shall be provided for mobile equipment with a   WR-740.415.1
current rating not exceeding 32 A for use outdoors.

### 5.5.6 Temporary electrical installations

An adequate number of socket-outlets shall be installed   WR-740.55.7
to allow the user's requirements to be met safely.

 **Note:** In booths, stands and for fixed installations, one socket-outlet for each square metre or linear metre of wall is generally considered adequate.

Socket-outlets dedicated to lighting circuits placed   WR-740.55.7
out of arm's reach shall be encoded or marked
according to their purpose.

When used outdoors, plugs, socket-outlets and   WR-740.55.7
couplers shall comply with BS EN 60309.

## 5.6 Protection by residual current devices

For installations and locations where there is an increased risk of shock (such as agricultural and horticultural premises, building sites, bathrooms, swimming pools etc.), additional measures may be required, such as automatic disconnection of supply by means of an RCD with a rated residual operating current not exceeding 30 mA.

If an installation includes an RCD, then it shall have a notice (fixed in a prominent position) that reads as shown in Figure 5.2.

Other requirements from the Regulations include:

This installation, or part of it, is protected by a device that automatically switches off the supply if an earth fault develops. Test quarterly by pressing the button marked 'T' or 'Test'. The device should switch off the supply and should then be switched on to restore the supply. If the device does not switch off the supply when the button is pressed, seek expert advice.

**Figure 5.2** Inspection and testing notice.

Every installation shall be divided into circuits, as necessary, to reduce the possibility of unwanted tripping of RCDs due to excessive protective conductor currents produced by equipment.   WR-314.1

In a.c. systems, additional protection by means of an RCD shall be provided for:   WR-411.3.3

- socket-outlets with a rated current not exceeding 20 A that are used by ordinary persons unless:

  - they are used under the supervision of skilled or instructed persons, or
  - a suitably labelled and/or identified socket-outlet is provided for connection of a particular item of equipment;

- mobile equipment with a current rating not exceeding 32 A for use outdoors.

### 5.6.1 Construction

Where the installation is not intended to be under the supervision of a skilled or instructed person (and the installation is liable to impact), consideration shall be given to providing additional protection by means of an RCD.

The magnetic circuit of the transformer of an RCD shall enclose all the live conductors of the protected circuit. The associated protective conductor shall be outside the magnetic circuit.   WR-531.2.2

The rated residual operating current of the protective device shall comply with the requirements appropriate to the type of system earthing.   WR-531.2.3

Where an RCD is used for fault protection with, but separate from, an OCPD, it shall be verified that the residual current operated device is capable of withstanding (without damage) the thermal and mechanical stresses to which it is likely to be subjected if a fault occurs on the load side of the point at which it is installed.   WR-531.2.8

Where two or more RCDs are in series (and where discrimination in their operation is necessary to prevent danger) their characteristics shall ensure that the intended discrimination is achieved.   WR-531.2.9

An RCD shall be so selected and the electrical          WR-531.2.4
circuits so subdivided that any protective
conductor current that may be expected to occur
during normal operation of the connected load(s)
will be unlikely to cause unnecessary tripping of
the device.

The use of an RCD associated with a circuit normally    WR-531.2.5
expected to have a protective conductor **shall not** be
considered sufficient for fault protection if there is no
such conductor.

## 5.6.2 Devices for protection against the risk of fire

Where it is necessary to limit the consequence          WR-532.1
of fault currents in a wiring system from the
point of view of fire risk, the circuit shall
either:

- be protected by an RCD for fault protection; and
- the RCD shall be installed at the origin of the
  circuit to be protected; and
- the RCD shall switch all live conductors; and
- the rated residual operating current of the RCD
  shall not exceed 300 mA; or
- the circuit will need to be continuously
  monitored by an insulation monitoring device
  that initiates an alarm on the occurrence of an
  insulation fault.

Where fault protection and/or additional protection     WR-612.13.1
is to be provided by an RCD, the effectiveness of
any test facility incorporated in the device shall be
verified.

## 5.6.3 Electrode water heaters and boilers

If an electrode water heater or electrode boiler is     WR-554.1.4
directly connected to a supply exceeding low voltage,
the installation shall include an RCD.

## 5.6.4 Installation

| | |
|---|---|
| An RCD shall be capable of disconnecting all the line conductors of the circuit at substantially the same time. | WR-531.2.1 |
| An RCD shall be located so that its operation will not be harmed by magnetic fields caused by other equipment. | WR-531.2.7 |
| Where an RCD could be operated by a person other than a skilled or instructed person, it shall be designed and installed so that it is not possible to modify or adjust the setting or the calibration of its rated residual operating current (or time-delay mechanism) without a deliberate act involving the use of either a key or a tool which results in a visible indication of its setting or calibration. | 531.2.10 |
| Where an installation includes an RCD, a notice (see Figure 5.2) shall be fixed in a prominent position at or near the origin of the installation. | WR-514.12.2 |

## 5.6.5 Locations containing a bath or shower

| | |
|---|---|
| Additional protection shall be provided for all circuits in these locations, by the use of one or more RCDs. | WR-701.411.3.3 |
| Where the location containing a bath or shower is in a building with a protective equipotential bonding system, supplementary equipotential bonding may be omitted where **all** of the following conditions are met: | WR-701.415.2 |

- all final circuits of the location have additional protection by means of an RCD;
- all final circuits of the location comply with the requirements for automatic disconnection; and
- all extraneous-conductive-parts of the location are effectively connected to the protective equipotential bonding.

## 5.6.6 Medical locations

In medical locations of Group 1 and Group 2, where RCDs are required, only type A (according to BS EN 61008 and BS EN 61009) or type B (according to IEC 62423) shall be selected, depending on the possible fault current arising.

WR-710.411.3.2.1

 Type AC RCDs shall not be used.

In Group 1 and Group 2 medical locations, the following shall apply:

WR-710.411.3.2.5

- for TN, TT and IT systems, the voltage presented between simultaneously accessible exposed-conductive-parts and/or extraneous-conductive-parts shall not exceed 25 V a.c. or 60 V d.c.;
- for TN and TT systems, the requirements in Table 5.11 shall apply.

 **Note:** In TN systems, 25 V a.c. or 60 V d.c. may be met with protective equipotential bonding, by complying with the disconnection time in accordance with Table 5.11.

Table 5.11 Maximum disconnection times (in seconds) (data from BS 7671:2008)

| System | Maximum disconnection time (s) | | | | | | | | | |
|---|---|---|---|---|---|---|---|---|---|---|
| | $25V < U_o\ 50V$ | | $50V < U_o\ 120V$ | | $120V < U_o\ 5230V$ | | $230V < U_o\ 400V$ | | $1.10 > 400V$ | |
| | a.c. | d.c. | a.c. | d.c. | a.c. | d.c. | a.c. | d.c. | a.c. | d.c. |
| TN | 5 | 5 | 0.3 | 2 | 0.3 | 0.5 | 0.05 | 0.06 | 0.02 | 0.02 |
| TT | 5 | 5 | 0.15 | 0.2 | 0.05 | 0.1 | 0.02 | 0.06 | 0.02 | 0.02 |

Supplementary equipotential bonding connection points for the connection of medical electrical equipment shall be provided in each medical location, as follows:

WR-710.415.2.1

- Group 1: one per patient location;
- Group 2: one per medical IT socket-outlet.

**Note:** Fixed conductive non-electrical patient supports such as operating theatre tables, physiotherapy couches and dental chairs should be connected to the equipotential bonding conductor, unless they are intended to be isolated from Earth.

### 5.6.6.1 IT system

| | |
|---|---|
| Where a medical IT system is used, additional protection by means of an RCD shall not be used. | WR-710.411.3.3 |
| In Group 2 medical locations, an IT system shall be used for final circuits supplying medical electrical equipment and systems intended for life support and surgical applications, and also for all other electrical equipment located (or that may be moved into) the 'patient environment'. | WR-710.411.6.3.1 |

For each group of rooms serving the same function, at least one IT system is necessary. The IT system shall be equipped with an insulation monitoring device (IMD), with the following additional specific requirements:

- a.c. internal impedance shall be $100\,k\Omega$;
- internal resistance shall be $250\,k\Omega$;
- test voltage shall be 25 V d.c.;
- injected current (even under fault conditions) shall be 1 mA peak;
- indication shall take place at the latest when the insulation resistance has decreased to $50\,k\Omega$.

**Note:** If the response value is adjustable, the lowest possible set-point value shall be $50\,k\Omega$. A test device shall be provided.

### 5.6.6.2 TN system

| | |
|---|---|
| In final circuits of Group 1 rated up to 63 A, RCDs shall be used. | WR-710.411.4 |

In TN-S systems, the insulation level of all live conductors shall be monitored.

In Group 2 medical locations, protection by automatic disconnection of supply by means of RCDs shall only be used on the following circuits:

- circuits for the supply of movements of fixed operating tables; or
- circuits for X-ray units;

**Note:** This requirement is mainly applicable to mobile X-ray units brought **into** Group 2 locations.

- circuits for large equipment with a rated power greater than 5 kV A.

### 5.6.6.3 TT system

| | |
|---|---|
| In Group 1 and Group 2 medical locations, RCDs shall be used as disconnection devices. | WR-710.411.5 |

## 5.6.7 Outdoor lighting installation

| | |
|---|---|
| Lighting arrangements in places such as telephone kiosks, bus shelters and town plans shall be provided with additional protection by an RCD. | WR-559.10.3.2 |

## 5.6.8 Power supply

| | |
|---|---|
| An RCD that is powered from an independent auxiliary source and does not operate automatically in the case of failure of the auxiliary source shall **only** be used if:<br><br>• fault protection is maintained even in the case of failure of the auxiliary source; or<br>• the device is incorporated in an installation intended to be supervised by an instructed person or a skilled person, and inspected and tested by a competent person. | WR-531.2.6 |
| The generating set shall be connected so that any part of the installation that is protected by an RCD remains effective for every intended combination of sources of supply. | WR-551.4.2 |
| In a TN, TT or IT system, one or more RCDs with a rated residual operating current of not more than 30 mA shall be installed to protect every circuit. | WR-551.4.4.2 |
| Where RCDs are also used for protection against fire, the conditions for protection by automatic disconnection of the supply shall be verified. | WR-612.8 |

Where RCDs are required for additional protection, the effectiveness of automatic disconnection of supply by RCDs shall be verified using suitable test equipment according to BS EN 61557–6 to confirm that the relevant requirements are met.   WR-612.10

### 5.6.8.1 TN systems

An RCD **shall not** be used in a TN-C system.

In a TN system, the integrity of the earthing of the installation depends on the reliable and effective connection of the PEN or PE conductors to Earth.

The neutral conductor shall be protected against short-circuit current.   WR-431.2.1

An RCD may be used as a protective device for fault protection.   WR-411.4.4

**Note:** Where an RCD is used for Earth fault protection the circuit should also incorporate an OCPD.

Where an RCD is used, the maximum values of Earth fault loop impedance in Table 5.12 may be applied for non-delayed RCDs to BS EN 61008–1 and BS EN 61009–1 for a nominal voltage of $U_o$ of 230 V.   WR-411.4.9

Table 5.12 Maximum earth fault loop impedance $(Z_s)$ for non-delayed RCDs to BS EN 61008–1 and BS EN 61009–1 for a nominal voltage $U_o = 230$ V (data from BS 7671:2008)

| Rated residual operating current (mA) | Maximum earth fault loop impedance 4 $(\Omega)$ | | | |
|---|---|---|---|---|
| | $50V < U_o\ 120V$ | $120V < U_o\ 230V$ | $230V < U_o\ 5400V$ | $U_o > 400V$ |
| 30 | 1667* | 1667* | 1533* | 1667* |
| 100 | 500* | 500* | 460* | 500* |

Where an RCD is used in a TN-C-S system, a PEN conductor **shall not** be used on the load side.

| | |
|---|---|
| Except for mineral-insulated cables, busbar trunking or powertrack systems, a wiring system shall be protected against insulation faults in a TN system by an RCD that has a rated residual operating current ($I_{An}$) not exceeding 300 mA. | WR-422.3.9 |
| In a TN system where, for certain equipment or a certain part of the installation, the characteristics of a protective device do not satisfy the requirements, that part may be protected by an RCD. | WR-531.3.1 |
| In these sorts of circumstances, the exposed-conductive-parts of that part of the installation shall be connected to the TN earthing system's protective conductor, or to a separate Earth electrode that provides an impedance suitable for the operating current of the RCD. | |

 In the latter case, the circuit shall be treated as a TT system.

| | |
|---|---|
| For a TN-S system where the neutral is not isolated, any RCD that is used shall be positioned so as to avoid incorrect operation due to the existence of any parallel neutral–Earth path. | WR-551.6.2 |

### 5.6.8.2 TT system

 Where an RCD is used for Earth fault protection, the circuit should also incorporate an OCPD.

| | |
|---|---|
| Except for mineral-insulated cables, busbar trunking or powertrack systems, a wiring system shall be protected against insulation faults in a TT system by an RCD that has a rated residual operating current ($I_{A}$,) not exceeding 300 mA. | WR-422.3.9 |
| If an installation that is part of a TT system is protected by a single RCD, this shall be placed at the origin of the installation, unless the part of the installation between the origin and the device complies with the requirements for protection by the use of Class II equipment or equivalent insulation. | WR-531.4.1 |

**Note:** Where there is more than one origin, this requirement applies to each origin.

### 5.6.8.3 IT system

| | |
|---|---|
| Where fault protection is provided by an RCD, the product of the rated residual operating current ($I_{An}$) in amperes and the Earth fault loop impedance in ohms shall not exceed 50. | WR-411.8.3 |

In an IT system, the neutral conductor **shall not** be distributed unless:    WR-431.2.2

- the circuit is protected by an RCD with a rated residual operating current not exceeding 0.2 times the current-carrying capacity of the corresponding neutral conductor; or
- overcurrent detection is provided for the neutral conductor of every circuit; or
- the neutral conductor is effectively protected against short-circuit by a protective device installed on the supply side.

In an IT system without a neutral conductor, it is permitted to omit the over-load protective device in one of the line conductors **if** an RCD is installed in each circuit.

### 5.6.9 Swimming pools and other basins

Equipment for use in the interior of basins that is only intended to be in operation when people are not inside zone 0 shall be supplied by a circuit protected by automatic disconnection of the supply using an RCD, or SELV or electrical separation.    WR-702.410.3.4.1

**Note:** The same ruling applies to fountains in zone 1.

In zones 0 or 1:    WR-702.53

- switchgear or controlgear **shall not** be installed; and
- a socket-outlet **shall not** be installed.

> In zone 2, a socket-outlet or a switch is permitted only if the supply circuit is protected by an RCD or SELV – or by electrical separation.　　WR-702.53

## 5.7  Residual current monitors

 A residual current monitor (RCM) is **not** intended to provide protection against electric shock.

> An RCM is intended to permanently monitor any leakage current in the downstream installation, or part of it.　　WR-538.4
>
> An RCM is intended to alert the user of the installation before the protective device is activated.　　WR-538.4.1
>
> RCMs for use in a.c. systems shall comply with BS EN 62020.　　WR-538.4
>
> Where an RCM is used in an a.c. IT system, it is recommended that a directionally discriminating RCM is used so as to avoid inopportune signalling of leakage current.　　WR-538.4
>
> In supply systems, RCMs may be installed to reduce the risk of the protective device operating in the event of excessive leakage current of the installation or the connected appliances.　　WR-538.4.1

 **Note:** Where an RCD is installed upstream of the RCM, it is recommended that the RCM has a rated residual operating current not exceeding a third of that of the RCD.

> In all cases, the RCM shall have a rated residual operating current not higher than the first fault current level intended to be detected.
>
> In an IT system where interruption of the supply in case of a first insulation fault to Earth is not required or permitted, an RCM may be installed to assist in locating a fault. It is recommended an RCM is installed at the beginning of outgoing circuits.　　WR-538.4.2

## 5.8 Rotating machines and motors

All equipment, including cable, of every circuit carrying the starting, accelerating and load currents of a motor shall be suitable for a current at least equal to the full-load current rating of the motor.

Where the motor is intended for intermittent duty with frequent starting and stopping, account shall be taken of any cumulative effects of the starting or braking currents on the temperature rise of the equipment of the circuit.    WR-552.1.1

Every electric motor having a rating exceeding 0.37 kW shall be provided with control equipment that includes a system for protection against overload of the motor.    WR-552.1.2

Every motor shall be provided with means to prevent automatic restarting after a stoppage due to a drop in voltage (or failure) of supply, as unexpected restarting of the motor might cause danger.    WR-552.1.3

 Except, that is, where failure to start after a brief interruption would be likely to cause greater danger.

 **Note:** These requirements do not exclude arrangements for starting a motor at intervals by an automatic control device, provided that other precautions are taken against danger from unexpected restarting.

Lighting for premises where machines with moving parts are in operation should consider stroboscopic effects, which can give a misleading impression of moving parts being stationary. Such effects may be avoided by selecting luminaires with suitable lamp controlgear, such as high-frequency controlgear, or by distributing lighting loads across all the phases of a three-phase supply.    WR-559.9

## 5.9 Supplies

### 5.9.1 Consumer units

A consumer unit (sometimes known as a 'consumer control unit' or 'electricity control unit') is a particular type of distribution board for the control and

distribution of electrical energy, primarily in domestic premises. It includes a manually operated method for isolation (both poles of the incoming circuit(s)) and assemblies of one or more fuses, circuit-breakers, RCDs, signalling, and other devices that have been purposely manufactured for this purpose.
One of the requirements of the Regulations is that:

All circuits and final circuits **shall** be provided with a means of switching for interrupting the supply on load.

**Note:** This regulation particularly applies to circuits and parts of an installation that (for safety reasons) need to be switched independently of other circuits and/or installations. It does not apply to short connections between the origin of the installation and the consumer's main switchgear.

### 5.9.2 Batteries

In 1957 a battery was discovered in Bagdad. It was made by the Parthians, who ruled Bagdad from 250 BC to 224 BC, and was used to electroplate silver!

| | |
|---|---|
| Stationary batteries shall be installed so that they are accessible **only** to skilled or instructed persons. | WR-551.8.1 |

**Note:** This generally requires the battery to be installed in a secure location or, for smaller batteries, a secure enclosure (which shall be adequately ventilated).

Battery connections shall have:      WR-551.8.2

- basic protection by insulation and/or enclosures; or
- shall be arranged so that two bare conductive parts having between them a potential difference exceeding 120 V cannot be inadvertently touched simultaneously.

Basic protection and fault protection is deemed to be      WR-414.2
provided where:

- the nominal voltage cannot exceed the upper limit of voltage Band I;
- and the supply is from a recognized source such as a battery.

A battery may be used as a source for SELV and PELV      WR-414.3
systems.

### 5.9.2.1 Central power supply sources

Batteries shall be of the vented (or valve-regulated)   WR-560.6.10
maintenance-free type, with a heavy-duty industrial
design that has a minimum declared life of 10 years.

### 5.9.2.2 Low-power supply sources

The power output of a low-power supply system is   WR-560.6.11
limited to 500 W for 3–hour duration or 1500 W for
1–hour duration. However, the batteries shall be:

- of the vented (or valve-regulated) maintenance-
  free type; and
- of a heavy-duty industrial design; with
- a minimum declared life of 5 years.

### 5.9.2.3 Protection against fault current

A device for protection against fault current need not   WR-434.3
be provided for a conductor connecting an accumulator
battery to the associated control panel where the
protective device is located, **provided** that the wiring:

- is carried out in such a way as to reduce the risk of
  fault to a minimum; and
- is installed in such a manner as to reduce to a
  minimum the risk of fire or danger to persons.

### 5.9.2.4 Temporary electrical installations

In a temporary electrical installation, the supply to a   WR-740.415.1
battery-operated emergency lighting circuit shall be
connected to the same RCD protecting the lighting
circuit.

### 5.9.2.5 Uninterruptible power supply (UPS) sources

A static, uninterruptible power supply (UPS) source   WR-560.6.12
(when operating in the emergency condition from the
inverter supplied by the battery) shall be able to:

- operate distribution circuit protective devices; and
- start the safety devices.

### 5.9.3 Electric motors

| | |
|---|---|
| An electric motor that is automatically or remotely controlled, or that is not continuously supervised shall be protected against excessive temperature by a protective device with a manual reset. | WR-422.3.7 |
| A motor with star-delta starting shall be protected against excessive temperature in both the star and delta configurations. | WR-422.3.7 |
| All equipment, including cable, of every circuit carrying the starting, accelerating and load currents of a motor shall be suitable for a current at least equal to the full-load current rating of the motor. | WR-552.1.1 |
| Every electric motor having a rating exceeding 0.37 kW shall be provided with control equipment to protect against overloading the motor. | WR-552.1.2 |
| Except where failure to start after a brief interruption would be likely to cause greater danger, every motor shall be provided with means to prevent automatic restarting after a stoppage due to a drop in voltage or failure of supply. | WR-552.1.3 |
| Fixed electric motors shall be provided with an efficient means of switching off that is readily accessible, easily operated and so placed as to prevent danger. | WR-132.15.2 |
| Motor control circuits shall be designed so as to prevent any motor from restarting automatically after a stoppage due to a fall in or loss of voltage – if such starting is liable to cause danger. | WR-537.5.4.1 |
| Where reverse-current braking of a motor is provided, provision shall be made for the avoidance of reversal of the direction of rotation at the end of braking – if such reversal may cause danger. | WR-537.5.4.2 |
| Where the motor is intended for intermittent duty with frequent starting and stopping, account shall be taken of any cumulative effects of the starting or braking currents on the temperature rise of the equipment of the circuit. | WR-552.1.1 |

> Where safety depends on the direction of rotation of a motor, provision shall be made for the prevention of reverse operation due to, for example, a reversal of phases.
>
> WR-537.5.4.3

### 5.9.4 Generating sets

 **Note:** Where the fault protection of an installation (or parts of an installation) is supplied by a static convertor that relies on the automatic closure of the bypass switch and operation of a protective device, which does not operate within the time required (see Table 41.1 of the Regulations), supplementary equipotential bonding shall be provided.

#### 5.9.4.1 As power sources

> Generator sets (independent of the normal supply) are a recognized supply source for safety services such as emergency escape lighting, fire detection and alarm systems, installations for fire pumps, fire rescue service lifts, and smoke- and heat-extraction equipment.
>
> WR-351
>
> The source of supply to a reduced low-voltage circuit shall be either:
>
> WR-411.8.4.1
>
> - a motor-generator with windings that provide isolation equivalent to that provided by the windings of an isolating transformer; or
> - a source independent of other supplies (such as an engine-driven generator).
>
> The neutral (star) point of the secondary windings of three-phase generators (or the midpoint of the secondary windings of single-phase transformers and generators) shall be connected to Earth.
>
> WR-411.8.4.2

#### 5.9.4.1.1 Basic and fault protection

Basic protection and fault protection is deemed to be provided where the nominal voltage cannot exceed the upper limit of voltage Band I and the supply is from a recognized source such as a diesel-driven generator or motor-generator.

> The following sources may be used for SELV and PELV    WR-414.3
> systems:
>
> - a source independent of a higher voltage circuit
>   (such as a diesel-driven generator);
> - a motor-generator with windings providing
>   equivalent isolation to a safety isolating transformer.

 Connection of live parts of the generator with Earth may affect the protective
measure.

> A device for protection against fault current need not    WR-434.3
> be provided for a conductor connecting a generator
> to the associated control panel, provided that the
> protective device is placed in the panel, and **provided**
> that both of the following conditions are simultaneously
> fulfilled:
>
> - the wiring is carried out in such a way as to reduce
>   the risk of fault to a minimum; and
> - the wiring is installed in such a manner as to reduce
>   to a minimum the risk of fire or danger to persons.
>
> The generating set shall be connected so that any RCD    WR-551.4.2
> used in the installation remains effective for every
> intended combination of sources of supply.

### 5.9.4.2 Low-voltage generating sets

Low-voltage generating sets can be:

- combustion engines;
- turbines;
- electric motors;
- photovoltaic cells;
- electrochemical accumulators;
- other suitable sources.

Where a generating set with an output not exceeding 16 A is to be connected
in parallel with a system for distribution of electricity to the public, procedures
for informing the electricity distributor are given in the Electricity Safety,
Quality and Continuity Regulations (ESQCR) (2002).

| | |
|---|---|
| In addition to the ESQCR requirements, where a generating set with an output exceeding 16 A is to be connected in parallel with a public distribution system, the requirements of the electricity distributor should be ascertained **before** the generating set is connected. | WR-551.1 |

 **Note:** The requirements of the distributor for the connection of units rated up to 16 A are given in BS EN 50438.

| | |
|---|---|
| The safety and proper functioning of other sources of supply shall not be impaired by the generating set. | WR-551.2.1 |
| The prospective short-circuit current and prospective Earth fault current shall be assessed for each source of supply (or combination of sources) that can operate independently of other sources or combinations. | WR-551.2.2 |
| The short-circuit rating of protective devices within the installation and, where appropriate, connected to a public distribution system, **shall not** be exceeded for any of the intended methods of operation of the sources. | WR-551.2.2 |
| Where the generating set is intended to provide a supply to an installation that is not connected to a public distribution system (or to provide a supply as a switched alternative to such a system), the capacity and operating characteristics of the generating set shall ensure that there will be no danger or damage to equipment after the connection or disconnection of any intended load as a result of any voltage or frequency deviation. | WR-551.2.3 |
| It shall be possible to automatically disconnect such parts of the installation as may be necessary if the capacity of the generating set is exceeded. | WR-551.2.3 |

## 5.9.4.3 Protection against overcurrent

| | |
|---|---|
| Where overcurrent protection of the generating set is required, it shall be located as near as is practicable to the generator terminals. | WR-551.5.1 |

> Where a generating set is intended to operate in      WR-551.5.2
> parallel with a public distribution system, or where
> two or more generating sets may operate in parallel,
> circulating harmonic currents shall be limited so that
> the thermal rating of conductors is not exceeded.

**Note:** The effects of circulating harmonic currents may be limited by one or more of the following:

- the selection of a generating set with compensated windings;
- the provision of a suitable impedance in the connection to the generator star points;
- the provision of switches that interrupt the circulatory circuit but which are interlocked so that at all times fault protection is not impaired;
- the provision of filtering equipment.

### 5.9.4.4 Standby systems and switched alternatives to public supplies

> Precautions for isolation shall be taken so that the      WR-551.6.1
> generator cannot operate in parallel with a public
> distribution system.

Protection by automatic disconnection of supply cannot rely on the connection to the earthed point of the public distribution system when the generator is operating as a switched alternative to a TN system.

## 5.9.5 Medical power supply systems

In medical locations, a power supply for safety purposes is required that will energize the installations needed for continuous operation in case of failure of the general power system, for a defined period, within a pre-set changeover time.

The safety power supply system shall automatically take over if the voltage of one or more incoming live conductors of the main distribution board of the building with the main power supply has dropped for more than 0.5 s and by more than 10% with regard to the nominal voltage (see Table 5.11 on p. 248 for a list of examples).

Primary cells are **not** allowed as safety power sources. An additional main incoming power supply, from the general power supply, is **not** regarded as a source of the safety power supply.

The availability (i.e. readiness for service) of safety power sources shall be monitored and indicated at a suitable location.

 The circuit that connects the power supply source for safety services to the main distribution board shall be considered a safety circuit.

### 5.9.5.1 General

| | |
|---|---|
| In medical locations, the distribution system shall be designed and installed to facilitate the automatic changeover from the main distribution network to the electrical safety source feeding essential loads. | WR-710.313.1 |
| Automatic changeover devices shall be arranged so that safe separation between supply lines is maintained. | WR-710.537.1 |
| In Group 1 and Group 2 medical locations, at least two different sources of supply shall be provided – one of which shall be connected to the electrical supply system for safety services. | WR-710.559 |

### 5.9.5.2 Failure of the general power supply source

| | |
|---|---|
| In case of a failure of the general power supply source, the power supply for safety services shall be energized to feed the equipment with electrical energy for a defined period of time, and within a predetermined changeover period. | WR-710.560.5.6 |
| In the event of a voltage failure on one or more line conductors at the distribution board, a safety power supply source shall be used and be capable of providing power for a period of at least 3 h for the following:<br><br>• the luminaires of operating theatre tables;<br>• medical electrical equipment containing light sources that are essential for the application of the equipment, such as. endoscopes and other associated, essential equipment, (e.g. monitors). | WR-710.560.6.1.1 |

The normal power supply to life-supporting medical electrical equipment shall be restored within a changeover period not exceeding 0.5 s.

| | |
|---|---|
| Equipment that is required for the maintenance of hospital services shall be connected either automatically or manually to a safety power supply source capable of maintaining it for a minimum period of 24 h. This equipment may include, for example:<br><br>• sterilization equipment;<br>• technical building installations, in services and waste disposal systems;<br>• cooling equipment;<br>• catering equipment;<br>• storage-battery chargers. | WR-710.560.6.1.3 |
| Other services that may require a safety service supply with a changeover period not exceeding 15 s include, for example, the following:<br><br>• selected lifts for fire-fighters;<br>• ventilation systems for smoke extraction;<br>• paging systems;<br>• medical electrical equipment used in Group 2 medical locations that is used for surgical or other procedures of vital importance;<br>• electrical equipment for medical gas supplies – including compressed air, vacuum supply and narcosis (anaesthetics) exhaustion, as well as their monitoring devices;<br>• fire detection and fire alarms to BS 5839;<br>• fire-extinguishing systems. | WR-710.560.11 |
| Where socket-outlets are supplied from the safety power supply source, they shall be readily identifiable according to their safety services classification. | WR-710.560.5.7 |

### 5.9.5.3 Power supplies for medical locations of Group 2

| | |
|---|---|
| In the event of a first fault to Earth, a total loss of supply in Group 2 locations shall be prevented. | WR-710.512.1.2 |

# 5.10 Switches

No means of isolation or switching shall be inserted in the outer conductor of a concentric cable.

| | |
|---|---|
| Switches shall be mounted in a suitable mounting box or enclosure. | WR-530.4.2 |
| A switch with a lock shall be provided for every electric discharge lighting installation that has an open circuit voltage exceeding low voltage. | WR-537.2.1.6 |
| Persons and livestock shall be protected against injury, and property shall be protected against damage, as a consequence of overvoltages such as those originating from switching. | WR-131.6.2 |
| Disconnecting devices shall be provided so as to allow electrical installations, circuits or individual items of equipment to be switched off or isolated for the purposes of operation, inspection, fault detection, testing, maintenance and repair. | WR-132.10 |

**Note:** Electrical equipment shall not cause harmful effects on other equipment or impair the supply during normal service, including switching operations.

| | |
|---|---|
| Every circuit shall be provided with a means of isolation from all live supply conductors by a linked switch or a linked circuit-breaker. | WR-422.3.13 |
| A label or other suitable means of identification shall be provided to indicate the purpose and identification of each item of switchgear. | WR-514.1.1 WR-514.9.1 |

## 5.10.1 Devices for isolation and switching

| | |
|---|---|
| Where it is intended that isolation and switching is carried out **only** by instructed persons, for TN systems the means of switching the supply on load and the means of isolation may be provided by a suitably rated fuse carrier. | WR-559.10.6.1 |

> Where the distributor's cut-out is used as the means     WR-559.10.6.2
> of isolation of a highway power supply, the approval
> of the distributor shall be obtained.

Combined protective and neutral (PEN) conductors **shall not** be isolated or switched.

### 5.10.2 Emergency switching

Emergency switching may be emergency switching ON or emergency switching OFF.

> Means shall be provided for emergency switching of     WR-537.4.1.1
> any part of an installation where it may be necessary
> to control the supply in order to remove an unex-
> pected danger.
>
> Other than where a risk of electric shock is involved,     WR-537.4.1.2
> the emergency switching device shall be an isolating
> device and shall interrupt **all** live conductors.

Except where the neutral conductor can be regarded as being reliably connected to Earth in a TN-S or TN-C-S system, the neutral conductor need not
be isolated or switched.

> The execution of emergency switching shall ensure     WR-537.4.1.3
> that only one single action is required to interrupt the
> appropriate supply conductors.
>
> The operation of an emergency switching device shall     WR-537.4.1.4
> **not** introduce a further danger or interfere with the
> complete operation necessary to remove the danger.

### 5.10.2.1 Devices for emergency switching

A device for emergency switching shall be capable of breaking the full load current of the relevant part(s) of the installation, taking account of stalled motor currents where appropriate.

Hand-operated switching devices for direct interruption of the main circuit shall be selected where practicable, and these shall be clearly identified, prefer-ably by colour.

An emergency switching device may consist of:   WR-537.4.2.2

- a switch in the main circuit (or pushbuttons in the control (auxiliary) circuit) that is capable of directly cutting off the appropriate supply; or
- a combination of equipment activated by a single action for the purpose of cutting off the appropriate supply.

The means of operating (handle, pushbutton, etc.) an emergency switching device:

- shall be readily accessible at places where a danger might occur and/or at any additional remote position from which that danger can be removed;   WR-537.4.2.5

- shall be capable of latching or being restrained in the OFF or STOP position;   WR-537.4.2.6

- **shall not** re-energize the relevant part of the installation;   WR-537.4.2.6

- shall be so placed and marked so as to be readily identifiable and convenient for the intended use.   WR-537.4.2.7

## 5.10.2.2 Functional switching devices

In general, all current-using equipment requiring control shall be controlled by an appropriate functional switching device.   WR-537.5.1.3

**Note:** A single functional switching device may control two or more items of equipment intended to operate simultaneously.

A functional switching device that is designed to ensure the changeover of supply from alternative sources:   WR-537.5.1.4

- shall affect all live conductors; and
- shall not be capable of putting the sources in parallel, unless the installation is specifically designed for this condition.

**Note:** In these cases, no provision shall be made for isolation of the PEN or protective conductors.

- shall be suitable for the most onerous duty they are intended to perform;    WR-537.5.2.1

- shall be provided for each part of a circuit that may require to be controlled independently of other parts of the installation;    WR-537.5.1.1

 But it need **not** necessarily control all live conductors of a circuit.

- may control the current without necessarily opening the corresponding poles.    WR-537.5.2.2

 Semiconductor switching devices are examples of devices capable of interrupting the current in the circuit but not opening the corresponding poles.

### 5.10.2.3 Fire-fighter's switches

A fire-fighter's switch shall be provided in the low-voltage circuit supplying:    WR-537.6.1

- exterior electrical installations operating at a voltage exceeding low voltage; and
- interior discharge lighting installations operating at a voltage exceeding low voltage.

Every fire-fighter's switch shall comply with the following:    WR-537.6.3
WR-537.6.2

- for an exterior installation, the switch shall be outside the building and adjacent to the equipment (or, alternatively, a notice indicating the position of the switch shall be placed adjacent to the equipment, and a notice shall be fixed near the switch so as to render it clearly distinguishable);
- for an interior installation, the switch shall be independent of the switch for any exterior installation and in the main entrance to the building;
- the switch shall be placed in a conspicuous position that is reasonably accessible to fire-fighters, and at not more than **2.75 m** from the ground or from a person standing beneath the switch;

I, personally, could **not** understand why the switch should be *'not more than 2.7 m from the ground'* as fire-fighters are not that tall as far as I am aware! I, therefore, sought advice from the local fire service and was told that the reason why the switch was positioned so high was to stop an unauthorized person from switching it off. Fire-fighters use a ceiling hook to operate the switch. I am now no longer confused!!

- where more than one switch is installed on any one building, each switch shall be clearly marked to indicate the installation or part of the installation that it controls.

A fire-fighter's switch shall:   WR-537.6.4

- be coloured **red** and have fixed on or near it a permanent nameplate marked with the words '**FIRE-FIGHTER'S SWITCH**';
- have its **ON** and **OFF** positions clearly indicated by lettering legible to a person standing on the ground at the intended site, with the OFF position at the top;
- be provided with a device to prevent the switch being inadvertently returned to the ON position; and
- be arranged to facilitate operation by a fire-fighter.

 **Note:** An insulation monitoring device (IMD) shall be used on a circuit comprising safety equipment that is normally de-energized by a switch that disconnects all live poles and which is only energized in the event of an emergency (provided that the IMD is automatically deactivated whenever the safety equipment is activated).

### 5.10.3 Isolation and switching

Every device provided for isolation or switching shall   WR-537.1.1 comply with the relevant requirements of this section.

 **Note:** Table 5.9 (see p. 239) provides information on selection.

### 5.10.4 Main switches

Where an installation is supplied from more than one     WR-537.1.6
source:

- a main switch shall be provided for each source of
  supply; and
- a warning notice shall be permanently fixed in a
  prominent position so that any person seeking to
  operate any of these main switches will be warned
  of the need to operate all such switches to achieve
  isolation of the installation; or
- a suitable interlock system shall be provided.

Where an installation is supplied from more than one     WR-537.1.5
source of energy (one of which requires a means of
earthing independent of the means of earthing of other
sources, and it is necessary to ensure that no more
than one means of earthing is applied at any time) a
switch may be inserted in the connection between the
neutral point and the means of earthing, **provided** that
the switch is a linked switch that is capable of disconnecting and connecting the earthing conductor for the
appropriate source, at substantially the same time as
the related live conductors.

### 5.10.5 Main linked switches

A main linked switch shall be provided as near as     WR-537.1.4
is practicable to the origin of every installation as a
means of switching the supply on load and as a means
of isolation.

 **Note:** In a TN-S or TN-C-S system the neutral conductor need not be isolated
or switched where it can be regarded as being reliably connected to Earth by
suitably low impedance.

A main switch intended to be operated by ordinary     WR-537.1.4
persons (e.g. a householder) shall interrupt both live
conductors of a single-phase supply.

## 5.10.6 Mechanical maintenance

> The capability of switching off for mechanical main-
> tenance shall be provided where mechanical mainte-
> nance could involve a risk of physical injury.
>
> WR-537.3.1.1
> WR-537.3.1.2

Suitable means should be provided to prevent electrically powered equipment from becoming unintentionally reactivated during mechanical
maintenance.

### 5.10.6.1 Devices for switching off for mechanical maintenance

> A device such as a:
>
> WR-537.3.2.1
>
> • multi-pole switch;
> • circuit-breaker;
> • control and protective switching device (CPS);
> • control switch operating a contactor;
> • plug and socket-outlet;
>
> may be inserted in the main supply circuit for switch-
> ing off for mechanical maintenance.

The open position of the contacts of the device shall be visible or be clearly
and reliably indicated by the use of the symbols 0 and I to indicate the open
and closed positions, respectively.

> A device for switching off for mechanical
> maintenance:
>
> • shall require manual operation;
> • shall be designed and/or installed so as to prevent     WR-537.3.2.3
>   inadvertent or unintentional switching on;
> • shall be so placed, readily identifiable and con-       WR-537.3.2.4
>   venient for the intended use.
>
>  The switch need not necessarily interrupt the     WR-537.3.2.5
> neutral conductor.

**Note:** A plug and socket-outlet or similar device of rating not exceeding 16 A
may be used as a device for switching off for mechanical maintenance.

### 5.10.6.1.1 Heaters and/or boilers

A heater or boiler shall be permanently connected to
the electricity supply through a double-pole linked
switch that is both separate from and within easy reach
of the heater or boiler, or is incorporated therein.

WR-554.3.3

The wiring from the heater or boiler shall be connected
directly to that switch without the use of a plug and
socket-outlet.

If the heater or boiler is installed in a room containing a fixed bath, the switch
shall comply with Section 701 of BS 7671:2008 (Incorporating Amendment
No. 1).

Where a step-up transformer is used, a linked switch
shall be provided for disconnecting the transformer
from all live conductors of the supply.

WR-555.1.3

### 5.10.7 Lighting installations

A lighting installation shall be appropriately
controlled (see Table 5.9 on p. 239).

WR-559.6.1.9

### 5.10.8 Protective devices

Switches shall only be inserted in the line conductor.

WR-132.14.1

Non-linked switches shall be inserted in an earthed
neutral conductor.

WR-132.14.2

**Note:** Any linked switch or linked circuit-breaker inserted in an earthed neutral conductor shall be capable of breaking all of the related line conductors.

### 5.10.9 Compatibility

Equipment shall be selected and erected so that it will
neither cause harmful effects to other equipment nor
impair the supply during normal service, including
switching operations.

WR-512.1.5

Switchgear, protective devices, accessories and other types of equipment shall be connected to conductors intended to operate at temperatures exceeding 70°C at the equipment in normal service, **unless:**

- the equipment manufacturer has confirmed that the equipment is suitable for such conditions, or the current (including any harmonic current) carried by the conductor does not exceed the appropriate temperature limit specified in Table 5.13.

WR-512.1.5
WR-523.1

Table 5.13 Maximum operating temperatures for types of cable insulation

| Type of insulation | Temperature limit |
|---|---|
| Thermoplastic | 70°C at the conductor |
| Thermosetting | 90°C at the conductor b |
| Mineral (thermoplastic covered, or bare exposed to touch) | 70°C at the sheath |
| Mineral (bare not exposed to touch, and not in contact with combustible material) | 105°C at the sheath b, c |

### 5.10.10 Single-pole switching devices

Single-pole switching devices:

- shall not be inserted in the neutral conductor of a multiphase circuit; or
- alone in the neutral conductor single-phase circuits.

WR-530.3.2

### 5.10.11 Switchboards

Passageways and working platforms that have access to an open-type switchboard or an equipment that has dangerous exposed live parts need to allow persons, without hazard, to:

- operate and maintain the equipment;
- pass one another as necessary with ease; and
- back away from the equipment.

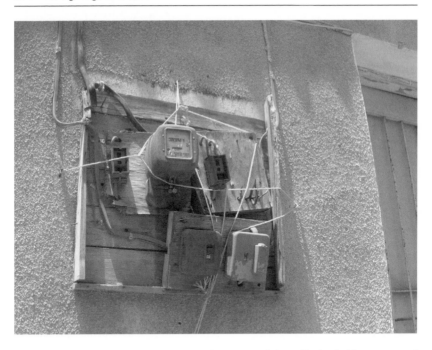

**Figure 5.4** Certainly NOT the right sort of installation! (Courtesy of StingRay.)

Where equipment carrying current of different types (or    WR-515.2
at different voltages) is grouped in a common assembly
such as a switchboard, all equipment belonging to any
one type of current or any one type of voltage shall be
effectively segregated so as to avoid mutual detrimental
influence.

Any identification of a switchboard busbar or conductor    WR-514.3.3
shall comply with the requirements of Table 5.14 insofar as these are applicable.

 **Notes:**
1. Power circuits include lighting circuits.
2. M identifies either the mid-wire of a three-wire d.c. circuit, or the earthed
   conductor of a two-wire earthed d.c. circuit.
3. Only the middle wire of three-wire circuits may be earthed.
4. An earthed PELV conductor is blue.

**Table 5.14** Identification of conductors

| Function | Alphanumeric | Colour |
|---|---|---|
| Protective conductors | | Green-and-yellow |
| Functional earthing conductor | | Cream |
| a.c. power circuit (Note 1) | | |
| Phase of single-phase circuit | L | Brown |
| Neutral of single- or three-phase circuit | N | Blue |
| Phase 1 of three-phase a.c. circuit | L1 | Brown |
| Phase 2 of three-phase a.c. circuit | L2 | Black |
| Phase 3 of three-phase a.c. circuit | L3 | Grey |
| Two-wire unearthed d.c. power circuit | | |
| Positive of two-wire circuit | L+ | Brown |
| Negative of two-wire circuit | L– | Grey |
| Two-wire earthed d.c. power circuit | | |
| Positive (of negative earthed) circuit | L+ | Brown |
| Negative (of negative earthed) circuit (Note 2) | M | Blue |
| Positive (of positive earthed) circuit (Note 2) | M | Blue |
| Negative (of positive earthed) circuit | L– | Grey |
| Three-wire d.c. power circuit | | |
| Outer positive of two-wire circuit derived from three-wire system | L+ | Brown |
| Outer negative of two-wire circuit derived from three-wire system | L– | Grey |
| Positive of three-wire circuit | L+ | Brown |
| Mid-wire of three-wire circuit (Notes 2 and 3) | M | Blue |
| Negative of three-wire circuit | L– | Grey |
| Control circuits, ELV and other applications | | |
| Phase conductor | L | Brown, black, red, orange, yellow, violet, grey, white, pink or turquoise |
| Neutral or mid-wire (Note 4) | N or M | Blue |

## 5.10.12  Switching devices

A switching device:

- shall be protected against overcurrent;   WR-536.5.1

- without integral overcurrent protection shall be co-ordinated with an appropriate OCPD;   WR-536.5.1

- **shall not** be inserted in a protective conductor unless the switch:   WR-543.3.3

o   has been inserted in the connection between the neutral point and the means of earthing;

o   is a linked switch arranged to disconnect and connect the earthing conductor at substantially the same time as the related live conductors;

o   is a multi-pole linked switch or plug-in device in which the protective conductor circuit has not been interrupted before the live conductors and re-established not later than when the live conductors are reconnected.

### 5.10.13 Switchgear

- Shall be accessible only to authorized persons;   WR-422.2.2
- that is placed in an escape route, shall be enclosed in a cabinet or an enclosure constructed of non-combustible or not readily combustible material;

- shall be installed outside the location unless:   WR-422.3.3

  o   it is suitable for the location; or
  o   it is installed in an enclosure providing a degree of protection of at least IP4X or, in the presence of dust, IP5X, or, in the presence of electrically conductive dust, IPX6.

- shall be subjected to a functional test to show it   WR-612.13.2
  is properly mounted, adjusted and installed in accordance with the relevant requirements of the Regulations.

For all installations with a 230 V single-phase supply   WR-530.3.4
rated up to 100 A that is under the control of ordinary persons:

- switchgear assemblies shall either comply with BS EN 60439–3; or
- be a consumer unit incorporating components and protective devices specified by the manufacturer and complying with BS EN 60439–3.

An auto-reclosing device for protection, isolation, switching or control may be installed only in an installation that is intended to be under the supervision of skilled or instructed persons and which is intended to be inspected and tested by competent persons.

| | |
|---|---|
| Means of access to all live parts of switchgear where different nominal voltages exist shall be marked to indicate the voltages present. | WR-514.10.1 |

Where a metal enclosure or frame of a low-voltage switchgear is used as a protective conductor:

- its electrical continuity should be protected against mechanical, chemical or electrochemical deterioration;
- its cross-sectional area should be in accordance with BS EN 60439–1;
- it should permit the connection of other protective conductors at every predetermined tap-off point.

### 5.10.14  Special installations and locations

#### 5.10.14.1  Agricultural and horticultural premises

| | |
|---|---|
| The electrical installation of each building or part of a building shall be isolated by a single isolation device. | WR-705.537.2 |
| These isolation devices: | WR-705.537.2 |
| • shall be clearly marked according to the part of the installation to which they belong; and | |
| • emergency stopping devices or emergency switching shall **not** be erected where they are accessible to livestock or in any position where access may be impeded by livestock. | WR-705.537.2 |

#### 5.10.14.2  Construction and demolition site installations

| | |
|---|---|
| Each assembly for construction site (ACS) shall incorporate suitable devices for switching and isolating the incoming supply. | WR-704.537.2.2 |

These devices shall be suitable for securing in the    WR-704.537.2.2
OFF position by padlock (or by location) inside a
lockable enclosure.

### 5.10.14.3 Electrical installations in caravans and motor caravans

In an installation consisting of only one final cir-    WR-721.537.2.1.1
cuit, the isolating switch may be the OCPD.

 A notice shall be permanently fixed near the main isolating switch inside the caravan.

### 5.10.14.4 Exhibitions, shows and stands

Switchgear shall be placed in closed cabinets    WR-711.51
that can only be opened by the use of a key
or a tool, except for those parts designed and
intended to be operated by ordinary persons.

A separate circuit shall be used to supply signs,    WR-711.559.4.4.3
lamps or exhibits, which shall be controlled by
an emergency switch.
The switch shall be easily visible, accessible and
clearly marked.

### 5.10.14.5 Marinas and similar locations

At least one means of isolation:    WR-709.537.2.1.1

• shall be installed in each distribution
  cabinet;
• shall disconnect all live conductors, including the neutral conductor.

 **Note:** There shall be one isolating switching device for a maximum of four socket-outlets.

### 5.10.14.6 Medical locations

Requirements for medical electrical equipment for use in conjunction with flammable gases and vapours are contained in BS EN 60601.

### 5.10.14.7 Rooms and cabins containing sauna heaters

| | |
|---|---|
| Switchgear that forms part of the sauna heater equipment (or of other fixed equipment installed in zone 2) may be installed within the sauna room or cabin. | WR-703.537.5 |
| Other switchgear (e.g. for lighting) shall be placed outside the sauna room or cabin. | WR-703.537.5 |

### 5.10.14.8 Solar photovoltaic power supply systems

| | |
|---|---|
| To allow maintenance of the photovoltaic (PV) convertor, means of isolating the PV convertor from the d.c. side and the a.c. side shall be provided. | WR-712.537.2.1.1 |
| Switchgear assemblies shall be in compliance with BS EN 60439–1. | WR-712.511.1 |

In the selection and erection of devices for isolation and switching to be installed between the PV installation and the public supply, the public supply shall be considered the source and the PV installation shall be considered the load.

| | |
|---|---|
| A switch disconnector shall be provided on the d.c. side of the PV convertor. | WR-712.537.2.2.5 |

**Note:** All junction boxes (PV generator and PV array boxes) shall carry a warning label indicating that parts inside the boxes may still be live after isolation from the PV convertor.

### 5.10.14.9 Swimming pools and other basins

| | |
|---|---|
| In zone 2, a switch is permitted **only** if the supply circuit is protected by one of the following protective measures: | WR-702.53 |

- SELV;
- automatic disconnection of supply using an RCD;
- electrical separation.

 In zones 0 or 1, switchgear or controlgear shall not be installed.

### 5.10.14.10 Temporary electrical installations

| | |
|---|---|
| Switchgear shall be placed in cabinets that can be opened only by the use of a key or a tool, except for those parts designed and intended to be operated by ordinary persons. | WR-740.51 |
| Every electrical installation in a booth, stand or amusement device shall have its own means of isolation and switching, which shall be readily accessible. | WR-740.537.1 |
| Temporary electrical installations for amusement devices (and distribution circuits supplying outdoor installations) shall be provided with their own readily accessible and properly identified means of isolation. | WR-740.537.2.1.1 |
| A device for isolation shall disconnect all live conductors (i.e. line and neutral conductors). | WR-740.537.2.2 |
| A separate circuit shall be used to supply luminous tubes, signs or lamps, which shall be controlled by an emergency switch. | WR-740.55.3.2 |

 **Note:** This switch shall be easily visible, accessible and marked in accordance with the requirements of the local authority.

# 5.11 Rectifiers

A device for protection against fault current need not be    WR-434.3
provided for a conductor connecting a rectifier where
the protective device is placed in the panel, **provided** that
both of the following conditions are simultaneously
fulfilled:

- the wiring is carried out in such a way as to reduce
  the risk of fault to a minimum; and
- the wiring is installed in such a manner as to
  reduce to a minimum the risk of fire or danger to
  persons.

## 5.11.1 Autotransformers and step-up transformers

Where an autotransformer is connected to a circuit    WR-555.1.1
having a neutral conductor, the common terminal
of the winding shall be connected to the neutral
conductor.

Where a step-up transformer is used, a linked switch    WR-555.1.3
shall be provided for disconnecting the transformer
from all live conductors of the supply.

 A step-up autotransformer **shall not** be connected to an IT system.

## 5.11.2 Electric dodgems

At amusement parks and fairgrounds:    WR-740.55.9

- electric dodgems shall only be operated at voltages
  not exceeding 50 V a.c. or 120 V d.c;
- the circuit shall be electrically separated from the
  supply mains by means of a transformer or a motor-
  generator set.

### 5.11.3 Extra-low-voltage (ELV) transformers and electronic converters

At exhibitions, shows and stands:　　　　　WR-711.55.6

- ELV (extra-low-voltage) transformers shall be mounted out of arm's reach of the public and shall have adequate ventilation;
- a manual reset protective device shall protect the secondary circuit of each transformer or electronic convertor;
- electronic converters shall conform to BS EN 61347–1.

### 5.11.4 IT systems

An IT system at mobile and/or transportable units　WR-717.411.6.2
can be provided by:

- an isolating transformer or a low-voltage generating set, with an insulation monitoring device installed; or
- a transformer offering simple separation and providing:

  ○ automatic disconnection of the supply in case of a first fault between live parts and the frame of the unit, or
  ○ an RCD.

### 5.11.5 Safety isolating transformers

A safety isolating transformer for an extra-low-volt-　WR-559.11.3.1
age lighting installation shall comply with BS EN
61558–2–6 and:

- either the transformer shall be protected on the primary side by a protective device; or
- the transformer shall be short-circuit proof (both inherently and non-inherently).

At booths in fairgrounds etc.:                                    WR-740.55.5

- safety isolating transformers shall comply with BS EN 61558–2–6 or provide an equivalent degree of safety;
- a manually reset protective device shall protect the secondary circuit of each transformer or electronic convertor;
- safety isolating transformers shall be mounted out of arm's reach or be mounted in a location that provides equal protection, and shall have adequate ventilation.

### 5.11.6 Temporary installations

All booths in fairgrounds etc. containing rectifiers shall be adequately ventilated, and the vents shall not be obstructed when in use.                                    WR-740.55.5

## 5.12 Transformers

The magnetic circuit of the transformer of an RCD shall enclose all the live conductors of the protected circuit. The associated protective conductor shall be outside the magnetic circuit.                                    WR-531.2.2

The source of supply to a reduced low-voltage circuit may be a double-wound isolating transformer (complying with BS EN 61558–1 and BS EN 61558–2–23).                                    WR-411.8.4.1

At booths in fairgrounds etc., enclosures containing transformers shall be adequately ventilated, and the vents shall not be obstructed when in use.                                    WR-740.55.5

### 5.12.1 Medical locations

Transformers should be installed in close proximity to the medical location, with the following additional requirements:

- The leakage current of the output winding to Earth and the leakage current of the enclosure, when measured in the no-load condition (and the transformer supplied at rated voltage and rated frequency), shall not exceed 0.5 mA.
- At least one single-phase transformer per room or functional group of rooms shall be used to form the IT systems for mobile and fixed equipment, and the rated output shall be not less than 0.5 kVA and shall not exceed 10 kVA. Where several transformers are needed to supply equipment in one room, they shall not be connected in parallel.
- If the supply of three-phase loads via an IT system is also required, a separate three-phase transformer shall be provided for this purpose.

 Capacitors shall not be used in transformers for medical IT systems.

| | |
|---|---|
| Overload current protection shall not be used in either the primary or secondary circuit of the transformer of a medical IT system. | WR-710.512.1.1 |
| Overcurrent protection against short-circuit and overload current is required for each final circuit. | |

# 6

# Cables and conductors

Within the Wiring Regulations there is frequent reference to different types of cables (e.g. single core, multi-core, fixed, flexible), conductors (e.g. live supply, protective, bonding) and conduits, cable ducting, cable trunking, and so on. Unfortunately, similar to equipment and components, the requirements for these items are liberally sprinkled throughout the BS 7671. The aim of this chapter, therefore, is to provide a catalogue of all the different types identified and referred to in the Wiring Regulations, under two main headings (i.e. 'cables', and 'conductors and conduits'), and then give a list of their essential requirements.

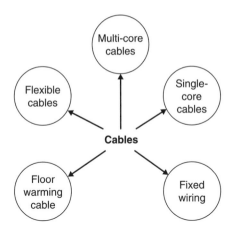

**Figure 6.1** Cables.

 Similar to other chapters, please remember that these lists of requirements are **only** the author's impression of the most important aspects of the Wiring Regulations, and electricians should **always** consult BS 7671 to satisfy compliance!

## 6.1 Cables

An electric power cable is defined as:

> *'an assembly of two or more electrical conductors consisting of a core protected by twisted wire strands held together with (and typically covered by) an overall*

*sheath. The conductors may be of the same or different sizes, each with their own insulation. A bare conductor is normally used for the equipment safety Earth'.*

There are five main types of cables found in electrical installations. These are:

- single-core cables;
- multi-core cables;
- flexible cables;
- heating cables;
- fixed wiring.

### 6.1.1 General

All cables should comply with the requirements of BS EN 50265–2–1 or 2–2.

### 6.1.2 Types of cable

#### 6.1.2.1 Single-core cables

The general requirements for single-core cables include the following.

Metallic sheaths and/or the non-magnetic armour of single-core cables that are in the same circuit must either be bonded together:

- at both ends of their run (solid bonding); or
- at one point in their run (single point bonding);

provided that, at full load, voltages from sheaths and/or armour to Earth;

- do not exceed 25 V; and
- do not cause corrosion; and
- do not cause danger or damage to property.

Single-core cables may be used as protective conductors, but shall be coloured **green-and-yellow** throughout their length.

The conductance of the outer conductor of a concentric single-core cable shall not be less than the internal conductor.

When non-twisted single-core cables with a cross-sectional area greater than $50 \, mm^2$ in copper (or $70 \, mm^2$ in aluminium) are connected in parallel, the load current shall be shared equally between them. (This ruling does not apply to final ring circuits.)

Owing to possible electromagnetic effects, single-core cables that are armoured with steel wire or tape **shall not** be used for a.c. circuits.

Notes:

1. The steel wire or steel tape armour of a single-core cable is regarded as a ferromagnetic enclosure.

2. For single-core armoured cables, the use of aluminium armour may be considered.

### 6.1.2.1.1 Armoured single-core cables

The metallic sheaths and/or non-magnetic armour of      WR-523.100
single-core cables in the same circuit shall normally
be bonded together at both ends of their run (solid
bonding).

### 6.1.2.2 Multi-core cables

The general requirements for multi-core cables include the requirements that:

For telecommunication circuits, data transfer circuits and similar, consideration shall also be given to electrical interference, both electromagnetic and electrostatic (see BS EN 50081 and BS EN 50082).

 A Band I circuit **shall not** be contained in the same wiring system as a Band II voltage circuit, unless it is in a multi-core cable and the cores of the Band I circuit are:

* insulated for the highest voltage present in the Band II circuit;
* separated from the cores of the Band II circuit by an earthed metal screen, and the cables are:
    o   insulated for their system voltage;
    o   installed in a separate compartment of a cable ducting or trunking system;
    o   installed on a tray or ladder separated by a partition;
    o   use a separate conduit, trunking or ducting system.

Separated extra-low-voltage (SELV) circuit conductors that are contained in a multi-core cable with other circuits should be insulated for the highest voltage present in that cable.

Separated circuits shall, preferably, use a separate wiring system. If this is not feasible, multi-core cables (without a metallic sheath or insulated conductors) may be used.

The conductance of the outer conductor of a concentric cable for a multi-core cable:

* serving a number of points contained within one final circuit (or where the internal conductors are connected in parallel) should not be less than that of the internal conductors;
* in a multi-phase or multi-pole circuit should not be less than that of one internal conductor.

A voltage Band I circuit shall not be contained in the     WR-528.1
same wiring system as a Band II circuit, unless:

- each conductor of a multi-core cable is insulated
  for the highest voltage present in the cable;
- for a multi-core cable, the cores of the Band I
  circuit are separated from the cores of the Band II
  circuit by an earthed metal screen of equivalent
  current-carrying capacity to that of the largest core
  of a Band II circuit.

Each part of a circuit shall be arranged such that the     WR-521.8.1
conductors are **not** distributed over different multi-core
cables, conduits, ducting systems, franking systems, tray
or ladder systems.

The line and neutral conductors of each final circuit     WR-521.8.2
shall be electrically separate from those of every other
final circuit, so as to prevent the indirect energizing of a
final circuit intended to be isolated.

Two or more circuits are allowed in the same cable (but     WR-521.7
see Section 528 for specific requirements).

Where multi-core cables are installed in parallel, each     WR-521.8.1
cable shall contain one conductor of each line.

In caravans and motor caravans, all protective con-     WR-721.543.2.3
ductors shall be incorporated in a multi-core cable
or in a conduit together with the live conductors.

### 6.1.2.3 Flexible cables

Flexible cables:

- shall be of a heavy-duty type with a voltage rating of not less than 45 V or
  50 V; or
- shall be suitably protected against mechanical damage;
- that are liable to mechanical damage shall be visible throughout their
  length;
- that are used as an overhead, low-voltage line shall comply with the relevant British or Harmonized Standard;
- shall only be used for fixed wiring where the relevant Regulations permit.

Insulated flexible cables may include a flexible metallic armour, braid or screen.

Non-flexible cables (and flexible cables not forming part of a portable appliance or luminaire) that are sheathed with lead, polyvinyl chloride (PVC) or an elastomeric material may include a catenary wire (or hard-drawn copper conductor) for aerial use or when suspended.

Provided that all flexible equipment cables (other than Class II equipment) have a protective conductor for use as an equipotential bonding conductor, then source supplies may supply more than one item of equipment.

| | |
|---|---|
| Flexible cables shall be visible throughout any part of their length that is liable to mechanical damage. | WR-413.3.4 |

For separated circuits, the use of separate wiring systems is recommended.

| | |
|---|---|
| All flexible cables (unless they supply equipment with double or reinforced insulation) shall include a protective bonding conductor. | WR-418.3.6 |
| A flexible cable shall be used for fixed wiring only where the relevant provisions of the Regulations are met. | WR-521.9.1 |
| Equipment that is intended to be moved while in use shall be connected by flexible cables, except equipment supplied by contact rails. | WR-521.9.2 |
| Stationary equipment that is moved temporarily for the purposes of connecting, cleaning etc. (e.g. a cooker or a flush-mounting unit for installations in a false floor) shall be connected with flexible cable. | WR-521.9.3 |
| Flexible cables shall be either:<br><br>• of a heavy-duty type (with a voltage rating of not less than 450/750 V); or<br>• suitably protected against mechanical damage. | WR-422.3.100 |

In locations subject to an external heat source (e.g. from solar gain of the wiring system or its surrounding medium):

| | |
|---|---|
| • the parts of a cable within an accessory, appliance or luminaire shall be suitable for the temperatures likely to be encountered, or shall be provided with additional insulation suitable for those temperatures. | WR-522.2.2 |

Where no vibration or movement can be expected, cables with non-flexible cores may be used.

> In caravans and motor caravans, flexible cables **shall not**  WR-711.52
> be laid in areas accessible to the public unless they are
> protected against mechanical damage.

### 6.1.2.4 Heating cables

The general requirements for heating cables include the following:
   Heating cables:

- passing through (or in close proximity to) a fire hazard:
    - shall be enclosed in material with an ignitability characteristic P as specified in BS 476 Part 5;
    - shall be protected from any mechanical damage;
- that are going to be laid (directly) in soil, concrete, cement screed, or other material used for road and building construction:
    - shall be capable of withstanding mechanical damage;
    - shall be constructed of material that will be resistant to damp and/or corrosion;
- that are going to be laid (directly) in soil, a road, or the structure of a building shall be installed so that they are:
    - completely embedded in the substance it is intended to heat;
    - not damaged by movement (by the substance in which it is embedded);
    - complies with the maker's instructions and recommendations.

The maximum loading of floor-warming cable under operating conditions is shown in Table 6.1.

**Table 6.1** Maximum conductor operating temperatures for a floor-warming cable

| Type of cable | Maximum conductor operating temperature (°C) |
| --- | --- |
| General-purpose PVC over conductor | 70 |
| Enamelled conductor, polychlorophene over enamel, PVC overall | 70 |
| Enamelled conductor PVC overall | 70 |
| Enamelled conductor, PVC over enamel, lead-alloy E sheath overall | 70 |
| Heat-resisting PVC over conductor | 85 |

| | |
|---|---|
| Nylon over conductor, heat-resisting PVC overall | 85 |
| Synthetic rubber or equivalent elastomeric insulation over conductor | 85 |
| Mineral insulation over conductor, copper sheath overall | Temperature dependent on type of seal employed, outer covering etc. |
| Silicone-treated woven-glass sleeve over conductor | 180 |

Where a heating cable is required to pass through, or be in close proximity to, material that presents a fire hazard, the cable:    WR-554.4.1

- shall be enclosed in material having the ignitability characteristic **P** as specified in BS 476–12; and
- shall be adequately protected from any mechanical damage that is reasonably foreseeable during installation and use.

A heating cable intended for laying directly in soil, concrete, cement screed or other material used for road and building construction shall be:    WR-554.4.2

- capable of withstanding mechanical damage under the conditions that can reasonably be expected to prevail during its installation; and
- constructed of material that will be resistant to damage from dampness and/or corrosion under normal conditions of service.

A heating cable laid directly in soil, a road or the structure of a building, shall be installed so that it:    WR-554.4.3

- is completely embedded in the substance it is intended to heat; and
- does not suffer damage in the event of movement normally to be expected in it or the substance in which it is embedded; and
- complies in all respects with the manufacturer's instructions and recommendations.

## 6.1.2.4.1 Electric floor heating systems

The load of every floor-warming cable (under operation) shall be limited to a value such that the manufacturer's stated conductor temperature is not exceeded.    WR-554.4.4

In locations containing a bath or a shower, only heat-    WR-701.753
ing cables or thin-sheet flexible heating elements of
electric floor heating systems, shall be erected, **provided** that they have either a metal sheath or a metal
enclosure or a fine-mesh metallic grid.

### 6.1.3 Cable Construction and manufacture

Cable conduits shall comply with the appropriate part of    WR-521-6
the BS EN 61386 series.

Cable tray and ladder systems shall comply with BS EN    WR-521-6
61537.

Cable trunking or ducting shall comply with the appro-    WR-521-6
priate part of the BS EN 50085 series.

#### 6.1.3.1 Cross-sectional areas of conductors of cables

The cross-sectional area of each conductor in an a.c.    WR-524.1
circuit or of a conductor in a d.c. circuit shall be not less
than the values given in Table 6.2.

**Table 6.2** Minimum nominal cross-sectional area of conductor (data from BS 7671:2008)

| Type of wiring system | Use of circuit | Conductor | |
|---|---|---|---|
| | | Material | Minimum cross-sectional area (mm$^2$) |
| Cables and insulated conductors | Power and lighting circuits | Copper | 1.0 |
| | | Aluminium | 16.0 |
| | Signalling and control circuits | Copper | 0.5 |
| Bare conductors | Power circuits | Copper | 10 |
| | | Aluminium | 16 |
| | Signalling and control circuits | Copper | 4 |
| Flexible connections with insulated conductors and cables | For a specific appliance | Copper | See relevant British Standard |
| | For any other application | Copper | 0.5 |
| | Extra-low-voltage circuits for special applications | Copper | 0.5 |

## 6.1.3.2 Identification of cables

| | |
|---|---|
| Cores of cables shall be identified at its terminations (and preferably throughout its length) by: | WR-514.3.1<br>WR-514.3.2 |

- colour (see BS 7671:2008 Regulation 514.4); and/or
- lettering and/or numbering (se BS 7671:2008 Regulation 514.5).

| | |
|---|---|
| Single-core cables that are coloured **green-and-yellow** throughout their length shall **only** be used as a protective conductor and **shall not** be over marked at their terminations. | WR-514.4.2 |

| | |
|---|---|
| Identification by colour or marking is not required for: | WR-514.6.1 |

- concentric conductors of cables;
- the metal sheath or armour of cables when used as a protective conductor.

### 6.1.4 Installation of cables

Every cable should have adequate strength and be so installed as to withstand the electromechanical forces that may be caused by any current, including fault current, it may have to carry in service.

A bare live conductor shall be installed on insulators.

| | |
|---|---|
| Non-sheathed cables are permitted in a cable trunking system that provides a minimum of IP4X or IPXXD protection, **and** if the cover can only be removed by means of a tool or a deliberate action. | WR-521.10.1 |
| Non-sheathed cables for fixed wiring shall be enclosed in conduit, ducting or trunking. | WR-521.10.1 |
| The installation method of a wiring system in relation to the type of conductor or cable used shall be in accordance with Table 4A1 of Appendix 4 of BS 7671:2008 (Incorporating Amendment No. 1). | WR-521.1 |

Where risks due to structural movement exist (CB3),     WR-522.15.1
the cable support and protection system employed
shall be capable of permitting relative movement so
that conductors and cables are not subjected to excessive mechanical stress.

### 6.1.4.1 Electromagnetic effects – a.c. circuits

Single-core cables that are armoured with steel wire or     WR-521.5.2
steel tape **shall not** be used for an a.c. circuit.

**Notes:**

1. The steel wire or steel tape armour of a single-core cable is regarded as a
   ferromagnetic enclosure.
2. For single-core armoured cables, the use of aluminium armour may be considered.

### 6.1.4.2 Busbar trunking systems and powertrack systems

A busbar trunking or a powertrack system shall take account of external influences (see Appendix 8 of BS 7671:2008 (Incorporating Amendment No. 1).

All:                                                        WR-521.4

- busbar trunking systems shall comply with BS EN
  60439–2; and
- powertrack systems shall comply with the appropriate
  part of the BS EN 61534 series.

### 6.1.4.3 Electrode water heaters and boilers

The shell of the electrode water heater or electrode       WR-554.1.3
boiler shall be bonded to the metallic sheath and
armour ( if any) of the incoming supply cable.

### 6.1.4.4 Impact

Wiring systems shall be selected and erected so as to minimize the damage
arising from mechanical stress (e.g. impact, abrasion, penetration, tension or
compression) during installation, use and/or maintenance.

The degree of protection of electrical equipment shall be maintained after installation of the cables and conductors.

WR-522.6.4

A cable concealed in a wall or partition at a depth of less than 50 mm from a surface of the wall or partition shall:

WR-522.6.101

- incorporate an earthed metallic covering; or
- be enclosed in earthed conduit; or
- be enclosed in earthed trunking or ducting complying; or
- be mechanically protected against damage sufficient to prevent penetration of the cable by nails, screws etc.; or
- be installed in a zone within 150 mm from the top of the wall or partition or within 150 mm of an angle formed by two adjoining walls or partitions.

A cable installed under a floor or above a ceiling shall be run in such a position that it is not liable to be damaged by contact with the floor or the ceiling or their fixings.

522.6.100

A cable passing through a joist within a floor or ceiling construction or through a ceiling support (e.g. under floorboards), shall:

522.6.5

- be at least 50 mm measured vertically from the top, or bottom as appropriate, of the joist or batten; or
- incorporate an earthed metallic covering; or
- be enclosed in earthed conduit; or
- be enclosed in earthed trunking or ducting; or
- be mechanically protected against damage sufficient to prevent penetration of the cable by nails, screws etc.; or
- form part of a separated extra-low voltage (SELV) or protective extra-low-voltage (PELV) circuit.

The cables of an installation **not** intended to be under the supervision of a skilled or instructed person, and which are concealed in a wall or partition (the internal construction of which includes metallic

WR-522.6.103

parts, other than metallic fixings such as nails, screws etc.), shall:

- incorporate an earthed metallic covering; or
- be enclosed in earthed conduit; or
- be enclosed in earthed trunking or ducting; or
- be sufficiently mechanically protected to avoid damage to the cable during construction of the wall or partition and during installation of the cable; or
- form part of an SELV or PELV circuit.

 Where the installation is not intended to be under the supervision of a skilled or instructed person, consideration shall be given to providing additional protection by means of an residual current device (RCD).

### 6.1.4.5 Cable couplers

| | |
|---|---|
| A cable coupler shall be arranged so that the connector of the coupler is fitted at the end of the cable that is **remote** from the supply. | WR-553.2.2 |
| Every cable coupler in a functional extra-low voltage (FELV) system: | WR-411.7.5 |
| • shall have a protective conductor contact; and<br>• **shall not** be dimensionally compatible with those used for any other system in use in the same premises. | |
| Except for an SELV or a Class II circuit, a cable coupler shall be non-reversible, and shall have provision for the connection of a protective conductor. | WR-553.2.1 |

### 6.1.4.6 Current-carrying capacities of cables

| | |
|---|---|
| The current, including any harmonic current, to be carried by any conductor for sustained periods during normal operation shall be such that the appropriate temperature limit specified in Table 6.3 is not exceeded. | WR-523.1 |

**Table 6.3** Maximum operating temperatures for types of cable insulation (data from BS 7671:2008)

| Type of insulation | Temperature limit' |
|---|---|
| Thermoplastic | 70°C at the conductor |
| Thermosetting | 90°C at the conductor b |
| Mineral (thermoplastic covered, or bare exposed to touch) | 70°C at the sheath |
| Mineral (bare not exposed to touch, and not in contact with combustible material) | 105°C at the sheath b, c |

## 6.1.4.7 Earth electrodes

As covered in Chapter 3 of this book (Earthing), the following types of Earth electrode are recognized as being suitable for the purposes of the Wiring Regulations:

- Earth rods or pipes;
- Earth tapes or wires;
- Earth plates;
- underground structural metalwork embedded in foundations;
- welded metal reinforcement of concrete (except pre-stressed concrete) embedded in the Earth;
- lead sheaths and other metal coverings of cables;
- other suitable underground metalwork.

 **Note:** Further information on Earth electrodes can be found in BS 7430.

The use, as an Earth electrode, of the lead sheath or other metal covering of a cable shall be subject to all of the following conditions:    WR-542.2.5

- adequate precautions have been taken to prevent excessive deterioration by corrosion;
- the sheath or covering shall be in effective contact with Earth;
- the consent of the owner of the cable shall be obtained;
- arrangements shall exist for the owner of the electrical installation to be warned of any proposed change to the cable that might affect its suitability as an Earth electrode.

### 6.1.4.8 Fire precautions

In the selection and erection of installations in locations of national, commercial, industrial or public significance, the following measures may be considered:

- installation of mineral-insulated cables according to BS EN 60702;
- installation of cables with improved fire-resisting characteristics in case of a fire hazard;
- installation of cables in non-combustible solid walls, ceilings and floors;
- installation of cables in areas with constructional partitions having a fire-resisting capability for a time of 30 minutes or 90 minutes.

 **Note:** Where these measures are not practicable, improved fire protection may be possible by the use of reactive fire-protection systems.

### 6.1.4.9 Temperature

| | |
|---|---|
| Cables and wiring accessories shall only be installed or handled at temperatures within the limits stated in the relevant product specification or as given by the manufacturer. | WR-522.1.2 |
| Parts of a cable or flexible cord within an accessory, appliance or luminaire shall be suitable for the temperatures likely to be encountered, or shall be provided with additional insulation suitable for those temperatures. | WR-522.2.100 |

### 6.1.4.10 Thermal insulation of cables

| | |
|---|---|
| A cable should preferably **not** be installed in a location where it is liable to be covered by thermal insulation. | WR-523.9 |
| Where a cable is to be run in a space to which thermal insulation is likely to be applied, it shall, wherever practicable, be fixed in a position such that it will **not** be covered by the thermal insulation. | WR-523.9 |
| For a single cable that is likely to be totally surrounded by thermally insulating material for over 0.5 m, the current-carrying capacity shall be assumed to be at least 0.5 times the current-carrying capacity for that cable clipped directly to a surface and open. | WR-523.9 |

Where a cable is to be totally surrounded by thermal insulation for less than 0.5 m, the current-carrying capacity of the cable shall be reduced appropriately, according to the size of cable, the length of cable in insulation and the thermal properties of the insulation.

WR-523.9

### 6.1.4.10.1 Precautions within a fire-segregated compartment

The risk of spread of fire shall be minimized by the selection of appropriate materials and erection.

 **Note:** In installations where particular risk is identified, cables shall meet the flame propagation requirements given in the relevant part of the BS EN 50266 series.

Cables complying with the requirements of BS EN 60332–1–2 may be installed without special precautions.

WR-527.1.3

Cables **not** complying with the flame propagation requirements of BS EN 60332–1–2 shall be limited to short lengths for connection of appliances to the permanent wiring system and shall not pass from one fire-segregated compartment to another.

WR-527.1.4

### 6.1.4.10.2 Locations with risks of fire due to the nature of processed or stored materials

Where BE2 conditions exist and where there is a risk of fire due to the manufacture, processing or storage of flammable materials, such as:

- barns (due to the accumulation of dust and fibres);
- woodworking facilities;
- paper mills and textile factories (due to the storage and processing of combustible materials);

a fire risk will be present and, in such locations, and in these circumstances:

- cables shall, as a minimum, satisfy the test under fire conditions specified in BS EN 60332–1–2;
- cables not completely embedded in non-combustible material, such as plaster or concrete, or

WR-422.3.4

otherwise protected from fire shall meet the
flame propagation characteristics as specified in
BS EN 60332–1–2;

- a cable trunking system or cable ducting system
  shall satisfy the test under fire conditions specified in BS EN 50085;
- precautions shall be taken such that a cable or
  wiring system cannot propagate flame;
- a cable tray system or cable ladder shall meet the
  requirements of BSEN 61537.

Except for mineral-insulated cables, a wiring system     WR-422.3.9
shall be protected against insulation faults:

- in a TN or TT system, by an RCD having a rated
  residual operating current (IA,) not exceeding
  300 mA;
- in an IT system, by an insulation monitoring
  device with audible and visual signals.

Flexible cables and flexible cords shall either be:     WR-422.3.100

- a heavy-duty type (with a voltage rating of not
  less than 450/750 V); or
- suitably protected against mechanical damage.

### 6.1.4.11 Connection of multi-wire, fine wire and very fine wire conductors

To avoid undesirable separation or spreading of individual wires, care should
be taken to ensure that suitable terminals are used (and that conductor ends
are suitably treated) for all connections of multi-wire, fine wire or very fine
wire conductors.

Cores of sheathed cables from which the sheath has     WR-526.8
been removed, and non-sheathed cables at the termination of conduit, ducting or trunking shall be enclosed.

Soldering (tinning) of the whole conductor end of      WR-526.9.2
multi-wire, fine wire and very fine wire conductors is **not**
permitted if screw terminals are used.

Soldered (tinned) conductor ends on fine wire and      WR-526.9.3
very fine wire conductors are **not** permissible at

connection and junction points that are subject (while in service) to a relative movement between the soldered and the non-soldered part of the conductor.

### 6.1.4.12 Lifts and/or the proximity to non-electrical services

No cable shall be run in a lift or hoist shaft unless it forms part of the lift installation.                    WR-528.3.5

### 6.1.4.13 Luminaires

Any flexible cable between the fixing means and the luminaire shall be installed so that any expected stresses in the conductors, terminals and terminations will not impair the safety of the installation.          WR-559.6.1.5

The connection of suspended current-using equipment (such as a luminaire) shall be made by cable with flexible cores.          WR-522.7.2

 **Note:** A cable for through wiring shall be selected in accordance with the temperature information on the luminaire or on the manufacturer's instruction sheet.

### 6.1.4.14 Electrical connections to bare connectors and/or busbars

Where a cable is to be connected to a bare conductor or busbar, its type of insulation and/or sheath shall be suitable for the maximum operating temperature of the bare conductor or busbar.          WR-526.4

### 6.1.4.15 Fault current protective devices

Every fault current protective device shall ensure that:

A fault occurring at any point in a circuit is quickly interrupted so that the fault current does not cause the permitted limiting temperature of any conductor or cable to be exceeded.          WR-434.5.2

### 6.1.4.16 Groups containing more than one circuit

The group rating factors (see Tables 4C1 to 4C5 of Appen-    WR-523.5
dix 4 to BS 7671:2008 (Incorporating Amendment No.
1)) are applicable to groups of non-sheathed or sheathed
cables having the same maximum operating temperature.

For groups containing non-sheathed or sheathed cables    WR-523.5
having different maximum operating temperatures, the
current-carrying capacity of all the non-sheathed or
sheathed cables in the group shall be based on the lowest maximum operating temperature of any cable in the
group, together with the appropriate group rating factor.

### 6.1.4.17 Reduced low-voltage system

Every cable coupler of a reduced low-voltage system:    WR-411.8.5

- shall have a protective conductor contact; and
- **shall not** be dimensionally compatible with those
  used for any other system in use in the same premises.

### 6.1.4.18 Requirements for SELV and PELV circuits

Protective separation of wiring systems of SELV or    WR-414.4.2
PELV circuits from the live parts of other circuits
(which have at least basic insulation) shall be achieved
by one of the following arrangements:

- SELV and PELV circuit conductors;
- circuit conductors contained in a multi-conductor
  cable or other grouping of conductors;
- the rated voltage of the cable(s) not being less
  than the nominal voltage of the system, mechanical protection, and basic insulation etc.

### 6.1.4.19 Rotating machines

All equipment (including cable) of every circuit carrying    WR-552.1.1
the starting, accelerating and load currents of a motor
shall be suitable for a current at least equal to the full-load
current rating of the motor when rated in accordance
with the appropriate British (or Harmonized) Standard.

### 6.1.4.20 Telecommunication cables

Protective equipotential bonding shall be applied to any metallic sheath of a telecommunication cable.

WR-411.3.1.2

### 6.1.4.21 Underground cables

In the event of a cable crossing (or being in the proximity of) underground telecommunication cables and underground power cables:

WR-528.2

- a minimum clearance of 100 mm shall be maintained;
- a fire-retardant partition shall be provided between the cables; and
- mechanical protection between the cables shall be provided.

### 6.1.4.22 Wiring systems

The installation of wiring systems will meet the requirements if:

WR-412.2.4.1

- the rated voltage of the cable(s) is not less than the nominal voltage of the system and at least 300/500 V; and
- adequate mechanical protection of the basic insulation is provided by (one or more) of the following:

  o the non metallic sheath of the cable;
  o non-metallic trunking or ducting (complying with the BS EN 50085);
  o a non-metallic conduit.

The minimum cross-sectional area of the extra-low-voltage conductors shall normally be 1.5 mm$^2$ copper, but:

WR-559.11.5.2

- for flexible cables with a maximum length of 3 m, a cross-sectional area of 1 mm$^2$ copper may be used;
- for suspended flexible cables (and especially for mechanical reasons), 4 mm$^2$ copper should be used;

- for composite cables consisting of a braided
  tinned copper outer sheath (with a material of
  high tensile strength inner core), 4 mm² copper
  should be used.

A voltage Band I circuit shall not be contained in    WR-528.1
the same wiring system as a Band II circuit, unless:

- every cable is insulated for the highest voltage
  present;
- each conductor of a multi-core cable is insulated
  for the highest voltage present in the cable;
- the cables are insulated for their system voltage
  and installed in a separate compartment of a
  cable ducting or cable trunking system;
- the cables are installed on a cable tray system
  where physical separation is provided by a
  partition;
- for a multi-core cable or cord, the cores of the
  Band I circuit are separated from the cores of
  the Band II circuit by an earthed metal screen of
  equivalent current-carrying capacity of the largest core of a Band II circuit.

### 6.1.5 Inspection of cables

As well as checking the routing of cables in safe zones (for protection against
mechanical damage), where relevant to the installation, and where necessary,
cables should also be checked during erection.

Inspection should always precede testing, and shall normally be done with that
part of the installation under inspection disconnected from the supply.

### 6.1.5.1 Mechanical stresses

A wiring system buried in a floor should be sufficiently protected to prevent
damage caused by the intended use of the floor.

A cable buried in the ground (i.e. not installed in a     WR-522.8.10
conduit or duct) shall incorporate an earthed armour
or metal sheath (or both) as a protective conductor.

A conduit system or cable ducting system (other than     WR-522.8.2
a pre-wired conduit assembly that has been specifically

designed for the installation) that is going to be
buried in the structure shall be completely
erected between access points before any cable
is drawn in.

A wiring system intended for the drawing in or out          WR-522.8.6
of conductors or cables shall have adequate means
of access to allow for this operation.

A wiring system shall be selected and erected to            WR-522.8.1
avoid (during installation, use or maintenance)
damage to the sheath or insulation of cables and
their terminations.

 The use of any lubricants that could have a detrimental effect on the cable or
wiring system is **not** permitted.

Buried cables shall be at a sufficient depth to avoid        WR-522.8.10
being damaged by any reasonably foreseeable distur-
bance of the ground.

Every cable shall be supported in such a way that it is      WR-522.8.5
not exposed to undue mechanical strain and so that
there is no appreciable mechanical strain on the terminations of the conductors.

The location of buried cables shall be clearly marked        WR-522.8.10
by cable covers or a suitable marking tape.

The radius of every bend in a wiring system shall be         WR-522.8.3
such that the cables do not suffer damage and termi-
nals are not stressed.

Where the conductors or cables are not supported             WR-522.8.4
continuously, they shall be supported by suitable
means at appropriate intervals in such a manner that
the conductors or cables do not suffer damage by their
own weight.

 **Note:** A wiring system buried in a floor should be sufficiently protected to
prevent damage caused by the intended use of the floor.

 See IEC 61386–24 for further details concerning underground conduits.

Cable supports and enclosures shall not have sharp edges that could damage the wiring system.   WR-522.8.11

A cable shall not be damaged by the means of fixing.   WR-522.8.12

Cables that pass across expansion joints shall be selected and/or erected so that any anticipated movement will not cause damage to the electrical equipment.   WR-522.8.13

### 6.1.5.2 Vibration

The cables and cable connections of a wiring system supported by (or fixed to a structure or equipment) that is subject to vibration of medium severity (AII2) or high severity (AH3 ) shall be suitable for such conditions   WR-522.7.1

## 6.1.6 Cables for special installations and locations

Where an electrical service is located in close proximity to one or more non-electrical services, it shall meet the following conditions:

- the wiring system shall be suitably protected against the hazards that are likely to arise from the presence of the other services in normal use;   WR-528.3.4
- fault protection shall be provided by automatic disconnection of supply.

### 6.1.6.1 Agricultural and horticultural premises

The particular requirements of this section apply to fixed electrical installations that are indoors and/or outdoors in agricultural and horticultural premises. Some of the requirements are also applicable to other locations that are in common buildings belonging to the agricultural and horticultural premises.

 **Note:** Section 705 does **not** cover electric fence installations.

Where vehicles and mobile agricultural machines are operated, the following methods of installation shall be applied:   WR-705.522

- cables shall be buried in the ground at a depth of at least 0.6 m, with added mechanical protection;

- cables in arable or cultivated ground shall be buried at a depth of at least 1 m;
- self-supporting suspension cables shall be installed at a height of at least 6 m.

Special attention shall be given to the presence of different kinds of fauna (e.g. rodents).

The following documentation shall be provided to      WR-705.514.9.3
the user of the installation:

- the routing of all concealed cables;
- a single-line distribution diagram.

### 6.1.6.2 Caravan and camping parks

**Note:** In order not to mix regulations on different subjects, such as those for electrical installation of caravan parks with those for electrical installation inside caravans, two sections have been created in BS 7671:2008 (Incorporating Amendment No. 1):

- Section 708, which concerns electrical installations in caravan parks, camping parks and similar locations; and
- Section 721, which concerns electrical installations in caravans and motor caravans.

In caravan and camping sites, underground      WR-708.521.1.1
cables shall be buried at a depth of at least
0.6 m and (unless having additional
mechanical protection) be placed outside any
caravan pitch or away from any surface where
tent pegs or ground anchors are expected to
be present.

No more than four socket-outlets should be grouped in one location, in order to avoid the supply cable crossing a pitch other than the one intended to be supplied.
In caravans and motor caravans:

| | |
|---|---|
| All cables (unless enclosed in rigid conduit) and all flexible conduits shall be supported at intervals not exceeding 0.4 m for vertical runs and 0.25 m for horizontal runs. | WR-721.522.8.1.3 |
| All cables shall, as a minimum, meet the requirements of BS EN 60332–1–2. | WR-721.521.2 |
| Cable management systems shall comply with BS EN 61386. | WR-721.521.2 |
| Low-voltage cable systems shall be run separately from the cables of extra-low-voltage systems so that there is no possibility of physical contact between the two wiring systems. | WR-721.528.1 |
| The wiring systems shall be installed using one or more of the following: | WR-721.521.2 |

- insulated single-core cables, with flexible class 5 conductors, in non-metallic conduit;
- insulated single-core cables, with stranded class 2 conductors (minimum of seven strands), in non-metallic conduit;
- sheathed flexible cables.

| | |
|---|---|
| Where cables have to run through such a compartment, they shall be protected against mechanical damage by installation within a conduit system or within a ducting system. | WR-721.528.3.4 |

### 6.1.6.3 Construction and demolition site installations

| | |
|---|---|
| Cables shall **not** be installed across a site road or a walkway unless the cable is adequately protected against mechanical damage. | WR-704.522.8.10 |
| For applications exceeding reduced low voltage, flexible cable shall be of HO7RN-F (BS 7919) type or equivalent, have a 450/750 V rating and be resistant to abrasion and water. | WR-704.522.8.11 |
| For reduced low-voltage systems, low temperature 300/500 V thermoplastic (BS 7919) or equivalent flexible cables shall be used. | WR-704.522.8.11 |

## 6.1.6.4 *Exhibitions, shows and stands*

The particular requirements of this section apply to the temporary electrical installations in exhibitions, shows and stands (including mobile and portable displays and equipment), to protect users.

| | |
|---|---|
| A cable intended to supply temporary structures shall be protected at its origin by an RCD whose rated residual operating current does not exceed 300 mA. | WR-711.410.3.4 |
| Armoured cables or cables protected against mechanical damage shall be used wherever there is a risk of mechanical damage. | WR-711.52 |
| Flexible cords **shall not** be laid in areas accessible to the public unless they are protected against mechanical damage. | WR-711.52 |
| Insulation piercing lampholders **shall not** be used unless the cables and lampholders are compatible, and provided that the lampholders are non-removable once fitted to the cable. | WR-711.559.4.3 |
| Joints shall not be made in cables, except where necessary as a connection into a circuit.<br>Where joints are made, these shall either use connectors that are in accordance with relevant standards or be in enclosures with a degree of protection of at least IP4X or IPXXD. | WR-711.526.1 |
| If no fire alarm system has been installed in a building that is going to be used for exhibitions etc., the cable systems shall be either:<br><br>• flame retardant to BS EN 60332–1–2 and low smoke to BS EN 61034–2; or<br>• single-core or multi-core unarmoured cables enclosed in metallic or non-metallic conduit or trunking, providing a degree of fire protection of at least IP4X. | WR-711.521 |
| Where strain can be transmitted to terminals, the connection shall incorporate suitable cable anchorage(s). | WR-711.526.1 |

Wiring cables shall be copper, have a minimum          WR-711.52
cross-sectional area of 1.5 mm², and shall comply
with an appropriate British Standard for either
thermoplastic or thermosetting insulated electric
cables.

### 6.1.6.5 Marinas and similar locations

The particular requirements of this section are applicable only to circuits that
are intended to supply pleasure craft or houseboats in marinas and similar
locations.

 They do **not** apply to the supply to houseboats if they are directly supplied from
the public network, or to the internal electrical installations of pleasure craft
or houseboats.

The following wiring systems are suitable for          WR-709.521.1.4
marina distribution circuits:

- underground cables;
- overhead cables or overhead insulated
  conductors;
- cables with copper conductors and thermoplastic or elastomeric insulation and sheath,
  installed within an appropriate cable management system (taking into account external
  influences such as movement, impact, corrosion and ambient temperature);
- mineral-insulated cables with a PVC protective covering;
- cables with armouring and serving of thermoplastic or elastomeric material;
- other cables and materials that are no less suitable than those listed above.

The following wiring systems **shall not** be used on     WR-709.521.1.5
or above a jetty, wharf, pier or pontoon:

- cables in free air suspended from or incorporating a support wire;
- non-sheathed cables in cable management systems, trunking etc.;
- cables with aluminium conductors;
- mineral-insulated cables.

| | |
|---|---|
| Underground distribution cables shall, unless provided with additional mechanical protection, be buried at a sufficient depth to avoid being damaged (e.g. by heavy-vehicle movement). | WR-709.521.1.7 |

 **Note:** A depth of 0.5 m is normally considered as the 'minimum depth' to fulfil this requirement.

| | |
|---|---|
| Cable management systems shall be installed to allow the drainage of water by drainage holes and/or installation of the equipment on an incline. | WR-709.521.1.6 |
| Cables shall be selected and installed so that mechanical damage due to tidal and other movement of floating structures is prevented. | WR-709.521.1.6 |

### 6.1.6.6 Mobile or transportable units

For the purposes of this section, the term 'unit' is intended to mean a vehicle and/or mobile (self-propelled or towed) or transportable structure (e.g. a container or cabin) in which all or part of an electrical installation is contained and which is provided with a temporary supply by means of, for example, a plug and socket-outlet.

| | |
|---|---|
| Flexible cables (for connecting the unit to the supply), or cables of equivalent design, having a minimum cross-sectional area of 2.5 mm$^2$ copper shall be used. | WR-717.52.1 |
| The flexible cable shall enter the unit by an insulating inlet in such a way as to minimize the possibility of any insulation damage or fault that might energize the exposed-conductive-parts of the unit. | WR-717.52.1 |
| The wiring system shall be installed using one or more of the following: <br><br> • unsheathed flexible cable with thermoplastic or thermosetting insulation installed in either a conduit or trunking or ducting; | WR-717.52.2 |

- sheathed flexible cable thermoplastic or thermosetting insulation.

Where cables have to run through such a compartment, they shall be protected against mechanical damage by installation within a conduit system or within a ducting system.    WR-717.528.3.4

Where installed, this conduit or duct shall be able to withstand an impact equivalent to AG3 without visible physical damage.    WR-717.528.3.5

### 6.1.6.7 Rooms and cabins containing sauna heaters

In zone 3 of a room (or cabin) containing a sauna heater, the insulation and sheaths of cables shall be capable of withstanding a minimum temperature of 170°C.

The particular requirements of this section apply to:

- sauna cabins erected on site (e.g. in a location or in a room);
- the room where the sauna heater is, or where the sauna heating appliance is installed.

### 6.1.6.8 Solar and photovoltaic power supply systems

On the a.c. side, the supply cable of a photovoltaic (PV) power supply system shall be connected to the supply side of the protective device for the automatic disconnection of circuits supplying current-using equipment.    WR-712.411.3.2.1.1

Overload protection may be omitted to the PV main cable if the continuous current-carrying capacity is equal to (or greater than) 1.25 times $I_{sc\ STC}$ of the PV generator.    WR-712.433.2

Overload protection may be omitted to PV string and PV array cables when the continuous current-carrying capacity of the cable is equal to (or greater than) 1.25 times $I_{sc\ STC}$ at any location.    WR-712.433.1

PV string cables, PV array cables and PV d.c. main cables shall be selected and erected so as to minimize the risk of Earth faults and short-circuits.

WR-712.522.8.1

The PV supply cable on the a.c. side shall be protected against fault current by an overcurrent protective device installed at the connection to the a.c. mains.

WR-712.434.1

Where protective bonding conductors are installed, they shall be parallel to, and in as close contact as possible with, d.c. cables and a.c. cables and accessories.

WR-712.54

Wiring systems shall be capable of withstanding expected external influences such as wind, ice formation, temperature and solar radiation.

WR-712.522.8.3

### 6.1.6.9 Swimming pools and other basins

In zones 0, 1 and 2 of a swimming pool (or similar basin) any metallic sheath or metallic covering of a wiring system shall be connected to the supplementary equipotential bonding.

WR-702.522.21

 **Note:** Cables should preferably be installed in conduits made of insulating material.

For a fountain, the following additional requirements shall be met:

WR-702.522.23

- a cable for electrical equipment in zone 0 shall be installed as far outside the basin rim as is reasonably practicable and run to the electrical equipment inside zone 0 by the shortest possible route;

> - in zone 1 a cable shall be selected, installed and provided with mechanical protection to medium severity (AG2) and the relevant sub-mersion in water depth (AD8).

### 6.1.6.10 Temporary electrical systems in amusement parks, circuses and fairgrounds

| | |
|---|---|
| All cables shall meet the requirements of BS EN 60332–1–2. | WR-740.521.1 |
| Armoured cables or cables protected against mechanical damage shall be used wherever there is a risk of mechanical damage due to external influence (e.g. to AG2 or better). | WR-740.521.1 |
| Buried cables shall be protected against mechanical damage. | WR-740.521.1 |
| Cables shall have a minimum rated voltage of 450/750 V. | WR-740.521.1 |
| Insulation-piercing lampholders shall not be used unless the cables and lampholders are compatible and the lampholders cannot be removed once fitted to the cable. | WR-740.55.1.2 |
| Joints shall not be made in cables except where necessary as a connection into a circuit. Where joints are made, these shall either use connectors in accordance with the relevant British Standard, or the connection shall be made in an enclosure with a degree of protection of at least IP4X or IPXXD. | WR-740.526 |
| Supply cables shall be flexible and have adequate protection against mechanical damage. | WR-740.55.1.4 |
| The routes of cables buried in the ground shall be marked at suitable intervals. | WR-740.521.1 |
| Where strain can be transmitted to terminals the connection shall incorporate cable anchorage(s). | WR-740.526 |

 Luminaires and floodlights shall be so fixed and protected that a focusing or concentration of heat is not likely to cause ignition of any material.

## 6.2 Conductors and conduits

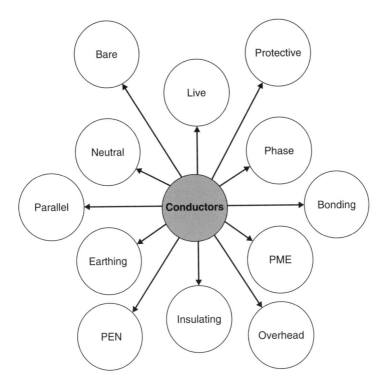

**Figure 6.2** Conductors.

Conductors:

- intended to operate at temperatures above 70°C **shall not** be connected to switchgear, protective devices, accessories or other types of equipment;
- **shall not** be subjected to excessive mechanical stress;
- **shall** be capable of withstanding all foreseen electromechanical forces (including fault current) during service.

### 6.2.1 General

| | |
|---|---|
| The number of conductors to be considered in a circuit are those carrying load current less those conductors that only serve the purpose of protective conductors. | WR-523.6.1 WR-523.6.4 |
| Combined protective and neutral (PEN) conductors need to be taken into consideration in the same way as neutral conductors. | WR-523.6.4 |

> Where risks due to structural movement exist (CB3),      WR-522.15.1
> the cable support and protection system employed
> shall be capable of permitting relative movement
> so that conductors are not subjected to excessive
> mechanical stress.

## 6.2.2 Types of protective conductor

A protective conductor may consist of one or more of the following:

- a single-core cable;
- a conductor in a cable;
- an insulated or bare conductor in a common enclosure with insulated live conductors;
- a fixed bare or insulated conductor;
- a metal covering (e.g. the sheath, screen or armouring of a cable);
- a metal conduit or other enclosure or electrically continuous support system for conductors.

 **Note:** If a protective conductor is formed by a conduit, trunking, ducting or the metal sheath and/or armour of a cable, the earthing terminal of each accessory shall be connected by a separate protective conductor to an earthing terminal incorporated in the associated box or enclosure.

 A gas pipe, an oil pipe, flexible or pliable conduit, support wires or other flexible metallic parts, or constructional parts that are subject to mechanical stress in normal service, **shall NOT** be selected as a protective conductor.

### 6.2.2.1 Bare conductors

The minimum cross-sectional area of phase conductors in a.c. circuits and of live conductors in d.c. circuits shall be as shown in Table 6.4.

**Table 6.4** Minimum cross-sectional area of bare conductors

| Type of wiring system | Use of circuit | Conductor | |
|---|---|---|---|
| | | Material | Minimum cross-sectional area (mm²) |
| Bare conductors | Power circuits | Copper | 10 |
| | | Aluminium | 16 |
| | Signalling and control circuits | Copper | 4 |

Bare conductors shall be painted or identified by a   WR 514.4.6
coloured tape, sleeve or disc as per Table 6.5.

**Table 6.5** Identification of conductors (data from BS 7671: 2008)

| Function | Alphanumeric | Colour |
| --- | --- | --- |
| Protective conductors | | Green-and-yellow |
| Functional earthing conductor | | Cream |
| a.c. power circuit (Note 1) | | |
| Phase of single-phase circuit | L | Brown |
| Neutral of single- or three-phase circuit | N | Blue |
| Phase 1 of three-phase a.c. circuit | L1 | Brown |
| Phase 2 of three-phase a.c. circuit | L2 | Black |
| Phase 3 of three-phase a.c. circuit | L3 | Grey |
| Two-wire unearthed d.c. power circuit | | |
| Positive of two-wire circuit | L+ | Brown |
| Negative of two-wire circuit | L− | Grey |
| Two-wire earthed d.c. power circuit | | |
| Positive (of negative earthed) circuit | L+ | Brown |
| Negative (of negative earthed) circuit (Note 2) | M | Blue |
| Positive (of positive earthed) circuit (Note 2) | M | Blue |
| Negative (of positive earthed) circuit | L− | Grey |
| Three-wire d.c. power circuit | | |
| Outer positive of two-wire circuit derived from three-wire system | L+ | Brown |
| Outer negative of two-wire circuit derived from three-wire system | L− | Grey |
| Positive of three-wire circuit | L+ | Brown |
| Mid-wire of three-wire circuit (Notes 2 and 3) | M | Blue |
| Negative of three-wire circuit | L− | Grey |
| Control circuits, ELV and other applications | | |
| Phase conductor | L | Brown, black, red, orange, yellow, violet, grey, white, pink or turquoise |
| Neutral or mid-wire (Note 4) | N or M | Blue |

**Notes:**

1. Power circuits include lighting circuits.
2. M identifies either the mid-wire of a three-wire d.c. circuit, or the earthed conductor of a two-wire earthed d.c. circuit.
3. Only the middle wire of three-wire circuits may be earthed.
4. An earthed PELV conductor is blue.

Colour or marking is not required for bare conductors (where permanent identification is not practicable).

| | |
|---|---|
A bare conductor or busbar used as a protective conductor shall be identified, where necessary, by equal **green** and **yellow** stripes (each not less than 15 mm and not more than 100 mm wide) close together, either throughout the length of the conductor, or in each compartment and unit and at each accessible position. If adhesive tape is used, it shall also be bi-coloured. | WR-514.4.2 |

### 6.2.2.2 Earthing conductors

 Earthing conductors **shall** be capable of being disconnected to enable the resistance of the earthing arrangements to be measured.

| | |
|---|---|
In every installation a main earthing terminal shall be provided to connect the following to the earthing conductor: | WR-542.4.1 |

- the circuit protective conductors;
- the protective bonding conductors;
- functional earthing conductors (if required);
- lightning protection system bonding conductor (if any).

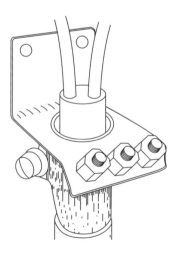

**Figure 6.3** Typical earthing conductor. (Courtesy of Stingray.)

If a protective conductor forms part of a cable, then this **shall** only be earthed in the installation containing the associated protective device.

| | |
|---|---|
| When the earthing conductor is buried in the ground, it shall have a cross-sectional area not less than that stated in Table 6.6. | WR-542.3.1 |

**Table 6.6** Minimum cross-sectional area of a buried earthing conductor

| | Protected against mechanical damage | Not protected against mechanical damage |
|---|---|---|
| Protected against corrosion by a sheath | See  Note A below | 16 mm² copper<br>16 mm² coated steel |
| Not protected against corrosion | 25 mm² copper<br>50 mm² steel | 25 mm² copper<br>50 mm² steel |

**Note A:** The cross-sectional area of protective conductor shall not be less than:

$$S = \frac{\sqrt{I^2 t}}{k} \quad \text{where:}$$

$S$ = the nominal cross-sectional area of the conductor in mm²
$I$ = the fault current in amperes (rms for a.c.) for a fault of negligible impedance, which can flow through the associated protective device
$t$ = the operating time in seconds of the disconnecting device
$k$ = is a factor taking account of the resistivity, temperature coefficient and heat capacity of the conductor material.

| | |
|---|---|
| For a tape or strip conductor, the thickness shall be such as to withstand mechanical damage and corrosion. | WR-542.3.2 |
| The connection of an earthing conductor to an Earth electrode or other means of earthing shall be: | WR-542.3.2 |

- soundly made;
- electrically and mechanically satisfactory;
- labelled; and
- suitably protected against corrosion.

| | |
|---|---|
| The earthing conductor of a street electrical fixture shall have a minimum copper equivalent cross-sectional area not less than that of the supply neutral conductor at that point or not less than 6 mm², whichever is the smaller. | WR-559.10.3.4 |

| | |
|---|---|
| Where overcurrent protective devices are used for fault protection, the protective conductor shall be incorporated in the same wiring system as the live conductors, or in their immediate proximity. | WR-543.6.1 |

### 6.2.2.3 Line conductors

| | |
|---|---|
| Detection of overcurrent shall be provided for all line conductors, and shall cause the disconnection of the conductor in which the overcurrent is detected. | WR-431.1.1 |
| In a TN or TT system, for a circuit supplied between line conductors and in which the neutral conductor is not distributed, overcurrent detection need not be provided for one of the line conductors, provided that: | WR-431.1.2 |

- there exists, in the same circuit or on the supply side, differential protection intended to detect unbalanced loads and cause disconnection of all the line conductors; **and**
- the neutral conductor is not distributed from an artificial neutral point of the circuits situated on the load side of this differential protective device.

### 6.2.2.4 Live conductors

- Live supply conductors **shall** be capable of being isolated from circuits.
- Bare live conductors **shall** be installed on insulators.
- Conductors **shall** be able to carry fault current without overheating.
- The supply to all live conductors **shall** be automatically interrupted in the event of an overload or fault current.
- Persons and livestock **shall** be protected against injury, and property **shall** be protected against damage, due to excessive temperatures (or electro-mechanical stresses) caused by any overcurrents likely to arise in live conductors.

### 6.2.2.5 Multi-wire, fine wire and very fine wire conductors

| | |
|---|---|
| In order to avoid inappropriate separation or spreading of individual wires of multi-wire, fine wire or very fine wire conductors, suitable terminals shall be used or the conductor ends shall be suitably treated. | WR-526.9.1 |

Soldering (tinning) of the whole conductor end of multi-wire, fine wire and very fine wire conductors is **not** allowed if screw terminals are used.

WR-526.9.2

Soldered (tinned) conductor ends on fine wire and very fine wire conductors are **not allowed** at connection and junction points that are subject in service to a relative movement between the soldered and the non-soldered part of the conductor.

WR-526.9.3

## 6.2.3 Construction

### 6.2.3.1 Automatic disconnection in case of a fault

A protective device shall automatically interrupt the supply to the line conductor of a circuit or equipment in the event of a fault of negligible impedance between the line conductor and an exposed-conductive-part (or a protective conductor) in the circuit or equipment, within the disconnection time required in Table 6.7.

WR-411.3.2.1

**Table 6.7** Maximum disconnection times (in seconds)

| System | Maximum disconnection time (s) | | | | | | | |
|---|---|---|---|---|---|---|---|---|
| | $50\,V < U_o\ 120\,V$ | | $120\,V < U_o\ 230\,V$ | | $230\,V < U_o\ 5400\,V$ | | $U_o > 400\,V$ | |
| | a.c. | d.c. | a.c. | d.c. | a.c. | d.c. | a.c. | d.c. |
| TN | 0.8 | See Note | 0.4 | 5 | 0.2 | 0.4 | 0.1 | 0.1 |
| TT | 0.3 | See Note | 0.2 | 0.4 | 0.07 | 0.2 | 0.04 | 0.1 |

 **Note:** Disconnection is not required for protection against electric shock, but may be required for other reasons, such as protection against thermal effects.

In a TN system, a disconnection time not exceeding 5 s is permitted for a distribution circuit and for a circuit not covered by Table 6.7.

WR-411.3.2.3

In a TT system, a disconnection time not exceeding 1 s is permitted for a distribution circuit and for a circuit not covered by Table 6.7.

WR-411.3.2.4

Where automatic disconnection cannot be achieved in the time required, supplementary equipotential bonding shall be provided.

### 6.2.3.2 Busbar trunking

| | |
|---|---|
| Where a busbar trunking system is used as a protective conductor: | WR-543.2.4 |

- its electrical continuity shall be assured, either by construction or by suitable connection, in such a way as to be protected against mechanical, chemical or electrochemical deterioration;
- its cross-sectional area shall be in accordance with BS EN 60439–1;
- it shall permit the connection of other protective conductors at every predetermined tap-off point.

### 6.2.3.3 Cable couplers

A cable coupler shall be arranged so that the connector of the coupler is fitted at the end of the cable that is remote from the supply.

| | |
|---|---|
| Except for a SELV or a Class II circuit, a cable coupler shall be non-reversible and shall be capable of being connected to a protective conductor. | WR-553.2.1 |
| Every cable coupler in an FELV system: | WR-411.7.5 |

- shall have a protective conductor contact; and
- **shall not** be dimensionally compatible with those used for any other system utilized in the same premises.

### 6.2.3.4 Cross-sectional area of conductors

| | |
|---|---|
| The cross-sectional area of conductors shall be determined for both normal operating conditions and, where appropriate, for fault conditions, according to: | WR-132.6 |

- the admissible maximum temperature;
- the admissible voltage drop limit;
- the electromechanical stresses likely to occur due to short-circuit and Earth fault currents;
- other mechanical stresses to which the conductors are likely to be exposed;

- the maximum impedance for correct operation of short-circuit and Earth fault protection;
- the method of installation;
- harmonics;
- thermal insulation.

The cross-sectional area of a phase conductor in an a.c. circuit or of a live conductor in a d.c. circuit shall be as shown in Table 6.8.

WR-524.1

Table 6.8 Minimum nominal cross-sectional area of conductor

| Type of wiring system | Use of circuit | Conductor | |
|---|---|---|---|
| | | Material | Minimum cross-sectional area (mm²) |
| Cables and insulated conductors | Power and lighting circuits | Copper | 1.0 |
| | | Aluminium | 16.0 |
| | Signalling and control circuits | Copper | 0.5 |
| Bare conductors | Power circuits | Copper | 10 |
| | | Aluminium | 16 |
| | Signalling and control circuits | Copper | 4 |
| Flexible connections with insulated conductors and cables | For a specific appliance | Copper | See relevant British Standard |
| | For any other application | Copper | 0.5 |
| | Extra-low-voltage circuits for special applications | Copper | 0.5 |

The neutral conductor (if any) shall have a cross-sectional area not less than that of the line conductor.

WR-524.2.1

For a polyphase circuit where each line conductor has a cross-sectional area greater than 16 mm² for copper or 25 mm² for aluminium, the neutral conductor is permitted to have a smaller cross-sectional area than that of the line conductors, providing that:

- the expected maximum current, including harmonics (if any), in the neutral conductor during normal service is not greater than the current-carrying capacity of the reduced cross-sectional area of the neutral conductor; and

- the neutral conductor is protected against overcurrents; and
- the size of the neutral conductor is at least equal to 16 mm² for copper or 25 mm² for aluminium.

WR-524.3

In caravans and motor caravans, the cross-sectional area of every conductor shall be not less than 1.5 mm².

WR-721.524.1

### 6.2.3.5 Current-carrying capacities of conductors

The current (including any harmonic current) to be carried by any conductor for sustained periods during normal operation shall be such that the appropriate temperature limit specified in Table 6.9 is not exceeded.

WR-523.1

Table 6.9  Maximum operating temperatures for types of cable insulation

| Type of insulation | Temperature limit |
|---|---|
| Thermoplastic | 70°C at the conductor |
| Thermosetting | 90°C at the conductor b |
| Mineral (thermoplastic covered, or bare exposed to touch) | 70°C at the sheath |
| Mineral (bare not exposed to touch, and not in contact with combustible material) | 105°C at the sheath b, c |

### 6.2.3.6 Electromechanical stresses

Every conductor or cable shall have adequate strength and be installed so as to withstand the electromechanical forces that may be caused by any current, including fault current, it may have to carry while in service.

### 6.2.3.7 Identification of conductors – by colour

Table 6.10  Identification of conductors

| Function | Alphanumeric | Colour |
|---|---|---|
| Protective conductors | | Green-and-yellow |
| Functional earthing conductor | | Cream |
| a.c. power circuit (Note 1) | | |
| Phase of single-phase circuit | L | Brown |
| Neutral of single- or three-phase circuit | N | Blue |

| | | |
|---|---|---|
| Phase 1 of three-phase a.c. circuit | L1 | Brown |
| Phase 2 of three-phase a.c. circuit | L2 | Black |
| Phase 3 of three-phase a.c. circuit | L3 | Grey |
| Two-wire unearthed d.c. power circuit | | |
| Positive of two-wire circuit | L+ | Brown |
| Negative of two-wire circuit | L− | Grey |
| Two-wire earthed d.c. power circuit | | |
| Positive (of negative earthed) circuit | L+ | Brown |
| Negative (of negative earthed) circuit (Note 2) | M | Blue |
| Positive (of positive earthed) circuit (Note 2) | M | Blue |
| Negative (of positive earthed) circuit | L− | Grey |
| Three-wire d.c. power circuit | | |
| Outer positive of two-wire circuit derived from three-wire system | L+ | Brown |
| Outer negative of two-wire circuit derived from three-wire system | L− | Grey |
| Positive of three-wire circuit | L+ | Brown |
| Mid-wire of three-wire circuit (Notes 2 and 3) | M | Blue |
| Negative of three-wire circuit | L- | Grey |
| Control circuits, ELV and other applications | | |
| Phase conductor | L | Brown, black, red, orange, yellow, violet, grey, white, pink or turquoise |
| Neutral or mid-wire (Note 4) | N or M | Blue |

**Notes:**

1. Power circuits include lighting circuits.
2. M identifies either the mid-wire of a three-wire d.c. circuit, or the earthed conductor of a two-wire earthed d.c. circuit.
3. Only the middle wire of three-wire circuits may be earthed.
4. An earthed PELV conductor is blue.

Conductors with **green-and-yellow** colour identification **shall not** be numbered other than for the purpose of circuit identification.

The single colour **green shall not** be used.

| | |
|---|---|
| Unambiguous marking shall be provided at the interface between conductors. | WR-514.1.3 |

Identification by colour or marking is not required for:

- concentric conductors of cables;
- metal sheath or armour of cables when used as a protective conductor;

- bare conductors where permanent identification is not practicable;
- extraneous-conductive-parts used as a protective conductor;
- exposed-conductive-parts used as a protective conductor

### 6.2.3.7.1 Bare conductors

| | |
|---|---|
| A bare conductor shall be identified by the application of tape, sleeve or disc of the appropriate colour prescribed in Table 6.10 or by painting it with such a colour. | WR-514.4.6 |

 Colour or marking is not required for bare conductors (where permanent identification is not practicable).

### 6.2.3.7.2 Neutral or midpoint conductors

| | |
|---|---|
| Where a circuit includes a neutral or midpoint conductor, the colour used shall be **blue**. | WR-514.4.1 |

### 6.2.3.7.3 PEN conductors

| | |
|---|---|
| A PEN conductor shall be marked by one of the following methods:<br><br>• **green-and-yellow** throughout its length and with **blue** markings at the terminations;<br>• **blue** throughout its length with **green-and-yellow** markings at the terminations. | WR-514.4.3 |

### 6.2.3.7.4 Other conductors

| | |
|---|---|
| All other conductors shall be identified by colour in accordance with Table 6.10. | WR-514.4.4 |

### 6.2.3.7.5 Main protective bonding conductors

| | |
|---|---|
| Except where protective multiple earthing (PME) conditions apply, a main protective bonding conductor shall have a cross-sectional area not less than half the cross-sectional area required for the earthing conductor of the installation, and certainly not less than 6 mm². | WR-544.1.1 |

The cross-sectional area need not exceed 25 mm$^2$ if the bonding conductor is of copper or of a cross-sectional area affording equivalent conductance in other metals.

> Except for highway power supplies and street       WR-544.1.1
> furniture (where PME conditions apply), the main
> protective bonding conductor shall be selected in
> accordance with the neutral conductor of the supply
> and Table 6.11.

**Table 6.11** Minimum cross-sectional area of the main protective bonding conductor in relation to the neutral of the supply (data from BS 7671: 2008)

| Copper equivalent cross-sectional area of the supply neutral conductor | Minimum copper equivalent* cross-sectional area of the main protective bonding conductor |
| --- | --- |
| 35 mm$^2$ or less | 10 mm$^2$ |
| Over 35 mm$^2$ up to 50 mm$^2$ | 16 mm$^2$ |
| Over 50 mm$^2$ up to 95 mm$^2$ | 225 mm$^2$ |
| Over 95 mm$^2$ up to 150 mm$^2$ | 235 mm$^2$ |
| Over 150 mm$^2$ | 20 mm$^2$ |

## 6.2.3.7.6 Non-standard colours

> If wiring additions or alterations are made to an      WR-514.14.1
> installation so that some of the wiring complies with
> the current Regulations but there is also wiring to pre-
> vious versions of these Regulations, a warning notice
> shall be affixed at or near the appropriate distribution
> board with the wording shown in Figure 6.4.

---

**CAUTION**

This installation has wiring colours to two versions of BS 7671. Great care should be taken before undertaking extension, alteration or repair that all conductors are correctly identified.

---

**Figure 6.4** Warning notice – non-standard colours.

### 6.2.3.8 Identification of conductors by letters and/or numbers

The lettering or numbering system applied to the identification of individual conductors and of conductors in a group:

- shall be clear, legible and durable;
- all numerals shall be in strong contrast to the colour of the insulation;
- shall be given in letters or Arabic numerals (in order to avoid confusion, unattached numerals 6 and 9 shall be underlined).

### 6.2.3.8.1 Numerical

| | |
|---|---|
| Conductors may be identified by numbers, the number 0 being reserved for the neutral or midpoint conductor. | WR-514.5.4 |

### 6.2.3.9 Plugs and socket-outlets

| | |
|---|---|
| Except for SELV, every plug and socket-outlet shall be of the non-reversible type, with provision for the connection of a protective conductor. | WR-553.1.2 |

### 6.2.3.10 Protective multiple earthing

Where protective multiple earthing (PME) exists, the cross-sectional area of the main equipotential bonding conductor) shall be in accordance with Table 6.12.

**Table 6.12** Minimum cross-sectional area of the main equipotential bonding conductor in relation to the neutral of the supply

| Copper equivalent cross-sectional area of the supply neutral conductor | Minimum copper equivalent cross-sectional area of the main equipotential bonding conductor |
|---|---|
| $35\,mm^2$ or less | $10\,mm^2$ |
| Over $35\,mm^2$ up to $50\,mm^2$ | $16\,mm^2$ |
| Over $50\,mm^2$ up to $95\,mm^2$ | $25\,mm^2$ |
| Over $95\,mm^2$ up to $150\,mm^2$ | $35\,mm^2$ |
| Over $150\,mm^2$ | $50\,mm^2$ |

 **Note:** Local distributor's network conditions may require a larger conductor.

### 6.2.3.11 Protective conductors

- A gas pipe, an oil pipe, flexible or pliable conduit, support wires or other flexible metallic parts (or constructional parts subject to mechanical stress in normal service) **shall not** be selected as a protective conductor.
- Exposed-conductive parts of equipment **shall not** be used as a protective conductor for other equipment.

- In installations and locations where the risk of an electric shock is increased by a reduction in body resistance and/or by contact with Earth potential, all plugs, socket-outlets and cable couplers of a reduced low-voltage system **shall** have a protective conductor contact.

If the protective conductor:                                    WR-543.1.1

- is not an integral part of a cable; or
- is not formed by conduit, ducting or trunking; or
- is not contained in an enclosure formed by a wiring system;

then the cross-sectional area shall be not less than:

- $2.5\,mm^2$ copper equivalent if protection against mechanical damage is provided; and
- $4\,mm^2$ copper equivalent if mechanical protection is not provided.

Where a protective conductor is common to two or        WR-543.1.2
more circuits, its cross-sectional area shall be:

- calculated for the most onerous of the values of fault current and operating time encountered in each of the various circuits; or
- selected so as to correspond to the cross-sectional area of the largest line conductor of those circuits.

A protective conductor with a cross-sectional area up   WR-543.3.2
to and including $6\,mm^2$ shall be protected throughout
by a covering at least equivalent to that provided by
the insulation of a single-core non-sheathed cable having a voltage rating of at least 450/750 V unless it is:

- a protective conductor forming part of a multi-core cable;
- cable trunking or conduit used as a protective conductor.

If the protective conductor:                                    WR-543.1

- is not an integral part of a cable; or
- is not formed by conduit, ducting or trunking; or
- is not contained in an enclosure formed by a wiring system;

then the cross-sectional area shall be not less than 2.5 mm² copper equivalent if protection against mechanical damage is provided, and 4 mm² copper equivalent if mechanical protection is not provided.

A separate metal enclosure for cable shall not be used as a PEN conductor.    WR-543.2.10

### 6.2.3.11.1 Types of protective conductor

**Note:** The metal covering (including the sheath – bare or insulated) of a cable, trunking, ducting and metal conduit may be used as a protective conductor for the associated circuit.

A protective conductor may consist of one or more of the following:    WR-543.2.1

- a single-core cable;
- a conductor in a cable;
- an insulated or bare conductor in a common enclosure with insulated live conductors;
- a fixed bare or insulated conductor;
- a metal covering (e.g. the sheath, screen or armouring of a cable);
- a metal conduit, metallic cable management system or other enclosure or electrically continuous support system for conductors;
- an extraneous-conductive-part.

An extraneous-conductive-part may be used as a protective conductor if:    WR-543.2.6

- electrical continuity can be ensured and either constructed or connected so that it is protected against mechanical, chemical or electrochemical deterioration;
- precautions have been taken against its removal;
- it has been considered for such a use and, if necessary, suitably adapted.

A protective conductor of the types listed above and cross-sectional area of which is 10 mm² or less, shall be of copper.

Where a metal enclosure or frame of a low-voltage     WR-543.2.4
switchgear or controlgear assembly or busbar trunking
system is used as a protective conductor:

- its electrical continuity shall be assured, either by
  construction or by suitable connection, in such a
  way as to be protected against mechanical, chemical
  or electrochemical deterioration;
- its cross-sectional area shall be in accordance with
  BS EN 60439-1;
- it shall permit the connection of other
  protective conductors at every predetermined
  tap-off point.

---

The bi-colour combination **green-and-yellow** shall be     WR-514.4.2
used exclusively for identification of a protective conductor, and this combination **shall not** be used for any
other purpose.

Single-core cables that are coloured **green-and-yellow**     WR-514.4.2
throughout their length shall **only** be used as a protective conductor and **shall not** be overmarked at their
terminations.

One of the colours shall cover at least 30% and at most     WR-514.4.2
70% of the surface being coloured, while the other colour shall cover the remainder of the surface.

A bare conductor or busbar used as a protective     WR-514.4.2
conductor shall be identified, where necessary, by
equal **green** and **yellow** stripes, each not less than
15 mm and not more than 100 mm wide, close together,
either throughout the length of the conductor or in
each compartment and unit and at each accessible
position.

## 6.2.3.12 RCDs

An RCD shall be capable of disconnecting all the     WR-531.2.1
line conductors of the circuit at substantially the
same time.

### 6.2.3.13 Suspended conductors

| | |
|---|---|
| Suspension devices for extra-low-voltage luminaires, including supporting conductors, shall be capable of carrying five times the mass of the luminaires (including their lamps) intended to be supported, but not less than 5 kg. | WR-559.11.6 |
| Terminations and connections of conductors shall be made by screw terminals or screwless clamping devices. | WR-559.11.6 |

### 6.2.3.14 Warning notices

#### 6.2.3.14.1 Safety Earth

A warning notice (as shown in Figure 6.5) shall be permanently fixed at:

| | |
|---|---|
| • the point of connection of every earthing conductor to an Earth electrode; and<br>• the point of connection of every bonding conductor to an extraneous-conductive-part; and<br>• the main Earth terminal, where separate from the main switchgear. | WR-514.13.1 |

#### 6.2.3.14.2 Electrical separation

**Safety Electrical Connection – Do Not Remove**

**Figure 6.5** Warning notice – earthing and bonding.

| | |
|---|---|
| Where electrical separation to the supply to more than one current-using equipment is used (Regulation 418.2.5 or 418.3), the warning notice shall read as shown in Figure 6.6. | WR-514.13.2 |

> The protective bonding conductors associated with the electrical installation in this location MUST NOT BE CONNECTED TO
>
> # EARTH
>
> Equipment having exposed-conductive-parts connected to earth must not be brought into this location

**Figure 6.6** Warning notice – protective bonding conductors.

### 6.2.3.14.3 Alternative supplies

Where an installation includes alternative or additional sources of supply, warning notices shall be affixed at the following locations in the installation:

> - at the origin of the installation;          WR-514.15.1
> - at the meter position, if remote from the origin;
> - at the consumer unit or distribution board to which the alternative or additional sources are connected;
> - at all points of isolation of all sources of supply.

These warning notices shall have the following shown in Figure 6.7.

> **WARNING — MULTIPLE SUPPLIES**
>
> **Isolate all electrical supplies before carrying out work.**
>
> **Isolate the mains supply at ....................................**
>
> **Isolate the alternative supplies at .........................**

**Figure 6.7** Warning notice – multiple supplies.

### 6.2.4 Installation

#### 6.2.14.1 Bare conductors

> If the nominal voltage does not exceed 25 V a.c. or          WR-559.11.5.3
> 60 V d.c., bare conductors may be used for extra-
> low-voltage lighting installations, provided that:

- the lighting installation has been designed,
  installed or enclosed in such a way that the
  risk of a short-circuit is reduced to a
  minimum;
- the conductors used have a cross-sectional area
  of at least $4\,mm^2$;
- the conductors are not placed directly on combustible material.

For suspended bare conductors, at least one                    WR-559.11.5.3
conductor and its terminals shall be insulated for
that part of the circuit between the transformer and
the short-circuit protective device to prevent
a short-circuit.

### 6.2.4.2 Bonding conductors

In agricultural and horticultural premises, protec-            WR-705.544.2
tive bonding conductors shall be protected against
mechanical damage and corrosion, and shall be
selected to avoid electrolytic effects.

In solar photovoltaic (PV) power supply                        WR-712.54
systems (where protective bonding conductors are
installed) they shall be parallel to and in as close
contact as possible with d.c. cables and a.c. cables and
accessories.

A permanent label/warning notice – with the words shown in Figure 6.8 – shall
be permanently fixed at or near the bonding conductor's connection point to
an extraneous part.

> **Safety Electrical Connection – Do Not Remove**

**Figure 6.8** Warning notice – earthing and bonding.

## 6.2.4.3 *Connecting conductors*

| | |
|---|---|
| A bare live conductor shall be installed on insulators. | WR-521.10.100 |
| A circuit supplying one or more items of Class II equipment shall have a circuit protective conductor run to (and terminated at) each point in wiring and at each accessory. | WR-412.2.3.2 |
| An insulation monitoring device (IMD) shall be connected between Earth and a live conductor of the monitored equipment. | WR-538.3 |
| The connection of conductors shall not affect the protection being supplied by an enclosure. | WR-412.2.3.1 |
| The 'line' terminal(s) of an IMD shall be connected as close as is practicable to the origin of the system, to either: | WR-538.1.2 |

- the neutral point of the power supply; or
- an artificial neutral point with impedances con-nected to the line conductors; or
- a line conductor or two or more line conductors.

| | |
|---|---|
| For d.c. installations, the 'line' terminal(s) of the IMD shall be connected either directly to the midpoint, if any, or to one or all of the supply conductors. | WR-538.1.2 |
| In some particular d.c. IT two-conductor installations, a passive IMD that does not inject current into the system may be used, provided that: | WR-538.1.4 |

- the insulation of all live distributed conductors is monitored; and
- all exposed-conductive-parts of the installation are interconnected; and
- circuit conductors are selected and installed so as to reduce the risk of an Earth fault to a minimum.

### 6.2.4.4 Conductors in parallel

Where two or more live conductors or PEN conductors     WR-523.7
are connected in parallel in a system, either:

- measures shall be taken to achieve equal load current
  sharing between them; and
- either the conductors in parallel are multi-core cables
  or twisted single-core cables or non-sheathed cables, or
- non-twisted single-core cables or non-sheathed
  cables in trefoil or flat formation and where the cross-
  sectional area is greater than $50\,\mathrm{mm}^2$ in copper or
  $70\,\mathrm{mm}^2$ in aluminium.

**Notes:**

1. This Regulation does not preclude the use of ring final circuits with or
   without spur connections.
2. Where adequate current sharing is not possible (or where four or more conductors have to be connected in parallel) consideration shall be given to
   the use of busbar trunking.

### 6.2.4.5 Earthing arrangements and protective conductors

**Note:** Where the heat dissipation differs from one part of a route to another,
the current-carrying capacity at each part of the route should be appropriate
for that part of the route.

The earthing arrangements of protective conductors     WR-542.1.3.1
shall be such that:

- the value of impedance from the consumer's main
  earthing terminal to the earthed point of the supply for TN systems (or to Earth for TT and IT
  systems) is in accordance with the protective and
  functional requirements of the installation, and
  are considered to be continuously effective;
- Earth fault currents and protective conductor
  currents that may occur are carried without danger (particularly from thermal, thermomechanical and electromechanical stresses);
- they are adequately robust or have additional
  mechanical protection appropriate to the
  assessed conditions of external influence.

For a TN-S system, means shall be provided for the     WR-542.1.2.1
main earthing terminal of the installation to be connected to the earthed point of the source of energy.
(Part of this connection may be formed by the distributor's lines and equipment.)

For a TN-C-S system, where PME is provided; means     WR-542.1.2.2
shall be provided for the main earthing terminal of
the installation to be connected by the distributor to
the neutral of the source of energy.

For a TT or IT system, the main earthing terminal     WR-542.1.2.3
shall be connected via an earthing conductor to an
Earth electrode.

Where the supply to an installation is at high voltage, protection against faults between the high-voltage supply and Earth shall be provided.

Where a number of installations have separate earth-     WR-542.1.3.3
ing arrangements, any protective conductor common
to any of these installations shall either:

* be capable of carrying the maximum fault current likely to flow through them; or
* be earthed within one installation only and
  insulated from the earthing arrangements of any
  other installation.

Precautions should be taken against the risk of damage to other metallic parts through electrolysis.

### 6.2.4.5.1 Earthing terminal

Where the protective conductor is formed by metal     WR-543.2.7
conduit, trunking, ducting or the metal sheath and/
or armour of a cable, the earthing terminal of each
accessory shall be connected by a separate protective
conductor to an earthing terminal incorporated in the
associated box or other enclosure.

**Note:** It is not recommended that an exposed-conductive-part of equipment is used to form a protective conductor for other equipment.

| | |
|---|---|
| Except where the circuit protective conductor is formed by a metal covering (or enclosure containing all the conductors of the ring), the circuit protective conductor of every ring final circuit shall also be run in the form of a ring having both ends connected to the earthing terminal at the origin of the circuit. | WR-543.2.9 |

| | |
|---|---|
| To enable the resistance of the earthing arrangements to be measured, the earthing conductor needs to be capable of being easily disconnected. | WR-542.4.2 |
| Any joint shall be capable of disconnection only by means of a tool. | WR-542.4.2 |

## 6.2.4.5.2 Earthing requirements for the installation of equipment having high protective conductor currents

| | |
|---|---|
| Equipment having a protective conductor current exceeding 3.5 mA, but less than 10 mA, shall be either permanently connected to the fixed wiring of the installation without the use of a plug and socket-outlet, or connected by means of a plug and socket-outlet complying with BS EN 60309–2. | WR-543.7.1.101 |
| Equipment having a protective conductor current exceeding 10 mA shall be connected to the supply:<br><br>• permanently via the wiring of the installation (e.g. a flexible cable); or | |
| Equipment having a protective conductor current exceeding 10 mA shall be connected to the supply:<br><br>• permanently via the wiring of the installation (e.g. a flexible cable); or<br>• via a flexible cable with a plug and socket-outlet; or<br>• via a protective conductor with an Earth monitoring system. | WR-543.7.1.102 |
| The wiring of every final circuit and distribution circuit where the total protective conductor current is likely to exceed 10 mA shall have a high-integrity protective connection complying with one or more of the following: | |

- a single protective conductor with a cross-sectional area greater than 10 mm$^2$;    WR-543.7.1.103
- a single copper protective conductor having a cross-sectional area of not less than 4 mm$^2$;
- two individual protective conductors;
- an Earth monitoring system that in the event of a continuity fault occurring in the protective conductor automatically disconnects the supply to the equipment;
- connection of the equipment to the supply by means of a double-wound transformer or equivalent unit, such as a motor-alternator set.

Where two protective conductors are used, the ends of the protective conductors shall be terminated independently of each other and at all connection points (e.g. at the distribution board, junction boxes and socket-outlets) throughout the circuit.    WR-543.7.1.104

At the distribution board, information shall be provided indicating those circuits having a high protective conductor current.    WR-543.7.1.105

**Note:** This information shall be positioned so as to be visible to a person who is modifying or extending the circuit.

### 6.2.4.6 Electrical connections

Connections between conductors or between a conductor and other equipment should provide robust electrical continuity and ample mechanical strength and protection.

The type of connector chosen shall take account of:    WR-526.2

- the cross-sectional area of the conductor;
- the material of the conductor and its insulation;
- the number and shape of the wires forming the conductor;
- the number of conductors to be connected together;
- the provision of adequate locking arrangements in situations subject to vibration or the thermal cycling temperature attained at the terminals in normal service.

Where a soldered connection is used, the design shall take account of creep, mechanical stress and temperature rise under fault conditions.   WR-526.2

Every electrical connection and joint shall be accessible for inspection, except for the following:   WR-526.3

- a joint designed to be buried in the ground;
- a compound-filled or encapsulated joint;
- a connection between a cold tail and the heating element, as in ceiling heating, floor heating or a trace heating system;
- a joint made by welding, soldering, brazing or an appropriate compression tool;
- joints or connections made in the equipment by the manufacturer of the product and not intended to be inspected or maintained;

equipment complying with the requirements of the Wiring Regulations for a maintenance-free accessory and marked with the symbol.

Where necessary, precautions shall be taken so that the temperature attained by a connection in normal service shall not impair the effectiveness of the insulation of the conductors connected to it or any insulating material used to support the connection.   WR-526.4

Where a cable is to be connected to a bare conductor or busbar, its type of insulation and/or sheath shall be suitable for the maximum operating temperature of the bare conductor or busbar.   WR-526.4

Every termination and joint in a live conductor or a PEN conductor shall be made within one (or a combination) of the following:   WR-526.5

- a suitable accessory;
- an equipment enclosure;
- an enclosure partially formed or completed with non-combustible building material.

There shall be no appreciable mechanical strain on the connections of conductors.   WR-526.6

| Where a connection is made in an enclosure, the enclosure shall provide adequate mechanical protection and protection against relevant external influences. | WR-526.7 |

### 6.2.4.7 Electric floor heating systems

The load of every floor-warming cable under operation shall be limited to a value such that the manufacturer's stated conductor temperature is not exceeded.

| In locations containing a bath and/or shower:<br><br>• only heating cables or thin-sheet flexible heating elements shall be erected (**provided** that they have either a metal sheath or a metal enclosure or a fine-mesh metallic grid);<br>• the fine-mesh metallic grid, metal sheath or metal enclosure shall be connected to the protective conductor of the supply circuit. | WR-701.753 |

 Compliance with the latter requirement is not required if the protective measure SELV is provided for the floor heating system.

| For electric floor heating systems in locations containing a bath or shower, the protective measure 'protection by electrical separation' is **not** permitted. | WR-701.753 |

### 6.2.4.8 Electrode water heaters and boilers

| If an electrode water heater or electrode boiler is connected to a three-phase low-voltage supply, the shell of the electrode water heater or electrode boiler shall be connected to the neutral of the supply as well as to the earthing conductor. | WR-554.1.5 |

| | |
|---|---|
| If the supply to an electrode water heater or electrode boiler is single-phase and one electrode is connected to a neutral conductor earthed by the distributor, the shell of the electrode water heater or electrode boiler shall be connected to the neutral of the supply as well as to the earthing conductor. | WR-554.1.6 |
| If the electrode water heater or electrode boiler is not piped to a water supply or in physical contact with any earthed metal (and where the electrodes and the water in contact with the electrodes are so shielded in insulating material that they cannot be touched while the electrodes are live), a fuse in the line conductor may be substituted for the circuit-breaker, and the shell of the electrode water heater or electrode boiler need not be connected to the neutral of the supply. | WR-554.1.7 |

### 6.2.4.8.1 Water heaters having immersed and uninsulated heating elements

| | |
|---|---|
| All metal parts of the heater or boiler that are in contact with the water (other than current-carrying parts) shall be solidly and metallically connected to a metal water pipe through which the water supply to the heater or boiler is provided, and that water pipe shall be connected to the main earthing terminal by means independent of the circuit-protective conductor. | WR-554.3.2 |
| Single-phase water heaters and boilers with an uninsulated heating element immersed in the water shall not have a single-pole switch, non-linked circuit-breaker or fuse fitted in the neutral conductor, in any part of the circuit between the heater or boiler, or in the origin of the installation. | WR-554.3.1 WR-554.3.4 |

### *6.2.4.9 Enclosures*

| | |
|---|---|
| No conductive part that is enclosed in an insulating enclosure shall be connected to a protective conductor. | WR-412.2.2.4 |

No exposed-conductive-part or intermediate part shall be connected to a protective conductor unless the specification for the equipment concerned allows for this to happen.

WR-412.2.2.4

### 6.2.4.10 Fault protection

Fault protection by automatic disconnection of supply shall be provided by means of an overcurrent protective device in each line conductor or by an RCD.

WR-411.8.3

Live parts of the separated circuit **shall not** be connected at any point to another circuit, or to Earth, or to a protective conductor.

WR-413.3.3

 For separated circuits the use of separate wiring systems is recommended.

The exposed-conductive-parts of the equipment of the FELV circuit shall be connected to the protective conductor of the primary circuit of the source . . .

WR-411.7.3

 . . . provided that the primary circuit is subject to protection by automatic disconnection of supply.

No exposed-conductive-part of the separated circuit shall be connected either to the protective conductor or exposed-conductive-parts of other circuits, or to Earth.

WR-413.3.6

Where overcurrent protective devices are used for fault protection, the protective conductor shall be incorporated in the same wiring system as the live conductors or in their immediate proximity.

WR-543.6.1

### 6.2.4.10.1 Protection against fault current

A device providing protection against fault current shall be installed at the point where a reduction in the cross-sectional area (or other change) causes a reduction in

WR-434.2

the current-carrying capacity of the conductors of the installation (except installations situated in locations presenting a fire risk or risk of explosion and where the requirements for special installations and locations specify different conditions).

The device protecting a conductor may be installed on the supply side of the point where a change occurs, **provided** that it possesses an operating characteristic such that it protects the wiring situated on the load side against fault current.

WR-434.2.2

A device for protection against fault current need not be provided for:

WR-434.3

- a conductor connecting a generator, transformer, rectifier or an accumulator battery to the associated control panel where the protective device is placed in the panel;
- a circuit where disconnection could cause danger for the operation of the installation concerned;
- certain measuring circuits;
- the origin of an installation where the distributor installs one or more devices that provide protection against fault current;

**provided** that both of the following conditions are simultaneously fulfilled:

- the wiring is carried out in such a way as to reduce the risk of a fault occurring to a minimum; **and**
- the wiring is installed in such a manner as to reduce to a minimum the risk of fire or danger to persons.

A single protective device may protect conductors in parallel against the effects of fault currents provided that the operating characteristic of the device results in its effective operation should a fault occur at the most onerous position in one of the parallel conductors.

WR-434.4

Conductors (other than live conductors) and any other parts intended to carry a fault current, shall be capable of carrying that current without attaining an excessive temperature.

WR-131.5

**Note:** Conductors should be provided with protection against electromechanical stresses of fault currents as necessary to prevent injury or damage to
persons, livestock or property.

> A fault occurring at any point in a circuit shall be inter-      WR-434.5.2
> rupted within a time such that the fault current does
> not cause the permitted limiting temperature of any
> conductor or cable to be exceeded.
>
> Conductors are considered to be protected against        WR-436
> overload current and fault current where they are sup-
> plied from a source incapable of supplying a current
> exceeding the current-carrying capacity of the conduc-
> tors (e.g. certain bell transformers, certain welding
> transformers and certain types of thermoelectric gener-
> ating sets).

### 6.2.4.11 Ferromagnetic enclosures: electromagnetic effects

> The conductors of an a.c. circuit installed in a ferro-      WR-521.5.1
> magnetic enclosure shall be arranged so that the line
> conductors, the neutral conductor (if any) and the
> appropriate protective conductor are all contained in
> the same enclosure.

 **Note:** Where such conductors enter a ferrous enclosure, they should be
arranged so that the conductors are only collectively surrounded by ferrous
material.

### 6.2.4.12 Final circuits

> A final circuit with a number of socket-outlets or      WR-543.7.2.101
> connection units that is intended to supply two or
> more items of equipment (and where it is known
> that the total protective conductor current in
> normal service will exceed 10 mA) shall be pro-
> vided with a high-integrity protective conductor
> connection.

The following arrangements of the final circuit are acceptable:

WR-543.7.2.101

- a ring final circuit with a ring protective conductor;
- a radial final circuit with a single protective conductor.

Each final circuit forming part of an electrical installation in a caravan and/or motor caravan shall be protected by an overcurrent protective device that disconnects all live conductors of that circuit.

WR-721.43.1

### 6.2.4.13 Fire risk

In agricultural and horticultural premises that are liable to fire risk, conductors of circuits supplied at extra-low voltage shall be protected:

WR-705.422.8

- either by barriers or enclosures affording a degree of protection of IPXXD or IP4X; or
- in addition to their basic insulation, by an enclosure of insulating material.

### 6.2.4.13.1 Protection against fire

It is recommended that, wherever possible, every termination of a live conductor or connection or joint between live conductors shall be contained within an enclosure.

A wiring system that passes through the location but is not intended to supply electrical equipment within that location shall:

WR-422.3.5

- have no connection or joints in the location, unless the connection or joint is installed in an enclosure, and
- is protected against overcurrent, and
- does **not** use bare live conductors.

A PEN conductor **shall not** be used unless it is a circuit traversing the location.

WR-422.3.12

Every circuit shall be provided with a means of isolation from all live supply conductors by a linked switch or a linked circuit-breaker.

WR-422.3.13

### 6.2.4.14 Fuses

A fuse base shall be arranged so as to exclude the possibility of the fuse carriers making contact between conductive parts belonging to two adjacent fuse bases.

WR-533.1.1.1

A fuse base using screw-in fuses shall be connected so that the centre contact is connected to the conductor from the supply, and the shell contact is connected to the conductor to the load.

WR-533.1.1.1

### 6.2.4.15 Functional switching

Functional switching devices ensuring the changeover of supply from alternative sources shall affect all live conductors and **shall not** be capable of putting the sources in parallel, unless the installation is specifically designed for this condition.

WR-537.5.1.4

Functional switching devices need not necessarily control all live conductors of a circuit.

WR-537.5.1.2

A functional switching device shall be provided for each part of a circuit that may require to be controlled independently of other parts of the installation.

WR-537.5.1.1

**Note:** In these cases, no provision shall be made for isolation of the PEN or protective conductors.

### 6.2.4.16 Harmonic currents

Overcurrent detection shall be provided for the neutral conductor in a multi-phase circuit where the harmonic content of the line currents is such that the current in the neutral conductor may exceed the current-carrying capacity of that conductor.

WR-431.2.3

**Note:** Overcurrent detection should cause disconnection of the line conductors but not necessarily the neutral conductor.

## 6.2.4.17 Isolation

Isolation is intended (for reasons of safety) to make dead a circuit by separating an installation or section from every source of electric energy.

| | |
|---|---|
| Every circuit shall be capable of being isolated from each of the live supply conductors. | WR-537.2.1.1 |

**Note:** In a TN-S or TN-C-S system, it is not necessary to isolate or switch the neutral conductor where it is regarded as being reliably connected to Earth by suitably low impedance.

| | |
|---|---|
| Where an installation, item of equipment or enclosure contains live parts that are connected to more than one supply, a warning notice shall be placed so that any person liable to gain access to live parts, **must** isolate those parts from the various supplies – unless an interlocking arrangement is provided to ensure that all the circuits concerned are isolated. | WR-537.2.1.3 |

**Note:** Where necessary, suitable means shall be provided for the discharge of stored electrical energy.

| | |
|---|---|
| Where an isolating device for a particular circuit is remote from the equipment to be isolated, the means of isolation should always be secured in the **open** position. | WR-537.2.1.5 |

Notes:

1. If this means of isolation is a lock or removable handle, the key or handle shall be non-interchangeable with any other used for a similar purpose within the premises.
2. If a switch is provided for this purpose:
   - it shall be capable of cutting off the full load current of the relevant part of the installation;
   - but, if used as a device for switching off for mechanical maintenance, the switch need not necessarily interrupt the neutral conductor.

Provision shall be made for disconnecting the          WR-537.2.1.7
neutral conductor. Where this is a joint it shall
be such that it is in an accessible position that
can only be disconnected by means of a tool, is
mechanically strong and will reliably maintain
electrical continuity.

In marinas and similar locations, this switch-          WR-709.537.2.1.1
ing device shall disconnect all live conductors,
including the neutral conductor.

In temporary electrical installations, devices for      WR-740.537.2.2
isolation shall disconnect all live conductors (line
and neutral).

### 6.2.4.17.1  Devices for isolation

An isolation device shall isolate all live supply con-   WR-537.2.2.1
ductors from the circuit concerned.

Where a link is inserted in the neutral conductor, the   WR-537.2.2.4
link shall:

- not be capable of being removed without the use
  of a tool;
- only be accessible to skilled persons.

 Semiconductor devices **shall not** be used as isolating devices.

### 6.2.4.18  Live conductors

- Live supply conductors **shall** be capable of being isolated from circuits.
- Bare live conductors **shall** be installed on insulators.
- Conductors **shall** be able to carry fault current without overheating.
- The supply to all live conductors **shall** be automatically interrupted in the
  event of overload current and fault current.
- Persons and livestock **shall** be protected against injury, and property
  shall be protected against damage, due to excessive temperatures or electromechanical stresses caused by any overcurrents likely to arise in live
  conductors.

A main switch that is intended to be operated by ordi-   WR-537.1.4
nary persons (such as within a householder or similar
individual) shall interrupt **both** live conductors of a
single-phase supply.

Where an installation is supplied from more than one source of energy (one of which requires a means of earthing independent of the means of earthing of other sources, and it is necessary to ensure that no more than one means of earthing is applied at any time), a switch may be inserted in the connection between the neutral point and the means of earthing, **provided** that the switch is a linked switch that has been arranged to disconnect and connect the earthing conductor for the appropriate source, at substantially the same time as the related live conductors.

WR-537.1.5

Where no neutral point or midpoint exists, a line conductor may be connected to Earth.

After the occurrence of a first fault – and in the event of a second fault occurring on a different live conductor – automatic disconnection of supply shall occur.

WR-411.6.4

**Note:** Where the exposed-conductive-parts are interconnected by a protective conductor and collectively earthed to the same earthing system, the conditions similar to a TN system shall apply.

Where both the live circuit conductors are uninsulated, they shall either:

• be provided with a protective device complying with the requirements of Regulation 559.11.4.2; or
• the system shall comply with BS EN 60598–2–23.

WR-559.11.4.1

In agricultural and horticultural premises:

• the electrical installation of each building or part of a building shall be isolated by a single isolation device.

WR-705.537.2

**Note:** Means of isolation of all live conductors, including the neutral conductor, should also be provided for circuits that are used occasionally (such as during harvest time).

- RCDs shall disconnect all live conductors.     WR-705.422.7

In zone 1 of swimming pools (and other basins), fixed     WR-702.55.4
equipment shall only be accessible via a hatch (or a
door) by means of a key or a tool, which shall disconnect all live conductors and the supply cable.

In addition, the main disconnecting facility shall be
installed in a way that provides protection of Class II
or equivalent insulation.

The supply circuit of the equipment shall be
protected by:

- SELV; or
- an RCD; or
- electrical separation.

In solar photovoltaic (PV) power supply systems,     WR-712.312.2
earthing of one of the live conductors of the d.c.
side is permitted if there is at least simple separation
between the a.c. side and the d.c. side.

**Note:** Any connections with Earth on the d.c. side should be electrically connected so as to avoid corrosion (see BS 7361–1:1991).

### 6.2.4.19 Low-voltage generating sets

Where a generating set is intended to operate in paral-     WR-551.5.2
lel with a system for distribution of electricity to the
public (or where two or more generating sets may
operate in parallel), circulating harmonic currents
shall be limited so that the thermal rating of conductors is not exceeded.

### 6.2.4.20 Luminaires

Any flexible cable between the fixing means and the luminaire should be
installed so that any expected stresses in the conductors, terminals and terminations will not affect the safety of the installation.

In circuits of a TN or TT system (except for E14 and E27 lampholders complying with BS EN 60238), the outer contact of every Edison-screw or single-centre bayonet-cap type lampholder shall be connected to the neutral conductor.                                  WR-559.6.1.8

 **Note:** This regulation **also** applies to track-mounted systems!

Groups of luminaires divided between the three line conductors of a three-phase system with only one common neutral conductor shall be provided with at least one device that simultaneously disconnects all line conductors.                                     WR-559.6.2.3

### 6.2.4.21 Mechanical stresses

The radius of every bend in a wiring system shall be such that conductors do not suffer damage and terminals are not stressed.                                        WR-522.8.3

Every conductor shall be supported in such a way that it is not exposed to undue mechanical strain and there is no appreciable mechanical strain on the terminations of the conductors.                                   WR-522.8.5

Where the conductors (or cables) are not supported continuously, they shall be supported at appropriate intervals in such a manner that the conductors do not suffer damage by their own weight.                       WR-522.8.4

Cables, busbars and other electrical conductors that pass across expansion joints shall be selected and/or erected so that any (anticipated) movement does not cause damage to the electrical equipment.              WR-522.8.13

A wiring system intended for the drawing in or out of conductors shall have adequate means of access to allow this operation.                                  WR-522.8.6

A cable buried in the ground (that is not installed in a conduit or duct) shall incorporate an earthed-armoured or metal sheath (or both) suitable for use as a protective conductor.                                       WR-522.8.10

Conductors shall not be damaged by the means of fixing.   WR-522.8.12

### 6.2.4.22 Multi-core cables, conduits, ducting systems, franking systems or tray or ladder systems

| | |
|---|---|
| Each part of a circuit shall be arranged such that the conductors are **not** distributed over different multi-core cables, conduits, ducting, trunking, tray or ladder systems. | WR-521.8.1 |
| Where multi-core cables are installed in parallel, each cable shall contain one conductor of each line. | WR-521.8.1 |
| The line and neutral conductors of each final circuit shall be electrically separate from those of every other final circuit, so as to prevent the indirect energizing of a final circuit intended to be isolated. | WR-521.8.2 |

### 6.2.4.23 Multi-phase circuit

| | |
|---|---|
| In multi-phase circuits an independently operated single-pole switching device or protective device shall not be inserted in the neutral conductor. | WR-530.3.2 |
| In single-phase circuits an independently operated single-pole switching or protective device shall not be inserted in the neutral conductor alone. | WR-530.3.2 |

### 6.2.4.24 Neutral conductor

Consideration should be given to the fact that:

- if the neutral conductor in a three-phase TN or TT system is interrupted, basic, double and reinforced insulation (as well as components rated for the voltage between line and neutral conductors) can be temporarily stressed with the line-to-line voltage;
- if a line conductor of an IT system is earthed accidentally, insulation or components rated for the voltage between line and neutral conductors can be temporarily stressed with the line-to-line voltage;
- if a short-circuit occurs in the low-voltage installation between a line conductor and the neutral conductor, the voltage between the other line conductors and the neutral conductor can reach the value of $1.45 \times U_o$ for a time of up to 5 s.

| | |
|---|---|
| If the total harmonic distortion due to third harmonic current (or multiples of the third harmonic) is greater than 15% of the fundamental line current, the neutral conductor shall not be smaller than the line conductors. | WR-523.6.3 |

| | |
|---|---|
| In a TN or TT system, the neutral conductor shall be protected against short-circuit current. | WR-431.2.1 |
| In an IT system, the neutral conductor shall not be distributed unless: | WR-431.2.2 |

- overcurrent detection is provided for the neutral conductor of every circuit; or
- the neutral conductor is effectively protected against short-circuit by a protective device installed on the supply side; or
- the circuit is protected by an RCD with a rated residual operating current not exceeding 0.2 times the current-carrying capacity of the corresponding neutral conductor.

| | |
|---|---|
| In a TN-S or TN-C-S system the neutral conductor need not be isolated or switched where it can be regarded as being reliably connected to Earth by suitably low impedance. | WR-537.1.2 |
| Where conductors in polyphase circuits carry balanced currents, the associated neutral conductor need not be taken into consideration. | WR-523.6.1 |
| Where the neutral conductor carries current without a corresponding reduction in load of the line conductors, the neutral conductor shall be considered when ascertaining the current-carrying capacity of the circuit. | WR-523.6.3 |
| The neutral conductor shall not be disconnected before the line conductors, and shall be reconnected at the same time as (or before) the line conductors. | WR-431.3 |
| In temporary electrical installations, the neutral conductor of the star-point of the generator shall, except for an IT system, be connected to the exposed-conductive-parts of the generator. | WR-740.551.8 |

### 6.2.4.25 Non-conducting location

This protective measure is intended to prevent simultaneous contact with parts that may be at different potentials through failure of the basic insulation of live parts. It is **not** recognized for general application.

In a non-conducting location there is no need    WR-418.1.3
for a protective conductor.

## 6.2.4.26 Non-earthed equipotential bonding conductor

Source supplies may supply more than one item of    WR-413–06–05
equipment, provided that:

* all exposed-conductive-parts of the separated
  circuit are connected together by an insulated and non-earthed equipotential bonding
  conductor;
* the non-earthed equipotential bonding
  conductor is not connected to a protective
  conductor (or to an exposed-conductive-
  part of any other circuit or to any
  extraneous-conductive-part).

## 6.2.4.27 Overhead conductors

In caravan and camping parks, marinas and similar    WR-708.521.1.2
locations, all overhead conductors:

* shall be insulated;

* shall be at a height above ground of not less
  than 6 m in all areas subject to vehicle movement and 3.5 m in all other areas.

## 6.2.4.28 PEN conductors

 • A separate metal cable enclosure **shall not** be used as a
  PEN conductor.

 • PEN conductors **shall not** be isolated or switched.

 • Automatic disconnection using an RCD **shall not** be
  applied to a circuit incorporating a PEN conductor.

 • PEN conductors **shall not** be isolated or switched.

| | |
|---|---|
| PEN conductors may only be used within an installation: | WR-543.4.1<br>WR-543.4.2 |

- where any necessary authorization for use of a PEN conductor has been obtained and where the installation complies with the conditions for that authorization; or
- where the installation is supplied by a privately owned transformer or converter in such a way that there is no metallic connection (except for the earthing connection) with the distributor's network; or
- where the supply is obtained from a private generating plant.

| | |
|---|---|
| For a fixed installation, a conductor of a cable not subject to flexing and having a cross-sectional area not less than $10\,mm^2$ for copper or $16\,mm^2$ for aluminium may serve as a PEN conductor, provided that the part of the installation concerned is not supplied through an RCD. | WR-543.4.100 |
| Other than a cable conforming to BS EN 60702–1, all PEN conductors of every cable shall be insulated or have an insulating covering suitable for the highest voltage to which it may be subjected. | WR-543.4.8 |

 **Note:** If, from any point of the installation, the neutral and protective functions are provided by separate conductors, those conductors should not then be reconnected beyond that point.

| | |
|---|---|
| In marinas (and similar locations) and electrical installations in caravans and camping parks, the Electricity Safety, Quality and Continuity Regulations 2002 prohibit the connections of a PME earthing facility to any metalwork of a boat or caravan etc. | WR-708.411.4 |

 In caravan and camping parks, socket-outlet protective conductors **shall not** be connected to any PEN conductor of the electricity supply.

In temporary electrical installations (where the type of system earthing is TN), a PEN conductor shall not be used downstream of the origin of the temporary electrical installation.

WR-740.411.4.3

**Note:** In Great Britain, Regulation 8(4) of the Electricity Safety, Quality and Continuity Regulations 2002 prohibits the use of PEN conductors in consumers' installations.

### 6.2.4.29 Protection against overcurrent

A protective device shall be provided to break any overcurrent in the circuit conductors before such a current could cause a danger due to thermal or mechanical effects detrimental to insulation, connections, joints, terminations or the surroundings of the conductors.

**Note:** Protection of conductors according to these Regulations does not necessarily protect the equipment connected to the conductors.

In an IT system, in the event of a second fault, an overcurrent protective device shall disconnect all corresponding live conductors, including the neutral conductor (if any).

WR-531.1.3

### 6.2.4.30 Protection against overload current

The rated current or current setting of a device protecting a conductor against overload shall not be less than the design current of the circuit, and:

WR-433.1.1

- shall not exceed the lowest of the current-carrying capacities of any of the conductors of the circuit; and
- the current causing effective operation of the protective device shall not exceed 1.45 times the lowest of the current-carrying capacities of any of the conductors of the circuit.

A device for protection against overload shall be installed at the point where a reduction occurs in the value of the current-carrying capacity of the conductors of the installation.

WR-433.2.1

**Note:** A reduction in current-carrying capacity may be due to a change in cross-sectional area, method of installation, type of cable or conductor, or in environmental conditions.

The device protecting a conductor against overload may be installed along the run of that conductor provided that part of the run (i.e. between the point where a change occurs) and the position of the protective device has neither branch circuits nor outlets for connection of current-using equipment and is protected against fault current; or    WR-433.2.2

- its length does not exceed 3 m;
- it is installed so as to reduce the risk of fault to a minimum; and
- it is installed so as to reduce to a minimum the risk of fire or danger to persons.

A device for protection against overload need not be provided:    WR-433.3.1

- for a conductor situated on the load side of the point where a reduction occurs in the value of current-carrying capacity, where the conductor is effectively protected against overload by a protective device installed on the supply side of that point;
- for a conductor which, because of the characteristics of the load or the supply, is not likely to carry overload current;
- at the origin of an installation where the distributor provides an overload device.

In an IT system without a neutral conductor it is permitted to omit the overload protective device in one of the line conductors if an RCD is installed in each circuit.    WR-433.3.2.2

 **Note:** Where a single protective device protects two or more conductors in parallel, there shall be no branch circuits or devices for isolation or switching in the parallel conductors.

### 6.2.4.31 Protection by Earth-free local equipotential bonding

Earth-free local equipotential bonding is intended to prevent the appearance of dangerous touch voltages, and shall **only** be used in special circumstances.

Protective bonding conductors shall interconnect every simultaneously accessible exposed-conductive-part and extraneous-conductive-part.

WR-418.2.2

Unless protection by automatic disconnection of supply can be applied, local protective bonding conductors **shall not**:

- be in direct electrical contact with Earth; nor
- through exposed-conductive-parts; nor
- through extraneous-conductive-parts.

WR-418.2.3

The exposed-conductive-parts of the separated circuit shall be connected together by insulated, non-earthed protective bonding conductors.

WR-418.3.4

 These conductors **shall not** be connected to the protective conductor or exposed-conductive-parts of any other circuit or to any extraneous-conductive-parts.

Every socket-outlet shall be provided with a protective conductor contact, which shall be connected to the equipotential bonding system.

WR-418.3.5

All flexible cables (unless they supply equipment with double or reinforced insulation) shall include a protective bonding conductor.

WR-418.3.6

If two faults affect two exposed-conductive-parts at the same time, and these parts are fed by conductors with a different polarity, a protective device shall disconnect the supply in a disconnection time conforming to Table 6.13.

WR-418.3.7

**Table 6.13** Maximum disconnection times (data from BS 7671: 2008)

| System | Maximum disconnection time (s) | | | | | | | |
|---|---|---|---|---|---|---|---|---|
| | $50V < U_o$ 120V | | $120V < U_o$ 230V | | $230V < U_o$ 5400V | | $U_o > 400V$ | |
| | a.c. | d.c. | a.c. | d.c. | a.c. | d.c. | a.c. | d.c. |
| TN | 0.8 | Not required for protection against electric shock | 0.4 | 5 | 0.2 | 0.4 | 0.1 | 0.1 |
| TT | 0.3 | | 0.2 | 0.4 | 0.07 | 0.2 | 0.04 | 0.1 |

## 6.2.4.32 Protective conductors

| | |
|---|---|
| Every plug, socket-outlet, luminaire supporting coupler (LSC), device for connecting a luminaire (DCL) and cable coupler of a reduced low-voltage system:<br><br>• shall have a protective conductor contact; and<br>• **shall not** be dimensionally compatible with any LSC, DCL etc. used for any other system that is in use in the same premises. | WR-411.8.5 |

 In a TN system, the integrity of the earthing of the installation depends on the reliable and effective connection of the PEN or PE conductors to Earth.

| | |
|---|---|
| Each exposed-conductive-part of the installation shall be connected by a protective conductor to the main earthing terminal of the installation, which shall be connected to the earthed point of the power supply system. | WR-411.4.2 |
| In a fixed installation, a single conductor may serve both as a protective conductor and as a neutral conductor (PEN conductor). | WR-411.4.3 |
| Where an RCD is used for fault protection in a TN system, the circuit should also incorporate a PEN conductor (in the case of a TN-C-S system, the PEN conductor shall be on the **source side** of the RCD). | WR-411.4.4 |
| For an outdoor lighting installation (where the protective measure for the whole installation is by double or reinforced insulation), no protective conductor shall be provided and the conductive parts of the lighting column shall not be intentionally connected to the earthing system. | WR-559.10.4 |
| An extra-low-voltage luminaire without provision for the connection of a protective conductor shall be installed **only as part of** a SELV system. | WR-559.11.2 |

| | |
|---|---|
| In caravans and motor caravans, all circuit protective conductors shall be incorporated in a multi-core cable or in a conduit, together with the live conductors. | WR-721.543.2.3 |

## 6.2.4.32.1 Preservation of electrical continuity of protective conductors

A protective conductor shall be suitably protected against mechanical and chemical deterioration and electrodynamic effects.    WR-543.3.1

A protective conductor with a cross-sectional area up to and including 6 mm² shall be protected throughout by a covering at least equivalent to that provided by the insulation of a single-core non-sheathed cable having a voltage rating of at least 450/750 V, unless it is:    WR-543.3.100

- a protective conductor forming part of a multi-core cable;
- a metal conduit, metallic cable management system or other enclosure used as a protective conductor cable trunking or conduit used as a protective conductor.

Where the sheath of a cable incorporating an uninsulated protective conductor of cross-sectional area up to and including 6 mm² is removed adjacent to joints and terminations, the protective conductor shall be protected by insulating sleeving complying with the BS EN 60684 series.    WR-543.3.100

Every connection and joint shall be accessible for inspection, testing and maintenance.

A switching device shall **not** be inserted in a protective conductor unless:    WR-543.3.3

- the switch has been inserted in the connection between the neutral point and the means of earthing, and that switch is a linked switch arranged to disconnect and connect the earthing conductor for the appropriate source, at substantially the same time as the related live conductors;
- it is a multi-pole linked switch or plug-in device in which the protective conductor circuit has not been interrupted before the live conductors and re-established not later than when the live conductors are reconnected.

| | |
|---|---|
| Joints intended to be disconnected for test purposes are permitted in a protective conductor circuit. | WR-543.3.3 |
| Where electrical monitoring of earthing is used, no dedicated devices (e.g. operating sensors, coils) shall be connected in series with the protective conductor. | WR-543.3.4 |
| Every joint in metallic conduit shall be mechanically and electrically continuous. | WR-543.3.6 |

### 6.2.4.33 Protective devices and switches

| | |
|---|---|
| Single-pole fuses, switches or circuit-breakers shall only be inserted in the line conductor. | WR-132.14.1 |
| Switches, circuit-breakers (except where linked) or fuses shall be inserted in an earthed neutral conductor. | WR-132.14.2 |
| Any linked switch or linked circuit-breaker inserted in an earthed neutral conductor shall be capable of breaking all of the related line conductors. | WR-132.14.2 |

### 6.2.4.34 Protective earthing

| | |
|---|---|
| Exposed-conductive-parts shall be connected to a protective conductor. | WR-411.3.1.1 |

 Simultaneously accessible exposed-conductive-parts shall be connected to the same earthing system individually, in groups or collectively.

| | |
|---|---|
| Conductors for protective earthing shall comply with Chapter 54 of BS 7671:2008. | WR-411.3.1.1 |
| A circuit protective conductor shall be run to and be terminated at each point in wiring and at each accessory (except a lampholder having no exposed-conductive-parts and suspended from such a point). | WR-411.3.1.1 |

### 6.2.4.35 *Protective equipotential bonding*

In each installation, main protective bonding con-     WR-411.3.1.2
ductors shall connect extraneous-conductive-parts to
the main earthing terminal, including the following:

- central heating and air-conditioning systems;
- exposed metallic structural parts of the building;
- gas installation pipes;
- water installation pipes;
- other installation pipework and ducting.

 Where an installation serves more than one building, the above requirement
shall be applied to each building.

In exhibitions and showgrounds, structural     WR-711.411.3.1.2
metallic parts that are accessible from within the
stand, vehicle, wagon, caravan or container shall
be connected through the main protective bonding conductors to the main earthing terminal
within the unit.

In mobile or transportable units:     WR-717.411.3.1.2

- accessible conductive parts of the unit, such
  as the chassis, shall be connected through
  the main protective bonding conductors to
  the main earthing terminal within the unit;

- the main protective bonding conductors
  shall be finely stranded.

In caravans and motor caravans, where protection by     WR-721.411.1
automatic disconnection of supply is used, an RCD
shall be and the wiring system shall include a circuit
protective conductor, which shall be connected to:

- the protective contact of the inlet; and
- the exposed-conductive-parts of the electrical
  equipment; and
- the protective contacts of the socket-outlets.

**Note:** Structural metallic parts that are accessible from within the caravan shall be connected through main protective bonding conductors to the main earthing terminal within the caravan.

### 6.2.4.36 RCDs

| | |
|---|---|
| An RCD shall be capable of disconnecting all the line conductors of the circuit at substantially the same time. | WR-531.2.1 |
| An RCD shall be so selected and the electrical circuits so subdivided that any protective conductor current that may be expected to occur during normal operation of the connected load(s) will be unlikely to cause unnecessary tripping of the device. | WR-531.2.4 |
| In an IT system, where protection is provided by an RCD (and disconnection following a first fault is not envisaged) the non-operating residual current of the device shall be at least equal to the current that circulates on the first fault to Earth of negligible impedance affecting a line conductor. | WR-531.5.1 |
| The use of an RCD associated with a circuit that would normally be expected to have a protective conductor shall **not** be considered sufficient for fault protection if there is no such conductor. | WR-531.2.5 |
| The magnetic circuit of the transformer of an RCD shall enclose all the live conductors of the protected circuit. The associated protective conductor shall be outside the magnetic circuit. | WR-531.2.2 |
| Where it is necessary to limit the consequence of fault currents in a wiring system from the point of view of fire risk, the circuit shall be either:<br><br>• protected by an RCD for fault protection; and<br>• the RCD shall be installed at the origin of the circuit to be protected; and<br>• the RCD shall switch all live conductors. | WR-532.1 |

### 6.2.4.37 SELV and PELV circuits

| | |
|---|---|
| The earthing of PELV circuits may be achieved by a connection to Earth or to an earthed protective conductor within the source itself. | WR-414.4.1 |

Protective separation of wiring systems of SELV          WR-414.4.2
or PELV circuits from the live parts of other circuits (which have at least basic insulation) shall be
achieved by one of the following arrangements:

- SELV and PELV circuit conductors:

    o   enclosed in a non-metallic sheath or insulat-
        ing enclosure in addition to basic insulation;
    o   separated from conductors of circuits at volt-
        ages higher than Band I by an earthed metal-
        lic sheath or earthed metallic screen;

- circuit conductors (at voltages higher than Band
  I) contained in a multi-conductor cable or other
  grouping of conductor (but only if the SELV and
  PELV conductors are insulated for the highest
  voltage present).

Plugs and socket-outlets in an SELV system shall **not**    WR-414.4.3
have a protective conductor contact.

Exposed-conductive-parts of an SELV circuit shall not    WR-414.4.4
be connected to protective conductors.

### 6.2.4.38  Supplementary equipotential bonding

The use of supplementary bonding does not exclude the need to disconnect
the supply for other reasons, such as protection against fire or thermal stresses
in equipment.

Supplementary equipotential bonding system shall be         WR-415.2.1
connected to the protective conductors of all equip-
ment, including those of socket-outlets.

In locations containing a bath and/or a shower,
local supplementary equipotential bonding shall be
established, connecting together the terminals of the
protective conductor of each circuit supplying Class I
and Class II equipment to the accessible extraneous-
conductive-parts, including the following:

- metallic pipes supplying services and metallic
  waste pipes (e.g. water, gas);

- metallic central heating pipes and air-condition-   WR-701.415.2
  ing systems.

 **Note:** Accessible metallic structural parts of
the building (metallic door architraves, window frames and similar parts) are not considered to be extraneous-conductive-parts unless
they are connected to metallic structural parts
of the building.

 Supplementary equipotential bonding may be installed outside or inside rooms
containing a bath or shower, preferably close to the point of entry of extraneous-conductive-parts into such rooms.

A supplementary bonding conductor connecting   WR-544.2.1
two exposed-conductive-parts shall have a
conductance (if sheathed or otherwise provided
with mechanical protection) not less than that of
the smaller protective conductor connected to the
exposed-conductive-parts.

 **Note:** If mechanical protection is not provided, its cross-sectional area shall
be not less than $4\,mm^2$.

A supplementary bonding conductor connecting an   WR-544.2.2
exposed-conductive-part to an extraneous conductive
part shall have a conductance (if sheathed or otherwise provided with mechanical protection) not less
than half that of the protective conductor connected
to the exposed conductive part.

A supplementary bonding conductor connecting two   WR-544.2.3
extraneous-conductive-parts shall have a cross-
sectional area not less than 2.5 mm$^2$ if sheathed, or
otherwise provided with mechanical protection or
$4\,mm^2$ if mechanical protection is not provided.

Supplementary bonding shall be provided by a   WR-544.2.4
supplementary conductor, a conductive part of a
permanent and reliable nature, or by a combination
of these.

Where supplementary bonding is to be applied to a fixed appliance (which is supplied via a short length of flexible cord from an adjacent connection unit or other accessory, incorporating a flex outlet), the circuit protective conductor within the flexible cord shall be deemed to provide the supplementary bonding connection to the exposed-conductive-parts of the appliance.

WR-544.2.5

All extraneous-conductive-parts in zones 0, 1 and 2 of a swimming pool (and/or other basins) shall be connected by supplementary protective bonding conductors to the protective conductors of exposed-conductive-parts of equipment situated in these zones.

WR-702.415.2

### 6.2.4.39 Suspended conductors

Insulation piercing connectors and termination wires that rely on counterweights hung over suspended conductors to maintain the electrical connection **shall not** be used.

WR-559.11.6

The suspended system shall be fixed to walls or ceilings by insulated distance cleats and shall be continuously accessible throughout the route.

WR-559.11.6

### 6.2.4.40 Supply source

In caravans and motor caravans, the means of connection to the caravan pitch socket-outlet shall be supplied with the caravan and shall comprise the following:

WR-721.55.2.6

- a plug complying with BS EN 60309–2; and
- a flexible cable of 25 m (±2 m) length incorporating a protective conductor.

 Annex A721 to BS 7671:2008 (Incorporating Amendment No. 1) provides guidance for extra-low-voltage d.c. installations.

In temporary electrical installations (irrespective of          WR-740.313.3
the number of supply sources) the line and neutral
conductors from different sources shall not be interconnected downstream of the origin of the temporary
electrical installation.

### 6.2.4.41 Transformers

Where an autotransformer is connected to a                      WR-555.1.1
circuit that has a neutral conductor, the common
terminal of the winding shall be connected to the
neutral conductor.

Where a step-up transformer is used, a linked switch            WR-555.1.3
shall be provided for disconnecting the transformer from
all live conductors of the supply.

A step-up autotransformer **shall not** be connected to an IT system.

### 6.2.4.42 Wiring systems

The installation method of a wiring system in relation          WR-521.1
to the type of conductor or cable used shall be in accordance with Table 4A1 of Appendix 4 of the Wiring
Regulations.

Examples of wiring systems are shown in Table 4A2 of the Wiring
Regulations.

A voltage Band I circuit shall not be contained in              WR-528.1
the same wiring system as a Band II circuit unless:

• every cable or conductor is insulated for the
  highest voltage present;
• each conductor of a multi-core cable is
  insulated for the highest voltage present in
  the cable.

Metallic structural parts of buildings (such as pipe            WR-559.11.5.1
systems or parts of furniture) **shall not** be used as live
conductors.

The minimum cross-sectional area of the extra-low-voltage conductors shall normally be 1.5 mm$^2$ copper, but:   WR-559.11.5.2

- in the case of flexible cables with a maximum length of 3 m, a cross-sectional area of 1 mm$^2$ copper may be used;
- in the case of suspended flexible cables or insulated conductors, for mechanical reasons, 4 mm$^2$ copper should be used;
- in the case of composite cables consisting of braided tinned copper outer sheath, having a material of high tensile strength inner core, 4 mm$^2$ copper should be used.

In marinas and similar locations, the following wiring systems are suitable for distribution circuits:   WR-709.521.1.4

- overhead cables or overhead insulated conductors;
- cables with copper conductors and thermoplastic or elastomeric insulation and sheath installed within an appropriate cable management system (taking into account external influences such as movement, impact, corrosion and ambient temperature).

In caravans and motor caravans, the wiring systems shall be installed using one or more of the following:   WR-721.521.2

- insulated single-core cables, with flexible class 5 conductors, in non-metallic conduit;
- insulated single-core cables, with stranded class 2 conductors (minimum of seven strands), in non-metallic conduit;
- sheathed flexible cables.

## 6.2.5 Inspection and testing

 **Note:** Inspection shall precede testing, and shall normally be done with that part of the installation under inspection disconnected from the supply.

The inspection shall include at least the checking of the   WR-611.3
following items, where relevant to the installation and,
where necessary, during erection:

- connection of conductors;
- identification of conductors;
- selection of conductors for current-carrying
  capacity and voltage drop, in accordance with
  the design;
- connection of single-pole devices for protection or
  switching in line conductors only.

A test shall be made to verify the continuity of each   WR-612.2.2
conductor, including the protective conductor, of every
ring final circuit.

The insulation resistance shall be measured between   WR-612.3.1
live conductors and between live conductors and
the protective conductor connected to the earthing
arrangement.

The insulation resistance measured with the test volt-   WR-612.3.2
ages indicated in Table 6.14 shall be considered satisfactory **if** the main switchboard and each distribution
circuit tested separately (with all its final circuits connected but with current-using equipment disconnected)
has an insulation resistance not less than the appropriate value given in Table 6.14.

 More stringent requirements are applicable for the wiring of fire alarm systems
in buildings (see BS 5839–1).

Table 6.14 Minimum values of insulation resistance

| Circuit nominal voltage (V) | Test voltage d.c. (V) | Minimum insulation resistance (Ω) (M.(2) |
|---|---|---|
| SELV and PELV | 250 | >0.5 |
| Up to and including 500 V, with the exception of the above systems | 500 | >1.0 |
| Above 500 V | 1000 | >1.0 |

In locations exposed to fire hazard, a measurement of          WR-612.3.2
the insulation resistance between the live conductors
should be applied.

Insulation resistance values are usually much higher than those in Table 6.14.

Where the circuit includes electronic devices that          WR-612.3.3
are likely to influence the results or be damaged,
only a measurement between the live conductors
connected together and the earthing arrangement
shall be made.

### 6.2.5.1 Electrical installations

The characteristics of the electrical equipment **shall not** be impaired by the
process of erection.

Every installation shall be divided into circuits, as nec-          WR-314.1
essary, to reduce the possibility of unwanted tripping of
RCDs due to excessive protective conductor (PE) currents that are not due to a fault.

The number of final circuits required, and the          WR-314.3
number of points supplied by any final circuit, shall
be such as to enable compliance with the requirements
for overcurrent protection, isolation and switching, and
with due regard to the current-carrying capacities of
conductors.

### 6.2.5.2 Emergency switching

Emergency switching may be emergency switching **ON** or emergency switching **OFF**.

Emergency switching shall act as directly as possible          WR-537.4.1.3
on the appropriate supply conductors.

Emergency switching shall be provided for any part of          WR-537.4.1.1
an installation where it may be necessary to control
the supply to remove an unexpected danger.

Other than where a risk of electric shock is involved, WR-537.4.1.2
the emergency switching device shall be an isolating
device and shall interrupt all live conductors.

Except where the neutral conductor can be regarded WR-537.4.1.2
as being reliably connected to Earth in a TN-S or
TN-C-S system, the neutral conductor need not be
isolated or switched.

### 6.2.5.2.1 Equipotential bonding

Where an installation has more than one source of WR-544.1.1
supply to which PME conditions apply, the main protective bonding conductor shall be selected according
to the largest neutral conductor of the supply.

### 6.2.5.3 Parallel conductors

Except for a ring final circuit (where spurs are permitted WR-433.4.1
and where a single device protects conductors in parallel and the conductors are sharing currents equally), the
value of the current-carrying capacity of the conductor
shall be the sum of the current-carrying capacities of the
parallel conductors.

Where the use of a single conductor is impractical and WR-433.4.2
the currents in the parallel conductors are unequal, the
design current and requirements for overload protection
for each conductor shall be considered individually.

 **Note:** Currents in parallel conductors are considered to be unequal if the difference between the currents is more than 10% of the design current for each
conductor.

### 6.2.5.4 PEN conductors

The outer conductor of a concentric cable shall not be WR-543.4.4
common to more than one circuit.

The conductance of the outer conductor of a concen- WR-543.4.5
tric cable (measured at a temperature of 20°C) shall:

- for a single-core cable, not be less than that of the internal conductor;
- for a multi-core cable serving a number of points contained within one final circuit (or having the internal conductors connected in parallel), not be less than that of the internal conductors connected in parallel.

At every joint in the outer conductor of a concentric    WR-543.4.6
cable and at a termination, the continuity of that joint
shall be supplemented by a conductor additional to
any means used for sealing and clamping the outer
conductor.

 No means of isolation or switching shall be inserted in the outer conductor of a concentric cable.

### 6.2.5.5 Polarity

A test of polarity shall be made, and it shall be verified    WR-612.6
that:

- every fuse and single-pole control and protective device is connected in the line conductor only; and
- circuits (except for E14 and E27 lampholders to BS EN 60238) that have an earthed neutral conductor, centre-contact bayonet and Edison-screw lampholders, have the outer or screwed contacts connected to the neutral conductor; and
- wiring has been correctly connected to socket-outlets and similar accessories.

### 6.2.5.6 Protective equipotential bonding conductors

- All socket-outlets shall be provided with a pro-    WR-413–06–05
tective conductor contact (that is connected to
the equipotential bonding conductor);
- all flexible equipment cables (other than
Class II equipment) shall have a protective
conductor for use as an equipotential bonding
conductor.

### 6.2.5.7 Verification of voltage drop

When required, compliance to the Regulations may be confirmed by using the following options:                                    WR-612.14

- the voltage drop evaluated by measuring the circuit impedance;
- the voltage drop evaluated by using calculations (e.g. by diagrams or graphs showing maximum cable length versus load current for different conductor cross-sectional areas, with different percentage voltage drops for specific nominal voltages, conductor temperatures and wiring systems).

 **Note:** Verification of voltage drop is not normally required during initial verification.

## 6.2.6 Conduits, cable ducting, cable trunking, busbar or busbar trunking

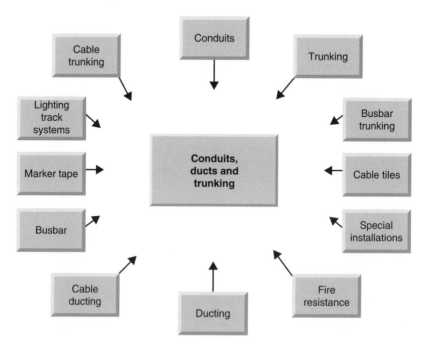

**Figure 6.9** Conduits and conduit fittings.

- Conduit and trunking **shall** comply with the resistance to flame propagation requirements of BS EN 50085 or BS EN 50086.
- Flexible or pliable conduit **shall not** be selected as a protective conductor.

**Note:** Conduits and conduit fittings shall comply with the appropriate British Standard shown in Table 6.15.

### 6.2.6.1 Busbars

**Table 6.15** Conduits and conduit fittings

| Conduit fitting | BS (or BS EN Standard) |
| --- | --- |
| Steel conduit and fittings | BS31, BSEN60423, BSEN50086–1 |
| Flexible steel conduit | BS731–1, BSEN60423, BSEN50086–1 |
| Steel conduit fittings with metric threads | BS 4568, BS EN 60423, BS EN 50086–1 |
| Non-metallic conduits and fittings | BS 4607, BS EN 60423, BS EN 50086–2–1 |

| | |
| --- | --- |
| A busbar trunking system or a powertrack system shall be installed in accordance with the manufacturer's instructions, taking account of external influences. | WR-521.4 |

**Note:** A busbar trunking system shall comply with BS EN 60439–2, and a powertrack system shall comply with the BS EN 61534 series.

| | |
| --- | --- |
| Busbars that pass across expansion joints shall be selected and/or erected so that any anticipated movement does not cause damage to the electrical equipment. | WR-522.8.13 |
| Except for mineral-insulated cables, busbar trunking systems or powertrack systems, a wiring system shall be protected against insulation faults: | WR-422.3.9 |

- in a TN or TT system, by an RCD having a rated residual operating current ($I_{An}$) not exceeding 300 mA;
- in an IT system, by an insulation monitoring device with audible and visual signals.

Where a busbar trunking system is used as a protective    WR-543.2.2
conductor:

- its electrical continuity shall be ensured, either by
  construction or by suitable connection, in such a
  way as to be protected against mechanical, chemical or electrochemical deterioration;
- its cross-sectional area shall be at in accordance
  with BS EN 60439–1;
- it shall permit the connection of other protective
  conductors at every predetermined tap-off point.

## 6.2.6.2 Ducting and trunking

A cable concealed in a wall or partition at a depth    WR-522.6.101
of less than 50 mm from a surface of the wall or partition shall:

- incorporate an earthed metallic covering; or
- be enclosed in earthed conduit; or
- be enclosed in earthed trunking or ducting; or
- be mechanically protected against damage sufficient to prevent penetration of the cable by
  nails, screws etc.

A cable installed under a floor or above a ceiling    WR-522.6.5.102
shall:

- be at least 50 mm measured vertically from the
  top (or bottom as appropriate) of the joist or
  batten; or
- incorporate an earthed metallic covering; or
- be enclosed in earthed conduit; or
- be enclosed in earthed trunking or ducting; or
- be mechanically protected against damage such
  that it is sufficient to prevent penetration of
  the cable by nails, screws etc.

 A cable trunking system or cable ducting system shall satisfy the test under fire
conditions specified in BS EN 50085.

A conduit system or cable ducting system (other than    WR-522.8.2
a pre-wired conduit assembly that has been

specifically designed for the installation) that is going to be buried in the structure shall be completely erected between access points before any cable is drawn in.

A conduit system, cable trunking system or cable ducting system classified as non-flame propagating according to the relevant product standard and having a maximum internal cross-sectional area of 710 mm$^2$ need not be internally sealed provided that:

WR-527.2.3

- the system satisfies the test of BS EN 60529 for IP33; and
- any termination of the system in one of the compartments, separated by the building construction being penetrated, satisfies the test of BS EN 60529 for IP33.

A voltage Band I circuit shall not be contained in the same wiring system as a Band II circuit unless:

WR-528.1

- the cables are insulated for their system voltage and installed in a separate compartment of a cable ducting or cable trunking system;
- the cables are installed on a cable tray system where physical separation is provided by a partition;
- a separate conduit, trunking or ducting system is employed.

A wiring system (such as a conduit system, cable ducting system, cable trunking system, busbar or busbar trunking system) that penetrates elements of building construction having specified fire resistance, shall be internally sealed to the degree of fire resistance of the respective element before penetration, as well as being externally sealed.

WR-527.2.2

Conduit and trunking systems shall be in accordance with BS EN 61386–1 and BS EN 50085–1, respectively, and shall meet the fire-resistance tests within these standards

WR-422.4.103

Cores of sheathed cables from which the sheath has been removed and non-sheathed cables at the termination of conduit, ducting or trunking shall be enclosed.

WR-526.8

If the cables of an installation not intended to be under the supervision of a skilled or instructed person are concealed in a wall or partition (the internal construction of which includes metallic parts, other than metallic fixings such as nails, screws and the like) they shall:

WR-522.6.101

- incorporate an earthed metallic covering; or
- be enclosed in earthed conduit; or
- be enclosed in earthed trunking or ducting; or
- be mechanically protected sufficiently to avoid damage to the cable during construction of the wall or partition and during installation of the cable.

In each installation, main protective bonding conductors shall connect extraneous-conductive-parts to the main earthing terminal.

WR-411.3.1.2

In multi-core cables, each part of a circuit shall be arranged so that the conductors are **not** distributed over different multi-core cables, conduits, ducting systems, trunking, tray or ladder systems.

WR-521.8.1

The line and neutral conductors of each final circuit shall be electrically separate from those of every other final circuit, so as to prevent the indirect energizing of a final circuit intended to be isolated.

WR-521.8.2

Non-sheathed cables are permitted in a cable trunking system that provides a minimum of IP4X or IPXXD protection, and if the cover can only be removed by means of a tool or a deliberate action.

WR-521.10.1

Non-sheathed cables for fixed wiring shall be enclosed in conduit, ducting or trunking.

WR-521.10.1

The installation of wiring systems will meet the requirements if non-metallic trunking or ducting (complying with the BS EN 50085) is used.

WR-412.2.4.1

The metal covering (including the sheath – bare or insulated) of a cable, trunking, ducting or metal conduit may be used as a protective conductor for the associated circuit.

WR-543.2.5

Two or more circuits are allowed in the same conduit, ducting or trunking system (but see Section 528 of the Wiring Regulations for specific requirements).

WR-521.7

Where the protective conductor is formed by metal conduit, trunking, ducting or the metal sheath and/or armour of a cable, the earthing terminal of each accessory shall be connected by a separate protective conductor to an earthing terminal incorporated in the associated box or other enclosure.

WR-543.2.7

Wherever equipment is fixed on or in cable trunking, skirting trunking or in mouldings, it should not be fixed on covers that can be inadvertently removed.

In agricultural and horticultural premises (where the wiring system may be exposed to impact and mechanical shock due to vehicles and mobile agricultural machines etc.), the external influences shall be classified AG3, and:

WR-705.522.16

- conduits shall provide a degree of protection against impact of 5 J according to BS EN 61386–2;
- cable trunking and ducting systems shall provide a degree of protection against impact of 5 J according to BS EN 50085–2–1.

In marinas and similar locations, the following wiring systems **shall not** be used on or above a jetty, wharf, pier or pontoon:

WR-709.521.1.5

- cables in free air suspended from, or incorporating, a support wire;
- non-sheathed cables in cable management systems;
- cables with aluminium conductors;
- mineral-insulated cables.

Where no fire alarm system is installed in a building    WR-711.521
used for exhibitions and shows etc., cable systems shall
be either:

- flame retardant to BS EN 60332–1–2 and low
  smoke to BS EN 61034–2; or
- single-core or multi-core unarmoured cables
  enclosed in metallic or non-metallic conduit or
  trunking, providing a degree of fire protection of at
  least IP4X.

The following applies to all temporary electrical systems in amusement parks,
circuses and fairgrounds:

Cable trunking systems and cable ducting systems    WR-740.521.1
shall comply with Part 2 of BS EN 50085.

Conduit systems shall comply with the BS EN 61386    WR-740.521.1
series.

# 7

# Special installations and locations

 These particular Regulations are **additional** to all of the other requirements, and **not** alternatives to them.

While the Regulations apply to all electrical installations in buildings, there are also some indoor and out-of-doors special installations and locations that are subject to special requirements due to the extra dangers they pose. This chapter considers the requirements for these special locations and installations.

In addition to the normal safety protection methods against direct and indirect contact listed in other parts of the Regulations, special installations and locations such as:

- agricultural and horticultural premises;
- conducting locations with restricted movement;
- construction and demolition sites;
- electrical installations in caravan/camping parks and similar locations;
- electrical installations in caravans and motor caravans;
- exhibitions, shows and stands;
- floor and ceiling heating systems;
- locations containing a bath or shower;
- marinas and similar locations;
- medical locations;
- mobile and transportable units;
- operating and maintenance gangways;
- rooms and cabins containing saunas;
- solar, PV power supply systems;
- swimming pools and other basins;
- temporary electrical installations for structures, amusement devices and booths at fairgrounds, amusement parks and circuses;

**must** also comply with the requirements for safety protection in respect of:

- electric shock;
- thermal effects;
- overcurrent;

- undervoltage;
- isolation and switching.

These requirements are described in more detail below.

# 7.1 General requirements

The following are intended to act as a reminder of the general requirements that are applicable to special installations and locations of conductors.

## 7.1.1 Accessibility

Connections and joints of heating appliances must be accessible for inspection, testing and maintenance, unless:

- they are in a compound-filled or encapsulated joint;
- the connection is between a cold tail and a heating element;
- the joint is made by welding, soldering, brazing or a compression tool.

## 7.1.2 Electrical heating units

The equipment, system design, installation and testing of an electric surface heating system **must** meet the requirements of BS 6351.

Heating appliances **must** be fixed so as to minimize the risks of burns to livestock and of fire from combustible material.

Electric heating units that are embedded in the floor (and intended for heating the location) may be installed below any zone in a bathroom, provided that they are covered by an earthed metallic grid or sheath that is connected to local supplementary equipotential bonding.

If an electric heating unit is embedded in the floor in zone B or C of a swimming pool, it must either:

- be connected to the local supplementary equipotential bonding by a metallic sheath; or
- be covered by an earthed metallic grid connected to the equipotential bonding.

Radiant heaters must be fixed not less than 0.5 m from livestock and from combustible material.

## 7.1.3 Equipotential bonding conductors

Main equipotential bonding conductors (for each installation) need to be connected to the main earthing terminal of that installation, which can include:

- water service pipes;
- gas installation pipes;
- other service pipes and ducting;
- central heating and air-conditioning systems;
- exposed metallic structural parts of the building;
- the lightning protective system.

 **Note:** Where an installation serves more than one building, the above requirement shall be applied to each building.

### 7.1.4 Heating appliances

 Heating appliances **must** always be fixed.

In the Regulations, the general requirements for heating appliances (e.g. water heaters, boilers, heating units, heating conductors and cables, surface and underfloor heating systems) are very important to special installations and locations, and the following, while perhaps not being a complete list, represents the most important requirements.

#### 7.1.4.1 Forced air heating systems

- Forced air heating systems **must** have two, independent, temperature-limiting devices.
- Electric heating elements of forced air heating systems (other than those of central-storage heaters) should:

   ○ not be capable of being activated until the prescribed air flow has been established;
   ○ deactivate when the air flow is reduced or stopped.

- Frames and enclosures of electric heating elements must be of non-ignitable material.

#### 7.1.4.2 Heating conductors and cables

- Heating cables passing through (or in close proximity to) a fire hazard:

   ○ must be enclosed in material with an ignitability characteristic P as specified in BS 476: Part 5;
   ○ must be protected from any mechanical damage.

- Heating cables that are going to be laid (directly) in soil, concrete, cement screed or other material used for road and building construction must be:

   ○ capable of withstanding mechanical damage;
   ○ resistant to damage from dampness or corrosion.

- Heating cables that are going to be laid (directly) in soil, a roadway, or the structure of a building must be installed so that they are:

- o    completely embedded in the substance it is intended to heat;
- o    not damaged by movement (either of the cable, or the substance in which it is embedded;
- o    compliant (in all respects) with the maker's instructions and recommendations.

### 7.1.5 Locations with risks of fire due to the nature of processed or stored materials

- Equipment enclosures (such as heaters and resistors) must not attain higher surface temperatures than:

  - o    90°C under normal conditions; and
  - o    115°C under fault conditions.

- Heat-storage appliances must not ignite combustible dust and/or fibres.
- Heating appliances mounted close to combustible materials must be protected by barriers.
- Where heating and ventilation systems containing heating elements are installed:

  - o    the dust or fibre content and the temperature of the air must not present a fire hazard;
  - o    temperature-limiting devices must have a manual reset.

### 7.1.6 Overhead wiring systems

While the only economic method of transmitting power from a grid station is by means of lines suspended from pylons, at lower voltages there is a choice between running them overhead or underground. The supply to most domestic buildings (particularly in towns) is predominantly underground, but for electrical installations such as in agricultural buildings the most cost-effective way is via overhead cables. The downside of this, of course, is the potential for the overhead cable to become a safety hazard, and protective methods must be used to guard against this possibility.

#### 7.1.6.1 Protection against indirect contact

Protective measures against indirect contact may only be dispensed with if:

- overhead line insulator brackets (and metal parts connected to them) are not within arm's reach;
- the steel reinforcement of concrete poles is not accessible;
- exposed-conductive-parts (including small isolated metal parts such as bolts, rivets, nameplates and cable clips) cannot be gripped or cannot be contacted by a major surface of the human body;

(a)

(b)

**Figure 7.1** Overhead wiring systems: (a) power lines; (b) urban installation. (Courtesy of Stingray.)

- there is no risk of fixing screws used for non-metallic accessories coming into contact with live parts;
- inaccessible lengths of metal conduit do not exceed 150 mm;
- metal enclosures mechanically protecting equipment comply with the requirements for Class II protection;
- unearthed street furniture that is supplied from an overhead line is inaccessible while in normal use.

### 7.1.6.2 Protection against direct contact

Bare live parts (other than overhead lines) must **not** be within arm's reach.

- Bare and/or insulated overhead lines being used for distribution between buildings and structures must be installed in accordance with the Electricity Safety, Quality and Continuity Regulations 2002 (ESQCR).

Conductors used as an overhead line, operating at low voltage, must comply with the relevant British and/or Harmonized Standard.

**Note:** If access to live equipment (from a normally occupied position) is restricted by an obstacle (e.g. a handrail, mesh or screen) with a degree of protection less than 1P2X or IPXXB, the extent of arm's reach must be measured **from** that obstacle.

When protection against direct contact is required for highway power supplies:

- protection by obstacles must not be used;
- protection by placing out of reach shall only apply to low-voltage overhead lines constructed in accordance with ESQCR;
- except when the maintenance of equipment is to be restricted to skilled persons who are specially trained; items of street furniture (or street located equipment) that are within 1.5 m of a low-voltage overhead line, protection against direct contact must be provided by some means other than placing out of reach.

### 7.1.6.3 Protection against overvoltages

No additional protection against overvoltages of atmospheric origin is necessary for:

- installations that are supplied by low-voltage systems that do not contain overhead lines;
- installations that are supplied by low-voltage networks that contain overhead lines and their location is subject to less than 25 thunderstorm days per year;
- installations that contain overhead lines and their location is subject to less than 25 thunderstorm days per year;

**provided** that they meet the required minimum equipment impulse withstand voltages shown in Table 7.1.

**Table 7.1 Required minimum impulse to withstand voltage (data from BS 7671:2008)**

| Nominal voltage of the installation (V) | Required minimum impulse, $U_w$ (kV) | | | |
|---|---|---|---|---|
| | Category IV (equipment with very high impulse voltage) | Category III (equipment with high impulse voltage) | Category II (equipment with normal impulse voltage) | Category I (equipment with reduced impulse voltage) |
| 230/240 277/480 | 6 | 4 | 2.5 | 1.5 |
| 400/690 | 8 | 6 | 4 | 2.5 |
| 1000 | 12 | 8 | 6 | 4 |

**Note:** Suspended cables having insulated conductors with earthed metallic coverings are considered to be *underground cables*.

Installations that are supplied by (or include) low-voltage overhead lines must incorporate protection against overvoltages of atmospheric origin if the location is subject to more than 25 thunderstorm days per year.

This protection must be provided either by:

- a surge protective device with a protection level not exceeding Category II; or
- other means providing at least an equivalent attenuation of overvoltages.

### 7.1.6.4 Overhead power lines and cables

All overhead conductors in caravan sites must be:

- protected by insulation of live parts;
- at least 2 m away from the boundary of any caravan pitch;
- not closer than 6 m in vehicle movement areas and 3.5 m in all other areas.

Mains operated electric fence controllers must:

- take account of the effects of induction when in the vicinity of overhead power lines;
- not be fixed to any supporting pole of an overhead power or telecommunication line.

**Note:** Poles and other overhead wiring supports shall be protected against any reasonably foreseeable vehicle movement.

### 7.1.7 Type of demand

The number and type of circuits required for lighting, heating, power, control, signalling, communication and information technology, etc. will generally depend on:

- the location and points of power demand;
- the loads to be expected on the various circuits;
- the daily and yearly variation in demand;
- any special conditions;
- requirements for control, signalling, communication and information technology, etc.

### 7.1.8 Water heaters and boilers

 Electric appliances producing hot water or steam must be protected against overheating.

Heaters that are intended for liquid or other substances must incorporate (or be provided with) an automatic device to prevent a dangerous rise in temperature.

Metal parts (other than the current-carrying parts of single-phase water heaters and boilers) that are in contact with the water shall be solidly and metallically connected to the metal water pipe supplying that heater/ boiler.

 The metal water pipe in question should be connected to the main earthing terminal by a circuit protective conductor that is independent of the heater/boiler.

The heater/boiler must be permanently connected to the electricity supply via a double-pole linked switch that is either:

- separate from and within easy reach of the heater/ boiler; or
- part of the boiler/heater (provided that the wiring from the heater or boiler is directly connected to the switch without use of a plug and socket-outlet).

## 7.2  Special installations and locations

### 7.2.1 Agricultural and horticultural premises

It is a mandatory requirement of the Regulations that all fixed agricultural and horticultural installations (outdoors and indoors) and locations where livestock is kept (such as stables, chicken houses, piggeries, feed-processing locations, lofts and storage areas for hay, straw and fertilizers) shall be inspected to confirm that they comply with Part 705 of the Regulations. (see Section 7.3.1 for a list of inspections and tests that need to be completed).

**Note:** If these premises include dwellings that are intended solely for human habitation, then the dwellings are excluded from the scope of these particular Regulations.

### 7.2.2  Conducting locations with restricted movement

Fixed equipment in conducting locations (particularly where the movement of persons is restricted by the location) and supplies to mobile equipment for use in such locations shall be inspected to confirm that they comply with Part 706 of the Regulations (see Section 7.3.2 for a list of inspections and tests that need to be completed).

### 7.2.3  Construction and demolition sites

Installations providing an electricity supply for: .

* new building construction;
* repairs, alterations, extensions or demolition of existing buildings;
* engineering construction;
* earthworks;

shall be inspected to confirm that they comply with Part 704 of the Regulations. (See Section 7.3.3 for a list of requirements, inspections and tests that need to be completed.)

These requirements do **not** apply to:

* construction site offices, cloakrooms, meeting rooms, canteens, restaurants, dormitories and toilets;
* installations covered by BS 6907.

### 7.2.4  Electrical installations in caravan/camping parks and similar locations

Electrical installations in caravan/camping parks and similar locations providing facilities for supplying leisure accommodation vehicles (including caravans) or tents, shall be inspected to confirm that they comply with Part 708 of the Regulations. (See Section 7.3.4 for a list of requirements, inspections and tests that need to be completed.)

### 7.2.5 Electrical installations in caravans and motor caravans

All electrical installations in caravans and motor caravans shall be inspected to confirm that they comply with Part 721 of the Regulations. (See Section 7.3.5 for a list of requirements, inspections and tests that need to be completed.)

It should be noted that the requirements of this section do **not** apply to:

• electrical circuits and equipment covered by the Road Vehicles Lighting Regulations 1989;
• installations covered by BS EN 1648-1 and BS EN 1648-2.

### 7.2.6 Exhibitions, shows and stands

Temporary electrical installations in exhibitions, shows and stands (including mobile and portable displays and equipment) shall be inspected to confirm that they comply with Part 711 of the Regulations. (See Section 7.3.6 for a list of inspections and tests that need to be completed.)

### 7.2.7 Floor and ceiling heating systems

Electric floor and ceiling heating systems that are erected as either thermal storage heating systems or direct heating systems shall be inspected to confirm that they comply with Part 753 of the Regulations. (See Section 7.3.7 for a list of requirements, inspections and tests that need to be completed.)

### 7.2.8 Locations containing a bath or shower

All locations containing a bath or shower shall be inspected to confirm that they comply with Part 701 of the Regulations. (See Section 7.3.8 for a list of requirements, inspections and tests that need to be completed.)

Locations containing baths or showers for medical treatment, or for disabled persons, may have special requirements.

### 7.2.9 Marinas and similar locations

Circuits intended to supply pleasure craft or houseboats in marinas and similar locations shall be inspected to confirm that they comply with Part 709 of the Regulations. (See Section 7.3.9 for a list of requirements, inspections and tests that need to be completed.)

### 7.2.10 Medical locations

Section 710 is a new section specifically aimed at electrical installations in medical locations. Although it mainly refers to hospitals, private clinics, medical and dental practices, healthcare centres and dedicated medical rooms in the workplace, this section also applies to electrical installations in locations designed for medical research and (where applicable) to veterinary clinics. (See Section 7.3.10 for a list of requirements, inspections and tests that need to be completed.)

## 7.2.11 Mobile and transportable units

A vehicle and/or mobile (self-propelled or towed) or transportable structure (such as a container or cabin) in which all or part of an electrical installation is contained, and which is provided with a temporary supply by means of (for example) a plug and socket-outlet, shall be inspected to confirm that it complies with Part 717 of the Regulations. (See Section 7.3.11 for a list of requirements, inspections and tests that need to be completed.)

## 7.2.12 Operating and maintenance gangways

Section 729 is a new (fairly small) addition to the standard, and centres on the operation and safe maintenance of switchgear and controlgear within areas that include gangways, and where access is restricted to skilled or instructed persons. Access areas and the requirements for operating and maintenance gangways shall comply with Part 729 of the Regulations. (See Section 7.3.12 for a list of requirements.)

## 7.2.13 Rooms and cabins containing saunas

Installations supplying electricity for locations containing hot air sauna heating equipment (in accordance with BSEN 60335-2-53) shall be inspected to confirm that they comply with Part 703 of the Regulations. (See Section 7.3.13 for a list of requirements, inspections and tests that need to be completed.)

## 7.2.14 Solar, photovoltaic (PV) power supply systems

Electrical installations of PV power supply systems (including subsystems with a.c. modules) shall be inspected to confirm that they comply with Part 712 of the Regulations. (See Section 7.3.14 for list of requirements, inspections and tests that need to be completed.)

## 7.2.15 Swimming pools and other basins

Installations associated with basins of swimming pools, paddling pools and other basins plus their surrounding zones shall be inspected to confirm that they comply with Part 702 of the Regulations. (See Section 7.3.15 for a list of requirements, inspections and tests that need to be completed.)

### 7.2.16 Temporary electrical installations for structures, amusement devices and booths at fairgrounds, amusement parks and circuses

The electrical installation and operation of temporarily erected mobile or transportable electrical machines and structures that incorporate electrical equipment shall be inspected to confirm that they comply with Part 740 of the Regulations. (See Section 7.3.16 for a list of requirements, inspections and tests that need to be completed.)

## 7.3 Requirements of the Regulations

Where an electrical service is located in close proximity to one or more non-electrical services, it shall meet the following conditions:

> * the wiring system shall be suitably protected against   WR-528.3.4
>   the hazards likely to arise from the presence of the
>   other services in normal use;
> * fault protection shall be provided by means of
>   automatic disconnection of supply.

### 7.3.1 Agricultural and horticultural premises

In contrast to normal domestic installations, an agricultural installation is usually prone to damp conditions, and so contact with Earth will be better and people and animals are more liable to electric shock. For animals (whose body resistance is much lower than that of humans) this situation is worsened as their contact with Earth will be greater, and so even a small voltage could prove lethal to them. Animals can also cause a lot of physical damage to electrical installations, and animal effluents present a risk of corrosion. Horticultural installations are also subject to the same wet/high Earth contact conditions, and for these reasons special requirements have been introduced for all agricultural or horticultural installations.

The Regulations list a number of requirements for fixed agricultural and horticultural installations (outdoors and indoors), and for locations where livestock is kept (e.g. cow sheds, stables, chicken houses, piggeries) plus food-processing stations and storage areas for hay, straw and fertilizers.

The particular requirements of this section apply to fixed electrical installations (indoors and outdoors) in agricultural and horticultural premises. Some of the requirements are also applicable to other locations that are in common buildings belonging to the agricultural and horticultural premises.

If these premises include dwellings that are intended **solely** for human habitation, then the dwellings are excluded from the scope of these particular Regulations.

## 7.3.1.1 General

### 7.3.1.1.1 Accessibility by livestock

As the possibility of animals unintentionally coming in direct contact with a live installation is greater than that for humans (e.g. they cannot read the warning notices!), and as livestock cannot be protected by earthed equipotential bonding and automatic disconnection (EEBAD) (because the voltages to which they would be subjected in the event of a fault would be unsafe for them), the following protective methods have to be used:

| | |
|---|---|
| Electrical equipment shall, generally speaking, be inaccessible to livestock. | WR-705.513.2 |
| Equipment that is unavoidably accessible to livestock (such as equipment for feeding and basins for watering) shall be adequately constructed and installed to avoid damage by, and to minimize the risk of injury to, livestock. | WR-705.513.2 |

### 7.3.1.1.2 Identification

| | |
|---|---|
| The following documentation shall be provided to the user of the installation:<br><br>• a plan indicating the location of all electrical equipment;<br>• the routing of all concealed cables;<br>• a single-line distribution diagram;<br>• an equipotential bonding diagram showing the locations of all bonding connections. | WR-705.514.9.3 |

## 7.3.1.2 Protective measures

### 7.3.1.2.1 Electric fence controllers

Electric fencing systems have been developed to stop the free movement of animals across pastures. They are semi-permanent solutions that can be extended, altered or removed to allow grazing to be divided up and/or protected from livestock access.

The system consists of plastic or wooden posts, insulators, conductive wire/rope/tape and an electrical energizer unit. This unit can be mains, battery or solar powered, and it sends short electrical impulses along a conductive wire (tape, rope etc.) so that, when the conductive wire or fence is touched by an animal, the current passes through it to the ground and causes the animal to feel a shock. The shock is sufficient to alarm an animal but not to harm it – and in time the animal will learn to stay away from it.

 **For** specific requirements for electric fence installations, see BS EN 60335-2, BS 7671:2008 and BS EN 6100-1.

### 7.3.1.2.2 External influences

| | |
|---|---|
| In agricultural and/or horticultural premises, electrical equipment shall have a minimum degree of protection of IP44. | WR-705.512.2 |
| Where equipment of IP44 rating is not available, it shall be placed in an enclosure complying with IP44. | WR-705.512.2 |
| Socket-outlets shall be installed in a position where they are unlikely to come into contact with combustible material. | WR-705.512.2 |
| Where there are conditions of external influences greater than AD4, AE3 and/or AG1, socket-outlets shall be provided with the appropriate protection. | WR-705.512.2 |
| Protection may also be provided by the use of additional enclosures or by installation in building recesses. | WR-705.512.2 |

 **Note:** These requirements do not apply to residential locations, offices, shops etc. belonging to agricultural and horticultural premises, or where (i.e. for socket-outlets) BS 1363-2 or BS 546 applies.

| | |
|---|---|
| Where corrosive substances are present (e.g. in dairies or cattle sheds) the electrical equipment shall be adequately protected. | WR-705.512.2 |

### 7.3.1.2.3 Protection against electric shock

| | |
|---|---|
| The protective measures of obstacles and placing out of reach are **not** permitted. | WR-705.410.3.5 |

The protective measures of non-conducting loca-   WR-705.410.3.6
tion and Earth-free local equipotential bonding) are
**not** permitted.

### 7.3.1.2.4 Protection against fire

Fire is a particular hazard in agricultural premises, as there are normally large
quantities of straw and other flammable material stored in these locations.

Electrical heating appliances used for the breeding   WR-705.422.6
and rearing of livestock shall be fixed so as to maintain an appropriate distance from livestock and combustible material, in order to minimize any risks of
burns to livestock and of fire.

For radiant heaters, the clearance shall be not less than 0.5 m or such other
clearance as recommended by the manufacturer.

For fire-protection purposes, RCDs shall be installed   WR-705.422.7
with a rated residual operating current not exceeding
300 mA.

RCDs shall disconnect all live conductors.   WR-705.422.7

Where improved continuity of service is required,   WR-705.422.7
RCDs not protecting socket-outlets shall be of the S
type or have a time delay.

In locations where a fire risk exists, conductors of cir-   WR-705.422.8
cuits supplied at extra-low voltage shall be protected:

- either by barriers or enclosures; or
- in addition to their basic insulation, by an enclosure of insulating material.

### 7.3.1.2.5 Supplementary equipotential bonding

In locations intended for livestock, supplementary   WR-705.415.2.1
bonding shall connect all exposed-conductive-
parts and extraneous conductive-parts that can be
touched by livestock.

Extraneous-conductive-parts in, or on, the floor    WR-705.415.2.1
(e.g. concrete reinforcement in general, or reinforcement of cellars for liquid manure) shall be
connected to the supplementary equipotential
bonding.

Where a metal grid is laid in the floor, it shall be    WR-705.415.2.1
included within the supplementary bonding of the
location (see Figure 7.2).

**Note:** It is recommended that:

- spaced floors made of prefabricated concrete elements are part of the supplementary equipotential bonding;
- the supplementary equipotential bonding and the metal grid (if any) shall
  be erected so that it is durably protected against mechanical stresses and
  corrosion.

Where a metal grid is not laid in the floor, the use of a protective multiple
earthing (PME) earthing facility is **not** recommended.

Protective bonding conductors shall be protected    WR-705.544.2
against mechanical damage and corrosion, and shall
be selected to avoid electrolytic effects.

### 7.3.1.3 Selection and erection of equipment

Only electrical heating appliances with visual indication    WR-705.53
of the operating position shall be used.

### 7.3.1.4 Supplies

A TN-C system **shall not** be used.

Where separated extra-low voltage (SELV) or pro-    WR-705.414.4.5
tective extra-low voltage (PELV) is used, whatever
the nominal voltage, protection shall be provided by:

- basic insulation; or
- barriers.

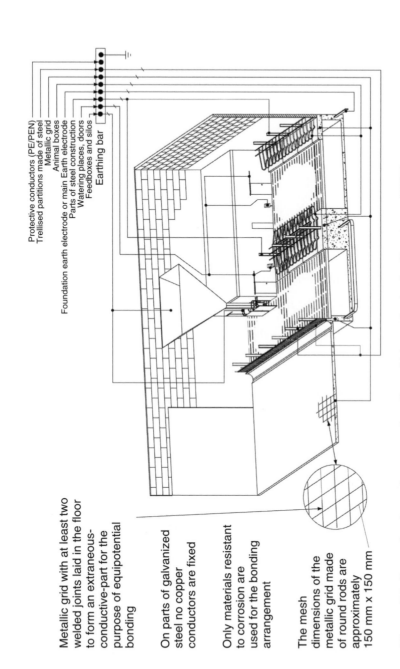

Protective conductors (PE/PEN)
Trellised partitions made of steel
Metallic grid
Animal boxes
Foundation earth electrode or main Earth electrode
Parts of steel construction
Watering places, doors
Feedboxes and silos
Earthing bar

Metallic grid with at least two welded joints laid in the floor to form an extraneous-conductive-part for the purpose of equipotential bonding

On parts of galvanized steel no copper conductors are fixed

Only materials resistant to corrosion are used for the bonding arrangement

The mesh dimensions of the metallic grid made of round rods are approximately 150 mm x 150 mm

**Figure 7.2** Example of supplementary bonding within cattle sheds. (Courtesy of BSI.)

### 7.3.1.4.1 Automatic disconnection of supply

In circuits, whatever the type of earthing system, the following disconnection device shall be provided:

WR-705.411.1

- in final circuits supplying socket-outlets with rated current not exceeding 32 A, an RCD with a rated residual operating current not exceeding 30 mA;
- in final circuits supplying socket-outlets with rated current more than 32 A, an RCD with a rated residual operating current not exceeding 100 mA;
- in all other circuits, RCDs with a rated residual operating current not exceeding 300 mA.

### 7.3.1.4.2 Isolation and switching

The electrical installation of each building or part of a building shall be isolated by a single isolation device.

WR-705.537.2

Means of isolation of all live conductors, including the neutral conductor, shall be provided for circuits used occasionally (e.g. during harvest time).

WR-705.537.2

The isolation devices shall be clearly marked according to the part of the installation to which they belong.

WR-705.537.2

Devices for isolation and switching and devices for emergency stopping or emergency switching shall not be erected where they are accessible to livestock or in any position where access may be impeded by livestock.

WR-705.537.2

### 7.3.1.4.3 Safety services

For high-density livestock rearing, systems operating for the life support of livestock shall be taken into account as follows:

WR-705.560.6

- where the supply of food, water, air and/or lighting to livestock is not ensured in the event of power supply failure, a secure source of supply shall be provided (e.g. an alternative or back-up supply) and separate final circuits for ventilation and lighting units shall be provided;
- where electrically powered ventilation is necessary in an installation, one of the following shall be provided:

  o a standby electrical source ensuring sufficient supply for ventilation equipment; or
  o temperature and supply voltage monitoring.

 **Note:** A notice should be placed adjacent to the standby electrical source, indicating that it should be tested periodically according to the manufacturer's instructions.

### 7.3.1.5 Wiring systems

### 7.3.1.5.1 Conduit systems, cable trunking systems and cable ducting systems

For locations where livestock is kept, external influences shall be classified AF4, and conduits shall have protection against corrosion of at least Class 2 (medium) for indoor use and Class 4 (high protection) for outdoors, according to BS EN 61386-21.

WR-705.522.16

For locations where the wiring system may be exposed to impact and mechanical shock due to vehicles and mobile agricultural machines etc., the external influences shall be classified AG3, and:

WR-705.522.16

- conduits shall provide a degree of protection against an impact of 5 J according to BS EN 61386-2;
- cable trunking and ducting systems shall provide a degree of protection against an impact of 5 J according to BS EN 50085-2-1.

### 7.3.1.5.2 Selection and erection of wiring systems

In locations that are accessible to (and are enclosing) livestock, wiring systems shall be erected so that they are inaccessible to livestock, or suitably protected against mechanical damage.      WR-705.522

Overhead lines shall be insulated.      WR-705.522

In areas of agricultural premises where vehicles and mobile agricultural machines are operated, the following methods of installation shall be applied:      WR-705.522

- cables shall be buried in the ground at a depth of at least 0.6 m, with added mechanical protection;
- cables in arable or cultivated ground shall be buried at a depth of at least 1 m;
- self-supporting suspension cables shall be installed at a height of at least 6 m.

Special attention shall be given to the presence of different kinds of fauna (e.g. rodents).

### 7.3.1.5.3 Socket-outlets

Socket-outlets used in agricultural and horticultural premises shall comply with:      WR-705.553.1

- BS EN 60309-1; or
- BS EN 60309-2 (when interchangeability is required); or

- BS 1363, BS 546 or BS 196 (provided the rated current does not exceed 20 A).

## 7.3.2 Conducting locations with restricted movement

A restrictive conductive location is one in which the surroundings consist mainly of metallic or conductive parts, such as a large metal container or boiler. People employed inside these locations (e.g. a person working inside the boiler while using an electric drill or grinder) would have their freedom of movement physically restrained, and a large proportion of their body would be in contact with the sides of that location and, therefore, prone to shock hazards.

The particular requirements of this section apply to:

- fixed equipment in conducting locations where movement of persons is restricted by the location; and
- to supplies for mobile equipment for use in such locations.

 This section does not apply to electrical systems used for private and public events, touring shows, theatrical, radio, TV or film productions, etc.

## 7.3.2.1 Protection against electric shock

| | |
|---|---|
| The protective measures of obstacles and placing out of reach are **not** permitted. | WR-706.410.3.5 |
| In a conducting location with restricted movement the following protective measures apply to circuits supplying the following current-using equipment: | WR-706.410.3.10 |

- for the supply to a hand-held tool or an item of mobile equipment:
  - electrical separation;
  - SELV;
- for the supply to handlamps:
  - SELV;
- for the supply to fixed equipment:
  - automatic disconnection of the supply with supplementary equipotential bonding; or
  - electrical separation: or
  - SELV; or
  - PELV.

## 7.3.2.2 Supplies

| | |
|---|---|
| As a protective measure, the unearthed source shall have simple separation, and shall be situated outside the conducting location. | WR-706.413.1.2 |
| A source for SELV or PELV shall be situated outside the conducting location. | WR-706.414.3(ii) |
| Whatever the nominal voltage, where SELV or PELV is used: | WR-706.414.4.5 |

- live parts shall be completely covered with
  insulation; or
- basic protection shall be provided by barriers
  or enclosures.

### 7.3.2.2.1 Automatic disconnection of supply

| | |
|---|---|
| If a functional Earth is required for certain equipment (e.g. measuring and control equipment), equipotential bonding shall be provided between all exposed-conductive-parts and extraneous-conductive-parts inside the conducting location with restricted movement and the functional Earth. | WR-706.411.1.2 |

## 7.3.3 Construction and demolition site installations

Electrical installations at construction sites are there primarily to provide lighting and power to enable work to proceed. As workmen will probably be working ankle deep in wet muddy conditions and using a selection of portable tools such as drills and grinders, they will be particularly susceptible to electric shock.

There are six levels of voltage normally associated with construction sites. These are:

| | |
|---|---|
| 25 V single-phase SELV | For portable handlamps in damp and confined situations. |
| 50 V single-phase centre-point earthed | For handlamps in damp and confined situations. |
| 400 V three-phase | For use with fixed or transportable equipment with a load of more than 3750 W. |
| 230 V single-phase | For site buildings and fixed lighting. |
| 110 V three-phase | For transportable equipment with a load up to 3750 W. |
| 110 V single-phase | For transportable tools and equipment, such as floodlighting. |

Equipment used must be suitable for the particular supply to which it is connected and for the application it will meet on site. Where more than one voltage is in use, plugs and sockets must be non-interchangeable to prevent misconnection.

Supplies will normally be obtained from the electrical supply company, but remote sites could need an IT supply (such as a generator), and care must be taken in complying with the safety requirements for this particular source.

The following requirements apply to temporary installations for construction and demolition sites during the period of the construction or demolition work, including, for example:

- construction work of new buildings;
- earthworks;
- engineering works;
- repair, alteration, extension, demolition of existing buildings or parts of existing buildings;
- work of similar nature;

and are applicable to:

- the main switchgear and protective devices;
- installations of mobile and transportable electrical equipment;
- the interface between the supply system and the construction site installations.

They do **not** apply to:

- installations covered by the IEC 60621 series 2, where equipment of a similar nature to that used in surface mining applications is involved;
- installations in administrative locations of construction sites (e.g. offices, cloakrooms, meeting rooms, canteens, restaurants, dormitories, toilets).

### 7.3.3.1 Protection against electric shock

 The protective measures of, non-conducting location and Earth-free local equipotential bonding are **not** permitted.

| | |
|---|---|
| A circuit supplying a socket-outlet with a rated current up to and including 32 A (and any other circuit supplying hand-held electrical equipment with a rated current up to and including 32 A) shall be protected by:<br><br>• reduced low voltage; or<br>• automatic disconnection of supply (with additional protection provided by an RCD); or | WR-704.410.3.10 |

- electrical separation of circuits (where each socket-outlet and item of hand-held electrical equipment is supplied by an individual transformer or by a separate winding of a transformer); or
- SELV or PELV.

### 7.3.3.2 Selection and erection of equipment

All equipment used for the distribution of electricity on construction and demolition sites shall meet the requirements of BS EN 60439-4.    WR-704.511.1

 A plug or socket-outlet with a rated current equal to or greater than 16 A shall comply with the requirements of BS EN 60309.

### 7.3.3.3 Supplies

Equipment shall be identified with (and be compatible with) the particular supply from which it is energized, and shall contain only components connected to one and the same installation, **except** for control or signalling circuits and inputs from standby supplies.    WR-704.313.3

Whatever the nominal voltage, where SELV or PELV is used:    WR-704.414.4.5

- live parts shall be completely covered with insulation; or
- basic protection shall be provided by barriers or enclosures.

### 7.3.3.3.1 Automatic disconnection of supply

A PME earthing facility **shall not** be used for the supply to a construction site, **unless** all extraneous parts are reliably connected to the main earthing terminal.    WR-704.411.3.1

Circuits supplying one or more socket-outlets with a rated current exceeding 32 A, shall be provided with an RCD having a rated residual operating current not exceeding 500 mA.

WR-704.411.3.2.1

## 7.3.3.3.2 Isolation devices

Each assembly for construction sites (ACS) shall incorporate suitable devices for the switching and isolation of the incoming supply.

WR-704.537.2.2

A device for isolating the incoming supply shall be suitable for securing in the OFF position (e.g. by padlock or location of the device inside a lockable enclosure).

WR-704.537.2.2

Current-using equipment shall be supplied by ACSs comprising:

WR-704.537.2.2

- overcurrent protective devices;
- devices affording fault protection; and
- socket-outlets, if required.

Safety and standby supplies shall be connected by means of devices arranged to prevent interconnection of the different supplies.

WR-704.537.2.2

## 7.3.3.4 Wiring systems

Cable **shall not** be installed across a site road or a walkway unless adequate protection of the cable against mechanical damage is provided.

WR-704.522.8.10

For reduced low-voltage systems, low-temperature 300/500 V thermoplastic or equivalent flexible cables shall be used.

WR-704.522.8.11

For applications exceeding reduced low-voltage, flexible cable shall be of HO7RN-F (BS 7919) type or equivalent, and have a 450/750 V rating and be resistant to abrasion and water.

WR-704.522.8.11

**Note:** A wiring system buried in a floor shall be sufficiently protected to prevent damage caused by the intended use of the floor.

The use of any lubricants that could have a detrimental effect on the cable or wiring system is **not** permitted.

### 7.3.4 Electrical installations in caravan/camping parks and similar locations

 **Note:** In order not to mix regulations on different subjects, such as those for electrical installation of caravan parks with those for electrical installation inside caravans, two sections have been created:

- Section 721 concerns electrical installations in caravans and motor caravans; and
- Section 708 concerns electrical installations in caravan parks, camping parks and similar locations.

The requirements of this particular section, therefore, apply to that portion of the electrical installation in caravan/camping parks and similar locations providing facilities for supplying leisure accommodation vehicles (including caravans) or tents. They do **not** apply to the internal electrical installations of leisure accommodation vehicles or mobile or transportable units.

The requirements of the Electricity Supply Regulations do not allow the supply neutral to be connected to any metalwork in a caravan (which means that only TT or TN-S systems may be used) and, in general, the supply of electrical energy to caravan and tent sites must ensure that:

- wherever possible the supply is via underground cables;
- overhead supplies use insulated as opposed to bare cables;
- cables are installed outside the area of the caravan pitch and be at least 3.5 m above ground level (increased to 6 m where vehicle movements are possible);
- each socket must have its own individual overcurrent protection in the form of a fuse or circuit-breaker;
- cables that are run below caravan pitches must be provided with additional protection, as shown in Figure 7.3;
- all sockets must be protected by an RCD complying with BS 4293, BS EN 61008-1 or BS EN 61009-1, with a 30 mA rating, either individually or in groups **not** exceeding three sockets.

The actual requirements (which only apply to installations supplying electricity to leisure accommodation vehicles in caravan/camping parks and similar locations) are as follows.

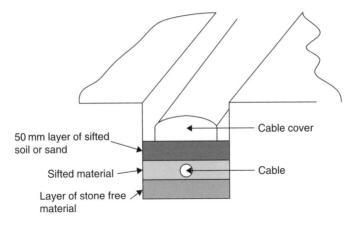

**Figure 7.3** Cable covers.

## 7.3.4.1 General

### 7.3.4.1.1 External influences

Electrical equipment installed outside shall comply at least with the following external influences:    WR-708.512.2

- presence of water;
- presence of foreign solid bodies;
- mechanical stress.

## 7.3.4.2 Protective measures

 The protective measures of:

- obstacles and placing out of reach; and
- non-conducting location and Earth-free local equipotential bonding;

are **not** permitted.

## 7.3.4.3 Supplies

 In the UK, the ESQCR prohibit the use of a TN-C-S system for the supply to a caravan or similar construction.

The nominal supply voltage of the installation for the supply of leisure accommodation vehicles shall be 230 V a.c. single-phase or 400 V a.c. three-phase.    WR-708.313.1.2

> Caravan pitch electrical supply equipment shall be located adjacent to the pitch and not more than 20 m from the connection facility on the leisure accommodation vehicle or tent when on its pitch.      WR-708.530.3

No more than four socket-outlets should be grouped in one location, in order to avoid the supply cable crossing a pitch other than the one intended to be supplied.

### 7.3.4.4 Wiring systems

### 7.3.4.4.1 Busbar trunking systems and powertrack systems

> All:      WR-521.4
>
> • busbar trunking systems shall comply with BS EN 60439-2; and
> • powertrack systems shall comply with the appropriate part of the BS EN 61534 series.

A busbar trunking system or a powertrack system shall take account of external influences. (See Appendix 8 of BS 7671:2008 (Incorporating Amendment No. 1).

### 7.3.4.4.2 Overhead distribution circuits

> All overhead conductors shall be insulated.      WR-708.521.1.2
>
> Poles and other supports for overhead wiring shall be located (or protected) so that they are unlikely to be damaged by vehicle movement.      WR-708.521.1.2
>
> Overhead conductors shall be not less than 6 m above ground in all areas subject to vehicle movement and 3.5 m in all other areas.      WR-708.521.1.2

### 7.3.4.4.3 Plugs and socket-outlets

> Each socket-outlet and its enclosure forming part of the caravan electricity supply equipment shall meet the degree of protection of at least IP44.      WR-708.553.1.8

| Socket-outlets shall be placed at a height of 0.5 m to 1.5 m from the ground, measured to the lowest part of the socket-outlet. | WR-708.553.1.9 |

 In **special cases**, due to environmental conditions such as risk of flooding or heavy snowfall, the maximum height is permitted to exceed 1.5 m.

| The current rating of socket-outlets shall be not less than 16 A. | WR-708.553.1.10 |

| At least one socket-outlet shall be provided for each caravan pitch. | WR-708.553.1.11 |

Socket-outlets shall be provided:

| • with individual overcurrent protection; | WR-708.553.1.12 |

| • individually with an RCD. | WR-708.553.1.13 |

| Socket-outlet protective conductors shall not be connected to any combined protective and neutral (PEN) conductor of the electricity supply. | WR-708.553.1.14 |

### 7.3.4.4.4 Underground distribution circuits

| Underground cables shall:<br><br>• be buried at a depth of at least 0.6 m; and<br>• unless they have additional mechanical protection, be placed outside any caravan pitch or away from any surface where tent pegs or ground anchors are expected to be present. | WR-708.521.1.1 |

 Unless mechanically protected, all underground cables **must** be installed outside of the caravan pitch and areas where tent pegs or ground anchors may be driven.

| The following wiring systems are suitable for distribution circuits feeding caravan or tent pitch electrical supply equipment: | WR-708.521.1 |

- underground distribution circuits;
- overhead distribution circuits.

 **Note:** The preferred method of supply is by means of underground distribution circuits.

### 7.3.5 Electrical installations in caravans and motor caravans

Caravans and motor caravans are designed as leisure accommodation vehicles, and are either towed (e.g. by a car) or self-propelled to a caravan site. They will often contain a bath or a shower, and special requirements for such installations will apply. In addition to the normal dangers associated with fixed electrical installations, there is the potential hazard of totally unskilled people moving the caravan/motor caravan, connecting and disconnecting the mains supply, and ensuring that it is correctly earthed.

 Caravans that are used as mobile workshops will **also** be subject to the requirements of the Electricity at Work Regulations 1989, and locations containing baths or showers for medical treatment, or for disabled persons, may have additional special requirements.

The Regulations include a number of special requirements for caravans and mobile homes (as listed below), but these do not deal with:

- electrical circuits and equipment covered by the Road Vehicles Lighting Regulations 1989;
- installations covered by BS EN 1648-1 and BS EN 1648-2 (for 12 V d.c. extra-low-voltage installations in leisure accommodation vehicles).

 **Note:** A mobile home is defined as a "*transportable leisure accommodation vehicle' that does not meet the requirements for use as a road vehicle*", and is usually a permanent fixture on a caravan park. They normally have recognized power supplies and earthing, and the internal electrical installation is outside the scope of the Regulations. Nevertheless, potential safety hazards still have to be considered.

### 7.3.5.1 General

#### 7.3.5.1.1 External influences – vibration (AH) and mechanical stresses (AJ)

All wiring shall be protected against mechanical damage (e.g. from vibration) either by location or by enhanced mechanical protection.     WR-721.522.7.1

| | |
|---|---|
| Wiring passing through metalwork shall be protected by bushes or grommets, securely fixed in position. | WR-721.522.7.1 |
| Precautions shall be taken to avoid mechanical damage due to sharp edges or abrasive parts. | WR-721.522.7.1 |
| All cables (unless enclosed in rigid conduit) and all flexible conduit shall be supported at intervals not exceeding 0.4 m for vertical runs and 0.25 m for horizontal runs. | WR-721.522.8.1.3 |

### 7.3.5.2 Protective measures

 The protective measures of:

- electrical separation (with the exception of shaver socket-outlets);
- obstacles and placing out of reach;
- non-conducting location and Earth-free local equipotential bonding;

are **not** permitted.

### 7.3.5.2.1 Protection against overcurrent – final circuits

| | |
|---|---|
| Each final circuit shall be protected by an overcurrent protective device that disconnects all live conductors of that circuit. | WR-721.43.1 |

### 7.3.5.2.2 Protective conductors

| | |
|---|---|
| All protective conductors shall be incorporated in a multi-core cable or in a conduit together with the live conductors. | WR-721.543.2.3 |

### 7.3.5.2.3 Protective equipotential bonding

| | |
|---|---|
| Structural metallic parts that are accessible from within the caravan shall be connected through main protective bonding conductors to the main earthing terminal within the caravan. | WR-721.411.3.1.2 |

### 7.3.5.2.4 Protective measure – automatic disconnection of supply

> Where protection by automatic disconnection of sup-  WR-721.411.1
> ply is used, an RCD shall be provided and the wiring
> system shall include a circuit protective conductor,
> which shall be connected to:
>
> - the protective contact of the inlet; and
> - the exposed-conductive-parts of the electrical
>   equipment; and
> - the protective contacts of the socket-outlets.

### 7.3.5.3 Selection and erection of equipment

The means of connection to the caravan pitch socket-outlet shall be supplied
with the caravan and shall comprise the following:

- a plug complying with BS EN 60309-2; and
- a flexible cord or cable of 25 m (±2 m) length and incorporating a protective conductor.

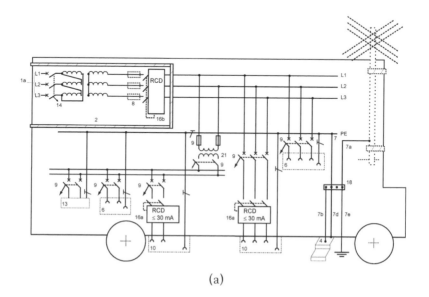

(a)

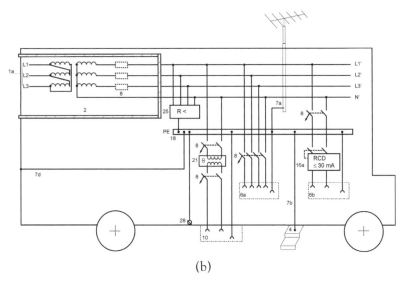

(b)

**Figure 7.4** (a) A two-pole and protective conductor supply system between the caravan pitch supply equipment. (b) Through the caravan and/or motor caravan. (Courtesy of BSI.)

### 7.3.5.3.1 Electrical inlets

| | |
|---|---|
| a.c. electrical inlets to caravans shall be in compliance with BS EN 60309. | WR-721.55.1.1 |
| The inlet shall be installed:<br><br>• not be more than 1.8 m above ground level; and<br>• be located in a readily accessible position; and<br>• have a minimum degree of protection of IP44 (with or without a connector engaged); and<br>• shall not protrude significantly beyond the body of the caravan. | WR-721.55.1.2 |

### 7.3.5.3.2 Isolation

| | |
|---|---|
| Each installation shall be provided with a main disconnector that will disconnect all live conductors and which is suitably located for ready operation within the caravan. | WR-721.537.2.1.1 |
| In an installation consisting of only one final circuit, the isolating switch may be the overcurrent protective device. | WR-721.537.2.1.1 |

| | |
|---|---|
| A notice shall be permanently fixed near the main isolating switch inside the caravan, bearing the text shown in Figure 721 on page 255 of BS 7671:2008. | WR-721.537.2.1.1.1 |

### 7.3.5.4 Supplies

| | |
|---|---|
| The nominal supply system voltage shall be chosen from IEC 60038. | WR-721.313.1.2 |
| The nominal a.c. supply voltage of the installation of the caravan shall not exceed 230 V single-phase, or 400 V three-phase. | WR-721.313.1.2 |
| The nominal d.c. supply voltage of the installation of the caravan shall not exceed 48 V. | WR-721.313.1.2 |
| Where there is more than one electrically independent installation, each independent installation shall be supplied by a separate connecting device and shall be segregated in accordance with the relevant requirements of the Regulations. | WR-721.510.3 |

 Annex A721 to BS 7671:2008 (Incorporating Amendment No. 1) provides guidance for extra-low-voltage d.c. installations.

### 7.3.5.5 Wiring systems

### 7.3.5.5.1 Types of wiring system

| | |
|---|---|
| The wiring systems shall be installed using one or more of the following:<br><br>• insulated single-core cables, with flexible class 5 conductors, in non-metallic conduit;<br>• insulated single-core cables, with stranded class 2 conductors (minimum of seven strands), in non-metallic conduit;<br>• sheathed flexible cables. | WR-721.521.2 |
| All cables shall, as a minimum, meet the requirements of BS EN 60332-1-2. | WR-721.521.2 |

Non-metallic conduits shall comply with BS EN 61386-21.   WR-721.521.2

Cable management systems shall comply with BS EN 61386.   WR-721.521.2

### 7.3.5.5.2 Accessories

Low-voltage socket-outlets (other than those supplied by an individual winding of an isolating transformer) shall incorporate an Earth contact.   WR-721.55.2.1

Socket-outlets supplied at extra-low voltage shall have the supply voltage visibly marked.   WR-721.55.2.2

If an accessory is exposed to the effects of moisture it shall be constructed (or enclosed) so as to provide a degree of protection not less than IP44.   WR-721.55.2.3

Luminaires in a caravan shall preferably be fixed directly to the structure or lining of the caravan.   WR-721.55.2.4

If a pendant luminaire is installed in a caravan, it shall be capable of being secured to prevent damage when the caravan is in motion.   WR-721.55.2.4

Accessories used to suspend a pendant luminaire shall be suitable for the mass suspended and the forces associated with vehicle movement.   WR-721.55.2.4

### 7.3.5.5.3 Identification

Instructions for use shall be provided with the caravan so that the caravan can be used safely. These instructions shall comprise (as a minimum):   WR-721.514.1

- a description of the installation;
- a description of the RCD(s) and its function;
- a description of the main isolating switch and its function;
- the use of the test button(s).

**Note:** The text for these instructions can be found in Figure 721 on page 255 of BS 7671:2008.

## 7.3.5.5.4 Proximity of wiring systems to other services

Cables of low-voltage systems shall be run separately from the cables of extra-low-voltage systems, so that there is no risk of physical contact between the two wiring systems.            WR-721.528.1

No electrical equipment (including wiring systems, but not extra-low-voltage equipment for gas-supply control) shall be installed in any gas cylinder storage compartment.            WR-721.528.3.4

Extra-low-voltage cables and electrical equipment may only be installed within the liquid petroleum gas (LPG) cylinder compartment **if** the installation indicates the operation of the gas cylinder (e.g. indication of empty gas cylinders) or is for use within the compartment.

 Such electrical installations and components shall be constructed and installed so that they are not a potential source of ignition.

Where cables have to run through such a compartment, they shall be protected against mechanical damage by installation within a conduit system or within a duct passing through the compartment.            WR-721.528.3.4

 This conduit or duct must be able to withstand an impact equivalent to AG3 without visible physical damage.

## 7.3.6 Exhibitions, shows and stands

The particular requirements of this section apply to the temporary electrical installations in exhibitions, shows and stands (including mobile and portable displays and equipment) in order to protect users.

### 7.3.6.1 General

### 7.3.6.1.1 External influences

Any external influences that could affect where a temporary electrical installation is erected (e.g. the presence of water or mechanical stresses) shall be taken into account.            WR-711.32

## 7.3.6.1.2 Isolation

| | |
|---|---|
| Separate, temporary structures, such as vehicles, stands or units, that are intended supply outdoor installations shall be provided with their own readily accessible and properly identifiable means of isolation. | WR-711.537.2.3 |

### 7.3.6.2 Protective measures

## 7.3.6.2.1 Protection against electric shock

| | |
|---|---|
| A cable supplying temporary structures shall be protected at its origin by an RCD the rated residual operating current of which does not exceed 300 mA. | WR-711.410.3.4 |
| The protective measures of obstacles and placing out of reach are **not** permitted. | WR-711.410.3.5 |
| The protective measures of non-conducting location and Earth-free local equipotential bonding are **not** permitted. | WR-711.410.3.6 |

## 7.3.6.2.2 Protection against fire and heat generation

| | |
|---|---|
| Lighting equipment, such as incandescent lamps, spotlights and small projectors, and other equipment or appliances with high-temperature surfaces shall be suitably guarded, installed and located. | WR-711.422.4.2 |
| Showcases and signs shall be constructed of material having an adequate heat resistance, mechanical strength, electrical insulation and ventilation, taking into account the combustibility of exhibits in relation to the heat generation. | WR-711.422.4.2 |
| Stand installations containing a concentration of electrical equipment, luminaires or lamps liable to generate excessive heat shall not be installed unless adequate ventilation provisions are made (e.g. a well-ventilated ceiling constructed of incombustible material). | WR-711.422.4.2 |

### 7.3.6.2.3 Protection against thermal effects

Luminaires mounted below 2.5 m from the floor level (or otherwise accessible to accidental contact) shall be firmly and adequately fixed, and so sited or guarded as to prevent risk of injury to persons or ignition of materials. WR-711.559.5

### 7.3.6.2.4 Protective equipotential bonding

Structural metallic parts that are accessible from within the stand, vehicle, wagon, caravan or container shall be connected through the main protective bonding conductors to the main earthing terminal within the unit. WR-711.411.3.1.2

### *7.3.6.3 Selection and erection of equipment*

Switchgear and controlgear shall be placed in closed cabinets that can only be opened by the use of a key or a tool, except for those parts designed and intended to be operated by ordinary persons.

### 7.3.6.3.1 Electrical connections

Joints **shall not** be made in cables (except where necessary) as a connection into a circuit. Where joints have to be made, then either connectors or enclosures (with a degree of protection of at least IP4X or IPXXD) shall be used. WR-711.526.1

Where strain can be transmitted to terminals, the connection shall include (i.e. incorporate) suitable cable anchorage(s). WR-711.526.1

### 7.3.6.3.2 Electric discharge lamp installations

All luminous tubes, signs or lamps that are used as an illuminated unit on a stand, or as an exhibit, with a nominal power supply voltage higher than 230/400 V a.c. shall comply with the following:

**Location** – the sign or lamp shall be installed out of arm's reach or shall be adequately protected to reduce the risk of injury to persons. WR-711.559.4.4

**Installation** – the facia or stand fitting material behind luminous tubes, signs or lamps shall be non-ignitable.

WR-711.559.4.5
WR-711.559.4.6

**Emergency switching devices** – a separate circuit shall be used to supply signs, lamps or exhibits, and shall be controlled by an emergency switch.

WR-711.559.4.7

 The switch shall be easily visible, accessible and clearly marked.

### 7.3.6.3.3 Luminaires and lighting installations

 Insulation-piercing lampholders shall **not** be used unless the cables and lampholders are compatible, and providing the lampholders are non-removable once fitted to the cable.

Extra-low-voltage lighting systems for filament lamps shall comply with BS EN 60598-2-23.

WR-711.559.4.2

### 7.3.6.3.4 Socket-outlets and plugs

An adequate number of socket-outlets shall be installed to allow user requirements to be met safely.

WR-711.55.7

Each socket-outlet circuit not exceeding 32 A and all final circuits (other than for emergency lighting) shall be protected by an RCD.

WR-711.411.3.3

Where a floor-mounted socket-outlet is installed, it shall be protected from accidental ingress of water and have sufficient strength to be able to withstand the expected traffic load.

WR-711.55.7

### 7.3.6.4 Supplies

The nominal supply voltage of a temporary electrical installation in an exhibition, show or stand shall not exceed 230/400 V a.c. or 500 V d.c.

WR-711.313

Except where an installation is within a building, a
PME earthing facility **shall not** be used, except where
the installation is permanently under the supervision of
a skilled or instructed person.

### 7.3.6.4.1 Electric motors – isolation

Where an electric motor might give rise to a hazard,      WR-711.55.4
the motor shall be provided with a means of
isolation on all poles adjacent to the motor that it
controls.

### 7.3.6.4.2 Extra-low-voltage transformers and electronic converters

Extra-low-voltage transformers shall be mounted out      WR-711.55.6
of arm's reach of the public and shall have adequate
ventilation.

A manual reset protective device shall protect the       WR-711.55.6
secondary circuit of each transformer or electronic
convertor.

Access by a competent person for testing and by a        WR-711.55.6
skilled person competent in such work for mainte-
nance shall be provided.

Electronic converters shall conform to BS EN             WR-711.55.6
61347-1.

### 7.3.6.4.3 Extra-low voltage provided by SELV or PELV

Where SELV or PELV is used, whatever the nomi-          WR-711.414.4.5
nal voltage, basic protection shall be provided by:

• covering all live parts with an insulation that
  can only be removed by destruction; or
• by barriers or enclosures affording a degree of
  protection of at least IP4X or IPXXD.

## 7.3.6.5 *Wiring systems*

Flexible cables **shall not** be laid in areas accessible to the public unless they are protected against mechanical damage.

| | |
|---|---|
| Armoured cables or cables protected against mechanical damage shall be used wherever there is a risk of mechanical damage. | WR-711.52 |
| Wiring cables shall be copper, have a minimum cross-sectional area of 1.5 mm$^2$, and shall comply with an appropriate British Standard for either thermoplastic or thermosetting insulated electric cables. | WR-711.52 |

### 7.3.6.5.1 Types of wiring system

| | |
|---|---|
| Where no fire alarm system is installed in a building used for exhibitions etc., cable systems shall be either:<br><br>• flame retardant and low smoke resistant; or<br>• single-core or multi-core unarmoured cables enclosed in metallic or non-metallic conduit or trunking that provides a degree of fire protection of at least IP4X. | WR-711.521 |

## 7.3.7 **Floor and ceiling heating systems**

**Note:** This section applies to the installation of electric floor and ceiling heating systems that are erected as either thermal storage heating systems or direct heating systems. It does not apply to the installation of wall heating systems.

Flexible sheet heating elements shall comply with the requirements of BS EN 60335-2-96.

## 7.3.7.1 *General*

### 7.3.7.1.1 Ambient temperature (AA)

| | |
|---|---|
| For cold tails (circuit wiring) and control leads installed in the zone of heated surfaces, the increase in ambient temperature shall be taken into account. | WR-753.522.1.3 |

### 7.3.7.1.2  Presence of solid foreign bodies (AE)

| | |
|---|---|
| Where heating units are installed, there shall be heating-free areas where drilling and fixing by screws, nails and the like are permitted. | WR-753.522.4.3 |
| The installer shall inform other contractors that no penetrating means, such as screws for door stoppers, shall be used in the area where floor or ceiling heating units are installed. | WR-753.522.4.3 |

### 7.3.7.2  Protective measures

The protective measures of:

- electrical separation (with the exception of shaver socket-outlets);
- obstacles and placing out of reach;
- non-conducting location and Earth-free local equipotential bonding;

are **not** permitted.

### 7.3.7.2.1  Prevention of mutual detrimental influences

| | |
|---|---|
| Heating units shall not cross expansion joints of the building or structure. | WR-753.515.4 |

### 7.3.7.2.2  Protection against burns

| | |
|---|---|
| In floor areas where contact with skin or footwear is possible, the surface temperature of the floor shall be limited (e.g. to 35°C). | WR-753.423 |

### 7.3.7.2.3  Protection by RCDs

RCDs shall be used as disconnecting devices.

| | |
|---|---|
| A circuit supplying heating equipment of Class II construction or equivalent insulation shall be provided with additional protection by the use of an RCD. | WR-753.415.1 |

## 7.3.7.3 *Selection and erection of equipment*

### 7.3.7.3.1 Heating units

Heating-free areas shall be provided in such a way that the heat emission is not prevented by the attachment of room fittings.

| | |
|---|---|
| Heating units for installation in ceilings shall have a degree of protection of not less than IPX1. | WR-753.512.2.5 |
| Heating units for installation in a floor of concrete or similar material shall have a degree of ingress protection not less than IPX7 and shall have the appropriate mechanical properties. | WR-753.512.2.5 |
| To avoid the overheating of floor or ceiling heating systems in buildings, one or more of the following measures shall be applied within the zone where heating units are installed, to limit the temperature to a maximum of 80°C:<br><br>• appropriate design of the heating system;<br>• appropriate installation of the heating system in accordance with the manufacturer's instructions;<br>• use of protective devices. | WR-753.424.3.1 |
| Heating units shall be connected to the electrical installation via cold tails or suitable terminals. | WR-753.424.3.1 |
| Heating units shall be inseparably connected to cold tails (e.g. by a crimped connection). | WR-753.424.3.1 |
| As the heating unit may cause higher temperatures or arcs under fault conditions, special measures should be taken when the heating unit is installed close to easily ignitable building structures (such as placing on a metal sheet, in metal conduit or at a distance of at least 10 mm in air from the ignitable structure). | WR-753.424.3.2 |
| Precautions shall be taken not to stress the heating unit mechanically (e.g. the material by which it is to be protected in the finished installation shall cover the heating unit as soon as possible). | WR-753.512.1.6 |

## 7.3.7.3.2  Identification and notices

The designer of the installation/heating system or installer shall provide a plan for each heating system, containing the following details:  WR-753.514

- manufacturer and type of heating units;
- number of heating units installed;
- length/area of heating units;
- rated power;
- surface power density;
- layout of the heating units in the form of a sketch, a drawing or a picture;
- position/depth of heating units;
- position of junction boxes;
- conductors, shields and the like;
- heated area;
- rated voltage;
- rated resistance (cold) of heating units;
- rated current of the overcurrent protective device;
- rated residual operating current of the RCD;
- insulation resistance of the heating installation and the test voltage used;
- leakage capacitance.

 This plan shall be fixed to, or adjacent to, the distribution board of the heating system.

## 7.3.8  Locations containing a bath or shower

All locations containing a bath or shower shall be inspected to confirm that they comply with Part 701 of the Regulations.

When people use bathrooms and showers, most of the time they are naturally unclothed and wet, and thus very vulnerable to electric shock due to their reduced body resistance (i.e. absence of shoes means less protection from shock, while water on the skin will tend to short-circuit its natural protection). Special measures are, therefore, required to ensure that the possibility of direct and/or indirect contact is reduced.

The following requirements apply to locations containing a fixed bath (bath tub or birthing pool), showers and their surrounding zones. They do **not** apply to emergency facilities in industrial areas and laboratories.

The requirements are based on three zones, which take into account the limitations of walls, doors, fixed partitions, ceilings and floors. The three zones are as shown in Figure 7.5 and Table 7.2.

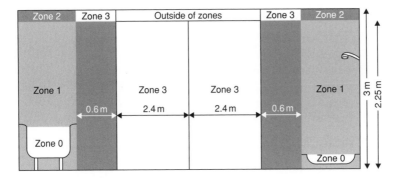

**Figure 7.5** Zone dimensions.

**Table 7.2 Limitations of zones**

Zone 0   The interior of the bath tub or shower basin.

  Note: For showers without a basin, the height of zone 0 is 0.10 m, with the same horizontal extent as zone 1.

Zone 1   Zone 1 does not include zone 0, and is limited by:

- the finished floor level and the highest fixed shower head or water outlet, or 2.25 m above the finished floor level (whichever is higher);
- the vertical surface;

  ○   circumscribing the bath tub or shower basin, and
  ○   1.20 m from the centre point of the fixed water outlet on the wall (or ceiling for showers without a basin).

  Note: The space under the bath tub or shower basin is considered to be zone 1. However, if the space under the bath tub or shower basin is only accessible with a tool, it is considered to be outside the zones.

Zone 2   Zone 2 is limited by:

- the finished floor level and the highest fixed shower head or water outlet, or 2.25 m above the finished floor level (whichever is higher);
- the vertical surface:

  ○   at the boundary of zone 1, and
  ○   0.60 m from the border of zone 1.

  Note: For showers without a basin, there is no zone 2, but the horizontal dimension of zone 1 is increased to 1.20 m.

 For locations containing a bath or shower for medical treatment, special requirements may be necessary.

### 7.3.8.1 General

| | |
|---|---|
| Horizontal or inclined ceilings, walls (with or without windows) doors, floors and fixed partitions may be taken into account where these effectively limit the extent of locations containing a bath or shower, as well as their zones. | WR-701.3 |

### 7.3.8.2 Protective measures

The protective measures of:

*   obstacles and placing out of reach; and
*   non-conducting location and Earth-free local equipotential bonding;

are **not** permitted.

**Note:** Protection by electrical separation shall **only** be used for:

*   circuits supplying one item of current-using equipment; or
*   one single socket-outlet.

#### 7.3.8.2.1 Extra-low voltage provided by SELV or PELV

| | |
|---|---|
| Where SELV or PELV is used (whatever the nominal voltage) basic protection for equipment in zones 0, 1 and 2 shall be provided by:<br><br>•  basic insulation under the supervision of a suitably qualified and/or trained person; or<br>•  barriers or enclosures affording a degree of protection of at least IPXXB or IP2X. | WR-701.414.4.5 |

#### 7.3.8.2.2 Protection by RCDs

| | |
|---|---|
| Additional protection shall be provided for all low-voltage circuits of the location, by the use of one or more RCDs. | WR-701.411.3.3 |

#### 7.3.8.2.3 Supplementary equipotential bonding

| | |
|---|---|
| Local supplementary equipotential bonding shall be established, connecting together the terminals of the protective conductor of each circuit supplying | WR-701.415.2 |

Class I and Class II equipment to the accessible extraneous-conductive-parts, within a room containing a bath or shower, including the following:

- metallic pipes supplying services and metallic waste pipes (e.g. water, gas);
- metallic central heating pipes and air-conditioning systems;
- accessible metallic structural parts of the building (metallic door architraves, window frames and similar parts are not considered to be extraneous-conductive-parts unless they are connected to metallic structural parts of the building).

Supplementary equipotential bonding may be installed outside or inside rooms containing a bath or shower, preferably close to the point of entry of extraneous-conductive-parts into such rooms.          WR-701.415.2

Where the location containing a bath or shower is in a building with a protective equipotential bonding system, supplementary equipotential bonding may be omitted where all of the following conditions are met:          WR-701.415.2

- all final circuits of the location comply with the requirements for automatic disconnection;
- all final circuits of the location have additional protection by means of an RCD;
- all extraneous-conductive-parts of the location are effectively connected to the protective equipotential bonding according.

 Special requirements exist for locations containing baths and showers for medical use or for disabled persons (see Building Regulations Part M, and Section 710 of BS 7671:2008 (Incorporating Amendment No. 1)).

### 7.3.8.3 Selection and erection of equipment

### 7.3.8.3.1 External influences

Installed electrical equipment shall have at least the following degrees of protection:          WR-701.512.2

- in zone 0, IPX7;
- in zones 1 and 2, IPX4.

**Note:** This requirement does not apply to shaver supply units installed in zone 2 and located where direct spray from showers is unlikely.

 However, electrical equipment exposed to water jets (e.g. for cleaning purposes) shall have a degree of protection of at least IPX5.

### 7.3.8.3.1 Electric floor heating systems

 The so-called protective measure 'protection by electrical separation' is **not** permitted for electric floor heating systems.

> Only heating cables or thin-sheet flexible heating elements may be installed for electric floor heating systems, provided that they have either a metal sheath or a metal enclosure or a fine-mesh metallic grid, which is connected to the protective conductor of the supply circuit.   WR-701.753

 Compliance with the latter requirement is not required if the protective measure SELV is provided for the floor heating system.

### 7.3.8.4 Supplies

### 7.3.8.4.1 Current-using equipment

In zone 0, current-using equipment shall **only** be installed provided that all the following requirements are met:

- the equipment is suitable for use in that zone;
- the equipment is fixed and permanently connected;
- the equipment is protected by SELV (the safety source being installed outside zones 0, 1 and 2).

In zone 1, the following fixed and permanently connected current-using equipment shall **only** be installed provided it is suitable for installation in zone 1 according to the manufacturer's instructions:

- electric showers;
- equipment protected by SELV or PELV (the safety source being installed outside zones 0, 1 and 2);
- luminaires;
- shower pumps;
- towel rails;
- ventilation equipment;
- water heating appliances;
- whirlpool units.

## 7.3.8.4.2 Erection of switchgear, controlgear and accessories according to external influences

**Note:** The following requirements do not apply to switches and controls that are incorporated in fixed current-using equipment or to insulating pull cords of cord-operated switches.

In zone 0:

- switchgear or accessories **shall not** be installed.

In zone 1:

- only switches of SELV circuits shall be installed, their safety source being installed outside zones 0, 1 and 2.

In zone 2:

- switchgear, accessories incorporating switches or socket-outlets shall not be installed, with the exception of:
  - ○ switches and socket-outlets of SELV circuits (the safety sources of which are installed outside zones 0, 1 and 2); and
  - ○ shaver supply units complying with BS EN 61558–2–5.

> Except for SELV socket-outlets and shaver supply units complying with BS EN 61558–2–5, socket-outlets are prohibited within a distance of 3 m horizontally from the boundary of zone 1.   WR-701.512.3

### 7.3.9 Marinas and similar locations

The particular requirements of this section are applicable only to circuits intended to supply pleasure craft or houseboats in marinas and similar locations. They do **not** apply to the supply to houseboats if they are supplied directly from the public network, **or** to the internal electrical installations of pleasure craft or houseboats.

### 7.3.9.1 General

#### 7.3.9.1.1 External influences

> For marinas, particular attention shall be given to the likelihood of corrosive elements, movement of structures, mechanical damage, presence of flammable fuel and the increased risk of electric shock due to:   WR-709.512.2

- presence of water;
- reduction in body resistance;
- contact of the body with Earth potential.

### 7.3.9.1.2 Impact (AG)

Equipment installed on or above a jetty, wharf, pier or pontoon shall be protected against mechanical damage (impact of medium severity AG2). Protection shall be afforded by one or more of the following:     WR-709.512.2.1.4

- the position or location, selected to avoid being damaged by any reasonably foreseeable impact;
- the provision of local or general mechanical protection;
- installing equipment complying with a minimum degree of protection for external mechanical impact IK08 (see BS EN 62262).

### 7.3.9.1.3 Presence of corrosive or polluting substances (AF)

Equipment installed on or above a jetty, wharf, pier or pontoon shall be unaffected by the presence of atmospheric corrosive or polluting substances (AF2).     WR-709.512.2.1.3

 **Note:** If hydrocarbons are present, AF3 is applicable.

### 7.3.9.1.4 Presence of water (AD)

In marinas, equipment installed on or above a jetty, wharf, pier or pontoon shall be unaffected by:     WR-709.512.2.1.1

- water splashes (AD4), to IPX4;
- water jets (AD5), to IPX5;
- water waves (AD6), to IPX6.

### 7.3.9.1.5  Presence of solid foreign bodies (AE)

| | |
|---|---|
| Equipment installed on or above a jetty, wharf, pier or pontoon shall be protected to at least IP3X in order to protect against the ingress of small objects (AE2). | WR-709.512.2.1.2 |

### 7.3.9.2 *Protective measures*

The protective measures of:

* obstacles and placing out of reach; and
* non-conducting location and Earth-free local equipotential bonding;

are **not** permitted.

### 7.3.9.2.1  Protection against overcurrent

| | |
|---|---|
| Each socket-outlet shall be protected by an individual overcurrent protective device. | WR-709.533 |
| A fixed connection for supply to each houseboat shall be protected individually by an overcurrent protective device. | WR-709.533 |

### 7.3.9.2.2  Protection by RCDs

| | |
|---|---|
| Socket-outlets shall be protected individually by an RCD. | WR-709.531.2 |
| Final circuits intended for fixed connection for the supply to houseboats shall be protected individually by an RCD. | WR-709.531.2 |

### 7.3.9.3 *Selection and erection of equipment*

### 7.3.9.3.1  Isolation

| | |
|---|---|
| At least one means of isolation (with a maximum of four outlet sockets) shall be installed in each distribution cabinet. | WR-709.537.2.1.1 |

**Note:** This switching device shall disconnect all live conductors, including the neutral conductor.

### 7.3.9.3.2 Plugs and socket-outlets

| | |
|---|---|
| Socket-outlets shall comply with BS EN 60309–1 if above 63 A, and BS EN 60309–2 if up to 63 A. | WR-709.553.1.8 |
| Every socket-outlet shall meet the degree of protection of at least IP44 or otherwise be protected by an enclosure. | WR-709.553.1.8 |
| Every socket-outlet shall be located as close as is practicable to the berth to be supplied. Socket-outlets shall be installed in the distribution board or in separate enclosures. | WR-709.553.1.9 |
| A maximum of four socket-outlets shall be grouped together in one enclosure. | WR-709.553.1.10 |
| One socket-outlet shall supply one pleasure craft or houseboat – only! | WR-709.553.1.11 |
| In general, single-phase socket-outlets (with rated voltage 200250 V and rated current 16 A) shall be provided. | WR-709.553.1.12 |
| Socket-outlets shall be placed at a height of not less than 1 m above the highest water level. In the case of floating pontoons or walkways only, this height may be reduced to 300 mm above the highest water level, provided that appropriate additional measures are taken to protect against the effects of splashing. | WR-709.553.1.13 |

### 7.3.9.3.3 Overhead cables or overhead insulated conductors

 All overhead conductors shall be insulated.

| | |
|---|---|
| Poles and other supports for overhead wiring shall be located or protected so that they are unlikely to be damaged by any foreseeable vehicle movement. | WR-709.521.1.8 |

Overhead conductors shall be at a height above
ground of not less than 6 m in all areas subjected to
vehicle movement, and 3.5 m in all other areas.

WR-709.521.1.8

### 7.3.9.3.4 Underground cables

Underground distribution cables shall, unless provided with additional mechanical protection, be
buried at a sufficient depth to avoid being damaged
(e.g. by heavy vehicle movement).

WR-709.521.1.7

**Note:** A depth of 0.5 m is generally considered as a minimum depth to fulfil
this requirement.

### 7.3.9.4 Supplies

The nominal supply voltage of the installation for
the supply to pleasure craft or houseboats shall be
230 V a.c. single-phase, or 400 V a.c. three-phase.

WR-709.313.1.2

In the UK, the ESQCR prohibit the connection of a PME earthing device to
any metalwork of a boat.

### 7.3.9.5 Wiring systems

The following wiring systems are suitable for distribution circuits of marinas:

WR-709.521.1.4

- underground cables;
- overhead cables or overhead insulated
  conductors;
- cables with copper conductors and thermoplastic or elastomeric insulation and sheath
  installed within an appropriate cable management system, taking into account external
  influences such as movement, impact, corrosion and ambient temperature;
- mineral-insulated cables with a polyvinyl
  chloride (PVC) protective covering;
- cables with armouring and serving of thermoplastic or elastomeric material;

- other cables and materials that are no less suitable than those listed above.

The following wiring systems shall not be used on or above a jetty, wharf, pier or pontoon:          WR-709.521.1.5

- cables in free air suspended from or incorporating a support wire;
- non-sheathed cables in cable management systems;
- cables with aluminium conductors;
- mineral insulated cables.

Cables shall be selected and installed so that mechanical damage due to tidal and other movement of floating structures is prevented.          WR-709.521.1.6

Cable management systems shall be installed to allow the drainage of water by drainage holes and/or installation of the equipment on an incline.          WR-709.521.1.6

### 7.3.10 Medical locations

Section 710 is a new section specifically aimed at electrical installations in medical locations. Although it mainly refers to hospitals, private clinics, medical and dental practices, healthcare centres and dedicated medical rooms in the workplace, this section also applies to electrical installations in locations designed for medical research and (where applicable) to veterinary clinics.

The requirements of Section 710 do not apply to **medical** electrical equipment.

#### 7.3.10.1 Definitions

To save you hunting through BS 7671:2008 and other documents for the meaning of some of the medical terms used below and elsewhere in this book, the following is provided as an aide memoire to the most important ones:

**Intracardiac procedure** is a procedure whereby an electrical conductor is placed within the heart of a patient, or is likely to come into contact with the heart, such conductor being accessible outside the patient's body. In this context, an electrical conductor includes insulated wires such as cardiac pacing electrodes or intracardiac ECG electrodes, or insulated tubes filled with conducting fluids.

**Medical electrical equipment (ME equipment)** is equipment having an applied part for transferring energy to or from the patient or detecting such energy transfer to or from the patient and which is:

(a) provided with not more than one connection to a particular supply mains; and

(b) intended by the manufacturer to be used

- in the diagnosis, treatment or monitoring of a patient, or
- for compensation or alleviation of disease, injury or disability.

ME equipment includes those accessories as defined by the manufacturer that are necessary to enable the normal use of the ME equipment.

**Medical electrical system (ME system)** is a combination, as specified by the manufacturer, of items of equipment, at least one of which is medical electrical equipment to be interconnected by functional connection or by use of a multiple socket-outlet.

**Note:** The system includes those accessories that are needed for operating the system and are specified by the manufacturer.

**Medical IT system** is an IT electrical system fulfilling specific requirements for medical applications.

**Note:** These supplies are also known as 'isolated power supply systems'.

**Medical location** is a location intended for purposes of diagnosis, treatment (including cosmetic treatment), monitoring and care of patients.

Patient is a living being (person or animal) undergoing a medical, surgical or dental procedure.

**Note:** A person under treatment for cosmetic purposes may also be considered as a patient.

**Patient environment** is any volume in which intentional or unintentional contact can occur between a patient and parts of the medical electrical equipment or medical electrical system, or between a patient and other persons touching parts of the medical electrical equipment or medical electrical system.

## 7.3.10.1.1 Safety

Electrical installations in medical locations (and also electrical installations in locations designed for medical research and (where applicable) veterinary clinics) must be capable of ensuring the continued safety of patients and medical staff. For convenience, these medical locations are classified into groups according to the level of safety required. For example:

- the type of contact between applied parts and the patient;
- the threat to the safety of the patient that represents a discontinuity (failure) of the electrical supply; and
- the purpose for which the location is used.

**Note:** 'Applied part' refers to that part of the medical electrical equipment that, while in normal use, comes into physical contact with the patient.

To ensure the protection of patients from possible electrical hazards, additional protective measures need to be applied in medical locations.

Care should also be taken to ensure that other installations do not compromise the level of safety provided by installations meeting the requirements of this section.

### 7.3.10.1.2 Medical locations

Medical locations are split into groups as follows (Table 7.3):

**Table 7.3 Examples of the allocation of group numbers and classification for the safety services of medical locations**

| Medical location | Group | | | Class | |
|---|---|---|---|---|---|
| | 0 | 1 | 2 | ≤0.5 s | >0.5 s ≤15 s |
| 1   Massage room | X | X | | | X |
| 2   Bedrooms | | X | | | X |
| 3   Delivery room | | X | | X[a] | X |
| 4   ECG, EEG, EHG room | | X | | | X |
| 5   Endoscopic room | | X[b] | | X | X[b] |
| 6   Examination or treatment room | | X | | X | X |
| 7   Urology room | | X[b] | | X | X[b] |
| 8   Radiological diagnostic and therapy room | | X | X | X | X |
| 9   Hydrotherapy room | | X | | | X |
| 10  Physiotherapy room | | X | | | X |
| 11  Anaesthetic area | | | X | X[a] | X |
| 12  Operating theatre | | | X | X[a] | X |
| 13  Operating preparation room | | | X | X[a] | X |
| 14  Operating plaster room | | | X | X[a] | X |
| 15  Operating recovery room | | | X | X[a] | X |
| 16  Heart catheterization room | | | X | X[a] | X |
| 17  Intensive care room | | | X | X[a] | X |
| 18  Angiographic examination room | | | X | X[a] | X |
| 19  Haemodialysis room | | | X | | X |
| 20  Magnetic resonance imaging (MRI) room | | X | X | X | X |
| 21  Nuclear medicine | X | | | | X |
| 22  Premature baby room | | | X | X[a] | X |
| 23  Intermediate Care Unit (IMCU) | | | X | X | X |

a  Luminaires and life-support medical electrical equipment which needs power supply within 0.5 s or less.
b  Not being an operating theatre.

**Group 0** – Medical location where no applied parts are intended to be used and where discontinuity (failure) of the supply cannot cause danger to life.

**Group 1** – Medical location where discontinuity of the electrical supply does not represent a threat to the safety of the patient, and applied parts are intended to be used:

- externally;
- invasively to any part of the body, except where group 2 applies.

**Group 2** – Medical location where applied parts are intended to be used, and where discontinuity (failure) of the supply can cause danger to life, in applications such as:

- intracardiac procedures;
- vital treatments and surgical operations.

### 7.3.10.1.3 Classes and types of medical electrical equipment

All medical electrical equipment is categorized into classes according to the type of protection against electric shock that it uses.

#### 7.3.10.1.3.1 Class I equipment

Class I equipment has a protective Earth, where the basic means of protection is the insulation between live parts and exposed-conductive-parts such as the metal enclosure. In the event of a fault (that would otherwise cause an exposed-conductive-part to become live) the supplementary protection (i.e. the protective Earth) comes into effect.

 **Note:** Large fault current flows from the mains part to Earth via the protective Earth conductor, and this will cause a protective device (usually a fuse) in the mains circuit to disconnect the equipment from the supply.

#### 7.3.10.1.3.2 Class II equipment

Class II equipment uses either double insulation or reinforced insulation. In double-insulated equipment the basic protection is provided by the first layer of insulation. If the basic protection fails, then supplementary protection is provided by a second layer of insulation – thus preventing contact with live parts

#### 7.3.10.1.3.3 Class III equipment

Class III equipment is either battery operated or supplied by an SELV transformer and, as the voltages do not exceed 25 V a.c. or 60 V d.c., protection against electric shock is a minimal requirement.

### 7.3.10.1.4 Patient safety

As the hazard to people will depend on the treatment being administered, hospital locations are divided into groups as follows:

- Group Zero: where no treatment or diagnosis using medical electrical equipment is administered (e.g. consulting rooms).

- Group One: where medical electrical equipment is in use, but not for the treatment of heart (intracardiac) conditions.
- Group Two: where medical electrical equipment is in use for heart (intracardiac) conditions.

### 7.3.10.1.5 Risk of explosion

The gases used as anaesthetics in operating theatres are flammable if present in high concentrations. However, provided that there is adequate ventilation (e.g. 20 air changes per hour) no special precautions are necessary for the electrical installation in medical locations.

| | |
|---|---|
| All electrical devices (such as socket-outlets and/or switches) shall be installed at least 0.2 m below any medical-gas outlet, so as to minimize the risk of ignition of flammable gases. | WR-710.512.2.1 |

### *7.3.10.2 Protective measures*

### 7.3.10.2.1 RCDs

| | |
|---|---|
| Care shall be taken to ensure that simultaneous use of many items of equipment connected to the same circuit cannot cause unwanted tripping of the RCD | WR-710.411.3.2.1 |

In Group 1 and Group 2 medical locations, the following shall apply:

- where RCDs are required, **only** type A (according to BS EN 61008 and BS EN 61009) or type B (according to IEC 62423) shall be selected, depending on the possible fault current arising.    WR-710.411.3.2.1

 Type AC RCDs **shall not** be used.

- For TN, TT and IT systems, the voltage presented between simultaneously accessible exposed-conductive-parts and/or    WR-710.411.3.2.5

extraneous-conductive-parts shall not exceed 25 V a.c. or 60 V d.c.

- For TN and TT systems, the requirements of Table 7.4 shall apply.

**Table 7.4  Maximum disconnection times (data from BS 7671:2008)**

| System | Maximum disconnection time (s) | | | | | | | | | |
| | 25V < U₀ 50V | | 50V < U₀ 120V | | 120V <U₀ 5230V | | 230V < Uo 400V | | 1.10 > 400V | |
| | a.c. | d.c. | a.c. | d.c. | a.c. | d.c. | a.c. | d.c. | a.c. | d.c. |
|---|---|---|---|---|---|---|---|---|---|---|
| TN | 5 | 5 | 0.3 | 2 | 0.3 | 0.5 | 0.05 | 0.06 | 0.02 | 0.02 |
| TT | 5 | 5 | 0.15 | 0.2 | 0.05 | 0.1 | 0.02 | 0.06 | 0.02 | 0.02 |

 **Note:** In TN systems, 25 V a.c. or 60 V d.c. may be met with protective equipotential bonding, by complying with the disconnection time in accordance with Table 7.4.

### 7.3.10.2.1.1  TN system

In Group 1 final circuits that are rated up to 63 A, RCDs **shall** be used.  WR-710.411.4

In TN-S systems, the insulation level of all live conductors shall be monitored.  WR-710.411.4

In Group 2 medical locations (except for a medical IT system), protection by automatic disconnection of supply by means of RCDs **shall** only be used on the following circuits:  WR-710.411.4

- circuits for the supply of movements of fixed operating tables; or
- circuits for X-ray units; or
- circuits for large equipment with a rated power greater than 5 kV A.

### 7.3.10.2.1.2  TT system

In Group 1 and Group 2 medical locations using a TT system, RCDs shall be used.  WR-710.411.5

### 7.3.10.2.1.3 IT system

 Where a medical IT system is used, additional protection by means of an RCD **shall not** be used.

In Group 2 medical locations, an IT system shall always be used for:

> - final circuits supplying medical electrical equipment and systems intended for life support;
> - surgical applications; and
> - other electrical equipment located – or that may be moved into – the 'patient environment'.
>
> WR-710.411.6.3.1

For each group of rooms serving the same function, at least one IT system is necessary and this shall be equipped with an insulation monitoring device (IMD) that has:

> - an acoustic and visual alarm system that is located in a suitable place so that it can be permanently monitored (audible and visual signals) by the medical staff and the technical staff.
>
> WR-710.411.6.3.1

This alarm system shall have:

> - a **green** signal lamp to indicate normal operation;
> - a **yellow** signal lamp that lights when the minimum value set for the insulation resistance is reached;
>
>  It shall **not** be possible for this light to be cancelled or disconnected – but the **yellow** signal should automatically go out when the normal condition is restored.
>
> - an audible alarm that sounds when the minimum value set for the insulation resistance is reached.
>
>  **Note:** This audible alarm may be capable of being be silenced.

- Documentation of all faults occurring in a medical location shall be maintained, and shall include:
  - o the meaning of each type of signal; and
  - o the procedure to be followed in case of an alarm at first fault.

 **Note:** See Figure 7.6 for an illustration of a typical theatre layout.

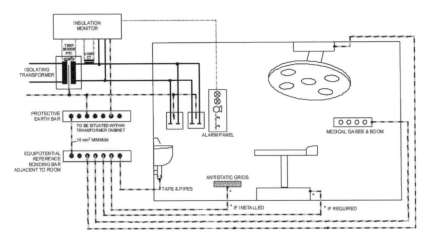

**Figure 7.6** Typical theatre layout. (Courtesy of BSI.)

### 7.3.10.2.1.4 Socket-outlets

For each circuit that is protected by an RCD, the possible unwanted tripping of the RCD due to excessive protective conductor currents produced by equipment in normal operation shall be considered.

WR-710.531.2.4

It is a mandatory requirement that in IT systems in Group 2 medical locations:

 Socket-outlets intended to supply medical electrical equipment **shall** be unswitched.

WR-710.553.1

Socket-outlets used on medical IT systems **shall** be coloured **blue**, and clearly and permanently marked 'Medical equipment only'.

In addition:

At each patient's place of treatment (e.g. bedheads):     WR-710.553.1

- each socket-outlet shall be supplied by an individually protected circuit; or
- several socket-outlets shall be separately supplied by a minimum of two circuits.

### 7.3.10.2.2 SELV and PELV

When using SELV and/or PELV circuits in Group     WR-710.414
1 and Group 2 medical locations, the nominal
voltage applied to current-using equipment shall
not exceed 25 V a.c. rms or 60 V ripple-free d.c.

Protection by basic insulation of live parts or by     WR-710.414
barriers or enclosures shall be provided.

In Group 2 medical locations, where PELV is     WR-710.414.4.1
used, exposed-conductive-parts of equipment
(e.g. operating theatre luminaire) shall be connected to the circuit protective conductor.

### 7.3.10.2.3 Supplementary equipotential bonding

For Group 1 and Group 2 medical locations,     WR-710.415.2.1
supplementary equipotential bonding shall
be installed for the following parts which are
located (or that may be moved into) the 'patient
environment':

- protective conductors;
- extraneous-conductive-parts;
- screening against electrical interference fields, if installed;

- connection to conductive floor grids, if installed;
- metal screens of isolating transformers, via the shortest route to the earthing conductor.

Supplementary equipotential bonding connection points for medical electrical equipment shall be provided in each medical location, as follows:  WR-710.415.2.1

- Group 1 – one per patient location;
- Group 2 – one per medical IT socket-outlet.

Unless they are intended to be isolated from Earth, fixed conductive non-electrical patient supports such as operating theatre tables, physiotherapy couches and dental chairs should be connected to the equipotential bonding conductor.  WR-710.415.2.1

The equipotential bonding busbar shall be located in or near the medical location, and all connections shall be accessible, labelled, clearly visible, and be capable of being easily disconnected individually.  WR-710.415.2.3

 It is recommended that radial wiring patterns are used to avoid 'Earth loops', which may cause electromagnetic interference.

### 7.3.10.2.4 Protection against electric shock

The protective measures:  WR-710.410.3.5

- obstacles and placing out of reach;
- non-conducting location;
- Earth-free local equipotential bonding; and
- electrical separation for the supply of more than one item of current-using equipment;

are **not** permitted in medical locations.

#### 7.3.10.2.4.1 Protective measure – functional extra-low voltage

In medical locations, functional extra-low voltage (FELV) is **not** permitted as a method of protection against electric shock.  WR-710.411.7

### 7.3.10.2.4.2 Protective measure – extra-low voltage provided by SELV or PELV

When using SELV and/or PELV circuits in medical locations of Group 1 and Group 2, protection by basic insulation of live parts (or by barriers or enclosures) shall be provided.  |  WR-710.414.1

In medical locations of Group 2, where PELV is used, exposed-conductive-parts of equipment (e.g. operating theatre luminaires) shall be connected to the circuit protective conductor.  |  WR-710.414.4.1

### 7.3.10.2.4.3 Supplementary equipotential bonding

In each Group 1 or Group 2 medical location, supplementary equipotential bonding shall be installed to equalize the potential differences between:  |  WR-710.415.2.1

- extraneous-conductive-parts;
- screening against electrical interference fields, if installed;
- connection to conductive floor grids, if installed;
- metal screens of isolating transformers, via the shortest route to the Earthing conductor;

which are located in, or that may be moved into, the 'patient environment'.

### 7.3.10.3 Supplies

### 7.3.10.3.1 Transformers for IT systems

In medical locations, IT transformers shall be installed in close proximity to the medical location, and shall ensure that:

- the leakage current of the output winding to Earth and the leakage current of the enclosure (when measured in no-load condition) and the transformer do not exceed 0.5 mA;  |  WR-710.512.1.1

- at least one single-phase transformer per room or functional group of rooms is used to form the IT systems for mobile and fixed equipment, and the rated output shall be no less than 0.5 kVA and shall not exceed 10 kVA;

   Where several transformers are needed to supply equipment in one room, they should not be connected in parallel.

- if the supply of three-phase loads via an IT system is also required, a separate three-phase transformer is provided for this purpose.

 Capacitors **shall not** be used in transformers for medical IT systems.

### 7.3.10.3.2 Safety services

| | |
|---|---|
| Medical locations shall have safety standby power supply available that will energize the installations required for continuous operation in the case of failure of the general power system | WR-710.56 |
| The safety power supply system shall be capable of automatically taking over if the main power supply has dropped for more than 0.5 s and by more than 10%, and shall be capable of providing power for a period of at least 3 h for: | WR-710.56<br>WR-710.560.6.1.1 |

- operating theatre table luminaires;
- medical electrical equipment containing light sources (endoscopes and monitors etc.);
- life-supporting medical electrical equipment.

 **Note:** Safety services that have been provided for a number of locations with different classifications should meet that classification which gives the highest security of supply.

### 7.3.10.3.2.1 General requirements for safety power supply sources of Group 1 and Group 2

| | |
|---|---|
| - Primary cells are not allowed as safety power sources; | WR-710.560.5.5 |

- an additional main incoming power supply,
  from the general power supply, is not regarded
  as a source of the safety power supply;
- the availability (readiness for service) of safety
  power sources shall be monitored;
- where socket-outlets are supplied from the
  safety power supply source they shall be
  identified according to their safety services
  classification.

### 7.3.10.3.2.2 Power supply sources with a changeover period greater than 15 seconds

Hospital maintenance services equipment such as:    WR-710.560.6.1.3

- sterilization equipment;
- technical building installations, in services
  and waste disposal systems;
- cooling equipment;
- catering equipment;
- storage battery chargers;

shall be connected either automatically (or manually) to a safety power supply source capable of maintaining it for a minimum period of 24 h.

### 7.3.10.3.2.3 Emergency lighting systems

In the event of mains power failure, the change-    WR-710.560.9.1
over period to the standby the safety services
sources such as:

- emergency lighting and exit signs;
- switchgear and control gear for emergency
  generating sets;
- normal power supply and the standby safety
  services power supply distribution boards;
- essential service rooms;
- locations of central fire alarm and monitoring
  systems;

shall not exceed 15 s.

- Group 1 medical location rooms shall be supplied with at least one luminaire in case of emergency;
- Group 2 medical locations rooms shall have at least 90% of their normal lighting requirements supplied from the standby safety service.

 The luminaires of the escape routes shall be arranged on alternate circuits.

#### 7.3.10.3.2.4 Other services

Other services such as: WR-710.560.11

- fire-fighter lifts;
- smoke extraction ventilation systems;
- paging systems;
- vitally important medical electrical equipment used in Group 2 medical locations;
- electrical equipment of medical gas supply (e.g. compressed air, vacuum supply and narcosis (anaesthetics)) including exhaustion as well as their monitoring devices;
- fire detection and fire alarms;
- fire extinguisher systems;

shall be provided with a changeover system to a standby safety service within 15 s.

### 7.3.10.3.3 Earthing

 PEN conductors **shall not** be used in medical locations and medical buildings downstream of the main distribution board.

- It is recommended that radial wiring patterns are used to avoid 'Earth loops', which may cause electromagnetic interference.
- In the event of a first fault to Earth, a total loss of supply in medical Group 2 locations shall be prevented.

### 7.3.10.4 Inspection and testing

The International Electrotechnical Commission (IEC) and/or British Standards (BS) manufacturers' standards for medical electrical equipment consist of two types of testing: type testing and routine testing.

### 7.3.10.4.1 Type testing

Type testing is carried out by an approved test house (under closely specified and closely controlled environmental conditions) on a single representative sample of a piece of equipment for which certification of compliance with a standard is being sought. These tests are not intended for routine use – indeed, it has been documented that repetition of many of the tests would certainly cause deterioration in performance and safety of the equipment under test.

### 7.3.10.4.2 Routine testing

Routine testing, on the other hand, is intended to provide an indication of the inherent safety of the equipment, without subjecting it to undue stress that would be liable to cause deterioration.

### 7.3.10.4.1 Initial verification

In addition to the requirements of Chapter 61 and Health Technical Memorandum 06–01 (Part A), the following tests **shall** be carried out, both prior to commissioning and after alteration or repairs and before re-commissioning:

> - complete functional tests of all IMDs associated     WR-710.6.1
>   with the medical IT system, including insulation
>   failure, transformer high temperature, overload,
>   discontinuity and the acoustic/visual alarms
>   linked to them;
> - measurements of leakage current from the IT
>   transformers of the output circuit and enclosure in
>   the no-load condition;
> - measurements to verify that the resistance of the
>   supplementary equipotential bonding is within
>   stipulated limits.

The dates and results of each verification shall be recorded.

### 7.3.10.4.2 Periodic inspection and testing

**Note:** Periodic inspection and testing should be carried out in accordance with Health Technical Memorandum 06–01 (Part B) and local health authority requirements as follows, and at the given intervals:

> - **Annually** – complete functional tests of all IMDs     WR-710.6.2
>   associated with the medical IT system, including
>   insulation failure, transformer high temperature,

overload, discontinuity and the acoustic/visual alarms linked to them.

- **Annually** – measurements to verify that the resistance of the supplementary equipotential bonding is within the stipulated limits.
- **Every 3 years** – measurements of leakage current of the output circuit and of the enclosure of the medical IT transformers in the no-load condition.

The dates and results of each verification **shall** be recorded.

## 7.3.11 Mobile and transportable units

For the purposes of this section, the term 'unit' is intended to mean a vehicle and/or mobile (self-propelled or towed) or transportable structure (such as a container or cabin) in which all or part of an electrical installation is contained, and which is provided with a temporary supply by means of, for example, a plug and socket-outlet.

### 7.3.11.1 Protective measures

The protective measures of obstacles, placing out of reach, non-conducting location and Earth-free local equipotential bonding are **not** permitted.

Automatic disconnection of the supply shall be provided by means of an RCD.

#### 7.3.11.1.1 Protective equipotential bonding

| | |
|---|---|
| Accessible conductive parts of the unit, such as the chassis, shall be connected through the main protective bonding conductors to the main earthing terminal within the unit. | WR-717.411.3.1.2 |
| The main protective bonding conductors shall be finely stranded. | WR-717.411.3.1.2 |

#### 7.3.11.1.2 IT system

| | |
|---|---|
| Additional protection by an RCD shall be provided for every socket-outlet intended to supply current-using equipment outside the unit, with the exception of | WR-717.415 |

socket-outlets that are supplied from circuits with protection by:

- SELV; or
- PELV; or
- electrical separation.

### 7.3.11.1.3 TN System

A PME earthing facility shall not be used as a means     WR-717.411.4
of earthing, except:

- where the installation is continuously under
  the supervision of a skilled or instructed person;
  and
- the suitability and effectiveness of the means of
  earthing has been confirmed before the connection is made.

### 7.3.11.2 *Selection and erection of equipment*

### 7.3.11.2.1 Identification of equipment

Identification – a permanent notice shall be fixed to     WR-717.514
the unit in a prominent position, preferably adjacent to
each supply inlet connector. The notice should state in
clear and unambiguous terms the following:

- the type of supplies that may be connected to the
  unit;
- the voltage rating of the unit;
- the number of supplies, phases and their
  configuration;
- the on-board earthing arrangement;
- the maximum power requirement of the unit.

### 7.3.11.2.2 Plugs and connectors

Connecting devices used to connect the unit to the     WR-717.55.1
supply shall be:

- within an enclosure of insulating material (with at least a degree of IP55); and
- afford a degree of protection not less than IP44.

### 7.3.11.2.3 Proximity to non-electrical services

| | |
|---|---|
| No electrical equipment (including wiring systems), except extra-low-voltage equipment for gas-supply control, shall be installed in any gas cylinder storage compartment. | WR-717.528.3.5 |
| Where cables have to run through such a compartment, they shall be protected against mechanical damage by installation within a conduit or a ducting system. | WR-717.528.3.4 |

Extra-low-voltage cables and electrical equipment may only be installed within the LPG cylinder compartment **if** the installation indicates the operation of the gas cylinder (e.g. indication of empty gas cylinders) or is for use within the compartment.

 Such electrical installations and components shall be constructed and installed so that they are not a potential source of ignition.

| | |
|---|---|
| Where installed, this conduit or ducting system shall be able to withstand an impact equivalent to AG3 without visible physical damage. | WR-717.528.3.5 |

### 7.3.11.2.4 Socket-outlets

| | |
|---|---|
| Socket-outlets located outside the unit shall be provided with an enclosure affording a degree of protection not less than IP44. | WR-717.55.2 |

### 7.3.11.3 *Supplies*

 Generating sets that are able to produce voltages other than SELV or PELV (and which are mounted in a mobile unit) shall be automatically switched off in case of an accident to the unit.

### 7.3.11.3.1  IT system

| | |
|---|---|
| An IT system can be provided by:<br><br>• an isolating transformer or a low-voltage generating set, with an IMD or an insulation fault location system installed; or<br>• a transformer providing simple separation, with an RCD and an Earth electrode installed to provide automatic disconnection in the case when the transformer fails. | WR-717.411.6.2 |

### 7.3.11.4  Wiring systems

| | |
|---|---|
| Flexible cables (for connecting the unit to the supply), or cables of equivalent design, having a minimum cross-sectional area of 2.5 mm² copper shall be used. | WR-717.52.1 |
| The flexible cable shall enter the unit by an insulating inlet in such a way as to minimize the possibility of any insulation damage or fault that might energize the exposed-conductive-parts of the unit. | WR-717.52.1 |
| The wiring system shall be installed using one or more of the following:<br><br>• unsheathed flexible cable with thermoplastic or thermosetting insulation installed in either a conduit or trunking or ducting;<br>• sheathed flexible cable thermoplastic or thermosetting insulation. | WR-717.52.2 |
| Where cables have to run through such a compartment, they shall be protected against mechanical damage by installation within a conduit system or within a ducting system. | WR-717.528.3.4 |
| Where installed, this conduit or duct shall be able to withstand an impact equivalent to AG3 without visible physical damage. | WR-717.528.3.5 |

### 7.3.12  Operating and maintenance gangways

Section 729 is a new (fairly small) addition to the standard which centres on the operation and safe maintenance of switchgear and controlgear within areas that included gangways and where access is restricted to skilled or instructed persons.

### 7.3.12.1 Restricted access areas

Restricted access areas shall:                                    WR-729.3

- be clearly and visibly marked by appropriate signs;
- not provide access to unauthorized persons; and
- provide closed doors that, although normally in the closed position, nevertheless will allow easy evacuation in case of danger without the use of a key, tool or any other device that is not part of the opening mechanism.

### 7.3.12.2 Requirements for operating and maintenance gangways

The width of gangways and access areas shall be adequate for work, operational access, emergency access, emergency evacuation and for transport of equipment.

Gangways shall permit at least a 90° opening of          WR. 729.513.2
equipment doors or hinged panels.

### 7.3.12.3 Restricted access areas where basic protection is provided by barriers or enclosures

Where basic protection is provided by barriers or enclosures, the minimum dimensions given in Table 7.5 apply (Figure 7.7).

**Table 7.5 Minimum gangway dimensions – barriers or enclosures**

| Gangway | Dimensions |
|---|---|
| Gangway width between barriers or enclosures and switch handles or circuit-breakers:<br><br>• in the most onerous position; and<br>• in the most onerous position and the wall | 700 mm |
| Gangway width between barriers or enclosures or other barriers or enclosures and the wall | 700 mm |
| Height of gangway to barrier or enclosure above floor | 2000 mm |
| Live parts placed out of reach | 2500 mm |

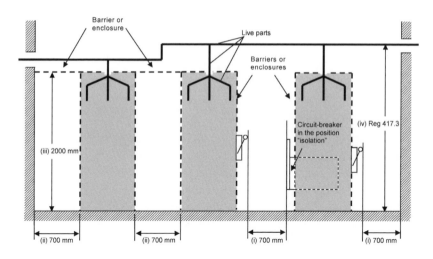

**Figure 7.7** Gangways in installations with protection by barriers or enclosures. (Courtesy of BSI.)

 **Note:** Where additional workspace is needed (e.g. for special switchgear and controlgear assemblies) larger dimensions may be required.

 **Note:** The above dimensions apply after barriers and enclosures have been fixed and with circuit-breakers and switch handles in the most onerous position, including 'isolation'.

### 7.3.12.3.1 Restricted access areas where the protective measure of obstacles is applied

Where basic protection is provided by obstacles, the minimum dimensions given in Table 7.6 apply (Figure 7.8).

**Table 7.6 Minimum gangway dimensions – obstacles (See note below)**

| Gangway | Dimensions |
|---|---|
| Gangway width between obstacles and switch handles or circuit-breakers: | 700 mm |
| • in the most onerous position; and <br> • in the most onerous position and the wall | |
| Gangway width between obstacles or other obstacles and the enclosures and the wall | 700 mm |
| Height of gangway to obstacles, barrier or enclosure above floor | 2000 mm |
| Live parts placed out of reach | 2500 mm |

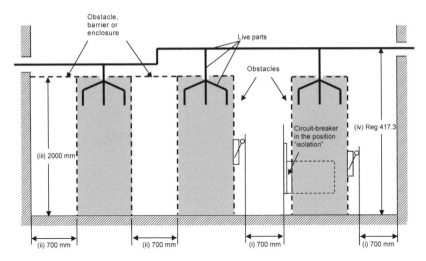

**Figure 7.8** Gangways in installations with protection by obstacles. (Courtesy of BSI.)

**Note:** Although these two figures and associated tables look very similar, there is a subtle change in the wording to indicate that one concerns protection by barriers and enclosures and the other for protection by obstacles (and barriers and enclosures)!

### 7.3.12.3.2 Access of gangways

| Gangways longer than 10 m shall be accessible from both ends. | WR-729.513.2.3 1 |
|---|---|

See Figure 729.3 of BS 7671:2008 (Incorporating Amendment No. 1) for examples of positioning of doors in closed restricted access.

## 7.3.13 Rooms and cabins containing saunas

Saunas, similar to bathrooms and showers, are primarily used by people who are unclothed and wet, and thus very vulnerable to electric shock due to their reduced body resistance (i.e. absence of shoes means less protection from shock, while water on the skin will tend to short-circuit its natural protection). Special measures are, therefore, needed to ensure that the possibility of direct and/or indirect contact is reduced.

The Regulations requirements for hot air saunas are based on four zones, which take into account the limitations of walls, doors, fixed partitions, ceilings and floors and the electric heater itself. The four zones  are shown in Figure 7.9.

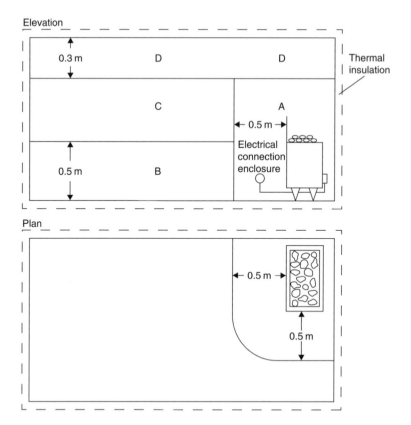

**Figure 7.9** Zones of ambient temperature.

**Zone 1**  Zone 1 is the volume containing the sauna heater, and is limited by the floor, the cold side of the thermal insulation of the ceiling, and a vertical surface circumscribing the sauna heater at a distance 0.5 m from the surface of the heater. If the sauna heater is located closer than 0.5 m to a wall, then zone 1 is limited by the cold side of the thermal insulation of that wall. In zone 1, only the sauna heater and equipment belonging to the sauna heater shall be installed.

**Zone 2**  Zone 2 is the volume outside zone 1, and is limited by the floor, the cold side of the thermal insulation of the walls, and a horizontal surface located 1.0 m above the floor. In zone 2 there is no special requirement concerning the heat resistance of equipment.

**Zone 3**  Zone 3 is the volume outside zone 1, and is limited by the cold side of the thermal insulation of the ceiling and walls, and a horizontal surface located 1.0 m above the floor. In zone 3 the equipment shall withstand a minimum temperature of 125°C and the insulation and sheaths of cables shall withstand a minimum temperature of 170°C.

### 7.3.13.1 Protective measures

The protective measures of:

- obstacles and placing out of reach; and
- non-conducting location and Earth-free local equipotential bonding;

are **not** permitted.

### 7.3.13.1.1 External influences

| | |
|---|---|
| The equipment shall have a degree of protection of at least IPX4. | WR-703.512.2 |
| If water jets are going to be used to clean the sauna, electrical equipment shall have a degree of protection of at least IPX5. | WR-703.512.2 |

### 7.3.13.1.2 Protection by RCDs

| | |
|---|---|
| Additional protection shall be provided for all circuits of the sauna by the use of one or more RCDs. | WR-703.411.3.3 |

### 7.3.13.2 Selection and erection of equipment

### 7.3.13.2.1 Isolation, switching, control and accessories

Socket-outlets **shall not** be installed within the location containing the sauna heater.

| | |
|---|---|
| Switchgear and controlgear that forms part of the sauna heater equipment or of other fixed equipment installed in zone 2, may be installed within the sauna room or cabin. | WR-703.537.5 |
| Other switchgear and controlgear (e.g. for lighting) shall be placed outside the sauna room or cabin. | WR-703.537.5 |

### 7.3.13.3 Supplies

### 7.3.13.3.1 Extra-low voltage provided by SELV or PELV

| | |
|---|---|
| Where SELV or PELV is used, whatever the nominal voltage, basic protection shall be provided by: | WR-703.414.4.5 |

- covering all live parts with an insulation that can only be removed by destruction; or
- by barriers or enclosures affording a degree of protection of at least IP4X or IPXXD.

### 7.3.13.4 Wiring systems

 Metallic sheaths and metallic conduits **shall not** be accessible in normal use.

| | |
|---|---|
| The wiring system should preferably be installed outside the zones (i.e. on the cold side of the thermal insulation). | WR-703.52 |
| If the wiring system is installed on the warm side of the thermal insulation in zones 1 or 3, then it must be heat-resisting. | WR-703.52 |

## 7.3.14  Solar, photovoltaic (PV) power supply systems

Photovoltaics is a technology that converts light directly into electricity by using photons from sunlight to knock electrons into a higher state of energy, thereby creating electricity. Solar cells are packaged in PV modules (often electrically connected in multiples as solar PV arrays) and convert energy from the sun into electricity.

### 7.3.14.1 Operational conditions and external influences

PV modules:

- may be connected in series up to the maximum allowed operating voltage of the PV modules and the PV convertor, whichever is lower;
- shall be installed in such a way that there is adequate heat dissipation under conditions of maximum solar radiation for the site.

### 7.3.14.2 Protective measures

Where protective bonding conductors are installed, they shall be parallel to (and in as close contact as possible with) the d.c. cables, a.c. cables and accessories.

### 7.3.14.2.1 Double or reinforced insulation

| | |
|---|---|
| Protection by the use of Class II or equivalent insulation is the preferred option on the d.c. side. | WR-712.412 |

## 7.3.14.2.2 Extra-low voltage provided by SELV or PELV

For SELV and PELV systems, $U_{oc\,STC}$ replaces $U_o$    WR-712.414.1.1
and shall not exceed 120 V d.c.

### 7.3.14.3 Selection and erection of equipment

The selection and erection of equipment shall enable safe maintenance.

## 7.3.14.3.1 Protection against electromagnetic interference in buildings

To minimize voltages induced by lightning, the    WR-712.444.4.4
area of all wiring loops shall be as small as possible.

### 7.3.14.4 Supplies

Electrical equipment on the d.c. side shall be    WR-712.512.1.1
suitable for direct voltage and direct current.

If blocking diodes are used, they shall be con-    WR-712.512.1.1
nected in series with the PV strings and their
reverse voltage rated for $2U_{oc\,STC}$ of the PV
string.

PV equipment on the d.c. side shall be consid-    WR-712.410.3
ered to be energized, even when the system is
disconnected from the a.c. side.

On the a.c. side, the PV supply cable shall be    WR-712.411.3.2.1.1
connected to the supply side of the protective
device for automatic disconnection of circuits
supplying current-using equipment.

To allow maintenance of the PV convertor, the    WR-712.537.2.1.1
PV convertor should be capable of being iso-
lated both from the d.c. side and the a.c. side.

Where an electrical installation includes a PV    WR-712.411.3.2.1.2
power supply system that does not possess a sim-
ple separation between the a.c. side and the d.c.
side, a type B RCD shall be installed to provide
fault protection by automatically disconnecting
the supply.

> Note: If the PV convertor is not con-
> structed to be able to feed d.c. fault cur-
> rents into the electrical installation, an
> RCD of type B is not required.

The protective measures of non-conducting location and Earth-free local equipotential bonding are **not** permitted on the d.c. side

### 7.3.14.4.1 Devices for isolation

| | |
|---|---|
| In the selection and erection of devices for isolation and switching to be installed between the PV installation and the public supply, the public supply shall be considered the source and the PV installation shall be considered the load. | WR-712.537.2.2.1 |
| A switch disconnector shall be provided on the d.c. side of the PV convertor. | WR-712.537.2.2.5 |
| All junction boxes (PV generator and PV array boxes) shall carry a warning label indicating that parts inside the boxes may still be live after isolation from the PV convertor. | WR-712.537.2.2.5.1 |

### 7.3.14.4.2 Earthing arrangements

| | |
|---|---|
| Earthing of one of the live conductors of the d.c. side is permitted, provided that there is at least simple separation between the a.c. side and the d.c. side. | WR-712.312.2 |

**Note:** Any connections with Earth on the d.c. side should be electrically connected so as to avoid corrosion (see BS 7361–1:1991).

### 7.3.14.4.3 Protection against fault current

| | |
|---|---|
| The PV supply cable on the a.c. side shall be protected against fault current by an overcurrent protective device installed at the connection point to the a.c. mains. | WR-712.434.1 |

## 7.3.14.4.4 Protection against overload on the d.c. side

Overload protection may be omitted:

- for PV string and PV array cables when the con-   WR-712.433.1
tinuous current-carrying capacity of the cable
is equal to or greater than $1.25I_{sc\ STC}$ at any
location;

- for the PV main cable if the continuous cur-   WR-712.433.2
rent-carrying capacity is equal to or greater than
$1.25I_{sc\ STC}$ of the PV generator.

### 7.3.14.5 Wiring systems

### 7.3.14.5.1 Selection and erection of wiring systems

PV string cables, PV array cables and PV d.c. main   WR-712.522.8.1
cables shall be selected and erected so as to minimize the risk of Earth faults and short-circuits.

Wiring systems shall withstand the expected exter-   WR-712.522.8.3
nal influences such as wind, ice formation, temperature and solar radiation.

## 7.3.15 Swimming pools and other basins

The particular requirements of this section apply to the basins of swimming
pools, the basins of fountains and the basins of paddling pools, and to the
surrounding zones of these basins. Swimming pools are, by design, wet areas,
and people using them are normally wet, which will increase their vulnerability
to electric shock. Special measures are, therefore, needed to ensure that all
possibility of direct and/or indirect contact is reduced.

**Note:** Except for areas especially designed as swimming pools, the requirements
of this section do not apply to natural waters, lakes in gravel pits, coastal areas
and the like.

Special requirements may be necessary for swimming pools for medical
purposes.

### 7.3.15.1 Classification of external influences

As shown in Figure 7.10, these requirements concern the dimensions of three
zones .

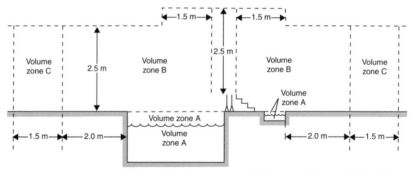

Note: The dimensions are measured taking account of walls and fixed partitions

**Figure 7.10** Swimming and paddling pool zone dimensions.

 Zones 1 and 2 may be limited by fixed partitions having a minimum height of 2.5 m.

**Zone 0**   This zone is the interior of the basin of the swimming pool or fountain, including any recesses in its walls or floors, basins for foot cleaning, and waterjets or waterfalls and the space below them.

**Zone 1**   This zone is limited by:

- zone 0;
- a vertical plane 2 m from the rim of the basin;
- the floor or surface expected to be occupied by persons;
- the horizontal plane 2.5 m above the floor or the surface expected to be occupied by persons.

Where the swimming pool or fountain contains diving boards, springboards, starting blocks, chutes or other components expected to be occupied by persons, zone 1 comprises the zone limited by:

- a vertical plane situated 1.5 m from the periphery of the diving boards, springboards, starting blocks, chutes and other components such as accessible sculptures, viewing bays and decorative basins;
- the horizontal plane 2.5 m above the highest surface expected to be occupied by persons.

**Zone 2**   This zone is limited by:

- the vertical plane external to zone 1 and a parallel plane 1.5 m from the former;
- the floor or surface expected to be occupied by persons;

- the horizontal plane 2.5 m above the floor or surface expected to be occupied by persons.

 There is no zone 2 for fountains.

 **Note:** For a swimming pool where it is not possible to locate a socket-outlet or switch outside zone 1, a socket-outlet or switch (preferably with a non-conductive cover or coverplate) is permitted in zone 1 **if**:

- it is installed at least 1.25 m horizontally from the border of zone 0;
- is placed at least 0.3 m above the floor, and is protected by:

  o SELV; or
  o automatic disconnection of supply; or
  o electrical separation.

### 7.3.15.2 Protective measures

 The protective measures of:

- obstacles;
- placing out of reach;
- non-conducting location; and
- Earth-free local equipotential bonding

are **not** permitted.

### 7.3.15.2.1 Protective measures against electric shock

### 7.3.15.2.1.1 Zones 0 and 1 (swimming pools and other basins)

 Except for fountains in zone 0 and zone 1, **only** protection by SELV is permitted, with the SELV source installed outside of zones 0, 1 and 2.

| | |
|---|---|
| All extraneous-conductive-parts in zones 0, 1 and 2 shall be connected by supplementary protective bonding conductors to the protective conductors of exposed-conductive-parts of equipment situated in these zones. | WR-702.415.2 |
| Any equipment that is located in the interior of a basin that is only intended to be in operation when people are not inside zone 0, shall be supplied by a circuit protected by: | WR-702.410.3.4.1 |

- SELV;
- automatic disconnection of the supply using an RCD; or
- electrical separation.

The socket-outlet of a circuit supplying such equipment (and similarly the control device of such equipment) shall have a notice warning users that this equipment shall be used only when the swimming pool is not occupied by persons.

WR-702.410.3.4.1

### 7.3.15.2.1.2 Zones 0 and 1 of fountains

In zones 0 and 1 of fountains, one or more of the following protective measures shall be employed:

WR-702.410.3.4.2

- SELV;
- automatic disconnection of supply using an RCD;
- electrical separation.

### 7.3.15.2.1.3 Zone 2 (swimming pools and other basins)

One or more of the following protective measures shall be employed:

WR-702.410.3.4.3

- SELV;
- automatic disconnection of supply;
- electrical separation.

### 7.3.15.2.2 Requirements for SELV and PELV circuits

Where SELV or PELV is used, whatever the nominal voltage, basic protection shall be provided by:

WR-702.414.4.5

- covering all live parts with an insulation that can only be removed by destruction; or
- by barriers or enclosures affording a degree of protection of at least IP4X or IPXXD.

### 7.3.15.3 *Selection and erection of equipment*

### 7.3.15.3.1 Current-using equipment of swimming pools

| | |
|---|---|
| In zones 0 and 1, it is only permitted to install fixed current-using equipment specifically designed for use in a swimming pool. | WR-702.55.1 |
| Equipment that is intended to be in operation only when people are outside zone 0 may be used in all zones, provided that it is supplied by a protected circuit. | WR-702.55.1 |
| It is permitted to install an electric heating unit embedded in the floor, provided that it: | WR-702.55.1 |

- is protected by SELV;
- incorporates an earthed metallic sheath connected to the supplementary equipotential bonding, and its supply circuit is additionally protected by an RCD; or
- is covered by an embedded earthed metallic grid connected to the supplementary equipotential bonding, and its supply circuit is additionally protected by an RCD.

### 7.3.15.3.2 Erection according to the zones

| | |
|---|---|
| In zones 0, 1 and 2, any metallic sheath (or metallic covering) of a wiring system shall be connected to the supplementary equipotential bonding. | WR-702.522.21 |

 **Note:** Cables should preferably be installed in conduits made of insulating material.

### 7.3.15.3.3 External influences

| | |
|---|---|
| Electrical equipment shall have at least the following degree of protection: | WR-702.512.2 |

- zone 0: IPX8;
- zone 1: IPX4, IPX5 (where water jets are likely to occur for cleaning purposes);

- zone 2:
  - IPX2 for indoor locations;
  - IPX4 for outdoor locations;
  - IPX5 where water jets are likely to occur for cleaning purposes.

### 7.3.15.3.4 Junction boxes

A junction box:

- shall not be installed in zone 0;                    WR- 702.522.24
- shall not be installed in zone 1 unless it is an     WR-702.522.24
  SELV circuit.

### 7.3.15.3.5 Special requirements for the installation of electrical equipment in zone 1 of swimming pools and other basins

Fixed equipment (such as filtration systems, jet stream pumps etc.) that is designed for use in swimming pools and other basins and which is supplied at low voltage, is permitted in zone 1, subject to all the following requirements:

- the equipment shall be located inside an insulat-    WR-702.55.4
  ing enclosure providing at least Class II or equivalent insulation and providing protection against
  mechanical impact of medium severity (AG2);

**Note:** This Regulation applies irrespective of the classification of the equipment.

- the equipment shall only be accessible via a hatch    WR-702.55.4
  (or a door) by means of a key or a tool which shall
  disconnect all live conductors and the supply
  cable (and the main disconnecting means shall
  be installed in a way which provides protection of
  Class II or equivalent insulation);
- the supply circuit of the equipment shall be protected by:
- SELV; or
- an RCD; or
- electrical separation.

Switchgear, controlgear and socket-outlets **shall not** be installed in zones 0 or

| | |
|---|---|
| In zone 2, a socket-outlet or a switch is permitted only if the supply circuit is protected by one of the following protective measures:<br><br>• SELV;<br>• automatic disconnection of supply using an RCD;<br>• electrical separation. | WR-702.53 |

 For a swimming pool where it is not possible to locate a socket-outlet or switch outside zone 1, a socket-outlet or switch, preferably having a non-conductive cover or coverplate, is permitted in zone 1 if it is installed outside (1.25 m) the border of zone 0, is placed at least 0.3 m above the floor, and is protected by:

- SELV; or
- automatic disconnection of supply using an RCD; or
- electrical separation.

### 7.3.15.3.6 Underwater luminaires for swimming pools

| | |
|---|---|
| A luminaire for use in the water or in contact with the water shall be fixed and shall comply with BS EN 60598–2–18. | WR-702.55.2 |
| Underwater lighting located behind watertight portholes, and serviced from behind, shall be installed in such a way that no intentional or unintentional conductive connection between any exposed-conductive-part of the underwater luminaires and any conductive parts of the portholes can occur. | WR-702.55.2 |

### 7.3.15.4 Wiring systems

### 7.3.15.4.1 Additional requirements for the wiring of fountains

| | |
|---|---|
| For a fountain, the following additional requirements shall be met:<br><br>• a cable for electrical equipment in zone 0 shall be installed as far outside the basin rim as is reasonably practicable, and run to the electrical | WR-702.522.23 |

> equipment inside zone 0 by the shortest practi-
> cable route;
> - in zone 1, a cable shall be selected, installed
>   and provided with mechanical protection to
>   medium severity (AG2) and the relevant sub-
>   mersion in water depth (AD8).

### 7.3.15.4.2 Electrical equipment of fountains

| | |
|---|---|
| Electrical equipment in zones 0 or 1 shall be provided with mechanical protection to medium severity (AG2) – for example, by using a mesh glass or grids that can only be removed by the use of a tool. | WR-702.55.3 |
| A luminaire installed in zones 0 or 1 shall be fixed and shall comply with BS EN 60598–2–18. | WR-702.55.3 |
| An electric pump shall comply with the requirements of BS EN 60335–2–41. | WR-702.55.3 |

### 7.3.15.4.3 Limitation of wiring systems according to the zones

| | |
|---|---|
| In zones 0 and 1, a wiring system shall be limited to that necessary to supply equipment situated in these zones. | WR-702.522.22 |

## 7.3.16 Temporary electrical installations for structures, amusement devices and booths at fairgrounds, amusement parks and circuses

This section specifies the minimum electrical installation requirements for the safe design, installation and operation of temporarily erected mobile or trans-portable electrical machines and structures that incorporate electrical equipment. These machines and structures are typically installed repeatedly and temporarily, at fairgrounds, amusement parks, circuses and/or similar places.

### 7.3.16.1 Protective measures

The protective measures of:

- obstacles;
- placing out of reach;

- non-conducting location; and
- Earth-free local equipotential bonding;

are **not** permitted.

### 7.3.16.1.1  RCDs

| | |
|---|---|
| All final circuits for:<br><br>• lighting;<br>• socket-outlets rated up to 32 A; and<br>• mobile equipment connected by means of a flexible cable or cord with a current-carrying capacity up to 32 A;<br><br>shall be protected by RCDs | WR-740.415.1 |
| The supply to a battery-operated emergency lighting circuit shall be connected to the same RCD protecting the lighting circuit. | WR-740.415.1 |
|  This requirement does not apply to:<br><br>• circuits protected by SELV or PELV; or<br>• circuits protected by electrical separation; or<br>• lighting circuits placed out of arm's reach. | |

### 7.3.16.1.2  Protection against electric shock

| | |
|---|---|
| Automatic disconnection of supply to the temporary electrical installation shall be provided at the origin of the installation by one or more RCDs with a rated residual operating current not exceeding 300 mA. | WR-740.410.3 |
| Placing out of arm's reach is acceptable for electric dodgems. | WR-740.410.3 |

### 7.3.16.1.3  Protection against thermal effects

| | |
|---|---|
| A motor that is automatically or remotely controlled and that is not continuously supervised shall be fitted with a manually reset protective device against excess temperature. | WR-740.422.3.7 |

### 7.3.16.1.4 Supplementary equipotential bonding

- Extraneous-conductive-parts in, or on, the floor (such as concrete reinforcement in general, or reinforcement of cellars for liquid manure) shall be connected to the supplementary equipotential bonding.    WR-740.415.2.1
- In locations intended for livestock, supplementary bonding shall connect all exposed-conductive-parts and extraneous-conductive-parts that can be touched by livestock.
- It is recommended that spaced floors made of prefabricated concrete elements be part of the equipotential bonding.
- The supplementary equipotential bonding and the metal grid, if any, shall be erected so that it is durably protected against mechanical stresses and corrosion.
- Where a metal grid is laid in the floor, it shall be included within the supplementary bonding of the location (see Figure 7.2).

### 7.3.16.2 Selection and erection of equipment

### 7.3.16.2.1 Electrical connections

Joints shall not be made in cables except where necessary as a connection into a circuit.    WR-740.526

If joints are required, they shall either use connectors in accordance with the relevant British Standard, or made in an enclosure with a degree of protection of at least IP4X or IPXXD.

Where strain can be transmitted to terminals, the connection shall incorporate cable anchorage(s).    WR-740.526

 The electrical installation between its origin and the electrical equipment shall be inspected and tested after it has been assembled on site.

### 7.3.16.2.2 Electric discharge lamp installations

The location of a luminous tube, sign or lamp shall be installed out of arm's reach or shall be adequately protected to reduce the risk of injury to persons.    WR-740.55.3.1

### 7.3.16.2.3 Emergency switching device

A separate circuit shall be used to supply luminous tubes, signs or lamps, which shall be controlled by an emergency switch.    WR-740.55.3.2

The switch shall be easily visible, accessible and marked in accordance with the requirements of the local authority.    WR-740.55.3.2

### 7.3.16.2.4 Floodlights

Where transportable floodlights are used, they shall be mounted so that the luminaire is inaccessible.    WR-740.55.1.4

Supply cables shall be flexible and have adequate protection against mechanical damage.    WR-740.55.1.4

 Luminaires and floodlights shall be so fixed and protected that a focusing or concentration of heat is not likely to cause ignition of any material.

### 7.3.16.2.5 Lampholders

Insulation-piercing lampholders shall not be used unless the cables and lampholders are compatible and the lampholders are non-removable once fitted to the cable.    WR-740.55.1.2

### 7.3.16.2.6 Lamps in shooting galleries

All lamps in shooting galleries and other sideshows where projectiles are used shall be suitably protected against accidental damage.    WR-740.55.1.3

### 7.3.16.2.7 Luminaires

| | |
|---|---|
| Every luminaire and decorative lighting chain shall: | WR-740.55.1.1 |
| • have a suitable IP rating; | |
| • be installed so as not to impair its ingress protection; and | |
| • be securely attached to the structure or support intended to carry it. | |
| Its weight shall not be carried by the supply cable, unless it has been selected and erected for this purpose. | WR-740.55.1.1 |
| Luminaires and decorative lighting chains mounted less than 2.5 m (arm's reach) above floor level or otherwise accessible to accidental contact, shall be firmly fixed and so sited or guarded as to prevent risk of injury to persons or ignition of materials. | WR-740.55.1.1 |
| Access to the fixed light source shall only be possible after removing a barrier or an enclosure which shall require the use of a tool. | WR-740.55.1.1 |
| Lighting chains shall use HO5RN-F (BS 7919) cable or equivalent. | WR-740.55.1.1 |

### 7.3.16.2.8 Selection and erection of equipment

| | |
|---|---|
| Switchgear and controlgear shall be placed in cabinets that can be opened only by the use of a key or a tool, except for those parts designed and intended to be operated by ordinary persons (see Appendix 5 to BS 7671:2008 (Incorporating Amendment No. 1)). | WR-740.51 |
| Electrical equipment shall have a degree of protection of at least IP44. | WR-740.512.2 |

### 7.3.16.2.9 Socket-outlets and plugs

| | |
|---|---|
| An adequate number of socket-outlets shall be installed to allow the user's requirements to be met safely. | WR-740.55.7 |

**Note:** In booths, stands and for fixed installations, one socket-outlet for each square metre or linear metre of wall is generally considered adequate.

### 7.3.16.2.10 Switchgear and controlgear - isolation

Every electrical installation of a booth, stand or amusement device shall have its own means of isolation, switching and overcurrent protection,

WR-740.537.1

 which shall be readily accessible.

Every separate temporary electrical installation for amusement devices and each distribution circuit supplying outdoor installations shall be provided with its own readily accessible and properly identified means of isolation.

WR-740.537.2.1.1

A device for isolation shall disconnect all live conductors (line and neutral conductors).

WR-740.537.2.2

### 7.3.16.3 Supplies

### 7.3.16.3.1 Electrical supply

At each amusement device, there shall be a connection point readily accessible and permanently marked to indicate the following essential characteristics:

WR-740.55.8

*   rated voltage;
*   rated current;
*   rated frequency.

 Where an alternative system is available, an IT system **shall not** be used.

### 7.3.16.3.2 Automatic disconnection of supply

For supplies to a.c. motors, RCDs (where used) should be of the time-delayed type or of the S-type, to prevent unwanted tripping.

WR-740.411

### 7.3.16.3.3 Electric dodgems

| | |
|---|---|
| Electric dodgems shall only be operated at voltages not exceeding 50 V a.c. or 120 V d.c. | WR-740.55.9 |
| The circuit used for electric dodgems shall be electrically separated from the supply mains by means of a transformer or a motor-generator set. | WR-740.55.9 |

### 7.3.16.3.4 Low-voltage generating sets – generators

| | |
|---|---|
| All generators shall be so located or protected to prevent any danger and injury to people by inadvertent contact with hot surfaces and dangerous parts. | WR-740.551.8 |
| Electrical equipment associated with the generator shall be mounted securely and, if necessary, on anti-vibration mountings. | WR-740.551.8 |
| Where a generator supplies a temporary installation that is part of a TN, TT or IT system, care shall be taken to ensure that the earthing arrangements are in accordance with the Regulations. | WR-740.551.8 |
| The neutral conductor of the star-point of the generator shall, except for an IT system, be connected to the exposed conductive-parts of the generator. | WR-740.551.8 |

### 7.3.16.3.5 Safety isolating transformers and electronic convertors

| | |
|---|---|
| • A manually reset protective device shall protect the secondary circuit of each transformer or electronic convertor.<br>• Access by competent persons for testing (or by a skilled person competent in such work for protective device maintenance) shall be provided.<br>• Enclosures containing rectifiers and transformers shall be adequately ventilated, and the vents shall not be obstructed when in use.<br>• Electronic converters shall conform to BS EN 61347–2–2. | WR-740.55.5 |

- Safety isolating transformers shall comply with BS EN 61558–2–6 or provide an equivalent degree of safety.
- Safety isolating transformers shall be mounted out of arm's reach or be mounted in a location that provides equal protection, and shall have adequate ventilation.

### 7.3.16.3.6 Supply from the public network

Irrespective of the number of sources of supply, the line and neutral conductors from different sources shall not be interconnected downstream of the origin of the temporary electrical installation.    WR-740.313.3

### 7.3.16.3.7 TN system

 The ESQCR prohibit the use of a PME earthing facility as the means of earthing a caravan or similar construction.

A PME earthing facility shall not be used as a means of earthing in a TN system.    WR-740.411.4

### 7.3.16.3.8 Voltage

The nominal supply voltage of temporary electrical installations in booths, stands and amusement devices shall not exceed 230/400 V a.c. or 440 V d.c.    WR-740.313.1.1

### *7.3.16.4 Wiring systems*

### 7.3.16.4.1 Cables and cable management systems

- All cables shall meet the requirements of BS EN 60332–1–2.    WR-740.521.1
- Armoured cables or cables protected against mechanical damage shall be used wherever there is a risk of mechanical damage due to external influence (e.g. above AG2).

- Buried cables shall be protected against mechanical damage.
- Cables shall have a minimum rated voltage of 450/750 V.
- Cable trunking systems and cable ducting systems shall comply with the relevant part 2 of BS EN 50085.
- Conduit systems shall comply with the relevant part of the BS EN 61386 series.
- Flexible conduit systems shall comply with BS 7671:2008 (Incorporating Amendment No. 1) and BS EN 61386–23.
- Mechanical protection shall be used in public areas and in areas where wiring systems are crossing roads or walkways.
- The routes of cables buried in the ground shall be marked at suitable intervals.
- Tray and ladder systems shall comply with BS EN 61537.
- Where subjected to movement, wiring systems shall be of flexible construction.

# 8

# External influences

While Chapter 51 of BS 7671:2008 (Incorporating Amendment No. 1) still contains the requirements for external influences (see below), Amendment No. 1 has now been further developed in accordance with new ISO standards and the BS EN 60721 and BS EN 61000 series on environmental conditions:

> "512.2   External influences
>
> 512.2.1   Equipment shall be of a design appropriate to the situation in which it is to be used or its mode of installation shall take account of the conditions likely to be, encountered.
>
> 512.2.2   If the equipment does not, by its construction, have the characteristics relevant to the external influences of its location, it may nevertheless be used be used on condition that it is provided with appropriate additional protection in the erection of the installation. Such protection shall not adversely affect the operation of the equipment thus protected.
>
> 512.2.3   Where different external influences occur simultaneously, they may have independent or mutual effects and the degree of protection shall be provided accordingly.
>
> 512.2.4   The selection of equipment according to external influences is necessary not only for proper functioning, but also to ensure the reliability of the measures of protection for safety complying with these Regulations generally. Measures of protection afforded by the construction of the equipment are valid only for the given conditions of external influence if the corresponding equipment specifications are made in these conditions of external influence."

It has also been determined that, for the purpose of these Regulations, the following classes of external influence are conventionally regarded as normal (i.e. that the requirement must generally satisfy applicable standards):

| | | |
|---|---|---|
| AA | Ambient temperature | AA4 |
| AB | Atmospheric humidity | AB4 |
| AC to AS | Other environmental conditions | XX1 of each parameter |
| B and C | Utilization and construction of buildings | XX1 of each parameter, except XX2 for the parameter BC |

**Note:** A list of external influences and their characteristics has been included as Appendix 5 in BS 7671:2008 (Incorporating Amendment No. 1), and the following notes concerning external influences are offered as guidance. Also included in this chapter are extracts from the current Regulations that have an impact on the environment.

# 8.1 Environmental factors and influences

The actual environment to which equipment is likely to be exposed is normally complex and comprises a number of environmental conditions. When defining the conditions for a certain application it is, therefore, necessary to consider all environmental influences that may be as a result of:

- conditions from the surrounding medium;
- conditions caused by the structure in which the equipment is situated or attached;
- influences from external sources or activities.

## 8.1.1 Combined environmental factors

Equipment may, of course, be simultaneously exposed to a large number of environmental factors and corresponding parameters. Some of the parameters are statistically dependent (e.g. low air velocity and low temperature; sun radiation and high temperature). Other parameters are statistically independent (e.g. vibration and temperature). The effect of a combination of environmental factors is, therefore, extremely important, and has to be considered during manufacture and operation.

## 8.1.2 Sequences of environmental factors

Certain effects of exposing electrical equipment and electrical installations to environmental conditions are a direct result of two or more factors, or parameters, happening either simultaneously or after each other (e.g. thermal shock caused when exposing equipment to a high temperature immediately after exposing it to a low temperature). These possibilities must always be taken into account when designing and installing electrical equipment.

## 8.1.3 Environmental application

Although the conditions affecting electrical equipment mainly consist of the environment (ambient and created), consideration must also has to be given to where the equipment will be operating from and how it will be used.

For simplicity this can be broken down into two basic categories:

Conditions     The environmental conditions that have been identified as
               having an effect on equipment (Table 8.1).

Situations     The main uses of electronic equipment (Table 8.2).

## 8.1.4 Environmental conditions

There are eight basic conditions that have a direct effect on electrical equipment and electrical installations. These are listed in Figure 8.1 and Table 8.1.

Table 8.1 Basic conditions

| Climatic | Externally generated influences |
|---|---|
| • Altitude ambient temperature<br>• Ambient temperature<br>• Atmospheric pressure<br>• Condensation<br>• Precipitation (i.e. rain, snow and hail)<br>• Relative atmospheric humidity<br>• Solar radiation<br>• Wind | • Air movement<br>• Dust<br>• Temperature<br>• Precipitation (e.g. water spray)<br>• Pressure changes (e.g. tunnels) |

| Mechanical | Ergonomic aspects |
|---|---|
| • Shock (sinusoidal and random)<br>• Vibration | • Achieving maximum task effectiveness<br>• Protecting the health of the engineer and end user<br>• The comfort of the operator and end user |

| Electrical | Chemical |
|---|---|
| • Earthing and bonding<br>• Electromagnetic environment (electromagnetic compatibility (EMC) and electromagnetic interference (EMI))<br>• Power supplies<br>• Susceptibility and generation<br>• Transients (spikes and surges) | • Corrosion<br>• Dangerous substances<br>• Pollution and contamination<br>• Resistance to solvents |

| Biological | General |
|---|---|
| • Animals<br>• Humans (vandalism)<br>• Vegetation | • Components<br>• Design of equipment<br>• Earthquakes<br>• Flammability and fire hazardous areas<br>• Maintainability<br>• Safety<br>• Reliability<br>• Waste |

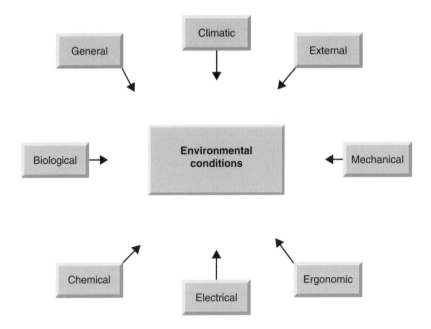

**Figure 8.1** Environmental conditions.

**Table 8.2 Environmental conditions**

| Operational | Storage | Transportation |
|---|---|---|
| • When installed and operational<br>• When installed and not in use | • When in storage | • When being transported |

### 8.1.5 Equipment situations

Obviously not all equipment will be fully operational all of the time, and so various equipment 'situations' also have to be considered (Table 8.2).

As a reference on the whole topic of environmental requirements, the following book, *Environmental Requirements for Electromechanical and Electronic Equipment* (by the same author no less!) is recommended. This book has become known as the definitive reference for designers and manufacturers of electrical and electromechanical equipment worldwide.

### 8.1.6  Requirements from the Regulations

#### 8.1.6.1  General environmental conditions

| | |
|---|---|
| Electrical equipment shall be selected so as to withstand safely the stresses, the environmental conditions and the characteristics of its location. | WR-133.3 |

A coating of paint, varnish or similar product is generally **not** considered to comply with these requirements.

| | |
|---|---|
| The design of electrical equipment shall take into account the environmental conditions to which it will be subjected. | WR-132.5.1 |
| Equipment in surroundings susceptible to risk of fire or explosion shall be so constructed or protected, and such other special precautions shall be taken, as to prevent danger. | WR-132.5.2 |
| The installation shall have an adequate level of immunity against electromagnetic disturbances so as to function correctly in the specified environment. | WR-131.6.4 |

## 8.1.6.2 General external influences

| | |
|---|---|
| Equipment installed shall be appropriate to the external influences foreseen. | WR-530.3 |

## 8.1.6.2.1 Type of wiring and method of installation

| | |
|---|---|
| The choice of the type of wiring system and the method of installation shall include consideration of the following: | WR-132.7 |

- the nature of the location;
- the nature of the structure supporting the wiring;
- accessibility of wiring to persons and livestock;
- voltage;
- the electromechanical stresses likely to occur due to short-circuit and Earth fault currents;
- electromagnetic interference (EMI);
- other external influences (e.g. mechanical, thermal and those associated with fire) to which the wiring is likely to be exposed during the erection of the electrical installation or in service.

### 8.1.6.2.2 Electrical connections

> Where a connection is made in an enclosure, the enclo-    WR-526.7
> sure shall provide adequate mechanical protection and
> protection against relevant external influences.

### 8.1.6.2.3 Sealing of wiring system penetrations

Any sealing arrangement shall resist external influences to the same degree as the wiring system with which it is used and, in addition, it shall meet all of the following requirements:

- it shall be resistant to the products of combustion to the same extent as the elements of building construction that have been penetrated;
- it shall provide the same degree of protection from water penetration as that required for the building construction element in which it has been installed;
- it shall be compatible with the material of the wiring system with which it is in contact;
- it shall permit thermal movement of the wiring system without reduction of the sealing quality;
- it shall be of adequate mechanical stability to withstand the stresses that may arise through damage to the support of the wiring system due to fire.

## 8.2 Ambient temperature

Temperature, humidity, rainfall, wind velocity and the duration of sunshine all affect the climate of an area. These elements are, in turn, the result of the interaction of a number of determining causes, such as latitude, altitude, wind direction, distance from the sea, relief and vegetation. The elements and their determining causes are similarly interrelated, and also contribute to temperature changes. For example, the length of day is a factor that helps to determine temperature; however, the duration of actual sunshine is an element with far-reaching effects on plant and animal life.

Of all the elements that have an effect on man and equipment, none is more vital to living organisms than temperature. Temperature has a large influence on where humans live and the areas where they work. Protective housing or artificial heat sources may overcome low temperatures (and high altitudes) and, similarly, cooling devices and reflective coatings protect equipment from high temperatures. Temperature is, therefore, a particularly important aspect of the environment, and its accurate measurement and definition requires careful consideration.

The ambient temperature at any given time is the temperature of the air measured under standardized conditions, and with certain recognized precautions against errors introduced by radiation from the sun or other heated body. Temperature figures with respect to climate are generally 'shade' temperatures (i.e. the temperature of the air measured with due precautions taken to exclude the influence of the direct rays of the sun), and it is usual for the temperature to be much higher in the direct sunshine. Many mountain areas have air temperatures in the region of zero in winter, but the presence of bright sunshine will produce a feeling of warmth and permits the wearing of light clothing (Figure 8.2).

**Figure 8.2** Enjoying the environment.

Seasonal fluctuations in temperature do not pass below ground deeper than 60–80 ft. Below that depth, borings and mine-shafts show that the temperature increases (downwards), depending on the geographical position, location and depth.

On average, however, a rise of about 1°C may be taken for each 64 ft of descent. Assuming that this rate of increase is maintained, it stands to reason that the interior of the Earth must be excessively hot and, therefore, it must warm the surface to some extent (Figure 8.3). It is not possible to determine the precise influence of this temperature increase, but it has an effect on tunnels at a depth greater than 80 ft. As the heat from the core is virtually negligible on the surface of the Earth (compared with that of the sun), it has not been considered in this book, and the only source of heat that has been taken into account is the sun.

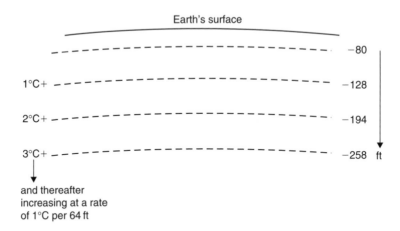

Earth's surface

1°C+ ---- −128

2°C+ ---- −194

3°C+ ---- −258 ft

and thereafter
increasing at a rate
of 1°C per 64 ft

**Figure 8.3** Temperature changes below the Earth's crust.

The difference between summer and winter temperatures for any locality is known as the 'annual range of temperature' or the 'absolute range of temperature' of that particular locality, and it is the difference between the highest and lowest temperatures ever experienced at the place in question. The maximum and minimum temperatures are obviously not the same every year, and, should their average over a series of years be taken, it would be known as the 'mean annual extreme range'.

Although air near the surface (especially at night) may be cooler than the air just above it, there is, generally speaking, a gradual falling off of temperature from the ground level up. Over thousands of feet this cooling averages 1°C per 300 ft, and thus at approximately 25,000–55,000 ft (5–10 miles) above the ground the temperature will be down to 55–60°C below zero. Above this the air temperature ceases to fall off regularly, in fact it may even rise for a bit. Usually, however, it remains fairly constant and, because of this, it is sometimes referred to as the 'isothermal layer' – but to meteorologists and airmen it is known as the stratosphere. The lower layers of the atmosphere (i.e. where temperature falls off with height) is known as the troposphere and the boundary layer between the two is called the tropopause (Figure 8.4).

### 8.2.1 Electrical installations

When equipment has been installed without any protection it can be expected to be exposed to more extreme air temperatures and more severe combinations of air temperatures and relative humidities than a similar piece of equipment that has been installed or housed in a temperature-controlled environment.

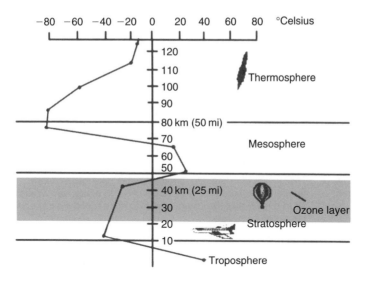

**Figure 8.4** Temperature changes above the Earth's surface.

In addition to open air temperature, the temperature stresses on equipment depend on a number of other environmental parameters (e.g. solar radiation, heating from adjacent equipment and air velocity), and these must be taken into account when designing, manufacturing and installing equipment.

The performance of equipment is also influenced and limited by the internal temperature of the equipment. Internal temperatures, in turn, depend on the external ambient conditions and the heat generated within the device itself. Indeed, whenever a temperature gradient exists within a system formed by a device and its surrounding environment, a process of heat transfer will follow.

Thus, in any generic (or system specific) specification or standard relating to ambient temperature it is necessary to consider the:

| | |
|---|---|
| **Operating temperature range** | The specified operating temperature for the equipment, which must always be the lowest and the highest ambient temperature expected to be experienced by the equipment during its normal operation. |
| **Storage temperature range** | The specified storage temperature, which is always the lowest and the highest ambient temperature that the equipment is expected to experience (with the power turned off) during storage or from exposure to climatic extremes. |

 **Note:** The equipment is not normally expected to be capable of operating at these extreme temperatures, merely survive them without damage.

## 8.2.2 Requirements from the Regulations – ambient temperature

The most common environmental requirements concerning ambient temperature are listed in Table 8.3.

**Table 8.3 Typical requirements – ambient temperature**

| | |
|---|---|
| Temperature ranges | Equipment will need to be designed and manufactured to meet the full performance specification requirement for the selected temperature category. |
| **Temperature increases** | The design of equipment should always take into account temperature increases within cubicles and equipment cases so as to ensure that the components do not exceed their specified temperature ranges. |
| **Temperature stresses** | In addition to open air temperature, temperature stresses on equipment caused by other environmental parameters (e.g. solar radiation, air velocity and heating from adjacent equipment) will need to be considered. |
| Operational requirements | • The specified operating temperatures should be the lowest and the highest ambient temperatures expected to be experienced by equipment during normal operation.<br>• When equipment is turned on it should be expected to operate within the temperature ranges stipulated and be fully operational within a specified time after initial turn on – unless otherwise specified.<br>• The permissible limit temperatures of the operating equipment are not allowed to be exceeded as a result of the temperature rise occurring in operation (including temporary acceleration). |
| Storage | The specified storage temperatures are normally the lowest and highest ambient temperatures that the sample is expected to experience (with the power turned off) during storage or exposure to climatic extremes. The equipment is not expected to be capable of operating at these extreme temperatures, but to survive them without damage. |
| Peripheral units | For peripheral units (measuring transducers etc.) or situations where equipment is in a decentralized configuration, ambient temperature ranges are frequently exceeded. In these cases, the actual temperature occurring at the location of the equipment concerned needs to be considered during the design and installation. |
| Installation | When equipment is installed in a controlled climatic environment, provided the equipment is not required to operate outside of those conditions, the temperature range can normally be agreed between purchaser and supplier. |

For the purpose of these Regulations, the AA2 and AA4 classes of ambient temperature: (from –40°C to + 40°C) are generally recommended, with AA4 conventionally being regarded as normal.

## 8.2.2.1 Mandatory safety requirements

Persons and livestock **shall** be protected against injury, and   WR-131.4
property shall be protected against damage, due to excessive temperatures or electromechanical stresses caused by
any overcurrent likely to arise in live conductors.

Where the temperature of an exposed part of electrical   WR-134.1.6
equipment is likely to cause injury to persons or livestock, that part **shall** be so located or guarded in order to
prevent accidental contact.

 **Note:** Where necessary, suitable safety warning signs and/or notices shall be
provided.

Persons, livestock, fixed equipment and fixed materi-   WR-131.3.2
als adjacent to electrical equipment **shall** be protected
against harmful effects of heat or thermal radiation
emitted by electrical equipment.

 **Note:** This requirement particularly applies to:

- combustion, ignition, or degradation of materials;
- the risk of burns;
- disturbing the safe function of the equipment.

## 8.2.2.2 Electrical installations

Electrical installations shall be so arranged that:   WR-131.3.1

- the risk of ignition of flammable materials due to
  high temperature or electric arc is minimized;
- during normal operation of the electrical equipment, there shall be minimal risk of burns to persons or livestock.

 Electrical equipment shall not present a fire hazard to adjacent materials.

## 8.2.2.3 Conductors

Conductors, other than live conductors, and any other parts   WR-131.5
intended to carry a fault current, shall be capable of carrying
that current without attaining an excessive temperature.

| | |
|---|---|
| The cross-sectional area of conductors shall be determined for both normal operating conditions and, where appropriate, for fault conditions, according to the admissible maximum temperature. | WR-132.6 |

### 8.2.2.4 Electrical equipment

| | |
|---|---|
| Electrical equipment shall be so selected and erected that its normal temperature rise and foreseeable temperature rise during a fault cannot cause a fire. | WR-422.1.2 |

 A temperature cut-out device shall have manual reset – only!

| | |
|---|---|
| Electrical equipment shall be installed so that design temperatures are not exceeded. | WR-134.1.5 |
| Electrical equipment that is likely to cause high temperatures or electric arcs shall be placed (or guarded) so as to minimize the risk of ignition of flammable materials. | WR-134.1.6 |

 Switchgear, protective devices, accessories and other types of equipment **shall not** be connected to conductors intended to operate at a temperature exceeding 70°C at the equipment in normal service.

### 8.2.2.5 Fixed electrical equipment

| | |
|---|---|
| Fixed electrical equipment shall be selected and erected such that its temperature in normal operation will not cause a fire. | WR-421.1.2 |
| Where fixed equipment may attain surface temperatures that could cause a fire hazard to adjacent materials, the equipment shall: | WR-421.2 |

• be mounted on a support that has low thermal conductance; or
• within an enclosure that will withstand such temperatures as may be generated; or

- be screened by materials of low thermal conductance that can withstand the heat emitted by the
  electrical equipment; or
- be mounted so as to allow safe dissipation of heat
  and at a sufficient distance from adjacent material
  on which such temperatures could have deleterious
  effects.

 **Note:** Any means of support shall be of low thermal conductance.

Where arcs, sparks or particles at high temperature          WR-421.1.3
may be emitted by fixed equipment in normal service,
the equipment shall be:

- totally enclosed in arc-resistant material; or
- screened by arc-resistant material from materials
  upon which the emissions could have harmful
  effects; or
- mounted so as to allow safe extinction of the
  emissions at a sufficient distance from materials
  upon which the emissions could have harmful
  effects.

Measures shall be taken to prevent an enclosure of           WR-422.3.2
electrical equipment such as a heater or resistor from
exceeding the following temperatures:

- 90°C under normal conditions; and
- 115°C under fault conditions.

Electrical equipment such as installation boxes and          WR-422.4.103
distribution boards, installed on or in a combusti-
ble wall shall comply with the relevant standard for
enclosure temperature rise.

Accessible parts of fixed electrical equipment within        WR-423.1
arm's reach shall not attain a temperature in excess of
the appropriate limit stated in Table 8.4.

Each such part of the fixed installation likely to attain    WR-423.1
under normal load conditions, even for a short period,
a temperature exceeding the appropriate limit in
Table 8.4 shall be guarded so as to prevent accidental
contact.

**Table 8.4 Temperature limit under normal load conditions for an accessible part of equipment within arm's reach (data from BS 7671:2008)**

| Accessible part | Material of accessible surfaces | Maximum temperature (°C) |
|---|---|---|
| A hand-held part | Metallic | 55 |
| | Non-metallic | 65 |
| A part intended to be touched but not hand-held | Metallic | 70 |
| | Non-metallic | 80 |
| A part which need not be touched for normal operation | Metallic | 80 |
| | Non-metallic | 90 |

### 8.2.2.6 Accessories

Parts of a cable or flexible cord within an accessory, appliance or luminaire shall be suitable for the temperatures likely to be encountered, or shall be provided with additional insulation suitable for those temperatures.

WR-522.2.100

### 8.2.2.7 Cables

For groups containing non-sheathed or sheathed cables that have different maximum operating temperatures, the current-carrying capacity of all the non-sheathed or sheathed cables in the group shall be based on the lowest maximum operating temperature of any cable in the group together with the appropriate group rating factor.

WR-523.5

### 8.2.2.8 Conductors

Connections between conductors or between a conductor and other equipment will need to take account of the temperature attained at the terminals in normal service.

Where a soldered connection is used, the design shall take account of creep, mechanical stress and temperature rise under fault conditions.

WR-526.2

Where necessary, precautions shall be taken so that the temperature attained by a connection in normal service shall not impair the effectiveness of the insulation of the conductors connected to it or any insulating material used to support the connection. — WR-526.4

Where a cable is to be connected to a bare conductor or busbar, its type of insulation and/or sheath shall be suitable for the maximum operating temperature of the bare conductor or busbar. — WR-526.4

The current, including any harmonic current, to be carried by any conductor for sustained periods during normal operation shall be such that the appropriate temperature limit specified in Table 8.5 is not exceeded. — WR-523.1

Table 8.5 Maximum operating temperatures for types of cable insulation, data from BS7671: 2008

| Type of insulation | Temperature limit |
| --- | --- |
| Thermoplastic | 70°C at the conductor |
| Thermosetting | 90°C at the conductor b |
| Mineral (thermoplastic covered, or bare exposed to touch) | 70°C at the sheath |
| Mineral (bare not exposed to touch, and not in contact with combustible material) | 105°C at the sheath b, c |

Connections between conductors or between a conductor and other equipment shall take account of the temperature attained at the terminals in normal service. — WR-526.2

Where a cable is to be connected to a bare conductor or busbar, its type of insulation and/or sheath shall be suitable for the maximum operating temperature of the bare conductor or busbar. — WR-526.4

## 8.2.2.9 Heating conductors and cables

The load of every floor-warming cable under operation shall be limited to a value such that the manufacturer's stated conductor temperature is not exceeded. — WR-554.4.4

### 8.2.2.10 Floor and ceiling heating systems

In floor areas where contact with skin or footwear     WR-753.423.1
is possible, the surface temperature of the floor shall
be limited (e.g. 35°C).

For cold tails (circuit wiring) and control leads     WR-753.522.1.3
installed in the zone of heated surfaces, the increase
in ambient temperature shall be taken into account.

### 8.2.2.11 Immersed heating elements

Heaters for liquid or other substance shall have an auto-     WR-554.2.1
matic device to prevent a dangerous rise in temperature.

### 8.2.2.12 Lighting equipment

When installed at exhibition, shows and stands,     WR-711.422.4.2
lighting equipment such as incandescent lamps,
spotlights and small projectors and other equipment or appliances with high-temperature surfaces,
shall be suitably guarded.

### 8.2.2.13 Luminaires

Luminaires marked ▽D̸ are designed to provide limited surface temperature.

Every luminaire shall:     WR-422.3.8

- be appropriate for the location;
- be provided with an enclosure providing a degree
  of protection of at least IP4X or, in the presence of
  dust, IP5X or, in the presence of electrically conductive dust, IPX6;
- have a limited surface temperature (in accordance
  with BS EN 60598-2-24);
- be of a type that prevents the lamp components
  from falling (i.e. from the luminaire).

In locations where there may be fire hazards due to dust or fibres, luminaires
shall be installed so that dust or fibres cannot accumulate in dangerous
amounts.

Electrical installations shall be so arranged that:    WR-422.4.2

- the risk of ignition of flammable materials due to high temperature or electric arc is minimized;
- during normal operation of the electrical equipment, there shall be minimal risk of burns to persons or livestock.

Luminaires should only be installed and used at a    WR-422.4.2
reasonable (i.e. sufficient) distance from combustible materials.

Unless otherwise recommended by the manufacturer,    WR-422.4.2
small spotlights or projectors shall be installed at the following minimum distances from combustible materials:

- with a rating up to 100 W – 0.5 m;
- over 100 W and up to 300 W – 0.8 m;
- over 300 W and up to 500 W – 1.0 m.

**Note:** All luminaire components (e.g. lamps and other apparatus) shall be protected against foreseeable mechanical stresses. Such protective means shall not be **fixed** to lampholders unless they form an integral part of the luminaire or are fitted in accordance with the manufacturer's instructions.

Bayonet lampholders B15 and B22 shall comply with    WR-559.6.1.7
BS EN 61184 and shall have the temperature rating T2 described in that standard.

Only independent lamp controlgear marked as suit-    WR-559.7
able for independent use (according to the relevant standard) shall be used external to a luminaire.

Only the following are permitted to be mounted on    WR-559.8
flammable surfaces:

- a class P thermally protected ballast(s)/transformer(s);
- a temperature declared thermally protected ballast(s)/transformer(s), with a marked value equal to (or below) 130°C.

**Note:** For an explanation of the symbols used, see Table 8.6.

**Table 8.6  Explanation of symbols used in luminaires, in controlgear for luminaires and in the installation of luminaires**

| BS EN 60598-1: 2004 | | BS EN 60598-1: 2008 | |
| --- | --- | --- | --- |
| F | Luminaire suitable for direct mounting on normally flammable surfaces | ☒ (flame symbols) | Recessed luminaire not suitable for direct mounting on normally flammable surfaces |
| F (crossed) | Luminaire suitable for direct mounting on non-combustible surfaces only | (flame symbol) | Surface mounted luminaire not suitable for direct mounting on normally flammable surfaces |
| F | Luminaire suitable for direct mounting in/on normally flammable surfaces when thermally insulating material may cover the luminaire | (crossed symbol) | Luminaire not suitable for covering with thermally insulating material |

NOTE  Luminaire suitable for direct on normally flammable surfaces may be marked with the symbol shown according to BS EN 60598-1:2004

With the publication of BS EN 60598-1 Ed. 7. luminaires suitable for direct mounting on normally flammable surfaces have no special marking and only luminaires not suitable for mounting on normally flammable surfaces are marked with a symbol (see annex N.4 of BS EN 60598-1:2008 for further explanations).

F

Recessed

Surface

| Symbol | Description | Symbol | Description |
| --- | --- | --- | --- |
| (cable symbol) t ..... °C | Use of heat-resistant supply cables, interconnecting cables, or external wiring (number of conductors of cable is optional) (BS EN 60598 series) | E | Luminaire for use with high pressure sodium lamps that require an external ignitor (BS EN 60598 series) |
| (bowl mirror lamp symbol) | Luminaire designed for use with bowl mirror lamps (BS EN 60598 series) | I | Luminaire for use with high pressure sodium lamps having an internal starting device (BS EN 60598 series) |
| $t_a$.... °C | Rated maximum ambient temperature (BS EN 60598 series) | D | Luminaire with limited surface temperature (BS EN 60598-2-24) |
| COOL BEAM | Warning against the use of cool-beam lamps (BS EN 60598 series) | (transformer symbol) | Short-circuit proof (inherently or non-inherently) safety isolating transformer (BS EN 61558-2-6) |
| ⊂--- m | Minimum distance from lighted objects (metres) (BS EN 60598 series) | ... | Temperature declared thermally protected lamp controlgear (... replaced by temperature) (BS EN 61347-1) |
| T | Rough service luminaire (BS EN 60598 series) | 110 | Electronic convertor for an extra-low voltage lighting installation |
| (cracked screen symbol) | Replace any cracked protective screen (BS EN 60598 series) | P | Thermally protected lamp controlgear (class P) (BS EN 61347-1) |
| (lamp symbol) | Luminaire designed for use with self-shielded tungsten halogen lamps or self-shielded metal halide lamps only (BS EN 60598 series) | (ballast symbol) | Independent ballast EN 60417 sheet No. 5138 |

## 8.2.2.14 Motors

A motor that is automatically or remotely control-led or that is not continuously supervised shall be protected against excessive temperature by a protective device with a manual reset.

WR-422.3.7

A motor with star-delta starting shall be protected against excessive temperature in both the star and delta configurations.

WR-422.3.7

Where the motor is intended for intermittent duty and for frequent starting and stopping, account shall be taken of any cumulative effects of the starting or braking currents on the temperature rise of the equipment of the circuit.

WR-552.1.1

In temporary electrical installations for structures (e.g. amusement devices and booths at fairgrounds, amusement parks and circuses) a motor that is automatically or remotely controlled (and which is not continuously supervised) shall be fitted with a manually reset protective device against excess temperature.

WR-740.422.3.7

## 8.2.2.15 Type of wiring and method of installation

The choice of the type of wiring system and the method of installation shall include consideration of the following:

WR-132.7

- the nature of the location;
- the nature of the structure supporting the wiring;
- accessibility of wiring to persons and livestock;
- voltage;
- the electromechanical stresses likely to occur due to short-circuit and Earth fault currents;
- EMI;
- other external influences (e.g. mechanical, thermal and those associated with fire) to which the wiring is likely to be exposed during the erection of the electrical installation or in service.

## 8.2.2.16 Wiring system

A wiring system shall be selected and erected so as to be suitable for the highest and the lowest local ambient temperatures.

WR-522.1.1

In order to avoid the effects of heat from external
sources, one or more of the following shall be used to
protect a wiring system:

WR-522.2.1

- shielding;
- placing sufficiently far from the source of heat;
- selecting a system with due regard for the additional temperature rise that may occur;
- local reinforcement or substitution of insulating material.

Heat from external sources may be radiated, conducted or convected, for example:

- from a manufacturing process;
- from hot water systems;
- from plant, appliances and luminaires;
- from solar gain of the wiring system or its surrounding medium;
- through heat conducting materials.

In solar photovoltaic (PV) power supply systems,
the wiring system shall withstand the expected
external influences such as wind, ice formation,
temperature and solar radiation.

WR-712.522.8.3

### 8.2.2.17 Wiring system components

Wiring system components, including cables and wiring
accessories, shall only be installed or handled at temperatures within the limits stated in the relevant product
specification or as given by the manufacturer.

WR-522.1.2

### 8.2.2.18 Agricultural and horticultural premises

For high-density livestock rearing, systems operating
for the life support of livestock (and where electrically powered ventilation is necessary) a system of
temperature and supply voltage monitoring shall
be used.

WR-705.560.6

## 8.2.2.19 *Temporary electrical installations*

> For cold tails (circuit wiring) and control leads   WR-753.522.1.3
> installed in the zone of heated surfaces of temporary
> electrical installations for structures (e.g. amuse-
> ment devices and booths at fairgrounds, amusement
> parks and circuses), the increase in ambient tem-
> perature shall be taken into account.

# 8.3 Solar radiation

Of all the factors that control the weather, the sun is by far the most power-ful, and practically everything that occurs on the Earth is controlled, directly or indirectly, by it. The sun affects the places humans inhabit, in the kind of homes that are built, the work that is done, and the equipment that is used.

Less than one-millionth of the energy emitted from the sun's surface travels the ninety-odd million miles to reach this planet. The sun's energy crosses those miles in the form of short electromagnetic radio waves, identical in nature to those used in broadcasting, which pass through the atmosphere and are absorbed by the Earth's surface. These waves warm the Earth's surface and are then re-radiated back to space. The wavelength of the energy emitted by the Earth is very much longer than that emitted by the sun (because the Earth

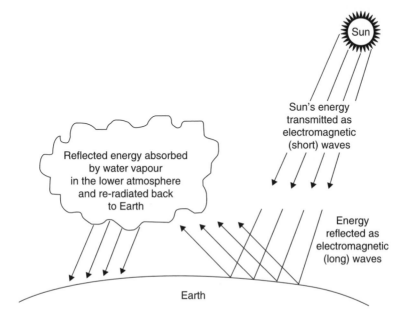

**Figure 8.5** Solar radiation – energy.

is much cooler than the sun), and these longer waves are not able to pass through the atmosphere as freely as short waves. A large proportion of the energy emitted by the Earth is absorbed by the water vapour and water droplets in the lower atmosphere, and this energy is then, in turn, re-radiated back to Earth. Thus the Earth plays the part of a receiving station absorbing short electromagnetic waves and converting them into longer electromagnetic waves, while the atmosphere acts as a trap containing most of the longer electromagnetic waves before they are lost to space.

Radiation from the sun consists of rays of three different wavelengths: heat rays, actinic rays and light rays. Heat rays and actinic rays are intercepted by solid bodies and produce peculiar effects in varying degrees according to the nature of the surface on which they fall. The light rays are responsible for daylight, and both light rays and actinic rays are necessary for the life processes of plants. The most important aspect of heat rays is temperature, and the amount of sunshine (and therefore the temperature) will depend on latitude and the length of day.

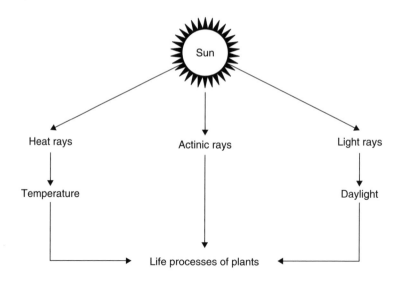

**Figure 8.6** Sun's radiation.

Radiant energy can be reflected from solid surfaces and intensified by that reflection. For example, reflection from walls is frequently used for ripening peaches and pears. Reflection from bare ground can also assist in the ripening of melons and other creeping plants, while reflection from water surfaces enhances the 'climatic reputation' of waterside resorts. However, radiant energy can also cause damage to equipment, as heat rays can warm the material or its environment to dangerous levels, and photochemical degradation of materials can be caused by the ultraviolet content of solar radiation.

## 8.3.1  What are the effects of solar radiation?

On cloudless nights when atmospheric radiation is very low, objects exposed to the night sky will attain surface temperatures below that of the surrounding air temperature. For example (and by experiment), a horizontal disk thermally insulated from the ground and exposed to the night sky during a clear night can attain a temperature of –14°C when the air temperature is 0°C and the relative humidity is close to 100%, and these values are of assistance when determining the 'under temperature' of components.

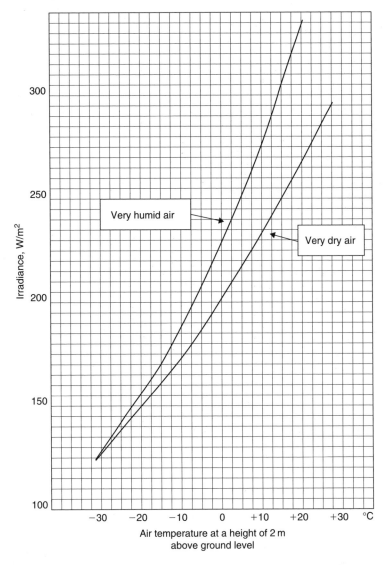

**Figure 8.7**  Lowest values of atmospheric radiation during clear nights. (Reproduced from BS 7527:1991, Section 2.4, by kind permission of the BSI.)

The sun's electromagnetic radiation consists of a broad spectrum of light ranging from ultraviolet to near infrared. Owing to the distance of the sun from the Earth, solar radiation appears on the Earth's surface as a parallel beam, and the highest (maximum) level of radiation occurs at noon, on a cloudless day, at a surface perpendicular to the sun.

Most of the sun's energy reaches the surface of the Earth in the 0.3–0.4 μm range, and the density of the solar radiated power (or irradiance – expressed in watts per square metre) is dependent on the content of aerosol particles, ozone and water vapour in the air. The actual amount of irradiance will vary considerably with geographical latitude and type of climate (i.e. temperature, humidity, air velocity etc.).

Having said that, an object subjected to solar radiation will obtain a temperature depending on the surrounding ambient air temperature, the intensity of radiation, the air velocity, the angle of incidence of the radiation on the object, the duration of exposure, the thermal properties of the object itself (e.g. surface reflectance, size, shape, thermal conductance and specific heat), together with other factors such as wind and heat conduction to mountings and surface absorbency etc.

### 8.3.2 Photochemical degradation of material

One of the biggest problems caused by solar radiation is the photochemical degradation of most organic materials, which in turn causes the elasticity and plasticity of certain rubber compounds and plastic materials to be affected and can, in exceptional cases, make optical glass opaque. Although solar radiation can bleach out colours in paints, textiles, paper, etc. (a major consideration when trying to read the colour coding of components!!), by far the most important effect is the heating of materials.

The combined effect of solar radiation, atmospheric gases, temperature and humidity changes etc., is often termed 'weathering', and results in the 'ageing' and ultimate destruction of most organic materials (e.g. plastics, rubbers, paints, timber). Typical defects caused by weathering are:

- rapid deterioration and breakdown of paints;
- cracking and disintegration of cable sheathing;
- fading of pigments;
- bleaching out of colours in paints, textiles and paper.

### 8.3.3 Effects of irradiance

To guard against the effects of irradiance, the following guidelines should be considered when locating electrical equipment:

- the sun should be allowed to shine only on the smallest possible casing surfaces;

- windows should be avoided on the sunny side of rooms housing electronic equipment;
- heat-sensitive parts must be protected by heat shields made, for instance, of polished stainless steel or aluminium plate;
- air-conditioning plant and cooling fans (when used) in rooms housing electronic equipment should be efficient and reliable;
- convection flow should sweep across the largest possible surfaces composed of materials with good conduction properties.

### 8.3.4  Heating effects

As noted previously, probably the most important effect of solar radiation heating is caused mainly by the short-term, high-intensity radiation around noon on cloudless days. Typical peak values of irradiance are shown in Table 8.7.

**Table 8.7  Typical peak values of irradiance from a cloudless sky**

| Area type | Irradiance (W/m$^2$) | | |
| --- | --- | --- | --- |
| | Large cities | Flat land | Mountainous areas |
| Subtropical climates and deserts | 700 | 750 | 1180 |
| Other areas | 1050 | 1120 | 1180 |

As equipment (if fully exposed to solar radiation) at an ambient temperature (e.g. 35–40°C) can attain temperatures in excess of 60°C, one has to consider the outside surface of the equipment. To a major extent the surface reflectance of an object affects its temperature rise from solar heating, and changing the finish from a dark colour to a gloss white can cause a considerable reduction in temperature. On the downside, a pristine finish designed specifically to reduce temperature can be expected to deteriorate over time and result in an increase in temperature.

Another problem with most of today's materials is that they are also selective reflectors (i.e. their spectral reflectance factor changes with wavelength). For example, paints are, in general, poor infrared reflectors, although they may be very efficient as a visible warning. Care should, therefore, be taken when selecting materials and finishes for equipment casings.

### 8.3.5  Requirements from the Regulations – solar radiation

Table 8.8 lists the most common environmental requirements concerning solar radiation.

**Table 8.8 Typical requirements – solar radiation**

| | |
|---|---|
| Survivability | Equipment that is exposed to the effect of solar radiation should remain unaffected. |
| Exposure | The sun should be allowed to shine only on the smallest possible casing surfaces, and the convection flow should sweep across the largest possible surfaces of materials with good conduction properties. |
| Windows | Windows should be avoided on the sunny side of rooms housing electronic equipment. |
| Heat shields | Heat-sensitive parts shall be protected by heat shields made of (for instance) polished stainless steel or aluminium plate. |
| Air-conditioning | Air-conditioning plant and cooling fans (when used) in rooms housing electronic equipment should be efficient and reliable. |

| | |
|---|---|
| The wiring system shall be selected and erected (or adequate shielding shall be provided) where and whenever significant solar radiation (AN2) or ultraviolet radiation is experienced or anticipated. | WR-522.11.1 |
| Special precautions may need to be taken for equipment that is subject to ionizing radiation. | WR-522.11.1 |
| Solar PV modules shall be installed in such a way that there is adequate heat dissipation under conditions of maximum solar radiation for the site. | WR-712.512.2.1 |

# 8.4 Humidity

The atmosphere is normally described as 'a shallow skin or envelope of gases surrounding the surface of the Earth which is made up of nitrogen, oxygen and a number of other gases which are present in very small quantities'. While the ratio of these components shows no appreciable variation either with latitude or altitude, the water vapour content of the atmosphere is subject to extremely wide fluctuations. The amount of water present in the air is referred to as 'humidity'.

When air and water come into contact, they will exchange particles with each other (i.e. the air particles will pass into the water and water particles will pass into the air in the form of vapour). There is always a certain amount of water vapour present in the air, and a certain amount of air present in water, and there is always a constant movement between the two mediums. If there is only a small amount of water vapour present in the air, then more particles of water will pass from the water into the air than from the air into the water, and so the water will gradually dry up (i.e. evaporate). Conversely, if the amount of water vapour in the air is large, then as many particles of vapour will pass from the air into the water as from the water into the air, and the water will not evaporate. In such a case the air is said to be 'saturated' – or, to put it another way, it holds as much water vapour as it can possibly contain.

Water vapour is collected in the air above the oceans and is carried by the wind towards the land masses. The amount of water vapour in the air varies greatly depending on the place and season but, in general, evaporation is most rapid at high temperature and slower at lower temperatures. We may, therefore, expect to find the greatest amount of water vapour over the oceans near the Equator and the smallest amount over the land in a cold region such as North-East Asia in winter. However, even when the surface is covered with snow and ice, water evaporation may take place, and, occasionally during a long frost, the snow will gradually disappear without melting.

Except for the water vapour being present, the composition of the atmosphere near the surface of the Earth up to a height of some 2000 ft is practically uniform throughout the globe. However, at greater altitudes (i.e. above the atmospheric boundary in the tropopause), there is practically no water vapour, or water, in any form.

### 8.4.1  What is humidity?

Temperature and the relative humidity of air (in varying combinations) are climatic factors that act upon electrical equipment and installations during storage, transportation or operation. Humidity and the electrolytic damage resulting from moisture, mostly affects plug points, soldered joints (in particular dry joints), bare conductors, relay contacts and switches. Humidity also promotes metal corrosion (see Section 8.7) owing to its electrical conductivity

In many cases, however, environmental influences such as mechanical and thermal stresses are merely the forerunner of the impending destruction of components by humidity – especially as the majority of electronic component failures are caused by water!

 **Note:** 'Humidity' (in the context of this book) has been taken to cover relative humidity, absolute humidity, condensation, adsorption, absorption and diffusion, and details of these 'subsets' are provided in the following paragraphs.

### 8.4.2  Relative and absolute humidity and their effect on equipment performance

The performance of virtually all electrical equipment is influenced and limited by its internal temperature, which, in turn, is dependent on the external ambient conditions and on the heat generated within the device itself. Fortunately, most electrical and electronic components (especially resistors) will normally remain dry when under load owing to the amount of internal/external heat dissipation. Indeed, many components either have to be de-rated in order to improve their reliability or, for reasons of circuit function, only energized intermittently.

### 8.4.2.1 Externally mounted equipment

Equipment and components that are mounted in external cabinets run the risk of coming into contact with water or water vapour (e.g. drifting snow, fog, dew, rain, spray water or water from hoses), and the equipment must, therefore, be adequately protected from such humidity in order to prevent the ingress of vapour into the system within the casing.

### 8.4.2.2 Housed equipment

In most locations (e.g. cabinets, equipment rooms, workshops and laboratories), although temperatures above 30°C may often occur, they are normally combined with a lower relative humidity than that found in the open air. In other rooms (e.g. offices), however, where several heat sources are present, the temperature and relative humidity can differ dramatically across the room.

The sun also plays its part because, in certain circumstances (such as when equipment is placed in an unventilated enclosure), the intense heat caused through solar radiation can generate relative humidities in excess of 95% when combined with:

- high relative humidity caused by the release of moisture from hygroscopic materials;
- the breathing and perspiration of human beings;
- open vessels containing water or other sources of moisture.

## 8.4.3 Condensation

Condensation occurs when the surface temperature of an item is lower than that of the dew point (i.e. the temperature with a relative humidity of 100% at which condensation occurs) and which can change electrical characteristics (e.g. decrease surface resistance, increase loss angle) between the absolute point at which atmospheric vapour condenses into droplets (i.e. the dew point), absolute humidity and vapour pressure.

For example, if a piece of equipment has a low thermal time constant, condensation (normally found on the surface of the equipment) will occur only if the temperature of the air increases very rapidly, or if the relative humidity is very close to 100%. Sudden changes in temperature may cause water to condense on parts of equipment, and leakage currents can occur.

## 8.4.4 Adsorption

Adsorption is the amount of humidity that may adhere to the surface of a material, and depends on the type of material, the surface structure and the vapour pressure. This layer of water (no matter how small) can cause electrical short-circuits and material distortion etc.

## 8.4.5 Absorption

The quantity of water that can be absorbed by a material depends largely on the water content of the ambient air, and the speed of penetration of the water molecules generally increases with increasing temperature.

## 8.4.6 Diffusion

Water vapour can penetrate encapsulations of organic material (e.g. into a capacitor or semiconductor) via the sealing compound, and enter the casing. This factor is frequently overlooked and can become a problem, especially as the moisture absorbed by an insulating material can cause a variation in a number of electrical characteristics (e.g. reduced dielectric strength, reduced insulation resistance, increased loss angle, increased capacitance).

## 8.4.7 Protection

The effects of humidity mainly depend on temperature, temperature changes and impurities in the air. As shown in Table 8.9, there are three basic methods of protecting the active parts of equipment and components from humidity.

Table 8.9 Protective methods – humidity

| | |
|---|---|
| Heating the surrounding air so that the relative humidity cannot reach high values | This method normally requires a separate heat source, which (especially in the case of equipment mounted in external cabinets) usually means a separate power supply must be provided. This method is disadvantaged by the reliability of the circuit being dependent on the efficiency of the heating. |
| Hermetically sealing components or assemblies using hydroscopic materials | This is an extremely difficult process, as the smallest crack or split can allow moisture to penetrate the component, particularly in the area of connecting wire entry points. Metal, glass and ceramic encapsulation do, nevertheless, produce some very satisfactory results. |
| Ventilation and the use of moisture-absorbing materials | Most water-retaining materials and paint etc. are suitable for the temporary absorption of excessive high air humidity in the casing, which, because of the risk of pollution and dust penetration, cannot be fully ventilated (i.e. air exchange with the outside is not possible). |

## 8.4.8 Typical requirements – humidity

Tables 8.10 and 8.11 give the most common environmental requirements with regard to humidity.

**Table 8.10  Typical requirements – solar radiation**

| | |
|---|---|
| Equipment interoperability | Equipment that is operated adjacent to the sea shore (and, therefore, subject to extreme humidity) must be able to function equally well as the same equipment housed in the low humidity of (for example) the desert. |
| External humidity levels | Equipment should be designed and manufactured to meet the following external humidity levels (limit values), over the complete range of ambient temperature values anticipated. |
| | Note: Meteorological measurements made over many years have shown that, within Europe, a relative humidity greater than 95% combined with a temperature above 30°C does not occur over long periods in free air conditions. |
| Condensation | Operationally caused infrequent and slight moisture condensation should not lead to malfunction or failure of the equipment. |
| Indoor installations | In all indoor installations, provision must be made for limiting the humidity of the ambient air to a maximum of 75% at –5°C. |
| Equipment in cubicles and cases | The design of equipment should take into account temperature rises within cubicles and equipment cases, in order to ensure that the components do not exceed their specified temperature ranges. |
| Peripheral units | For peripheral units (e.g. measuring transducers) or equipment employed in a decentralized configuration (i.e. where ambient temperature ranges are exceeded), the actual temperature occurring at the location of the equipment concerned should be utilized when designing equipment. |
| Product configuration | All proposed and date equipment, components or other articles must be tested in their production configuration without the use of any additional external devices that have been added expressly for the purpose of passing humidity testing. |

**Table 8.11  Humidity – external humidity levels**

| Duration | Limit value |
|---|---|
| Yearly average | 75% relative humidity |
| On 30 days of the year, continuously | 95% relative humidity |
| On the other days, occasionally | 100% relative humidity |
| On the other days, occasionally | 30 g/m³ occurring in tunnels |

## 8.4.9  Requirements from the Regulations – humidity

For the purpose of these Regulations, the AB2 and AB4 classes of ambient temperature (between 5% and 100%) are generally recommended.

A wiring system shall be selected and erected so that no    WR-522.3.1
damage is caused by condensation or ingress of water
during installation, use and maintenance.

Special consideration needs to be given to wiring systems that are liable to frequent splashing, immersion or submersion.

| | |
|---|---|
| Suitable means shall be provided for the escape of condensation that might form in a wiring system or where water might collect. | WR-522.3.2 |
| Wiring systems that could be subjected to waves (AD6) shall be protected from mechanical damage (e.g. impact, vibration and other mechanical stresses). | WR-522.3.3 |
| Where corrosive or polluting substances (including water) could cause corrosion and/or deterioration, the parts of the wiring system likely to be affected shall be suitably protected (e.g. by protective tapes, paints or grease) and/or manufactured from a material resistant to such substances. | WR-522.5.1 |
| If a wiring system is routed below services that are liable to cause condensation (e.g. water, steam or gas services), precautions shall be taken to protect the wiring system from harmful effects. | WR-528.3.2 |

## 8.5  Air pressure and altitude

Air pressure, also frequently referred to as 'atmospheric pressure', is 'the force exerted on a surface of a unit area caused by the Earth's gravitational attraction on the air vertically above that area'. Air pressure varies with altitude (i.e. elevation above mean sea level) and location. For instance, at the equator, where the tradewinds of both hemispheres converge, there is a low-pressure zone (known as the ITCZ, or 'international conveyance zone'), which is characterized by high humidity.

It is not widely appreciated that the location of equipment, especially with respect to its altitude above sea level, can affect the working of that equipment. But it is not just the height above sea level that has a significant effect! Even air pressure variations at ground level have to be considered.

### 8.5.1  Low air pressure

At altitudes above sea level, low air pressure can cause:

- leakage of gases or fluids from gasket-sealed containers;
- ruptures of pressurized containers;
- changes in physical or chemical properties;
- erratic breakdown or malfunction of equipment from arcing or corona;

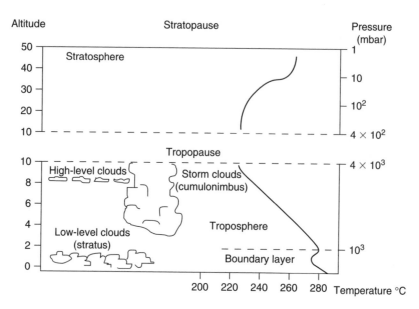

**Figure 8.8** Atmospheric structure.

**Table 8.12 Air pressure and altitude**

| Air pressure | | Approximate altitude above sea level (m) |
|---|---|---|
| (kPa) | (mbar) | |
| 1 | 10 | 31,200 |
| 2 | 20 | 26,600 |
| 4 | 40 | 22,100 |
| 8 | 80 | 17,600 |
| 15 | 150 | 13,600 |
| 25 | 250 | 10,400 |
| 40 | 400 | 7,200 |
| 55 | 550 | 4,850 |
| 70 | 700 | 3,000 |

- decreased efficiency of heat dissipation by convection and conduction in air, which will affect equipment cooling (e.g. an air pressure decrease of 30% has been found to cause an increase in temperature of 12%);
- acceleration of effects due essentially to temperature (e.g. volatilization of plasticizers, evaporation of lubricants).

## 8.5.2 Typical requirements – air pressure and altitude

Table 8.13 gives the most common environmental requirements concerning air pressure and altitude.

**Table 8.13 Typical requirements – air pressure and altitude**

| High air pressure | High air pressure occurring in natural depressions and mines can have a mechanical effect on sealed containers, and should always be borne in mind when designing and installing electrical equipment. |
| --- | --- |
| Installations up to 2000 m above sea level | Electrical equipment must be capable of working to an altitude (h) from –120 m to 2000 m above sea level, which corresponds to an air pressure range of 110.4–74.8 kPa. |

# 8.6 Weather and precipitation

Water is one of the most remarkable substances on the Earth. It is the substance that we most often see in all its three states: liquid (water), solid (ice) and gas (steam). No living organism can exist without water, and as much as half the weight of plants and animals is made up of water. Water in the oceans makes up approximately eleven times the volume of the solid part of the Earth, in addition, that is, to water frozen in ice flows, in lakes, rivers, within the ground, and in living plants and animals.

Water is a constantly moving in a cycle. As the sun beats down, some of the surface water evaporates; this water vapour rises as part of the air and is moved along by the wind. Should it pass over a land mass, it may become a cloud, and as more moisture is attracted to the cloud or the cloud passes over rising ground the water particles become larger and fall as rain, sleet or snow (Figures 8.9 and 8.10).

When the rain comes into contact with the ground there are several avenues open to it. It may be re-vaporized and return to the atmosphere, be absorbed by the ground, or remain on the surface of the ground and run downhill, forming streams and rivers, which run into the sea – and the cycle begins once more.

## 8.6.1 Water

Water (in all its three forms) is a major cause of failure in every application of electrical and electronic components. The humidity of the air and the possible formation of water particles must always be taken into consideration, especially as humidity possesses (almost without exception) a certain amount of electrical conductivity, which increases the possibility of corrosion of metals. Similarly, the ingress of water followed by freezing within electronic equipment can result in malfunction.

## 8.6.2 Salt water

Salt has an electrochemical effect on metallic materials (i.e. corrosion) which can damage and degrade the performance of equipment and/or parts that have been manufactured from metallic materials. Non-metallic materials can also

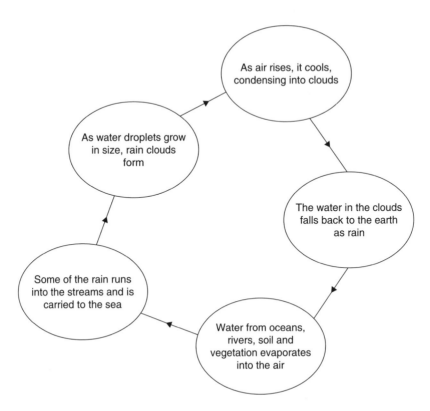

**Figure 8.9** The hydrological cycle.

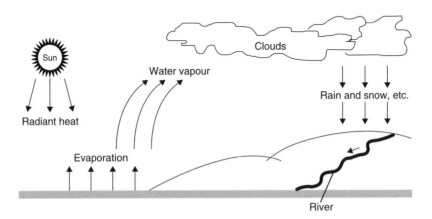

**Figure 8.10** Simplified water cycle.

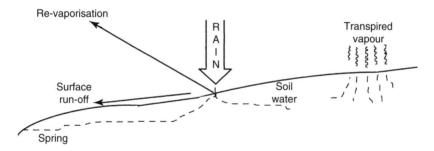

**Figure 8.11** The effect of rain.

be damaged by salt through a complex chemical reaction that is dependent on the supply of oxygenated salt solution to the surface of the material, its temperature, and the temperature and humidity of the environment. This is particularly a problem in areas close to the sea or mountain ranges.

### 8.6.3 Ice and snow

Water in the form of ice can cause problems in the cooling of equipment, or by freezing and thawing, which will result in cracks occurring, the breaking of cases, etc.

Powdered snow can easily be blown through ventilation ducts and then melt down in equipment compartments and cubicles, and this can cause damp problems in critical systems if not prevented in the original construction.

### 8.6.4 Weathering

As shown below, there are several types of 'weathering' (which is the collective term for the processes by which rock at or near the Earth's surface is disintegrated and decomposed by the action of atmospheric agents, water and living things).

### 8.6.5 Exfoliation

During the day rocks are warmed by the sun, and at night the surface can cool more rapidly than the underlying rocks. The outer skin of the rock then becomes tight and cracks, and thus layers of the rock peel off and the mountain becomes rounded or dome shaped. This exfoliation can have an effect on equipment that is sunk into rock faces or mounted on the surface of equipment.

### 8.6.6 Freeze–thaw

When water freezes it turns to ice, expanding by about one-twelfth of its volume. If this water is in the joint or a crack in a casing, then the space will become enlarged and the casing on either side will be forced apart. When the ice eventually thaws, more water will penetrate into the crack, and the cycle repeats itself, with the crack constantly enlarging.

### 8.6.7 Chemical weathering

Water can pick up quantities of sulphur dioxide from the atmosphere and will form a weak solution of acid. This acid can attack certain equipment housings, and the process whereby the housing is worn away is known as 'chemical weathering'.

### 8.6.8 Erosion

As the wind blows over dry ground it collects grit and 'throws' it vigorously against the surfaces nearby. This grit acts in a similar way as sandpaper, and gradually wears away the surface with which it comes into contact.

### 8.6.9 Mass movement

Once solids, such as sedimentary rocks, have been broken up, there is often a downwards movement of the particles that have broken off. This 'soil creep' can gather momentum and can, in certain circumstances, submerge equipment.

### 8.6.10 Requirements from the Regulations – weather and precipitation

Table 8.14 gives the most common environmental requirements concerning weather and precipitation.

**Table 8.14 Typical requirements – solar radiation**

| | |
|---|---|
| Weather protection | Equipment that is operated adjacent to the sea shore or on mountain ranges, and therefore subject to water and precipitation, must be able to function equally well as the same equipment housed in arid deserts. |
| Operation | All equipment should be capable of operating during rain, snow and hail, and be unaffected by ice, salt and water. |
| Rain | All equipment should be capable of operating in rain and be capable of preventing the penetration of rainfall at a minimum rate of 13 cm/h and an accompanying wind rate of 25 m/s. |
| Snow and hail | Consideration needs to be given to the effect of all forms of snow and/or hail. The maximum diameter of the hailstones is conventionally taken as 15 mm, but larger diameters can occur on occasion. |
| Salt water | Equipment should be capable of operating in (or be protected from) heavy salt spray, as would be experienced in seacoast areas and in the vicinity of salted roadways. |

## 8.6.10.1 Equipment

| | |
|---|---|
| The design of the electrical installation shall take into account the environmental conditions to which it will be subjected. | WR-132.5.1 |

## 8.6.10.2 Water

Where a wiring system passes through floors, walls, roofs, ceilings, partitions or cavity barriers, the openings remaining after passage of the wiring system should be sealed to provide the same degree of protection from water penetration as the building construction element in which it has been installed.

| | |
|---|---|
| A wiring system shall be selected and erected so that no damage is caused by condensation or ingress of water during installation, use and maintenance. | WR-522.3.1 |

 Special consideration needs to be given to wiring systems that are liable to frequent splashing, immersion or submersion.

| | |
|---|---|
| Where water may collect or condensation may form in a wiring system, provision shall be made for its escape. | WR-522.3.2 |
| Wiring systems that may be subjected to waves (AD6) shall be protected from mechanical damage (i.e. impact, vibration and other mechanical stresses). | WR-522.3.3 |
| Where the presence of corrosive or polluting substances, including water, is likely to give rise to corrosion or deterioration, parts of the wiring system likely to be affected shall be suitably protected (e.g. by protective tapes, paints or grease) and/or manufactured from a material resistant to such substances. | WR-522.5.1 |
| Where a wiring system is routed below services liable to cause condensation (such as water, steam or gas services), precautions shall be taken to protect the wiring system from harmful effects. | WR-528.3.2 |

## 8.6.10.3 Ice

| | |
|---|---|
| Wiring systems for Solar PV power supplies shall withstand the expected external influences such as ice formation. | WR-712.522.8.3 |

### 8.6.10.4 Marinas

| | |
|---|---|
| In marinas, equipment installed on or above a jetty, wharf, pier or pontoon shall be selected as follows, according to the external influences that may be present: | WR-709.512.2.1.1 |

- water splashes (AD4) – IPX4;
- water jets (AD5) – IPX5;
- water waves (AD6) – IPX6.

| | |
|---|---|
| For marinas, particular attention shall be given to the likelihood of corrosive elements, movement of structures, mechanical damage, presence of flammable fuel and the increased risk of electric shock due to the presence of water. | WR-709.512.2 |

**Note:** See Chapter 7 of this book for other requirements concerning weather and precipitation specific to special installations and locations.

### 8.6.10.5 Electrical installations in caravan and camping parks

| | |
|---|---|
| Socket-outlets of wiring systems shall be placed at a height of 0.5 m to 1.5 m from the ground, as measured to the lowest part of the socket-outlet. In special cases, due to environmental conditions such as risk of flooding or heavy snowfall, the maximum height is permitted to exceed 1.5 m. | WR-708.553.1.9 |

## 8.7 Pollutants and contaminants

Over the last 20 years environmental matters have become an area of wide-spread public concern, particularly those concerning the issues of pollution and contaminants. Pollutants and contaminants come in many forms, and can have an effect on the air, land or water courses. As pollutants move from one medium to another, they may be deposited on equipment and equipment housing – and they can cause extensive damage.

Pollution of the air can occur in both the troposphere and stratosphere, as shown in Figure 8.12. In the troposphere, pollutants from chimneys (for example) are carried by the air and can be deposited over time and distance, thus having a limited lifespan before they are washed out or deposited on the ground. If pollutants are injected straight into the stratosphere (as with a volcanic eruption) they will remain there for some time and result in noticeable effects over the whole region. On the other hand, the roughness of the ground will produce air turbulence, which will itself promote the mixing of pollutants, and even low wind speeds will result in high pollutant concentrations.

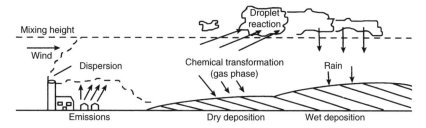

**Figure 8.12** The process between the emission of air pollutants and their being deposited on the ground.

Sources of natural pollutants include:

- sulphur – emitted by volcanoes and from biological processes;
- nitrogen – from biological processes in soil and lightning, and biomass burning;
- hydrocarbons – methane from fermentation of rice paddies, fermentation of the digestive tract of ruminants (e.g. cows), and released by insects, coal mining and gas extraction.

Sources of man-made pollutants include:

- carbon dioxide and carbon monoxide, produced during the burning of fossil fuels;
- soot formation accompanied by carbon monoxide, and generally due to inadequate or poor air supply;
- hydrocarbons – most boilers and central heating units burning fossil fuels result in very low emissions of gaseous hydrocarbons or oxygenated hydrocarbons such as aldehydes.

### 8.7.1 Pollutants

Although pollutant gases are normally only present in low concentrations, they can cause significant corrosion and a marked deterioration in the performance of contacts and connectors. The gases in operating environments that cause corrosion are oxygen, water vapour and the so-called 'pollutant gases', which include sulphur dioxide, hydrogen sulphide, nitrogen oxides and chlorine compounds.

Silver and some of its alloys are particularly susceptible to tarnishing by the minute quantities of hydrogen sulphide that occur in many environments.

The tarnish product is dark in colour, consists largely of b-silver sulphide, and separable electrical connections using silver and its alloys may, therefore, suffer from increased resistance and contact noise as a result.

The amount of tarnishing (of a metal) is dependent on the amount of humidity present. Less corrosion occurs below 70% relative humidity (RH), but above 80% RH the rate of tarnishing increases rapidly. Temperature also has an effect on the amount of tarnishing, as the nature of the corrosion mechanism has a tendency to change at temperatures above 30°C.

### 8.7.1.1 Sulphur dioxide

Sulphur dioxide is the pollutant gas most commonly found in the atmosphere, and is usually present in high concentrations in urban and industrial locations. In combination with other pollutants and moisture (e.g. humidity) it is responsible for the formation of high-resistance, visible corrosion layers on all but the most noble metals (e.g. silver and gold) and their alloys.

Although sulphur dioxide alone is less corrosive than other gases (such as sulphur trioxide, nitrogen oxides and chlorine compounds), the most extensive corrosion occurs when combustion products are present together with sulphur dioxide.

### 8.7.1.2 Hydrogen sulphide

Hydrogen sulphide is caused by bacterial reduction of sulphates in vegetation, soil, stagnant water and animal waste on a worldwide basis. In the atmosphere, hydrogen sulphide is oxidized to sulphur dioxide which, in turn, is brought to the ground by rain. In an aerobic soil, a bacterium turns the sulphur dioxide to sulphates. Sulphate-reducing bacteria complete the cycle, turning the sulphates to hydrogen sulphide – which is the principal natural sulphur input in the atmosphere and is, therefore, a widespread pollutant of air.

### 8.7.1.3 Nitrogen oxides

The production of nitrogen oxides is particularly significant because the rate of corrosion by sulphur dioxide is greatly accelerated in the presence of nitrogen dioxide, although the corrosion products have similar compositions.

## 8.7.2 Contaminants

Contaminants are composed of dust, sand, smoke and other particles that are contained within the air, and these can have an effect on electrical equipment in various ways, especially:

- ingress of dust into enclosures and encapsulations;
- deterioration of electrical characteristics (e.g. faulty contact, change of contact resistance);

- seizure or disturbance in motion bearings, axles, shafts and other moving parts;
- surface abrasion (erosion and corrosion);
- reduction in thermal conductivity;
- clogging of ventilating openings, bushes, pipes, filters and apertures that are necessary for operation etc.

The presence of dust and sand in combination with other environmental factors, such as water vapour, can also cause corrosion and promote mould growth. Damp heat atmospheres cause corrosion in connection with chemically aggressive dust, and similar effects are caused by salt mist. Effects of ion-conducting and corrosive dusts (e.g. de-icing salts) need also to be considered.

### 8.7.2.1 Dust and sand

The concentration of dust and sand in the atmosphere varies widely with geographical locality, local climatic conditions, and the type and degree of activity taking place. The amount of dust and sand found in the air is dependent on terrain, wind, temperature, humidity and precipitation. Under certain conditions enormous amounts of dust and sand may be temporarily released, and this suspended dust will drift away with the wind (see Table 8.15) depending on its concentration and the size of the particles.

**Table 8.15  Concentration of dust and sand**

| Region | Dust and sand concentration ($\mu g/m^3$) |
|---|---|
| Rural and suburban | 40–110 |
| Urban | 100–450 |
| Industrial | 500–2000 |

**(Extracted, with permission, from a paper by Herne European Consultancy.)**

Particles larger than 150 μm are generally confined to the air layer in the first metre above ground, and in this layer about half of the sand grains move within the first 10 mm above the surface.

The dust and sand appearing in enclosed and sheltered locations is generated by several sources (e.g. quartz, de-icing salts, fertilizers etc. penetrating into locations via ventilating ducts or badly fitting windows). The dust may also come from cloth or carpets in normal use in the working environment.

### 8.7.2.2 Dust

Dust may be defined as 'particulate matter of unspecified origin, composition and size ranging from 1 μm to 150 μm originating from quartz, flour, organic fibres etc.'. Particles of less than 75 μm can, because of their low terminal velocity, remain suspended in the atmosphere for very long periods through the natural turbulence of the air. In sheltered and enclosed locations, the maximum grain size tends to be smaller (e.g. less than 100 μm) than in non-weather-protected locations due to the filtering effect of the shelter.

The dust found in and around electrotechnical products may be generated by several different sources. The dust may be quartz, coal, de-icing salts, fertilizers, or small fibres from cotton or wool (real or artificial, that have been generated from cloth or carpets by normal use in living rooms and offices), and can penetrate into a piece of equipment by a number of mechanisms:

- carried in by forced air circulation (e.g. for cooling purposes);
- carried in by the thermal motion of the air;
- pumped in by variations in the atmospheric pressure caused by temperature changes;
- blown in by wind.

Dust itself can act as a physical agent or chemical component (or both), and can cause one or more of the following harmful effects:

- seizure of moving parts;
- abrasion of moving parts;
- adding mass to moving parts, thereby causing unbalance;
- deterioration of electric insulation;
- deterioration of dielectric properties;
- clogging of air filters;
- reduction of thermal conductivity;
- interference with optical characteristics;

and also

- corrosion and mould growth;
- overheating and fire hazard.

Dust adhering to the surface of materials may contain organic substances that provide a source of food for micro-organisms.

### 8.7.2.3 Sand

'Sand' is the term applied to 'segregated unconsolidated accumulation of detrital sediment, consisting mainly of tiny broken chips of crystalline quartz or other mineral, between 100 μm and 1000 μm in size'. Particles greater than 150 μm are unable to remain airborne unless continually subjected to strong winds, induced airflows or turbulence. Sand is generally harder than most fused silica glass compositions and can, quite naturally, scratch the surface of most glass optical devices. Pressure applied over trapped grains of sand can also cause fractures to occur in equipment.

The electrostatic charges produced by friction of the particles in sandstorms can interfere with the operation of equipment and sometimes be dangerous to personnel. The breakdown of insulators, transformers and lightning arresters and the failure of car ignition systems have also been known to occur as a result of such charges. The electrostatic voltages produced can be very large. Indeed, voltages as high as 150 kV have made telephone and telegraph communications inoperable during sandstorms.

Quartz, because of its hardness, can result in rapid wear or damage to products, particularly moving parts. However, erosion of material requires that the presence of dust and sand is combined with a high-velocity airstream over an extensive period of time.

Sand and the majority of dusts usually deposited on insulated surfaces are poor conductors in the absence of moisture. The presence of moisture, however, will result in the dissolving of the soluble particles and the formation of conducting electrolytes. For example, the leakage currents flowing over contaminated power line insulators can be of the order of one million times greater than those that flow through clean, dry insulators.

### 8.7.2.4 Smoke or fumes

Smoke or fumes are 'dispersive systems in the air consisting of particles below 1 µm'. As the particles are so small they do not usually affect equipment, provided that the equipment is properly designed.

### 8.7.2.5 Fauna and flora

With a few exceptions, fauna (rodents, insects, termites, birds etc.) and flora (plants, trees, seeds, fruit, blossom, mould, bacteria, fungi etc.) may be present at all locations where equipment is stored, transported or used. While fauna may be the cause of damage inside buildings as well as in open-air locations, damage by flora will predominantly occur in open-air conditions. Moulds and bacteria can, however, be present both inside buildings and in open-air conditions.

The frequency of flora and fauna depends on temperature and humidity. In warm damp climates, fauna and flora, especially insects and micro-organisms such as mould and bacteria, will find favourable conditions of life. Humid or wet rooms in buildings (or rooms in which processes produce humidity) are suitable living spaces for rodents, insects and micro-organisms. The range of temperature in which moulds may grow is 0–40°C, and the most favourable temperature for many cultures is between 20°C and 30°C.

If the surfaces of the products carry layers of organic substances (e.g. grease, oil, dust) or animal/vegetable deposits, the surfaces will become ideal locations for the growth of moulds and bacteria.

### 8.7.2.5.1 Effects of flora and fauna

The functioning of equipment and materials can be affected by physical attacks by fauna. Small animals and insects that feed from, gnaw at, eat into and chew on materials are particular problems, as are termites cutting holes into material.

Materials such as wood, paper, leather, textiles and plastics (including elastomers), and even some metals such as tin and lead, are all susceptible to attack.

Larger animals can also cause damage by stroke, impact or thrust. These attacks can cause:

- physical breakdown of material, parts, units or devices;
- mechanical deformation or compression;
- surface deterioration;
- electrical failure caused by mechanical deterioration.

Deposits from fauna (especially insects, rodents, birds etc.) can be caused by the presence of the animal itself, nest building, deposited feed stocks, or metabolic products such as excrement and enzymes etc.

Deposits from flora may consist of detached parts of plants (leaves, blossom, seeds, fruits etc.) and growth layers of cultures of moulds or bacteria. These attacks can lead to:

- deterioration of material;
- metallic corrosion;
- mechanical failure of moving parts;
- electrical failure due to:
  - increased conductivity of insulators;
  - failure of insulation;
  - increased contact resistance;
  - electrolytic and ageing effects in the presence of humidity or chemical substances;
  - moisture absorption and adsorption;
  - decreased heat dissipation.

These, in turn, can cause interruption of electrical circuits, malfunctioning of mechanical parts and clouding of optical surfaces (including glass).

### 8.7.2.6 Mould

Surface contamination in the form of dusts, splashes, condensed volatile nutrients or grease may be deposited on equipment. When that equipment is exposed (in use, storage or transportation) to the atmosphere, and without proper protective covering, mould growth will occur, and mould can cause unforeseen damage to equipment, whether constructed from mould-resistant materials or not!

A fungus grows in soil and in, or on, many types of common material, it propagates by producing spores, which become detached from the main growth and later germinate to produce further growth. The spores are very small and easily carried by the wind (or moving air). They also adhere to dust particles carried in the air. Contamination can also occur due to handling. Spores may be deposited by the hands or in the film of moisture left by the hands.

### 8.7.2.6.1 Germination and growth

Moisture is essential in allowing the spores to germinate, and when a layer of dust or other hydrophilic (i.e. moisture retaining) material is present on the surface, sufficient moisture may be abstracted by it from the atmosphere. In addition to high humidity, spores require (on the surface of the specimen) a layer of material that will absorb the moisture. Mould growth is also encouraged by stagnant air spaces and lack of ventilation.

When the relative humidity is below 65%, no germination or growth will occur. The higher the relative humidity above this value, the more rapid the growth will be. Spores can survive long periods of very low humidity and, even though the main growth may have died, they will germinate and start new growth as soon as the relative humidity becomes favourable again (i.e. in excess of 65%). The optimum temperature of germination for the majority of moulds is between 20°C and 30°C.

### 8.7.2.6.2 Effects of mould growth

Moulds can live on most organic materials, but some of these materials are much more susceptible to attack than others. Growth normally occurs only on surfaces exposed to the air, and those that absorb or adsorb moisture will generally be more prone to attack.

Even where only a slightly harmful attack on a material occurs, the formation of an electrically conducting path across the surface due to a layer of wet mycelium (i.e. the vegetative part of fungus) can drastically lower the insulation resistance between electrical conductors supported by an insulation material. When the wet mycelium grows in a position where it is within the electromagnetic field of a critically adjusted electronic circuit, it can cause a serious variation in the frequency-impedance characteristics of that circuit.

Among the materials that are highly susceptible to attack are leather, wood, textiles, cellulose, silk and other natural resources. Most plastic materials, although less susceptible, are also prone to attack, as they will probably contain oligomers (natural or synthetic compounds, usually of high molecular weight, that consist of millions of linked simple molecules), non-polymerized monomers (molecules that can combine with others to form a polymer) and/or additives which may radiate to the surface and be a nutrient for fungi. Some plastic materials depend, for a satisfactory lifespan, on the presence of a plasticizer (a substance added to plastics to make or keep them soft or pliable) which, if it is readily digested by fungi, will eventually give rise to failure of the main material.

Mould attack on materials usually results in a decrease in mechanical strength and/or changes in other physical properties, and the growing mould on the surface of a material can yield acid products and other electrolytes which will cause a secondary attack on the material. This attack can lead to electrolytic or ageing effects, and even glass can lose its transparency due to this process. Oxidation or decomposition may be facilitated by the presence of catalysts secreted by the mould.

### 8.7.2.6.3 Prevention of mould growth

All insulating materials used should be chosen to give as great a resistance to mould growth as possible, thus maximizing the time taken for mycelium to grow and minimizing any damage to the material consequent upon such growth. The use of lubricants during assembly (e.g. varnishes, finishes) is frequently necessary in order to obtain the required performance or durability of a product. Such materials should be chosen with regard to their ability to resist mould growth because, even though it can be shown that the lubricants do not support mould growth, they may collect dust, which in turn will support mould growth.

Moisture traps that could possibly be formed during the assembly of equipment and in which mould can grow should be avoided. Examples of such less obvious traps are between unsealed mating plugs and sockets, or between printed circuit cards and edge connectors. Other means of preventing mould growth include:

- complete sealing of the equipment in (and with) a dry, clean atmosphere;
- continuous heating within an enclosure can ensure a sufficiently low humidity;
- operation of equipment within a suitable controlled environment;
- regularly replaced desiccants (e.g. silica beads);
- periodic and careful cleaning of enclosed equipment.

Where the material and functioning of the equipment allows such treatment, ultraviolet radiation or ozone may be used for sterilization. Air currents flowing over the parts can retard the development of mould growth, and can be used to control the action of acaricides (i.e. mites and ticks).

### 8.7.3 Requirements from the Regulations – pollutants and contaminants

Table 8.16 gives the most common environmental requirements concerning pollutants and contaminants.

**Table 8.16 Typical requirements – pollutants and contaminants**

| | |
|---|---|
| Pollutants | • Although the severity of pollution will depend on the location of the equipment, the effects of pollution must be considered in the design of equipment and components.<br>• Means need to be provided to reduce pollution by the effective use of protective devices.<br>• The requirements of ISO 14001 regarding environmental protection and the prevention of pollution have to be met. |
| Contaminants | The following should be considered:<br><br>• chemical active substances;<br>• biologically active substances;<br>• flora and fauna;<br>• dust;<br>• sand. |
| Mould | • In an assembled state, equipment needs to operate when exposed to airborne mould spores and within climates that will be conducive to the growth of moulds.<br>• Insulating materials should be chosen to provide as much resistance to mould growth as possible, and all materials used should be chosen with regard to their ability to resist mould growth. |

## 8.7.3.1 Wiring systems

In escape routes that are likely to be BD2 (difficult), BD3 (crowded) and BD4 (difficult and crowded), the wiring system selected shall have a limited rate of smoke production.

WR-422.2.1

A wiring system **shall not** be installed in the vicinity of services that produce heat, smoke or fumes likely to be detrimental to the wiring, **unless** it is protected from harmful effects by shielding arranged so as not to affect the dissipation of heat from the wiring.

WR-528.3.1

Where no fire alarm system is installed in a building used for exhibitions etc., cable systems shall be flame retardant to BS EN 60332–1–2 and low smoke to BS EN 61034–2.

WR-711.521

## 8.7.3.2 Safety services

The location of a safety source shall be properly and adequately ventilated so that exhaust gases, smoke or fumes from the safety source cannot penetrate areas occupied by persons.

WR-560.6.4

### 8.7.3.3 Dust

Precautions shall be taken to prevent the accumulation of dust or other substances that could adversely affect the heat dissipation from a wiring system.

| | |
|---|---|
| Switchgear or controlgear shall be installed outside the location unless it is installed in an enclosure providing a degree of protection of at least IP4X or, in the presence of dust, IP5X, or in the presence of electrically conductive dust, IPX6. | WR-422.3.3 |
| A heat-storage appliance shall prevent the ignition of combustible dusts or fibres by the heat-storing core. | WR-422.3.102 |
| In a location where dust in significant quantity is present (AE4), additional precautions shall be taken to prevent the accumulation of dust or other substances in quantities that could adversely affect the heat dissipation from the wiring system. | WR-522.4.2 |

In agricultural and horticultural premises, luminaires shall be selected with regard to their degree of protection against the ingress of dust.

### 8.7.3.4 Mould

| | |
|---|---|
| Where previous or anticipated experiences of mould constitute a hazard (AK2), the wiring system shall be selected accordingly and/or special protective measures shall be adopted. | WR-522.9.1 |

**Note:** Possible preventive measures are closed types of installation (conduit or channel), maintaining distances to plants, and regular cleaning of the relevant wiring system.

### 8.7.3.5 Fauna

| | |
|---|---|
| Where previous or anticipated experience of fauna constitute a hazard (AL2), the wiring system shall be selected accordingly or special protective measures shall be adopted, for example, by:<br><br>• the mechanical characteristics of the wiring system; or | WR-522.101 |

- the location selected; or
- the provision of additional local or general protection against mechanical damage; or
- any combination of the above.

Where the conditions experienced or expected constitute a hazard (AK2), the wiring system shall be selected accordingly or special protective measures shall be adopted.　　　　WR-522.9.1

In locations accessible to, and enclosing, livestock, special attention shall be given to the presence of different kinds of fauna (e.g. rodents).　　　　WR-705.522

 **Note:** Possible preventive measures are closed types of installation (conduit or channel), maintaining distances to plants, and regular cleaning of the relevant wiring system.

# 8.8 Mechanical

Mechanics is the branch of physics concerned with the motions of objects and their response to forces. Modern descriptions of such behaviour begin with a careful definition of such quantities as displacement (distance moved), time, velocity, acceleration, mass and force.

There is often a tendency to underestimate the effect that the mechanical environment can have on the reliability of equipment, especially the effects of vibration and shock. Mechanical stresses are normally attributed to a moving mass, and there is frequently a tendency to underestimate the effect of the mechanical environment on the reliability of static installations. Experience suggests, however, that vibrations and shocks are a significant 'reliability, availability and maintainability' (RAM) factor, not only from the point of view of vehicle-mounted equipment, but also with respect to permanent installations.

If a spurious vibration acts on a printed circuit board, module or equipment, resonant oscillations will be induced in all components at their natural frequency. If, however, the frequency spectrum has some more or less distinctive frequency bands, the elements will perform forced oscillations at the cyclic frequency of the interference, and generally depend on both the characteristics of the oscillator (i.e. the component) and on the interference.

## 8.8.1 Shock

'Shock' is generally defined as 'an impact shock characterized by a simple acceleration and free impact on a firm base', and is usually the result of a violent collision or a heavy blow. While it is difficult to design and install electrical equipment so that components and systems are completely immune to shock, precautions should, nevertheless, be taken to guard against potential problem areas.

### 8.8.2 Vibration

Components, equipment and other articles during transportation or in service may be subjected to conditions involving vibration of a harmonic pattern, generated primarily by rotating, pulsating or oscillating forces caused by machinery and seismic incidents.

### 8.8.3 Acceleration

Equipment, components and electrotechnical products that are likely to be installed in moving bodies (e.g. rotating machinery) will be subjected to forces caused by steady accelerations. In general, the accelerations encountered in service will have different values along each of the major axes of the moving body and, in addition, usually have different values in the opposite senses of each axis.

### 8.8.4 Protection

- The resonant frequency of components is greatly influenced by the length of their connecting wires, and the actual length of these connecting wires may well be the decisive factor as to whether a component fails or remains functioning under given vibration and impact conditions.
- The amplitude of shocks on the equipment can be reduced by use of special mounting devices.
- Shock absorption is based on storing the impact and releasing it at a retarded rate. The peak acceleration is reduced and the high frequencies damped, thus providing protection for the components with their relatively higher natural frequencies (of some hundred hertz).
- Vibration dampers and shock absorbers are often used as a form of protection against mechanical stresses. The basic difference between vibration dampers and shock absorbers is that, with the former, the natural frequency lies below the interference frequency, while for the latter it is above it.

 Generally speaking, vibration dampers provide no protection whatsoever against shocks, and similarly shock absorbers offer no protection against vibrations. Only in exceptional cases can vibration dampers, for high frequencies, be used as shock absorbers.

- Elastic suspension of equipment can cause a critical increase in amplitude at certain frequencies and, where translatory and rotary displacements greater than six degrees of freedom are possible, very complex motions may arise. Wherever possible, therefore, one should try to ensure that none of the (possible) resonant frequencies fall within the range of the induced displacements.

- Sheathing circuits by means of cast resins can, in most cases, be a very effective means of counteracting mechanical stresses combined with temperature humidity.

## 8.8.5 Requirements from the Regulations – mechanical

Table 8.17 gives the most common environmental requirements concerning solar radiation.

**Table 8.17 Typical requirements – mechanical**

| | |
|---|---|
| Vibrations and shocks | Any dampers or anti-vibration mountings must be integral to the equipment so as to prevent the unit being accidentally installed without them. |
| Mechanical shock | Equipment should be capable of withstanding shock pulses (e.g. a minimum of 20,000 shocks at a shock level of 20 g.) |
| On or near the roadside | Equipment located on or near the roadside must be capable of withstanding vibrations and shocks. |
| Long-term exposure | Equipment must be capable of withstanding long-term exposure to shocks. |
| Encapsulated outdoor installations | Equipment contained in encapsulated outdoor installations must be capable of withstanding vibrations and shocks. |
| Closed rooms | Equipment located in closed room installations must be capable of withstanding self-induced vibrations. |
| In service | Equipment should be capable of withstanding, without deterioration or malfunction, all mechanical stresses that occur in service. |
| Random vibration | Equipment should be capable of withstanding random vibration. |

### 8.8.5.1 Mechanical stresses

A wiring system shall be selected and erected to avoid during installation, use or maintenance, damage to the sheath or insulation of cables and their terminations. — WR-522.8.1

A conduit system or cable ducting system (other than a pre-wired conduit assembly that has been specifically designed for the installation) that is going to be buried in the structure, shall be completely erected between access points before any cable is drawn in. — WR-522.8.2

| | |
|---|---|
| The radius of every bend in a wiring system shall be such that conductors or cables do not suffer damage and terminals are not stressed. | WR-522.8.3 |
| Where the conductors or cables are not supported continuously they shall be supported by suitable means at appropriate intervals in such a manner that the conductors or cables do not suffer damage by their own weight. | WR-522.8.4 |
| Every cable or conductor shall be supported in such a way that it is not exposed to undue mechanical strain and so that there is no appreciable mechanical strain on the terminations of the conductors. | WR-522.8.5 |
| A wiring system intended for the drawing in or out of conductors or cables shall have adequate means of access to allow this operation. | WR-522.8.6 |
| A cable buried in the ground (that is not installed in a conduit or duct) shall incorporate an earthed armour or metal sheath or both, suitable for use as a protective conductor. | WR-522.8.10 |
| The location of buried cables shall be marked by cable covers or a suitable marking tape. | WR-522.8.10 |
| Buried conduits and ducts shall be suitably identified. | WR-522.8.10 |
| Buried cables, conduits and ducts shall be at a sufficient depth to avoid being damaged by any reasonably foreseeable disturbance of the ground. | WR-522.8.10 |

 **Note:** See IEC 61386–24 for further details concerning underground conduits.

| | |
|---|---|
| Cable supports and enclosures shall not have sharp edges liable to damage the wiring system. | WR-522.8.11 |
| A cable or conductors shall not be damaged by the means of fixing. | WR-522.8.12 |
| Cables, busbars and other electrical conductors that pass across expansion joints shall be selected and/or erected such that anticipated movement does not cause damage to the electrical equipment. | WR-522.8.13 |

No wiring system shall penetrate an element of build-    WR-522.8.14
ing construction that is intended to be load bearing
unless the integrity of the load-bearing element can
be ensured after such penetration.

**Note:** A wiring system buried in a floor should be sufficiently protected to prevent damage caused by the intended use of the floor.

## 8.8.5.2 Shock

For locations where the wiring system may                WR-705.522.16
be exposed to impact and mechanical shock
due to vehicles, mobile agricultural machines
etc., the external influences shall be classified
AG3, and:

• conduits shall provide a degree of protection against impact of 5 J according to BS EN
  61386–2;
• cable trunking and ducting systems shall provide a degree of protection against impact of 5 J
  according to BS EN 50085–2–1.

## 8.8.5.3 Vibration

Stationary equipment that is moved temporarily           WR-521.9.2
for the purposes of connecting, cleaning etc. (e.g.
a cooker or a flush-mounting unit for installations
in a false floor) shall be connected with a non-flexible
cable; however, if they are subject to vibration
while in use they shall be connected by flexible
cables.

A wiring system supported by (or fixed to) a structure   WR-522.7.1
or equipment that is subject to vibration of medium
severity (AII2) or high severity (AH3) shall be suitable for such conditions, particularly where cables and
cable connections are concerned.

Where no vibration or movement can be expected, cable with non-flexible cores may be used.

### 8.8.5.4 Electrical connections

Connections between conductors or between a conductor and other equipment shall provide durable electrical continuity, and adequate mechanical strength and protection.                    WR-526.1

The selection of the means of connection shall take account of:                    WR-526.2

- the material of the conductor and its insulation;
- the number and shape of the wires forming the conductor;
- the cross-sectional area of the conductor;
- the number of conductors to be connected together;
- the temperature attained at the terminals in normal service;
- the provision of adequate locking arrangements in situations subject to vibration or thermal cycling.

### 8.8.5.5 Electrical connections in caravans and motor homes

In a caravan or motor caravan the wiring will be subjected to vibration, and so all wiring shall be protected against mechanical damage, either by location or by enhanced mechanical protection.                    WR-721.522.7.1

Wiring passing through metalwork shall be protected by suitable bushes or grommets, securely fixed in position.                    WR-721.522.7.1

Precautions shall be taken to avoid mechanical damage due to sharp edges or abrasive parts.                    WR-721.522.7.1

All cables, unless enclosed in rigid conduit, and all flexible conduit shall be supported at intervals not exceeding 0.4 m for vertical runs and 0.25 m for horizontal runs.                    WR-721.522.8.1.3

### 8.8.5.6 Low-voltage generating sets – generators

Electrical equipment associated with the generator shall be mounted securely and, if necessary, on anti-vibration mountings.                    WR-740.551.8

### 8.8.5.7 Protection against overcurrent

| | |
|---|---|
| Persons and livestock shall be protected against injury, and property shall be protected against damage, due to electromechanical stresses caused by any overcurrents likely to arise in live conductors. | WR-131.4 |

### 8.8.5.8 Protection against fault current

| | |
|---|---|
| Electrical equipment, including conductors, shall be provided with mechanical protection against electromechanical stresses of fault currents as necessary to prevent injury or damage to persons, livestock or property. | WR-131.5 |

### 8.8.5.9 Cross-sectional area of conductors

| | |
|---|---|
| The cross-sectional area of conductors shall be determined for both normal operating conditions and, where appropriate, for fault conditions according to:<br><br>• the electromechanical stresses likely to occur due to short-circuit and Earth fault currents;<br>• other mechanical stresses to which the conductors are likely to be exposed. | WR-132.6 |

### 8.8.5.10 Waves

| | |
|---|---|
| Wiring systems that may be subjected to waves (AD6) shall be protected from mechanical damage (i.e. impact, vibration and other mechanical stresses). | WR-522.3.3 |

## 8.9 Electromagnetic compatibility

Most car owners normally accept that when they drive near electric pylons their listening pleasure may be interrupted by loud crackles and/or buzzing noises. However, with the increased use of electronic equipment, the problem of interference has become one of our prime concerns.

Although most forms of interference are usually tolerated as being one of those things 'that you cannot do much about', the design of modern sophisticated equipment has become so susceptible to Electromagnetic Interference (EMI) that some form of regulation has had to be agreed.

Within Europe, this regulation is contained in the Electromagnetic Compatibility Directive 2004/108/EC (which repealed the original Directive 89/336/EEC), which clearly states that all electronic equipment shall be constructed so that:

- the electromagnetic disturbance it generates does not exceed a level allowing radio and telecommunications equipment and other apparatus to operate as intended;
- the apparatus has an adequate level of intrinsic immunity to electromagnetic disturbance.

Electromagnetic disturbances and EMI can seriously disrupt and even damage information technology (IT) systems and/or IT equipment, electronic components and circuits. Lightning, switching operations short-circuits and other electromagnetic phenomena can also cause overvoltages and EMI.

These effects are potentially more severe:

- where large metal loops exist;
- where different electrical wiring systems are installed in common routes, such as power supply, signalling and/or data communication cables connecting IT equipment within a building.

This is of particular relevance in (or near) rooms that are used for medical purposes, as electromagnetic disturbances can dramatically interfere with medical electrical equipment.

Potential sources of electromagnetic disturbances within an installation typically include:

- electric motors;
- fluorescent lighting;
- frequency convertors/regulators including variable-speed drives (VSDs);
- lifts;
- power distribution busbars and rectifiers;
- switchgear;
- switching devices for inductive loads;
- transformers;
- welding machines.

**Note:** For further information regarding electromagnetic disturbances, see the BS EN 50174 series of standards.

## 8.9.1 Requirements from the Regulations – electromagnetic compatibility

Tabe 8.18 gives the most common environmental requirements concerning electromagnetic compatibility (EMC).

**Table 8.18  Typical requirements – electromagnetic compatibility**

| | |
|---|---|
| Equipment | The use of electronic equipment shall not interfere with the operation of other equipment. |
| | All active electronic devices shall comply with the EMC Directive. |
| CE marking | Only CE (Conformity Europe) marked equipment may be offered for sale, and all active equipment connected to an electrical installation shall carry a CE mark. |
| Apparatus cases | Input/output connections from apparatus cases should always be of non-screened non-balanced signalling cable, and are normally restricted in length. |
| Atmospheric disturbances | To counteract the effects of storms, it is generally recommended that all equipment should be capable of withstanding (as a minimum) the following overvoltages: |
| | Magnitude        2000 V<br>Rise time        1.2 µs<br>Middle voltage time    50 µs |
| Equipment immunity levels | Equipment should be immune to induced common-mode voltages. |
| | Equipment should not experience a permanent loss of availability or suffer component damage for any induced common-mode voltage within this range. |
| Magnetic field | As low-frequency fields can influence cathode ray tubes, equipment should be capable of withstanding the following intensities: |
| | Hz          A/m<br>5         0.8<br>50       3.0<br>250     1.5 |
| Power supply lines | Equipment should be immune to the following high-frequency bursts: |
| | Initial peak-to-peak voltage    1 kV<br>Burst repetition rate        5 kHz |
| Transients | All electronic equipment should be capable of withstanding: |
| | • transients (either directly induced or indirectly coupled), so that no damage or failure occurs during operation;<br>• without damage or abnormal operation, transient non-repetitive surges. |

### 8.9.1.1 Measures to reduce EMI

To reduce the effects of EMI, the following measures shall be considered:

> Where screened signal or data cables are used, care WR-444.4.2.1
> should be taken to limit the fault current from power
> systems flowing through the screens and cores of
> signal cables, or data cables, which are earthed.

 **Note:** Additional conductors (such as a bypass conductor for screen reinforcement) may be necessary (see Figure 8.13).

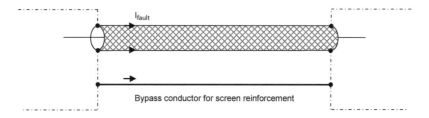

**Figure 8.13** Bypass conductor for screen reinforcement to provide a common equipotential bonding system. (Courtesy of the BSI.)

Other measures include:

> • Including surge protective devices and/or filters to WR-444.4.2.1
>   improve the EMC for electrical equipment sensitive
>   to electromagnetic disturbances.
> • Power cables (i.e. line, neutral and any protective Earth
>   conductors) should be installed close together in order
>   to minimize cable loop areas.
> • Adequate separation should be provided between
>   power and signal cables.
> • The installation of an equipotential bonding network.

### 8.9.1.2 TN system

To minimize electromagnetic disturbances, the following requirements shall be met:

>  A PEN conductor **shall not** be used downstream WR-444.4.3.1
> of the origin of the installation.

 **Note:** In Great Britain, Regulation 8(4) of the Electricity Safety, Quality and Continuity Regulations 2002 **prohibits** the use of PEN conductors in consumers' installations.

The installation shall have separate neutral and pro-tective conductors downstream of the origin of the installation (Figure 8.14).

WR-444.4.3.2

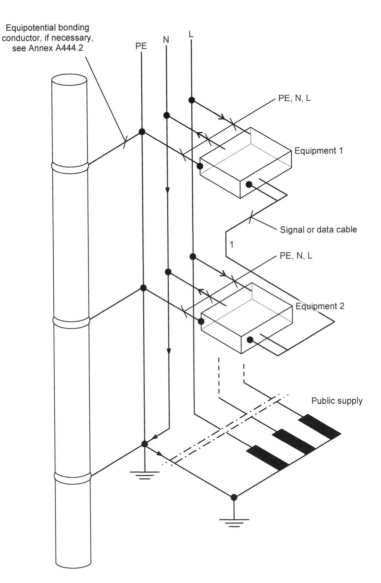

**Figure 8.14** Avoidance of neutral conductor currents in a bonded struc-ture by using an installation forming part of a TN-S system from the origin of the public supply up to and including the final circuit within a building. (Courtesy of the BSI.)

> Where the complete low-voltage installation    WR-444.4.3.3
> (including the transformer) is operated only by the
> user, an installation forming part of a TN-S system
> shall be installed (Figure 8.15).

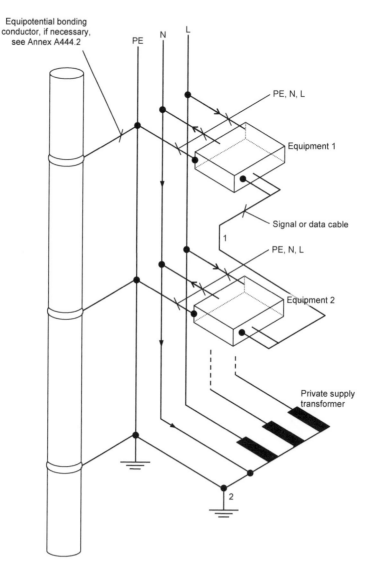

**Figure 8.15** Avoidance of neutral conductor currents in a bonded structure by using an installation forming part of a TN-S system downstream of
a consumer's private supply transformer. (Courtesy of the BSI.)

## 8.9.1.3 TT system

If an installation similar to that shown in Figure 8.16   WR-444.4.4
forms part of a TT system, consideration shall be given
to overvoltages that might exist between live parts and
any extraneous-conductive-parts of different buildings
that are connected to different Earth electrodes.

For a TN-S system, the use of an isolating transformer should also be considered.

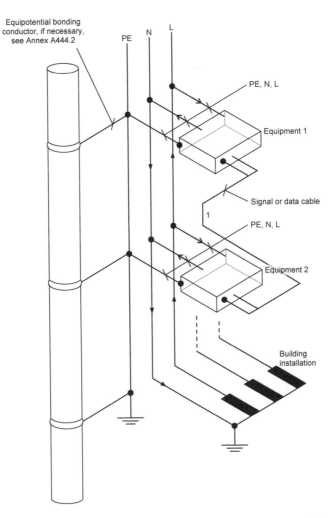

**Figure 8.16** Installation forming part of a TT system within a building installation. (Courtesy of the BSI.)

> Where screened signal and/or data cables are common    WR-444.4.4
> to several buildings that are supplied from an installa-
> tion forming part of a TT system, the use of a bypass
> equipotential bonding conductor (Figure 8.17) or
> single-point bonding shall be considered.

 **Note:** The bypass conductor shall have a minimum cross-sectional area of 16 mm² copper or equivalent.

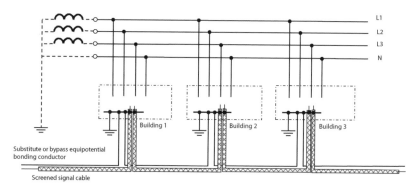

**Figure 8.17** Example of a substitute or bypass equipotential bonding conductor in an installation forming part of a TT system. (Courtesy of the BSI.)

> If the live conductors of the supply into any of the    WR-444.4.4
> buildings exceed 35 mm² in cross-sectional area, the
> bypass conductor shall have a minimum cross-sectional
> area in accordance with Table 8.19.

**Table 8.19** Minimum cross-sectional area of the main protective bonding conductor in relation to the neutral of the supply

| Copper-equivalent cross-sectional area of the supply neutral conductor | Minimum copper equivalent* cross-sectional area of the main protective bonding conductor |
| --- | --- |
| 35 mm² or less | 10 mm² |
| Over 35 mm² up to 50 mm² | 16 mm² |
| Over 50 mm² up to 95 mm² | 2 25 mm² |
| Over 95 mm² up to 150 mm² | 2 35 mm² |
| Over 150 mm² | 2 50 mm² |

### 8.9.1.4 Multiple-source TN or TT power supplies

| | |
|---|---|
| TN or TT multiple-source power supplies to an installation shall be earthed at one point only. | WR-444.4.6 |

For a TN system, a single point of connection only shall be made, as shown in Figure 8.18, so as to avoid having the neutral current flowing through the protective conductor.

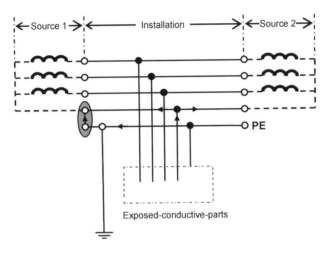

**Figure 8.18** TN multiple-source power supply with a single connection between the PEN and Earth. (Courtesy of the BSI.)

### 8.9.1.5 Transfer of supply

To prevent electromagnetic fields due to stray currents in the main supply system of an installation:

| | |
|---|---|
| The transfer from one supply to an alternative supply for any installation forming part of a TN system shall be via a multi-pole switching device that switches both the line conductors and the neutral conductor (if any). | WR-444.4.7 |

### 8.9.1.6 Separate buildings

| | |
|---|---|
| Where different buildings have separate equipotential bonding systems, metal-free optical fibre cables or other non-conducting systems (such as a microwave signal transformer) should be used for signal and data transmission. | WR-444.4.9 |

### 8.9.1.7 Earthing and equipotential bonding

#### 8.9.1.7.1 Within a single building

| | |
|---|---|
| Within a single building, all protective and functional earthing conductors of an installation shall normally be connected to the main earthing terminal. | WR-444.5.1.1 |

#### 8.9.1.7.2 Between several buildings

| | |
|---|---|
| Where a number of installations have separate earthing arrangements, any protective conductor common to any of these installations shall either be capable of carrying the maximum fault current that is likely to flow through them, or be earthed within one of the installations only (provided that that particular installation is insulated from the earthing arrangements of any of the other installations). | WR-444.5.1.2 |

 Where interconnection of the Earth electrodes is not possible or practicable, it is recommended that separation of communications networks is applied, for example, by using optical or radio links.

### 8.9.1.8 Equipotential bonding networks

To avoid or reduce the possibility of electromagnetic disturbances affecting an installation (in particular IT systems):

| | |
|---|---|
| • metal sheaths, screens or armouring of cables shall be bonded to the common bonding network (CBN) unless such bonding is required to be omitted for safety reasons;<br>• where screened signal or data cables are earthed, care shall be taken to limit the fault current from power systems flowing through the screens and cores of signal cables or data cables;<br>• the impedance of equipotential bonding connections intended to carry functional Earth currents having high-frequency components shall be as low as is practicable, and this should be achieved by the use of multiple, separated bonds that are as short as possible. | WR-444.5.2 |

**Note:** Where bonds of up to 1 m long are used, their inductive reactance and impedance of route can be reduced by choosing a conductive braid or a bonding strap/strip (with a width to thickness ratio of at least 5:1 and a length to width ratio no greater than 5:1).

 Bare conductors shall be protected against corrosion at their supports and on their passage through walls.

| | |
|---|---|
| The sizing and installation of an equipotential bonding ring network shall have the following minimum nominal dimensions:<br><br>• flat cross-section: 25 mm × 3 mm;<br>• round diameter: 8 mm. | WR-444.5.3 |

| | |
|---|---|
| The following parts shall be connected to the equipotential bonding network:<br><br>• metallic containment, conductive screens, conductive sheaths or armouring of data transmission cables or of information technology equipment;<br>• functional earthing conductors of antenna systems;<br>• conductors of the earthed pole of a d.c. supply for information technology equipment;<br>• functional earthing conductors;<br>• protective conductors. | WR-444.5.3.1 |

## 8.9.1.9 Information technology installations – earthing arrangements and equipotential bonding

| | |
|---|---|
| To enable information technology installations to be connected to the main earthing terminal by the shortest practicable route from any point in the building, consideration should be given to extending the main earthing terminal of the building by using one or more earthing busbars. | WR-444.5.7.1 |

**Note:** The earthing busbar should be accessible throughout its entire length, and bare conductors should be protected to prevent corrosion.

| | |
|---|---|
| In installations connected to a supply having a capacity of 200 A per phase or more, the cross-sectional area of the earthing busbar shall be not less than 50 mm² copper. | WR-444.5.7.2 |

 **Note:** For supplies with a capacity of less than 200 A per phase, the earthing busbar shall be selected in accordance with Table 8.20.

 Where the earthing busbar is used as part of a dc return current path, its cross-sectional area shall be selected according to the expected d.c. return currents.

### 8.9.1.10 Segregation of circuits

Cables that are used at voltage Band II (low voltage) and cables that are used at voltage Band I (extra-low voltage), and which share the same cable management system or the same route, shall be installed according to the following requirements:

**Table 8.20  Minimum cross-sectional area of the main protective bonding conductor in relation to the neutral of the supply**

| Copper-equivalent cross-sectional area of the supply neutral conductor | Minimum copper equivalent* cross-sectional area of the main protective bonding conductor |
|---|---|
| 35 mm² or less | 10 mm² |
| Over 35 mm² up to 50 mm² | 16 mm² |
| Over 50 mm² up to 95 mm² | 2 25 mm² |
| Over 95 mm² up to 150 mm² | 2 35 mm² |
| Over 150 mm² | 2 50 mm² |

| | |
|---|---|
| Each part of a circuit shall be arranged such that the conductors are **not** distributed over different multi-core cables, conduits, ducting systems, franking systems or tray or ladder systems. | WR-444.6.1 WR-521.8.1 |
| The line and neutral conductors of each final circuit shall be electrically separate from those of every other final circuit, so as to prevent the indirect energizing of a final circuit intended to be isolated. | WR-521.8.2 |
| Where multi-core cables are installed in parallel each cable shall contain one conductor of each line. | WR-444.6.1 WR-521.8.1 |

A voltage Band I circuit shall not be contained in the same wiring system as a Band II circuit unless: — WR-444.6.1 WR-528.1

- each conductor of a multi-core cable is insulated for the highest voltage present in the cable;
- for a multi-core cable, the cores of the Band I circuit are separated from the cores of the Band II circuit by an earthed metal screen of equivalent current-carrying capacity to that of the largest core of a Band II circuit.

In the event of crossing (or being in the proximity of) underground telecommunication cables and underground power cables, a minimum clearance of 100 mm shall be maintained, and: — WR-444.6.1 WR-528.2

- a fire-retardant partition shall be provided between the cables; and
- for crossings, mechanical protection between the cables shall be provided.

 **Note:** Electrical safety and EMC might produce different segregation or separation requirements. The design shall meet both requirements.

The minimum distance between information technology cables and discharge, neon and mercury vapour (or other high-intensity discharge) lamps shall be 130 mm. Data wiring racks and electrical equipment shall always be separated. — WR-444.6.2

### 8.9.1.11 Equipotential bonding networks in buildings with several floors

For buildings with several floors, it is recommended that, on each floor, an equipotential bonding system be installed and that the bonding systems of the different floors should be interconnected, at least twice, by protective conductors (selected in accordance with the requirements of Chapter 54 of BS 7671:2008 (Incorporating Amendment No. 1).

 Figure 8.19 shows an example of bonding networks in common use.

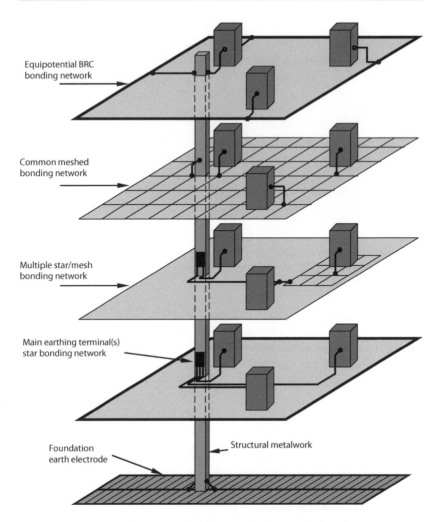

Equipotential BRC
bonding network

Common meshed
bonding network

Multiple star/mesh
bonding network

Main earthing terminal(s)
star bonding network

Foundation
earth electrode

Structural metalwork

**Figure 8.19** Example of equipotential bonding networks in a structure without a lightning protection system. (Courtesy of the BSI.)

 **Note:** Each floor represents a different type of bonding network.

### 8.9.1.12 Protection against voltage disturbances and measures against electromagnetic disturbances

Persons and livestock shall be protected against injury, WR-131.6.1
and property shall be protected against any harmful
effects, as a consequence of:

• a fault between live parts of circuits supplied at different voltages;

| | |
|---|---|
| • overvoltages such as those originating from atmos-pheric events or from switching; | WR-131.6.2 |
| • undervoltage and any subsequent voltage recovery. | WR-131.6.3 |
| The installation shall have an adequate level of immunity against electromagnetic disturbances so as to function correctly in the specified environment. | WR-131.6.4 |
| The installation design shall take into consideration the anticipated electromagnetic emissions, generated by the installation or the installed equipment. | WR-131.6.4 |

Also see the BS EN 62305 series of standards for further information concerning protection against lightning strikes.

### 8.9.1.13 Electrical installations

Electrical installations **shall** be arranged so that they do not mutually interfere (including EMI) with other electrical installations and non-electrical installations in a building.

| | |
|---|---|
| All fixed installations shall be in accordance with the EMC Regulations and with the relevant EMC standard. | WR-332.1 |

The previous EMC Directive (i.e. Directive 89/336/EC) was replaced by EMC Directive 2004/108/EC on 15 December 2004, and has been transposed into UK law by the EMC Regulations 2006 (SI 2006/3418), which came into force on 20 July 2007. These new Regulations have introduced a new regime for fixed installations, and require that **all** electrical and electronic apparatus marketed in the UK, including imports, satisfy the requirements of the EMC Directive.

| | |
|---|---|
| Immunity levels of equipment shall be chosen taking into account: <br> • the electromagnetic influences that can occur when connected and erected as for normal use; and <br> • the intended level of continuity of service necessary for the application. (See BS EN 50082.) | WR-515.3.1 |
| Equipment shall be chosen with sufficiently low emission levels so that it cannot cause unacceptable EMI with other electrical equipment. (See BS EN 50081.) | WR-515.3.2 |

Consideration shall be given by the planner and
designer of the electrical installation to measures reducing the effect of induced voltage disturbances and EMI).

WR-332.2

To minimize voltages induced by lightning, the area
of all wiring loops shall be as small as possible.

WR-712.444.4.4

### 8.9.1.14 Wiring installations

The choice of the type of wiring system and the method
of installation shall include consideration of the
following:

- the nature of the location;
- the nature of the structure supporting the wiring;
- accessibility of wiring to persons and livestock;
- voltage;
- the electromechanical stresses likely to occur due to
  short-circuit and Earth fault currents;
- EMI;
- other external influences (e.g. mechanical, thermal
  and those associated with fire) to which the wiring
  is likely to be exposed during the erection of the
  electrical installation or in service.

WR-132.7

Every installation shall be divided into circuits, as necessary, to:

WR-314.1

- avoid danger and minimize inconvenience in the
  event of a fault;
- ensure safe inspection, testing and maintenance;
- reduce the possibility of unwanted tripping of RCDs
  due to excessive protective conductor (PE) currents
  not due to a fault;
- prevent the indirect energizing of a circuit that is
  intended to be isolated;
- reduce the effects of EMI;
- take account of hazards that may arise from the failure of a single circuit (e.g. a lighting circuit).

The electrical installation shall be arranged in such a way
that no mutual detrimental influence will occur between
electrical installations and non-electrical installations.

WR-132.11

EMI shall be taken into account.

### 8.9.1.15 Cables and conductors

| | |
|---|---|
| The conductors of an a.c. circuit installed in a ferromagnetic enclosure shall be arranged so that the line conductors, the neutral conductor (if any) and the appropriate protective conductor are all contained in the same enclosure. | WR-521.5.1 |
| Where such conductors enter a ferrous enclosure, they shall be arranged such that the conductors are only collectively surrounded by ferrous material. | WR-521.5.1 |
| Single-core cables armoured with steel wire or steel tape **shall not** be used for an a.c. circuit. | WR-521.5.2 |

**Notes:**

1. The steel wire or steel tape armour of a single-core cable is regarded as a ferromagnetic enclosure.
2. For single-core armoured cables, the use of aluminium armour may be considered.

### 8.9.1.16 Medical locations

It is recommended that radial wiring patterns are used to avoid 'Earth loops', which may cause EMI.

| | |
|---|---|
| Special considerations have to be made concerning EMI and EMC in medical locations. | WR-710.444 |

## 8.10 Fire

A fire will normally start when sufficient thermal energy from, for example, a burning cigarette or an electric short-circuit, is supplied to a combustible material. Following ignition, the fire will then produce its own thermal energy, some of which will be used as feedback to maintain combustion, and some of which transferred via radiation and convection to other materials. These materials may also ignite and spread the fire.

The environmental conditions relating to the occurrence, development and spread of fire within a building and its effect on electrotechnical products exposed to fire is primarily covered by Section 8 (Fire Exposure) of IEC 721.2. This section provides background information for selecting the appropriate parameters and severities related to exposure of products to fire. More detailed information on fire condition characteristics and fire hazard testing is contained in specialist documentation.

The development of a fire generally consists of three processes:

- thermal;
- aerodynamic;
- chemical.

As a rule, radiation, convection and flame spread are the dominant physical factors.

### 8.10.1 Fire growth

Once a fire has started in a space (e.g. a room), its growth and spread is determined by:

- site;
- volume;
- arrangement of the fuel or fire load, its distribution, continuity, porosity and combustion properties;
- aerodynamic conditions of the space;
- shape and size of the space;
- thermal properties of the space.

During the growth of a fire, a hot layer of gas builds up under the ceiling of the space. Under certain conditions, this gas layer can give rise to a rapid fire growth, and flashover might occur.

### 8.10.2 Flashover

One normally defines flashover as the time when flames begin to emerge from the openings of the space, which correlates with a temperature of 500–600°C in the upper gas layer.

Flashover marks the transition from the growing fire (pre-flashover) to the fully developed fire (post-flashover).

#### 8.10.2.1 Pre-flashover

A pre-flashover fire primarily concerns the operation and function of products (e.g. detectors, alarm systems, associated cables and sprinklers) that are vital to maintaining the level of safety required for escape and/or the rescue of people caught in a fire.

### 8.10.2.1.1 Characteristics of pre-flashover fire

The ignitability properties of exposed material will depend on:

- the heat supplied;
- the exposure time;
- the presence, or not, of flames;
- the geometrical location;
- the thermal data;

together with time variations such as:

- rate of heat release;
- rate of flame spread;
- gas temperature.

### 8.10.2.1.2 Fire hazard of a pre-flashover

The fire hazard of a pre-flashover situation is normally considered in terms of a series of probabilities, which depend on:

- the presence of ignition sources;
- the presence of products;
- the product fire performance properties;
- the environmental factors;
- the presence of people;
- the presence/operation of detection and suppression devices;
- the availability of escape.

### 8.10.2.2 Post-flashover

Whilet most standards are normally concerned with conditions during the pre-flashover stage of a fire, conditions following flashover must also be considered. A post-flashover fire can seriously damage some of the structural and load-bearing elements of a building, and the fire can then, quite easily, spread from one fire space to another via partitions and ventilation systems. This can, of course, seriously damage electrical equipment located in these voids. For example, in a large space it is quite possible that a fire, small in relation to that space, could be large enough to damage some of the structural elements in the post-flashover state. An important factor of the post-flashover fire, which is often overlooked, is the amount of smoke and toxic gases that can affect people in escape routes and remote safety areas in a building. Smoke and toxic gases can also significantly affect equipment.

### 8.10.2.2.1 Characteristics of post-flashover fire

The main characteristics of a post-flashover fire are:

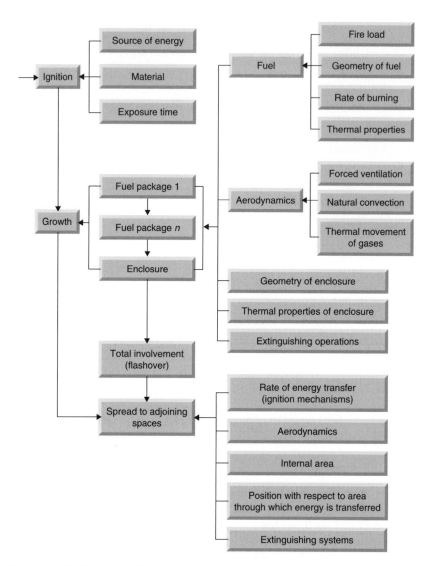

**Figure 8.20** Factors affecting the ignition, growth and spread of fire in a building.

- the rate of heat release;
- the gas temperature;
- the geometrical and thermal data for external flames;
- the smoke and its optical properties;
- the composition of the combustion products, particularly corrosive and toxic gases.

The possibility of a large external fire spreading from one storey to another in the same building (and eventually from one building to another) must also be considered. For these cases the first three characteristics – i.e. primarily gas temperature, geometrical and thermal data for the flames emerging from the window openings – are the most relevant.

## 8.10.3 Characteristics of smoke and gases as a fire product

Smoke is a mixture of heated gases, small liquid droplets and solid particles from the combustion. During a fire (pre- and post-flashover), smoke will be distributed within the building through the airflow between rooms and via ventilation ducts etc. In most circumstances this can have disastrous effects, because smoke can not only damage, and in some cases even destroy, property, it can also prevent the functioning of critical equipment. Most of the effects of smoke are of a chemical nature, and the most prevalent is destruction of or damage to electrotechnical products, in particular corrosion caused by hydrogen chloride, which is a substance in smoke.

Metal surfaces that are exposed to air under normal (non-fire) conditions often have a chloride deposit of up to $10 \, mg/m^2$. Such an amount is generally not harmful. However, after exposure to smoke from a fire involving polyvinyl chloride (PVC), a surface contamination of up to thousands of milligrams per square metre can be found, often causing significant damage. Chloride contamination of electrotechnical equipment can be removed by, for instance, detergents, solvents, neutralizing agents, ultrasonic vibrations and clean-air jets, but the procedures are not always effective, sometimes giving a temporary but not permanent cure.

Experiments, involving PVC-coated electrical wires and carried out on a scale large enough to be representative of real fires, are currently in hand.

## 8.10.4 Building designs

In the design of buildings, the fire design of load-bearing structural elements and partitions is normally considered as a national problem, and is directly related to the results of standard national and (when available) international fire resistance tests. In such tests the specimen is exposed, in a furnace, to a temperature rise that is varied over time and within specified limits, according to the particular test being used.

Over the last few decades, rapid progress has been made in the development of analytical and computational methods for determining the fire design of load-bearing and separating structures and structural elements. In the long term, it is foreseeable that this will develop into an analytical and/or computational design, directly based on a natural fire exposure. The design will be specified with regard to the combustion characteristics of a fire load and the geometrical, ventilation and thermal properties of the fire space.

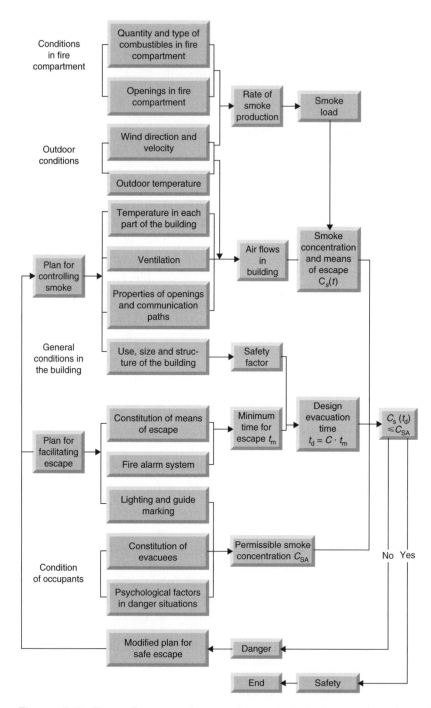

**Figure 8.21** Flow diagram of a smoke-control design system in a building.

## 8.10.5 Test standards

Fire tests on building materials, components and structures normally focus on the characteristics of pre-flashover fire. Simplified full-scale (i.e. room) tests for surface products' reaction to smoke, and in particular toxic combustion products, are already available, but considerable development work needs to be completed before a useful small-scale test is available.

If no mathematical model of a small-scale test is available, the test results should be statistically correlated directly to full-scale test data. If a validated mathematical model of a small-scale test exists, important material characteristics controlling the space fire growth can be given quantitative values, which can then be used as input data in mathematical models of a full-scale pre-flashover space fire for specified scenarios.

With a view to practical, long-term use, the results of small-scale reaction to fire tests to predict fire hazard should be based on a fundamental and scientific approach. Figure 8.22 outlines the structure of such an approach.

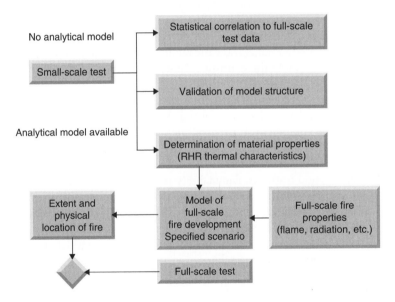

**Figure 8.22** Combination of basic property tests and mathematical models for assessing the contribution of a tested material or product to the overall fire safety.

## 8.10.6 Other related standards and specifications

IEC 60695 series   Fire hazard testing – Guidance, tests and specifications for assessing fire hazard of electrotechnical products

ISO 5657   Fire tests – Reaction to fire – Ignitability of building products

| ISO 5658 | Reaction to fire tests – Spread of flame on building products and vertical configuration |
| ISO 5660 | Fire tests – Reaction to fire – Rate of heat release from building products |
| ISO 9705 | Fire tests – Full scale room test for surface products |
| ISO TR 5924 | Fire tests – Reaction to fire – Smoke generated by building products (dual-chamber test) |
| ISO TR9112.1 | Toxicity testing of fire effluents – General |

### 8.10.7 Requirements from the Regulations – fire

In most contracts, reference is made to the IEC 60695 series of standards, which cover the assessment of electrotechnical products against a nominated fire hazard.

The CENELEC (Comité Européen de Normalisation Electrotechnique) standards, on the other hand, show the requirement for equipment to operate in fire hazardous areas as three distinct clauses:

**Class F0** – no special fire hazard envisaged. This is considered a normal service condition and, except for the characteristics inherent to the design of the equipment, no special measures need to be taken to limit flammability.

**Class F1** – equipment subject to fire hazard. This is considered an abnormal condition, and restricted flammability is required. Self-extinction of fire shall take place within a specified time period. Poor burning is permitted with negligible energy consumption. The emission of toxic substances shall be minimized. Materials and products of combustion shall, as far as possible, be halogen free and shall contribute with a limited quantity of thermal energy to an external fire.

**Class F2** – equipment subject to external fire. This is considered an abnormal condition and, in addition to the requirements of Class F1, the equipment shall (by means of special provisions) be able to operate for a given time period when subjected to an external fire. Materials are normally expected to confirm to those requirements defined in EN 60721.3.3 and EN 60721.3.4.

The 17th edition of BS 7671 was updated so as to maintain technical alignment with CENELEC harmonization documents. One of the main changes concerned the requirements for safety services (e.g. emergency escape lighting, fire alarm systems, installations for fire pumps, fire rescue service lifts, smoke and heat extraction equipment), which now need to be observed.

Safety services have also been expanded in line with International Electro-technical Commission (IEC) standardization.

### 8.10.7.1 Electrical installations

In electrical installations, risk of injury may result from excessive temperatures likely to cause burns, fires and other injurious effects. To guard against this happening:

> Equipment in surroundings susceptible to risk of fire       WR-132.5.2
> or explosion shall be constructed or protected so as to
> prevent danger.
>
>
> The choice of the type of wiring system and the method       WR-132.7
> of installation shall include consideration of the
> following:
>
> • the nature of the location;
> • the nature of the structure supporting the wiring;
> • accessibility of wiring to persons and livestock;
> • voltage;
> • the electromechanical stresses likely to occur due
>   to short-circuit and Earth fault currents;
> • EMI;
> • other external influences (e.g. mechanical, thermal
>   and those associated with fire) to which the wiring
>   is likely to be exposed during the erection of the
>   electrical installation or in service.

**Note:** In structures where the shape and dimensions are such as will facilitate the spread of fire, precautions shall be taken to ensure that the electrical instal-lation cannot propagate a fire (e.g. by the chimney effect).

### 8.10.7.1 Selection and erection of installations in locations of national, commercial, industrial or public significance

The following measures may be considered:

• installation of mineral insulated cables according to BS EN 60702;
• installation of cables with improved fire-resisting characteristics in case of a fire hazard;
• installation of cables in non-combustible solid walls, ceilings and floors;
• installation of cables in areas with constructional partitions having a fire-resisting capability for a time of 30 minutes or 90 minutes.

**Note:** Where these measures are not practicable improved fire protection may be possible by the use of reactive fire protection systems.

### 8.10.7.3 Precautions where a particular risk of fire exist

In locations that include buildings or rooms with assets of significant value (such as national monuments, museums and other public buildings.) or buildings such as railway stations and airports (that are generally considered to be of public significance) and or buildings or facilities such as laboratories, computer centres and certain industrial and storage facilities can be of commercial or industrial significance.

| | |
|---|---|
| Electrical equipment shall be so selected and erected so that its temperature in normal operation will not cause a fire. | WR-422.1.2 |

 A temperature cut-out device should have a **manual** reset.

Where BE2 conditions exist and where there is a risk of fire due to the manufacture, processing or storage of flammable materials such as:

- barns (due to the accumulation of dust and fibres);
- woodworking facilities;
- paper mills and textile factories (due to the storage and processing of combustible materials);

a fire risk will be present and the following precaution shall be observed:

| | |
|---|---|
| A cable shall, as a minimum, satisfy the test under fire conditions specified in BS EN 60332–1–2. | WR-422.3.4 |
| A cable not completely embedded in non-combustible material such as plaster or concrete or otherwise protected from fire shall meet the flame propagation characteristics as specified in BS EN 60332–1–2. | WR-422.3.4 |
| A conduit system shall satisfy the test under fire conditions specified in BS EN 61386–1. | WR-422.3.4 |
| A cable trunking system or cable ducting system shall satisfy the test under fire conditions specified in BS EN 50085. | WR-422.3.4 |
| A cable tray system or cable ladder shall satisfy the test under fire conditions specified in BS EN 61537. | WR-422.3.4 |

| | |
|---|---|
| A powertrack system shall satisfy the test under fire conditions specified in the BS EN 61534 series. | WR-422.3.4 |
| Precautions shall be taken such that a cable or wiring system cannot propagate flame. | WR-422.3.4 |
| Where the risk of flame propagation is high the cable shall meet the flame propagation characteristics specified in the appropriate part of the BS EN 50266 series. | WR-422.3.4 |
| Conduit and trunking systems shall be in accordance with BS EN 61386–1 and BS EN 50085–1, respectively, and shall meet the fire-resistance tests within these standards. | WR-422.4.103 |

### 8.10.7.4 Protection against the risk of fire

| | |
|---|---|
| Where it is necessary to limit the consequence of fault currents in a wiring system from the point of view of fire risk, the circuit shall be either: | WR-532.1 |

- protected by an RCD for fault protection; and
- the RCD shall be installed at the origin of the circuit to be protected; and
- the RCD shall switch all live conductors; and
- the rated residual operating current of the RCD shall not exceed 300 mA; or
- the circuit will need to be continuously monitored by an insulation monitoring device that initiates an alarm on the occurrence of an insulation fault.

 When selecting and erecting a luminaire, the thermal effects of radiant and convected energy on the surroundings shall be taken into account, including the fire resistance of adjacent material:

- at the point of installation; and
- in the thermally affected areas.

| | |
|---|---|
| A device for protection against fault current need not be provided where the wiring is installed in such a manner as to reduce to a minimum the risk of fire or danger to persons. | WR-434.3 |

The omission of devices for protection against overload is permitted for circuits supplying current-using equipment where unexpected disconnection of the circuit could cause danger or damage. Examples of such circuits are:

- a circuit supplying a fire-extinguishing device;
- a circuit supplying a safety service, such as a fire alarm or a gas alarm.

In such situations consideration should be given to the provision of an overload alarm.

### 8.10.7.5 Protection against thermal effects

Persons, livestock and property **shall** be protected against harmful effects of heat or fire that may be generated or propagated in electrical installations. These effects include:

- heat accumulation, heat radiation, hot components or equipment;
- failure of electrical equipment such as protective devices, switchgear, thermostats, temperature limiters, seals of cable penetrations and wiring systems;
- overcurrent;
- insulation faults or arcs, sparks and high-temperature particles;
- harmonic currents;
- external influences such as lightning surge.

**Note:** The use of supplementary bonding does not exclude the need to disconnect the supply for other reasons, such as protection against fire, thermal stresses in equipment, etc.

| | |
|---|---|
| Electrical heating appliances used for the breeding and rearing of livestock shall comply with BS EN 60335–2–71 and shall be fixed so as to maintain an appropriate distance from livestock and combustible material, to minimize any risks of burns to livestock and of fire. | WR-705.422.6 |

For radiant heaters the clearance shall be not less than 0.5 m or such other clearance as recommended by the manufacturer.

| | |
|---|---|
| For fire protection purposes, RCDs shall be installed with a rated residual operating current not exceeding 300 mA. | WR-705.422.7 |

RCDs shall disconnect all live conductors.    WR-705.422.7

Where improved continuity of service is required,    WR-705.422.7
RCDs not protecting socket-outlets shall be of the S
type or have a time delay.

In locations where a fire risk exists, conductors    WR-705.422.8
of circuits supplied at extra-low voltage shall be
protected:

- either by barriers or enclosures affording
  a degree of protection of IPXXD or
  IP4X; or
- in addition to their basic insulation, by an
  enclosure of insulating material.

## 8.10.7.6 Protection against fire heat generation

Lighting equipment such as incandescent lamps,    WR-711.422.4.2
spotlights and small projectors and other equip-
ment or appliances with high-temperature surfaces,
shall be suitably guarded, and installed and located
in accordance with the relevant standard.

Showcases and signs shall be constructed of mate-    WR-711.422.4.2
rial having an adequate heat resistance, mechani-
cal strength, electrical insulation and ventilation,
taking into account the combustibility of exhibits
in relation to the heat generation.

Stand installations containing a concentration of    WR-711.422.4.2
electrical equipment, luminaires or lamps liable to
generate excessive heat shall not be installed unless
adequate ventilation provisions are made (e.g. a
well-ventilated ceiling constructed of incombustible material).

## 8.10.7.7 Protection against fire caused by electrical equipment

Persons, livestock and property shall be protected    WR-421.1
against harmful effects of heat and fire which may be
generated or propagated in electrical installations.

Fixed electrical equipment shall be selected and erected such that its temperature in normal operation will not cause a fire.    WR-421.1.2

The heat generated by electrical equipment shall not cause danger or harmful effects to adjacent fixed material or to material.    WR-421.1.2

Where fixed equipment may attain surface temperatures that could cause a fire hazard to adjacent materials, the equipment shall be:

- mounted on a support that has low thermal conductance; or
- within an enclosure that will withstand such temperatures as may be generated; or
- screened by materials of low thermal conductance which can withstand the heat emitted by the electrical equipment; or
- mounted so as to allow safe dissipation of heat and    WR-421.1.2
  at a sufficient distance from adjacent material on which such temperatures could have deleterious effects.

 **Note:** Any means of support shall be of low thermal conductance.

Where arcs, sparks or particles at high temperature could be emitted by fixed equipment during normal service, the equipment shall be:    WR-421.1.3

- totally enclosed in arc-resistant material; or
- screened by arc-resistant material from materials upon which the emissions could have harmful effects; or
- mounted so as to allow safe extinction of the emissions at a sufficient distance from materials upon which the emissions could have harmful effects.

 **Note:** Arc-resistant material used for this protective measure shall be non-ignitable, of low thermal conductivity and of adequate thickness to provide mechanical stability.

| | |
|---|---|
| Fixed equipment that could cause a concentration and focus of heat shall be at a sufficient distance from any fixed object or building element. | WR-421.1.4 |
| Precautions shall be taken to prevent the spread of liquid, flame and other products of combustion for electrical equipment containing a significantly high amount of flammable liquid. | WR-421.1.5 |
| Materials used for the construction of enclosures for electrical equipment shall be capable of resisting heat and fire in accordance with an appropriate product standard. | WR-421.1.6 |

 It is recommended that, wherever possible, every termination of a live conductor or connection or joint between live conductors is contained within an enclosure.

### 8.10.7.8  Safety services

Safety services, by their very nature, need to be frequently regulated by statutory authorities – whose requirements are mandatory and have to be followed.

| | |
|---|---|
| Safety services may be required to operate at all material times that people or livestock are at risk, including during mains and local supply failure and through fire conditions. To meet this requirement, specific sources, equipment, circuits and wiring are necessary. | WR-560.5.1 |

 Some applications also have particular requirements.

| | |
|---|---|
| Circuits for safety services shall not pass through locations exposed to fire risk (BE2) unless they are fire-resistant. | WR-560.7.2 |
| For safety services required to operate in fire conditions:<br><br>• a safety source of supply shall be selected that will maintain a supply of adequate duration;<br>• equipment shall be provided (either by construction or by erection) with protection ensuring fire resistance of adequate duration. | WR-560.5.2 |

**Note:** The safety source is generally additional to the normal source (i.e. the public supply network), and examples include:

- emergency lighting;
- fire pumps;
- fire rescue service lifts;
- fire detection and alarm systems;
- carbon monoxide detection and alarm systems;
- fire evacuation systems;
- smoke ventilation systems;
- fire services communication systems;
- essential medical systems;
- industrial safety systems.

| | |
|---|---|
| The device protecting a conductor against overload should be installed so as to reduce to a minimum the risk of danger to persons. | WR-433.2.2 |

### 8.10.7.9 Protection against fault current

| | |
|---|---|
| A device for protection against fault current need not necessarily be provided if the wiring has been installed in such a manner as to reduce to a minimum the risk of fire or danger to persons. | WR-434.3 |

The omission of devices for protection against overload is permitted for circuits supplying current-using equipment where unexpected disconnection of the circuit could cause danger or damage.

Examples of such circuits are:

- a circuit supplying a fire-extinguishing device;
- a circuit supplying a safety service, such as a fire alarm or a gas alarm.

 In such situations, consideration should be given to the provision of an overload alarm.

### 8.10.7.10 Wiring systems

 A wiring system shall be installed so that the general building structural performance and fire safety are not reduced.

The risk of spread of fire shall be minimized by the selection of appropriate materials and erection.

WR-527.1.1

Cables not complying with the flame propagation requirements of BS EN 60332–1–2 shall be limited to short lengths for connection of appliances to the permanent wiring system, and shall not pass from one fire-segregated compartment to another.

WR-527.1.4

Where no fire alarm system is installed in a building used for exhibitions etc., cable systems shall be either:

WR-711.521

- flame retardant to BS EN 60332–1–2 and low smoke to BS EN 61034–2; or
- single-core or multi-core unarmoured cables enclosed in metallic or non-metallic conduit or trunking, providing a degree of fire protection of at least IP4X.

Where both the live circuit conductors are uninsulated, they shall either:

WR-559.11.4.1

- be provided with a protective device complying with the requirements of Regulation 559.11.4.2; or
- the system shall comply with BS EN 60598–2–23.

A device providing protection against the risk of fire shall meet the following requirements:

WR-559.11.4.2

- the device shall continuously monitor the power demand of the luminaires;
- the device shall automatically disconnect the supply circuit within 0.3 s in the case of a short-circuit or failure which causes a power increase of more than 60 W;
- the device shall provide automatic disconnection while the supply circuit is operating with reduced power or if there is a failure that causes a power increase of more than 60 W;
- the device shall provide automatic disconnection upon connection of the supply circuit if there is a failure that causes a power increase of more than 60 W;
- the device shall be fail-safe.

Wiring systems that are supplying safety circuits on escape routes that are likely to be BD2 (difficult), BD3 (crowded) or BD4 (difficult and crowded) shall have a resistance to fire of at least one hour.

WR-422.2.1

Where a heating cable is required to pass through, or be in close proximity to, material that presents a fire hazard, the cable:

WR-554.4.1

- shall be enclosed in material having the ignitability characteristic P as specified in BS 476–12; and
- shall be adequately protected from any mechanical damage reasonably foreseeable during installation and use.

### 8.10.7.11 Sealing of wiring system penetrations

Where a wiring system passes through elements of building construction (such as floors, walls, roofs, ceilings, partitions or cavity barriers), the openings remaining after passage of the wiring system shall be sealed according to the degree of fire resistance for the respective element of building construction.

WR-527.2.1

A wiring system (such as a conduit system, cable ducting system, cable trunking system, busbar or busbar trunking system) that penetrates elements of building construction having specified fire resistance shall be internally sealed to the degree of fire resistance of the respective element before penetration, and also beexternally sealed.

WR-527.2.2

In the event of a wiring system crossing (or being in the proximity of) underground telecommunication cables and underground power cables, a minimum clearance of 100 mm shall be maintained and a fire-retardant partition shall be provided between the cables.

WR-528.2

Temporary sealing arrangements shall be provided during the erection of a wiring system.

WR-527.2.1.1

During alteration work, sealing that has been disturbed shall be reinstated as soon as is practicable.

WR-527.2.1.2

All sealing work should resist external influences to the same degree as the wiring system with which it is used and, in addition, it shall:

- be compatible with the material of the wiring system with which it is in contact;
- be of adequate mechanical stability to withstand the stresses that may arise through damage to the support of the wiring system due to fire;
- be resistant to the products of combustion to the same extent as the elements of building construction that have been penetrated;
- permit thermal movement of the wiring system without reduction of the sealing quality;
- provide the same degree of protection from water penetration as that required for the building construction element in which it has been installed;

## 8.10.7.12 Fire-fighter's switches

A fire-fighter's switch shall be provided in the low-voltage circuit supplying: WR-537.6.1

- exterior electrical installations operating at a voltage exceeding low voltage; and
- interior discharge lighting installations operating at a voltage exceeding low voltage.

Temporary sealing arrangements shall be provided during the erection of a wiring system. WR-527.2.1.1

Similarly, every internal installation in each single premises shall be controlled by a single fire-fighter's switch that is independent of the switch for any other exterior installation. WR-537.6.2

Every fire-fighter's switch shall comply with the following: WR-537.6.3

- for an exterior installation, the switch shall be outside the building and adjacent to the equipment (or, alternatively, a notice indicating the position of the switch shall be placed adjacent to the equipment and a notice shall be fixed near the switch so as to render it clearly distinguishable);
- for an interior installation, the switch shall be in the main entrance to the building;
- the switch shall be placed in a conspicuous position, reasonably accessible to fire-fighters and at not more than 2.75 m from the ground or the standing beneath the switch;

I, personally, could not understand why the switch should be *'not more than 2.7 m from the ground'*, as fire-fighters are not that tall as far as I am aware! I therefore sought advice from the local fire service and was told that the reason why the switch is positioned so high is to stop unauthorized persons switching it off. Fire-fighters use a 'ceiling hook' to operate the switch. I am now no longer confused!!

- where more than one switch is installed on any one building, each switch shall be clearly marked to indicate the installation or part of the installation which it controls.

A fire-fighter's switch shall:                              WR-537.6.4

- be coloured **red** and have fixed on (or near it) a permanent nameplate marked with the words **'FIRE-FIGHTER'S SWITCH'**, the plate being of minimum size 150 mm by 100 mm, and have lettering that is easily legible from a distance appropriate to the site conditions, but not less than 36 point; and
- have its ON and OFF positions clearly indicated by lettering legible to a person standing at the intended site, with the OFF position at the top; and
- be provided with a device to prevent the switch being inadvertently returned to the ON position; and
- be arranged to facilitate operation by a fire-fighter.

### 8.10.7.13  Medical locations

In the event of mains power failure in medical locations, some fire equipment will require a minimum amount of luminance. For example:

- locations where central fire alarm and monitoring systems are installed;
- lifts for fire-fighters;
- fire detection and fire alarms;
- fire-extinguishing systems.

## 8.10.7.13.1 Protection against electric shock– medical locations

The protective measures of obstacles and placing out of reach are **not** permitted.

WR-710.410.3.5
WR 710.410.3.6

The protective measures of non-conducting location, Earth-free local equipotential bonding and electrical separation for the supply of more than one item of current-using equipment, are **not** permitted.

In medical locations, functional extra-low voltage (FELV) is not permitted as a method of protection against electric shock.

## 8.10.7.13.2 Patient environment

In Group 2 medical locations, an IT system shall be used for final circuits supplying medical electrical equipment and systems intended for life support, surgical applications and other electrical equipment located in, or that may be moved into, the 'patient environment'.

WR-710.411.6.3.1

In Group 1 and Group 2 medical locations supplementary equipotential bonding shall be installed so as to equalize potential differences between extraneous-conductive-parts etc. that are located in, or that may be moved into, the 'patient environment'.

WR-710.415.2.1

## *8.10.7.14 Inspection*

 **Note:** Inspection shall precede testing and shall normally be done with that part of the installation under inspection disconnected from the supply.

The inspection shall be made to verify that the installed electrical equipment is:

WR-611.2

- in compliance; and
- correctly selected and erected; and
- not visibly damaged or defective so as to impair safety.

The inspection shall include the checking (during erection) the presence of fire barriers, suitable seals and protection against thermal effects.

WR-611.3

> Where RCDs are also used for protection against fire, the conditions for protection by automatic disconnection of the supply shall be verified.
>
> WR-612.8

### 8.10.7.14.1 Periodic inspection and testing

> Periodic inspection comprising a detailed examination of the installation shall be carried out without dismantling . . .
>
>  **Note:** Having said that, the Regulations do permit a certain amount of 'partial dismantling', if required.
>
> . . . supplemented by appropriate tests to show that the requirements for disconnection times for protective devices are complied with, to provide for:
>
> WR-621.2
>
> • safety of persons and livestock against the effects of electric shock and burns;
> • protection against damage to property by fire and heat arising from an installation defect;
> • confirmation that the installation is not damaged or deteriorated so as to impair safety;
> • the identification of installation defects and departures from the requirements of these Regulations that may give rise to danger.

### *8.10.7.15 Testing*

> In locations exposed to fire hazard, a measurement of the insulation resistance between the live conductors should be made.
>
> WR-612.3.2

 Insulation resistance values are usually much higher than those in Table 8.21.

**Table 8.21 Minimum values of insulation resistance**

| Circuit nominal voltage (V) | Test voltage d.c. (V) | Minimum insulation resistance (MΩ) |
|---|---|---|
| SELV and PELV | 250 | >0.5 |
| Up to and including 500 V, with the exception of the above systems | 500 | >1.0 |
| Above 500 V | 1000 | >1.0 |

# 9
# Inspection and testing

Every installation (or alteration to an existing installation) shall, during erection and on completion before being put into service, be inspected and tested to verify, so far as is reasonably practicable, that the requirements of the Regulations have been met.

The verification shall be made by a competent person, and on completion of the verification a certificate shall be prepared.

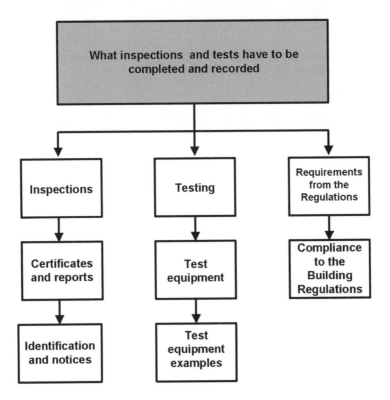

**Figure 9.1** Inspections and tests.

To meet these requirements it is essential for any electrician engaged in inspection, testing and certification of electrical installations to have a **full** working knowledge of the Institution of Engineering and Technology (IET) Wiring Regulations.

The electrician must also have above average experience and knowledge of the type of installation under test in order to carry out **any** inspection and testing. Without this prerequisite, it could be quite dangerous – particularly with regard to installations such as the one shown in Figure 9.2!

**Figure 9.2** Example of a non-conforming electrical installation! (Courtesy of Herne European Consultancy Ltd.)

## 9.1 What inspections and tests have to be completed and recorded?

Every installation must be inspected and tested during erection and on completion before being put into service:

- to verify that, so far as is reasonably practicable, the requirements of the Wiring Regulations have been met;
- to verify that precautions have been taken to avoid danger to persons and damage to property and installed equipment during inspection and testing;
- to make an assessment of the frequency and type of maintenance (e.g. periodic inspection, testing, maintenance and repair) that an installation can reasonably be expected to receive during its intended life.

## 9.2 Inspections

An inspection is (generally speaking) an official examination or formal evaluation exercise involving measurements, tests and gauges applied to certain characteristics with regard to an object or activity. The results of an inspection are usually compared with specified requirements (e.g. a British Standard) in order to determine whether an item or activity is in line with the relevant standard(s) and achieves certain criteria and characteristics. Inspections are usually non-destructive.

### 9.2.1 General

Inspection should **always** precede testing, and shall normally be done with that part of the installation under inspection disconnected from the supply.

Inspections shall be made to verify that the installed electrical equipment:

- complies with the requirements of the applicable British or Harmonized Standard appropriate to the intended use of the equipment;

**Note:** Equipment complying with a foreign national standard may be used **only** if it provides the same degree of safety afforded by a British or Harmonized Standard.

- is correctly selected and erected in accordance with the Regulations;
- is not visibly damaged or defective so as to impair safety.

### 9.2.2 Inspection check list

In accordance with the requirements of BS 7671:2008 (Incorporating Amendment No. 1), and for compliance with the Building Regulations 2010, the inspection shall include the following items:

- access to switchgear and equipment;
- cable routing;
- choice and setting of protective and monitoring devices;
- connection of accessories and equipment;
- connection of conductors;
- connection of single-pole devices for protection or switching in phase conductors;
- continuity of all protective conductors;
- continuity of all ring final circuit conductors;
- Earth electrode resistance;
- Earth fault loop impedance;
- erection methods;
- functional testing;
- identification of conductors;
- insulation of non-conducting floors and walls;
- insulation resistance;

- labelling of protective devices, switches and terminals;
- polarity;
- presence of danger notices and other warning signs;
- presence of diagrams, instructions and similar information;
- presence of fire barriers, suitable seals and protection against thermal effects;
- prevention of mutual (i.e. detrimental) influence;
- presence of undervoltage protective devices;
- prospective fault current;
- protection against electric shock:
  - capability of equipment to withstand mechanical, chemical, electrical and thermal influences and stresses normally encountered during service;
  - exposed-conductive-parts;
  - insulating enclosures;
  - insulation of operational electrical equipment;
  - verification of the quality of the insulation;

- protection against electric shock by direct and/or indirect contact:
  - separated extra-low voltage (SELV);
  - limitation of discharge of energy;

- protection against direct current:
  - barriers or an enclosure;
  - insulation of live parts;
  - obstacles;
  - protective extra-low voltage (PELV);
  - placing out of reach;

- protection against external influences;
- protection against indirect contact:
  - automatic disconnection of supply;
  - Earth-free local equipotential bonding;
  - earthed equipotential bonding;
  - earthing and protective conductors;
  - earthing arrangements for combined protective and functional purposes;
  - non-conducting location (absence of protective conductors);
  - electrical separation;
  - main equipotential bonding conductors;
  - use of Class II equipment or equivalent insulation;
  - supplementary equipotential bonding conductors;

- selection of conductors for current-carrying capacity and voltage drop;
- selection of equipment appropriate to external influences;

- site-applied insulation:
  - protection against direct contact;
  - protection against indirect contact;
  - supplementary insulation.

 The Building Regulations **specifically state** that inspections **shall** include "*the design, construction, inspection and testing of any new electrical installation or new work associated with an alteration or addition to an existing installation*".

In addition to the above list of mandatory inspections for compliance with the IET Wiring and the Building Regulations, the following are some of the extra inspections that electricians usually complete during initial and periodic inspections and tests of electrical installations:

- cables and conductors (current-carrying capacity, insulation and/or sheath);
- correct connection of accessories and equipment;
- electrical joints and connections (to ensure that they meet stipulated requirements concerning conductance, insulation, mechanical strength and protection);
- emergency switching;
- insulation;
- insulation monitoring devices (design, installation and security);
- inspection of another electrical installations;
- isolation and switching devices (and their correct location);
- locations with risks of fire due to the nature of processed and/or stored materials;
- plug and socket-outlets;
- protection against electric shock – special installations or locations;
- protection against Earth insulation faults;
- protection against mechanical damage;
- protection against overcurrent;
- protection by extra-low voltage systems (other than SELV);
- protection by non-conducting location;
- protection by residual current devices (RCDs);
- protection by separation of circuits;
- supplies;
- supplies for safety services;
- wiring systems (selection and erection, temperature variations).

### 9.2.3 Requirements from the Regulations – inspection

 Precautions shall be taken to avoid danger to persons and livestock, and to avoid damage to property and installed equipment, during inspection and testing.

Every electrical connection and joint shall be accessible
for inspection, except for the following:

WR-526.3

- a joint designed to be buried in the ground;
- a compound-filled or encapsulated joint;
- a connection between a cold tail and the heating
  element as in ceiling heating, floor heating or a trace
  heating system;
- a joint made by welding, soldering, brazing or an
  appropriate compression tool;
- joints or connections made in the equipment by the
  manufacturer of the product and not intended to be
  inspected or maintained;
- equipment complying with BS 7671:
  2008 (latest Amendment) for a
  maintenance free accessory and marked
  with the symbol

The inspection shall be made to verify that the installed
electrical equipment is:

WR-611.2

- in compliance;
- correctly selected and erected; and
- not visibly damaged or defective so as to impair
  safety.

The inspection shall include at least the checking of the
following items, where relevant to the installation and
where necessary, including any particular requirements
for special installations or locations (see Part 7 of BS
7671:2008 (Incorporating Amendment No. 1)):

- connection of conductors;
- identification of conductors;
- routing of cables in safe zones (or protection against
  mechanical damage);
- selection of conductors for current-carrying capacity
  and voltage drop, in accordance with the design;
- connection of single-pole devices for protection or
  switching in line conductors only;
- correct connection of accessories and equipment;
- presence of fire barriers, suitable seals and protection
  against thermal effects;
- methods of protection against electric shock;
- prevention of mutual detrimental influence;

- presence of appropriate devices for isolation and    WR-611.3
  switching correctly located;
- presence of undervoltage protective devices;
- labelling of protective devices, switches and
  terminals;
- selection of equipment and protective measures
  appropriate to external influences;
- adequacy of access to switchgear and equipment;
- absence of danger notices and other warning signs;
- absence of diagrams, instructions and similar
  information;
- erection methods.

# 9.3  Testing

Testing any electrical installation (even the simplest) can be very dangerous – not just to the tester himself, but also to bystanders and other people – unless it is carried out safely.

As a minimum, the electrician must:

- have an above average experience and knowledge of the type of installation under test;
- have a thorough understanding of the correct application and use of the relevant test instruments (and their associated leads, probes and accessories);
- ensure that the test equipment being used has recently been inspected, correctly maintained and (where necessary) calibrated either against a workshop standard or a national standard;
- observe the safety measures and procedures set out in Health and Safety Executive (HSE) Guidance Note GS38 concerning the safe use of instruments and their accessories.

The following are summarized details of the most important elements of the IET Wiring Regulations that an electrician must test for in order to confirm that the electrical installation meets the fundamental design requirements of the BS 7671:2008 (Incorporating Amendment No. 1) and is installed in conformance with the requirements of that British Standard.

## 9.3.1  Design requirements

Check to confirm that the number and type of circuits required for lighting, heating, power, control, signalling, communication and information technology etc. have taken consideration of:

- the location and points of power demand;
- the loads to be expected on the various circuits;
- the daily and yearly variation in demand;
- any special conditions;
- requirements for control, signalling, communication and information technology, etc.

### 9.3.2 Electricity distributor

Check to confirm that the electricity distributor has:

- evaluated and agreed proposals for new installations or significant alterations to existing ones;
- maintains the supply within defined tolerance limits;
- provided an earthing facility for all new connections;
- installed the cut-out and meter in a safe location;
- ensured that the cut-out and meter are mechanically protected and can be safely maintained;
- provided certain technical and safety information to the consumer to enable them to design their installations;
- ensured that their equipment on consumers' premises:
  - ○ is suitable for its purpose;
  - ○ is safe in its particular environment;
  - ○ clearly shows the polarity of the conductors.

### 9.3.3 Initial inspection and tests

In accordance with both the Wiring Regulations and the Building Regulations, all new installations (plus additions and/or alterations to existing circuits) need an initial verification to:

- ensure that equipment and accessories meet the requirements of the relevant standard;
- comply with the requirements of BS 7671;
- comply with the requirements of the Building Regulations;
- ensure that the installation is not damaged so as to impair safety.

 The following tests **shall** be carried out (and in the following order) before the installation is energized:

- a continuity test of all protective conductors (including main and supplementary equipotential bonding);
- a continuity test of all ring final circuit conductors;
- a measurement of the insulation resistance between live conductors and between each live conductor and Earth;

- a measurement of the insulation resistance of the main switchboard and each distribution circuit;
- confirmation that insulation for protection against direct and/or indirect contact meets requirements;
- verification that the separation of circuits is protected by SELV or PELV and/or that electrical separation meets requirements;
- a measurement of the insulation resistance of live parts from those of other circuits and those of other circuits and from Earth;
- confirmation that functional extra-low-voltage (FELV) circuits meet all the test requirements for low-voltage circuits;
- a test to ensure that the amount of protection against direct contact that is provided by a barrier or an enclosure (provided during erection) meets requirements;
- verification (by measurement) that the amount of protection against indirect contact provided by a non-conducting location meets requirements;
- a polarity test to verify that fuses, single-pole control and protective devices, lampholders and wiring meet requirements.

 The following tests shall then be carried out once the installation is energized:

- a measurement of the electrode resistance to Earth for earthing systems incorporating an Earth electrode;
- a measurement of Earth loop impedance;
- a measurement of prospective short-circuit and Earth fault;
- functional tests to verify the effectiveness of RCDs and test assemblies (e.g. switchgear, controlgear, drives, controls and interlocks) to show that they are properly mounted, adjusted and installed in accordance with the Regulations.

### 9.3.4 Insulation resistance

Check to confirm that:

- the insulation resistance (measured with all its final circuits connected but with current-using equipment disconnected) between live conductors and between each live conductor and Earth, is not be less than that shown in Table 9.1;
- the separation of live parts from those of other circuits and from Earth is in accordance with the values show in Table 9.1, by measuring the insulation resistance.

 **Note:** This test (more usually referred to as *meggering*) is, therefore, aimed at ensuring that the insulation of conductors, accessories and equipment is still capable of preventing dangerous leakage current between conductors and between conductors and Earth.

**Table 9.1 Minimum values of insulation resistance**

| Circuit nominal voltage (V) | Test voltage d.c. (V) | Minimum insulation resistance (MΩ) |
|---|---|---|
| SELV and PELV | 250 | 0.25 |
| Up to and including 500 V (with the exception of the above systems) | 500 | 0.5 |
| Above 500 V | 1000 | 1.0 |

To determine the insulation resistance between live conductors, the test is done between the live (phase and neutral) conductors at the distribution board (Figure 9.3).

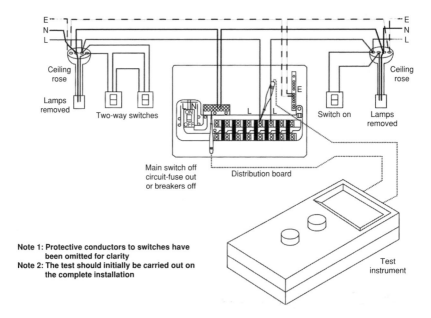

**Figure 9.3** Insulation resistance tests between live conductors of a circuit. (Courtesy of IET.)

The resistance readings obtained should be greater than the minimum values shown in Table 9.1.

To check the insulation resistance to Earth,

- single phase – test between the live conductors (phase and neutral) and the circuit protective conductors at the distribution board, as shown in Figure 9.4;
- three phase – test to Earth all live conductors (including the neutral) connected together; resistance values should be greater than the minimum figures shown in Table 9.1.

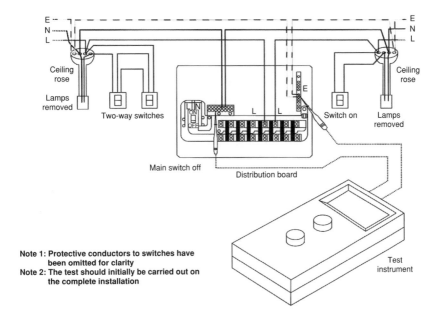

Note 1: Protective conductors to switches have
        been omitted for clarity
Note 2: The test should initially be carried out on
        the complete installation

**Figure 9.4** Insulation resistance tests to Earth. (Courtesy of IET.)

 **Notes:**

1. Measurements shall be carried out with d.c.
2. When the circuit includes electronic devices, only a measurement to protective Earth shall be made, with the phase and neutral connected together.

 Precautions may be necessary to avoid damage to electronic devices.

### 9.3.5 Site insulation

Confirm that:

- insulation applied on site to protect against direct contact is capable of withstanding, without breakdown or flashover, an applied test voltage as specified in the British Standard for similar type-tested equipment;
- supplementary insulation applied to equipment during erection to protect against indirect contact is tested to ensure that the insulating enclosure protects to at least IP2X or IPXXB; and
- is capable of withstanding, without breakdown or flashover, an applied test voltage as specified in the British Standard for similar type-tested equipment.

### 9.3.6 Protective measures

The Wiring Regulations stipulate that a continuity test of all protective conductors (including main and supplementary equipotential bonding) shall be made to ensure that the correct degree of protection is being provided by the following protective measures, by confirming, checking and testing:

| | |
|---|---|
| **Protection against overload current** | That the protective device is capable of breaking any overload current flowing in the circuit conductors before the current can damage the insulation of the conductors. |
| **Protection by Earth-free local equipotential bonding** | That Earth-free local equipotential bonding prevents the appearance of a dangerous voltage between simultaneously accessible parts in the event of failure of the basic insulation. |
| **Protection by electrical separation** | That equipment used as a fixed source of supply has been manufactured so that the output is separated from the input and from the enclosure by insulation for protection against indirect contact. |

 **Note:** This form of protection is intended for an individual circuit and is aimed at preventing shock current through contact with exposed-conductive-parts, which might be energized by a fault in the basic insulation of that circuit.

| | |
|---|---|
| **Protection by extra-low voltage systems (other than SELV)** | That if an extra-low-voltage system complies with the requirements for SELV, it is not be connected to a live part or a protective conductor forming part of another system, and is not connected to: |

- Earth;
- an exposed-conductive-part of another system;
- a protective conductor of any system; or
- an extraneous-conductive-part.

Protection against direct contact has been provided by either:

- insulation capable of withstanding 500 V a.c. rms for 60 s; or
- barriers or enclosures with a degree of protection of at least IP2X or IPXXB.

 This form of protection against direct contact is **not** required if the equipment is within a building in which main equipotential bonding is applied and the voltage does not exceed:

- 25 V a.c. rms or 60 V ripple-free d.c. when the equipment is normally used only in dry locations and large-area contact of live parts with the human body is not to be expected;
- 6 V a.c. rms or 15 V ripple-free d.c. in all other cases.

 **Note:** When an extra-low-voltage circuit is used to supply equipment whose insulation does not comply with the minimum test voltage required for the primary circuit, then the insulation of that equipment shall be reinforced to withstand a voltage of 1500 V a.c. rms for 60 s.

**Protection by insulation of live parts**

That the insulation protection has been designed to prevent contact with live parts.

 While, generally speaking, this method is for protection against direct contact, it also provides a degree of protection against indirect contact.

**Protection by non-conducting location**

That this form of protection prevents simultaneous contact with parts which may be at different potentials through failure of the basic insulation of live parts.

 While this protection is **not** recognized in the Regulations for general use, it may be applied in special situations provided that they are under effective supervision.

 Protection by non-conducting location **shall not** be used in installations and locations subject to increased risk of shock, such as agricultural and horticultural premises, caravans, swimming pools etc.

**Protection by RCDs**

That:

- parts of a TT system that are protected by a single RCD have been placed at the origin of the installation, unless that part between the origin and the device complies with the requirements for protection by using Class II equipment or an equivalent insulation;

 Where there is more than one origin, this requirement applies to each origin.

- installations forming part of an IT system have been protected by an RCD supplied by the circuit concerned or make use of an insulation monitoring device.

**Protection by SELV**

That circuit conductors for each SELV system have been physically separated from those of any other system. Where this proves impracticable, SELV circuit conductors have been:

- insulated for the highest voltage present;
- enclosed in an insulating sheath additional to their basic insulation.

**Protection by the use of Class II equipment or equivalent insulation**

That this form of protection prevents a fault in the basic insulation causing a dangerous voltage to appear on the exposed metalwork of electrical equipment.

A typical method for testing the continuity of protective conductors is illustrated in Figure 9.5, and basically this involves bridging the phase conductor to the protective conductor at the distribution board (so as to include all of the circuit) and then testing between phase and Earth terminals at each point of the circuit.

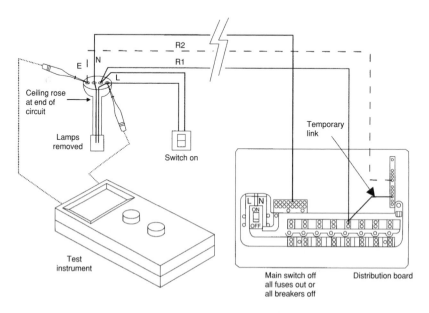

**Figure 9.5** Connections for testing the continuity of protective conductors. (Courtesy of IET.)

### 9.3.7 Protection against direct and indirect contact

The two methods for protecting against shock from both direct and indirect contact are:

- SELV – i.e. where the system voltage does not exceed extra low (e.g. 50 V a.c., 120V ripple free d.c.) and associated wiring etc. is separated from all other circuits of higher voltage;
- limitation of discharge of energy – i.e. where equipment is arranged so that the current that can flow through the body (or livestock) is limited to a safe level (e.g. electric fences).

#### *9.3.7.1  SELV*

The system is **not** deemed to be a SELV system if any exposed-conductive-part of an extra-low-voltage system is capable of coming into contact with an exposed-conductive-part of any other system. In addition, a system that does **not** use a device such as an autotransformer, potentiometer, semiconductor device etc. to provide electrical separation is also **not** deemed to be a SELV system.

Confirm that for protection by SELV:

- the nominal circuit voltage does **not** exceed extra-low voltage;
- the supply is from one of the following:

   ○ a safety isolating transformer complying with BS 3535;
   ○ a motor-generator with windings providing electrical separation equivalent to that of the safety isolating transformer specified above;
   ○ a battery or other form of electrochemical source;
   ○ a source independent of a higher voltage circuit (e.g. an engine-driven generator);
   ○ electronic devices that (even in the case of an internal fault) restrict the voltage at the output terminals so that it does not exceed extra-low voltage.

Confirm and test that:

- a mobile source for SELV has been selected and erected in accordance with the requirements for protection by the use of Class II equipment or by equivalent insulation;
- all live parts of a SELV system are:

   ○ electrically separated from that of any other higher voltage system;

**Note:** This electrical separation shall be not less than that between the input and output of a safety isolating transformer.

   ○ not connected to Earth;
   ○ not connected to a live part or a protective conductor forming part of another system;

- circuit conductors for each SELV system are physically separated from those of any other system and, where this proves impracticable, SELV circuit conductors are:

  o insulated for the highest voltage present; and (where this proves impracticable)
  o enclosed in an insulating sheath additional to their basic insulation;

- conductors of systems with a higher voltage than SELV are separated from the SELV conductors by an earthed metallic screen or an earthed metallic sheath;
- SELV circuit conductors that are contained in a multi-core cable with other circuits having different voltages are insulated, individually or collectively, for the highest voltage present in the cable or grouping);
- electrical separation between live parts of a SELV system (including relays, contactors and auxiliary switches) and any other system are maintained;
- exposed-conductive-parts of a SELV system are **not** connected to:

  o Earth;
  o an exposed-conductive-part of another system;
  o a protective conductor of any system;
  o an extraneous-conductive-part;

**Note:** except where that electrical equipment is mainly required to be connected to an extraneous-conductive-part (in which case, measures shall be incorporated so that the parts cannot attain a voltage exceeding extra low voltage).

- if the nominal voltage of a SELV system exceeds 25 V a.c. rms or 60 V ripple-free d.c., protection against direct contact is provided by one or more of the following:

  o a barrier (or an enclosure) capable of providing protection to at least IP2X or IPXXB;
  o insulation capable of withstanding a type-test voltage of 500 V a.c. rms for 60 s;

- the socket-outlet of a SELV system is:

  o incompatible with the plugs used for other systems in use in the same premises;
  o does not have a protective conductor contact;

- luminaire supporting couplers that have a protective conductor contact are **not** be installed in a SELV system.

### 9.3.7.2 Protection – by limiting discharge of energy

Protection against both direct and indirect contact shall be deemed to be provided when the equipment incorporates a means of limiting the amount of current that can pass through the body of a person or livestock to a value lower than that likely to cause danger.

### 9.3.8 Protection against direct contact

Electric shock caused by direct contact (i.e. when a body part directly touches live parts of equipment or systems that are intended to be live) is particularly dangerous, as the full voltage of the supply can be developed across the body. On the whole, however, if an electrical installation has been designed and installed correctly, there should not be too much risk from direct contact – but carelessness (such as changing a electric lightbulb without switching the mains off first) or overconfidence (such as working on a circuit with the power on) are the prime causes of injuries and death from electric shock.

The main protective methods against direct contact causing an electric shock are:

- barriers or an enclosure;
- insulation of live parts;
- by obstacles and placing out of reach;
- PELV and FELV;
- placing out of reach.

 The use of an RCD cannot prevent direct contact, but may be used to **supplement** other protective means that are used.

### 9.3.8.1 Protection by a barrier or an enclosure provided during erection

Test to ensure that the degree of protection against direct contact provided by a barrier or an enclosure (provided during erection) is not less than IP2X, or IPXXB or IP4X, as appropriate.

### 9.3.8.2 Protection by insulation of live parts

Complete a functional test to verify that protection by insulation of live parts **does** prevent contact with a live parts.

 **Note:** While, generally speaking, this basic form of insulation protection is for protection against direct contact, it also provides a degree of protection against indirect contact.

### 9.3.8.3 Protection by obstacles

Complete a functional test to verify that protection by obstacles prevents unintentional contact with a live part, but **not** intentional contact by deliberate circumvention of the obstacle.

The application shall be limited to protection against direct contact and in an area accessible only to skilled persons.

For some installations and locations where an increased risk of shock exists, this protective measure shall not be used.

### 9.3.8.4 PELV and FELV systems

PELV and FELV systems shall provide protection against electric shock, and meet the following requirements:

- barriers or enclosures with a degree of protection of at least IP2X or IPXXB; or
- insulation capable of withstanding 500 V a.c. rms for 60 s;

This form of protection against direct contact is **not** required if the equipment is within a building in which main equipotential bonding is applied and the voltage does not exceed:

- 25 V a.c. rms or 60 V ripple-free d.c. when the equipment is normally used only in dry locations and large-area contact of live parts with the human body is not to be expected;
- 6 V a.c. rms or 15 V ripple-free d.c. in all other cases.

- for extra-low-voltage systems that do not comply with the requirements for SELV in some respect, protection against direct contact shall be provided by one or more of the following:
  - barriers or enclosures;
  - insulation corresponding to the minimum voltage required for the primary circuit;

- when an extra-low-voltage circuit is used to supply equipment whose insulation does not comply with the minimum test voltage required for the primary circuit, the insulation of that equipment shall be reinforced to withstand a voltage of 1500 V a.c. rms for 60 s;
- if the primary circuit of the FELV source is protected by automatic disconnection, then the exposed-conductive-parts of equipment in that FELV system shall be connected to the protective conductor of the primary circuit;
- if the primary circuit of the FELV source is protected by electrical separation, the exposed-conductive-parts of equipment in that FELV system shall be connected to the non-earthed protective conductor of the primary circuit;
- all socket-outlets and luminaire supporting couplers in a FELV system shall use a plug that is dimensionally different from those used for any other system in use in the same premises.

### 9.3.8.5 Protection by placing out of reach

Check to ensure that:

- bare (or insulated) overhead lines being used for distribution between buildings and structures are installed in accordance with the Electricity Safety, Quality and Continuity Regulations 2002;
- bare live parts (other than overhead lines) are not be within arm's reach;

**Notes:**

1. If access to live equipment (from a normally occupied position) is restricted by an obstacle (such as a handrail, mesh or screen) with a degree of protection less than IP2X or IPXXB, the extent of arm's reach shall be measured from the obstacle.
2. If a bulky or long conducting object is normally handled in these areas, the distances shall be increased accordingly.

- bare live parts (other than an overhead line) are not within 2.5 m of:

  o   an exposed-conductive-part;
  o   an extraneous-conductive-part;
  o   a bare live part of any other circuit;

- if a bulky or long conducting object is normally handled in these areas, the distances required shall be increased accordingly;

No additional protection against overvoltages of atmospheric origin is necessary for:

- installations that are supplied by low-voltage systems that do not contain overhead lines;
- installations that are supplied by low-voltage networks that contain overhead lines and their location is subject to less than 25 thunderstorm days per year;
- installations that contain overhead lines and their location is subject to less than 25 thunderstorm days per year;

**provided** that they meet the required minimum equipment impulse withstand voltages shown in Table 9.2.

Suspended cables having insulated conductors with **earthed metallic coverings** are considered to be an 'underground cable'.

- check to ensure that installations that are supplied by (or include) low-voltage overhead lines incorporate protection against overvoltages of atmospheric origin, or (if the location is subject to more than 25 thunderstorm days per year) protection against overvoltages of atmospheric origin shall be provided in the installation of the building by:

**Table 9.2 Required minimum impulse to withstand voltage U$_w$**

| Nominal voltage of the installation (V) | Required minimum impulse (kV) | | | |
|---|---|---|---|---|
| | Category IV (equipment with very high impulse voltage) | Category III (equipment with high impulse voltage) | Category II (equipment with normal impulse voltage) | Category I (equipment with reduced impulse voltage) |
| 230/240 277/480 | 6 | 4 | 2.5 | 1.5 |
| 400/690 | 8 | 6 | 4 | 2.5 |
| 1000 | 12 | 8 | 6 | 4 |

- o a surge protective device with a protection level not exceeding Category II; or
- o by other means providing an equivalent attenuation of overvoltages;

- where protective measures against indirect contact only have been dispensed with, confirm that:

  - o overhead line insulator brackets (and metal parts connected to them) are not within arm's reach;
  - o the steel reinforcement of steel-reinforced concrete poles in not accessible;
  - o exposed-conductive-parts (including small isolated metal parts such as bolts, rivets, nameplates not exceeding 50 mm × 50 mm and cable clips) cannot be gripped or cannot be contacted by a major surface of the human body;
  - o there is no risk of fixing screws used for non-metallic accessories coming into contact with live parts;
  - o inaccessible lengths of metal conduit do not exceeding 150 mm$^2$;
  - o metal enclosures mechanically protecting equipment comply with the relevant British Standard;
  - o unearthed street furniture that is supplied from an overhead line is inaccessible while in normal use.

### 9.3.9 Protection against indirect contact

Indirect contact (i.e. touching conductive parts that are not meant to be live, but which have become live due to a fault) is the other main cause of electric shock. Again, this is particularly dangerous, and the main protection against indirect contact is for the electrical installation to be correctly earthed and for the circuit to be fitted with some form of overcurrent cut-out device.

The main protective methods against indirect contact causing an electric shock are:

- automatic disconnection of supply;
- Earth-free local equipotential bonding;
- earthed equipotential bonding;
- earthing and protective conductors;
- earthing arrangements for combined protective and functional purposes;
- non-conducting location (absence of protective conductors);
- electrical separation;
- main equipotential bonding conductors;
- supplementary equipotential bonding conductors;
- use of Class II equipment.

### 9.3.9.1 Automatic disconnection of supply

This intention of this form of protection is to prevent a dangerous voltage occurring between simultaneously accessible conductive parts. For installations and locations with increased risk of shock (e.g. those in Part 7 of the Regulations, such as agricultural and horticultural buildings, saunas etc.) additional measures may be required. For example:

- automatic disconnection of supply by means of an RCD with a rated residual operating current not exceeding 30 mA;
- supplementary equipotential bonding;
- reduction of maximum fault clearance time.

Confirm and test that:

- installations that are part of a TN system meet the requirements for Earth fault loop impedance and for circuit protective conductor impedance as specified in the Wiring Regulations regarding specified times for automatic disconnection of supplies;
- for circuits supplying fixed equipment that is outside of the earthed equipotential zone and that has exposed-conductive-parts which could be touched by a person who has direct contact directly with Earth, that the Earth fault loop impedance ensures that disconnection occurs within the time stated in Table 9.3;

**Table 9.3 Maximum disconnection times for TN systems**

| Installation nominal voltage, $U_o$ (V) | Maximum disconnection time, t (s) |
| --- | --- |
| 120 | 0.8 |
| 230 | 0.4 |
| 400 | 0.2 |
| >400 | 0.1 |

- if the installation is part of a TT system, all socket-outlet circuits are protected by an RCD;

- automatic disconnection using an RCD is not applied to a circuit incorporating a combined protective and neutral (PEN) conductor;
- installations that provide protection against indirect contact by automatically disconnecting the supply have a circuit protective conductor run to (and terminated at) each point in the wiring and at each accessory.

Except suspended lampholders that have no exposed-conductive-parts.

### 9.3.9.2 Earth-free local equipotential bonding

Earth-free local equipotential bonding is effectively a Faraday cage, where all metal is bonded together (but **not** to Earth!) so as to prevent the appearance of a dangerous voltage occurring between simultaneously accessible parts in the event of failure of the basic insulation.

Confirm that:

- Earth-free local equipotential bonding has **only** been used in special situations that are Earth-free;
- a warning notice (warning that Earth-free local equipotential bonding is being used) has been fixed in a prominent position adjacent to every point of access to the location concerned.

For some installations and locations with an increased shock risk (e.g. agricultural and horticultural, saunas), Earth-free local equipotential bonding shall to be used.

### 9.3.9.3 Earthing and protective conductors

A protective conductor may consist of one or more of the following:

- a single-core cable;
- a conductor in a cable;
- an insulated or bare conductor in a common enclosure with insulated live conductors;
- a fixed bare or insulated conductor;
- a metal covering (e.g. the sheath, screen or armouring of a cable);
- a metal conduit or other enclosure or electrically continuous support system for conductors;
- an extraneous-conductive-part.

Verify and test that:

- the thickness of tape or strip conductors is capable of withstanding mechanical damage and corrosion (see BS 7430);
- the connections of earthing conductors to the Earth electrode are:
  - soundly made;
  - electrically and mechanically satisfactory;

- ○ labelled in accordance with the Regulations;
- ○ suitably protected against corrosion;

- all installations have a main earthing terminal to connect the following to the earthing conductor:

  - ○ the circuit protective conductors;
  - ○ the main bonding conductors;
  - ○ functional earthing conductors (if required);
  - ○ lightning protection system bonding conductor (if any);

- earthing conductors are capable of being disconnected to enable the resistance of the earthing arrangement to be measured;
- any joint is:

  - ○ capable of disconnection only by means of a tool;
  - ○ mechanically strong;
  - ○ ensures the maintenance of electrical continuity;

 **Note:** For convenience (and if required) this may be combined with the main earthing terminal or bar.

- confirm that, unless a protective conductor forms part of a multi-core cable (or cable trunking or a conduit is used as a protective conductor), the cross-sectional area, up to and including 6 mm² has been protected, throughout, by a covering of at least equivalent in insulation to that of a single-core non-sheathed cable having a voltage rating of at least 450/750 V.

Where protective multiple earthing (PME) conditions apply, verify and test that:

- the main equipotential bonding conductor has been selected in accordance with the neutral conductor of the supply and Table 9.4;

**Table 9.4 Minimum cross-sectional area of the main equipotential bonding conductor in relation to the neutral**

| Copper-equivalent cross-sectional area of the supply neutral conductor | Minimum copper-equivalent cross-sectional area of the main equipotential bonding conductor |
|---|---|
| 35 mm² or less | 10 mm² |
| Over 35 mm² up to 50 mm² | 16 mm² |
| Over 50 mm² up to 95 mm² | 25 mm² |
| Over 95 mm² up to 150 mm² | 35 mm² |
| Over 150 mm² | 50 mm² |

Local distributor's network conditions may require a larger conductor.

- buried earthing conductors have a cross-sectional area not less than that stated in Table 9.5.

**Table 9.5 Minimum cross-sectional area of a buried earthing conductor**

| | Protected against mechanical damage | Not protected against mechanical damage |
|---|---|---|
| Protected against corrosion by a sheath | | 16 mm² copper<br>16 mm² coated steel |
| Not protected against corrosion | 25 mm² copper<br>50 mm² steel | 25 mm² copper<br>50 mm² steel |

All protective conductors – particularly main equipotential and supplementary bonding conductors – must be tested for continuity using a low-resistance ohmmeter.

Verify and test that:

- the cross-sectional area of all protective conductors (less equipotential bonding conductors) is not be less than

$$S = \frac{\sqrt{I^2 t}}{k}$$

- if the protective conductor:
  - is not an integral part of a cable; or
  - is not formed by conduit, ducting or trunking; or
  - is not contained in an enclosure formed by a wiring system;

then the cross-sectional area shall be not less than

  - 2.5 mm² copper equivalent if protection against mechanical damage is provided; or
  - 4 mm² copper equivalent if mechanical protection is not provided;

- protective conductors buried in the ground, shall have a cross sectional area not less than that stated in Table 9.5.

### 9.3.9.4 Earthing arrangements for combined protective and functional purposes

The following conductors may serve as a PEN conductor, provided that the part of the installation concerned is not supplied through an RCD:

- a conductor of a cable not subject to flexing and with a cross-sectional area not less than 10 mm² (for copper) or 16 mm² for aluminium (this applies to a fixed installation by the way);

- the outer conductor of a concentric cable, where that conductor has a cross-sectional area not less than 4 mm².

Verify and test (where necessary) that PEN conductors have only been used if:

- authorization to use a PEN conductor has been obtained by the distributor; or
- the installation is supplied by a privately owned transformer or convertor and there is no metallic connection (except for the earthing connection) with the distributor's network; or
- the installation is supplied from a private generating plant;
- the outer conductor of a concentric cable is not be common to more than one circuit;
- the conductance of the outer conductor of a concentric cable (and the terminal link or bar):
  - for a single-core cable, is not less than that of the internal conductor;
  - for a multi-core cable in a multi-phase or multi-pole circuit is not be less than that of one internal conductor;
  - for a multi-core cable serving a number of points contained within one final circuit (or where the internal conductors are connected in parallel) is not less than that of the internal conductors;
- the continuity of all joints in the outer conductor of a concentric cable (and at a termination of that joint) is supplemented by an additional conductor, additional (that is) to any means used for sealing and clamping the outer conductor;
- isolation devices or switching have **not** been inserted in the outer conductor of a concentric cable;
- PEN conductors of all cables have been insulated or have an insulating covering suitable for the highest voltage to which it may be subjected;
- if neutral and protective functions are provided by separate conductors, those conductors are not then be reconnected together beyond that point;
- separate terminals (or bars) have been provided for the PEN conductors at the point of separation;
- PEN conductors have been connected to the terminals or bar intended for the protective earthing conductor and the neutral conductor.

 **Note:** Where earthing is required for protective as well as functional purposes, the requirements for protective measures shall take precedence.

### 9.3.9.5 Electrical separation

This form of protection is intended for an individual circuit and is aimed at preventing shock current through contact with exposed-conductive-parts that might be energized by a fault in the basic insulation of that circuit.

Verify that:

- protection by electrical separation has been applied to the supply of individual items of equipment by means of a transformer complying with BS 3535 (the secondary of which is not earthed) or a source affording equivalent safety;
- protection by electrical separation has been used to supply several items of equipment from a single separated source (but **only** for special situations);
- equipment used as a fixed source of supply is either:

  - selected and/or installed with Class II or equivalent protection; or
  - manufactured so that the output is separated from the input and from the enclosure by insulation satisfying the conditions for Class II;

- the supply source to the circuit is either:

  - an isolating transformer complying with BS 3535; or
  - a motor generator;

- mobile supply sources (fed from a fixed installation) are selected and/or installed with Class II or equivalent protection;
- source supplies are only supplying more than one item of equipment provided that:

  - all exposed-conductive-parts of the separated circuit are connected together by an insulated and non-earthed equipotential bonding conductor;
  - the non-earthed equipotential bonding conductor is not connected to a protective conductor, or to an exposed-conductive-part of any other circuit or to any extraneous-conductive-part;
  - all socket-outlets are provided with a protective conductor contact (that is connected to the equipotential bonding conductor);
  - all flexible equipment cables (other than Class H equipment) have a protective conductor for use as an equipotential bonding conductor;
  - exposed-conductive-parts which are fed by conductors of different polarity (and which are liable to a double fault occurring) are fitted with an associated protective device;
  - any exposed-conductive-part of a separated circuit cannot come into contact with an exposed-conductive-part of the source;

- live parts of a separated circuit are not connected (at any point) to another circuit or to Earth;
- live parts of a separate circuit are electrically separated from all other circuits;

 **Note:** Live parts of relays, contactors etc. included in a separated circuit (and between a separated circuit and other live parts of other circuits) shall be similarly electrically separated.

- separated circuits, preferably, use a separate wiring system;

 If this is not feasible, multi-core cables (without a metallic sheath) or insulated conductors (in an insulating conduit) may be used.

- the voltage of an electrically separated circuit does not exceed 500 V;
- all parts of a flexible cable (or cord) that is liable to mechanical damage is visible throughout its length;
- for circuits supplying a single piece of equipment, no exposed-conductive-part of the separated circuit is connected:

  o    to the protective conductor of the source;
  o    to any exposed-conductive-part of any other circuit;

- a warning notice (warning that protection by electrical separation is being used) is fixed in a prominent position adjacent to every point of access to the location concerned.

### 9.3.9.6 Main equipotential bonding conductors

All main equipotential (and supplementary) bonding conductors must be tested for continuity.

 The normal approach is to just connect the leads from a low-resistance ohmmeter to the ends of the bonding conductor as shown in Figure 9.6 – making sure that one end is disconnected from its bonding clamp, otherwise the measurement may include the resistance of parallel paths of other earthed metalwork.

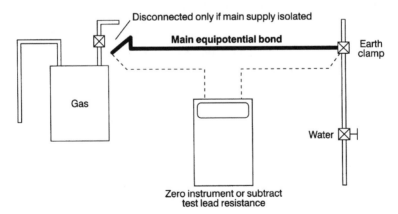

**Figure 9.6** Continuity testing of main and equipotential bonding conductors.

Confirm that main equipotential bonding conductors have (for each installation) been connected to the main earthing terminal of that installation. These conductors can include the following:

- water service pipes;
- gas installation pipes;
- other service pipes and ducting;
- central heating and air-conditioning systems;
- exposed metallic structural parts of the building;
- the lightning protective system.

 **Note:** Where an installation serves more than one building, the above requirement shall be applied to each building.

### 9.3.9.7 Non-conducting location

This form of protection (as the name implies) consists of an area in which the floor, walls and ceiling are all insulated, and within which protective conductors and socket-outlets do not have an earthing connection.

Test that the insulation of extraneous-conductive-parts:

- is not less than $0.5\,\text{M}\Omega$ (when tested at $500\,\text{V}$ d.c);
- is able to withstand a test voltage of at least $2\,\text{kV}$ a.c. rms;
- does not pass a leakage current exceeding $I\,\text{mA}$ in normal use.

Check that the:

- degree of protection against indirect contact provided by a non-conducting location is verified by measuring the resistance of the location's floors and walls to the installation's main protective conductor at not less than three points on each relevant surface;

 **Note:** One of these measurements should be not less than $I\,\text{m}$ and not more than $1.2\,\text{m}$ from any extraneous-conductive-part in the location. The other two measurements shall be made at greater distances.

- the insulation of extraneous-conductive-parts (to satisfy the requirements for protection be non-conducting location):
  ○ are not less than $0.5\,\text{M}\Omega$ when tested at $500\,\text{V}$ d.c.; and
  ○ are able to withstand a test voltage of at least $2\,\text{kV}$ a.c. rms; and
  ○ do not pass a leakage current exceeding $I\,\text{mA}$ in normal use.

### 9.3.9.8 Supplementary equipotential bonding

Using a low-resistance ohmmeter (similar to that employed for testing main equipotential bonding conductors described above), test all supplementary equipotential bonding conductors for continuity, particularly in locations intended for livestock, to confirm that:

- supplementary bonding connects all exposed and extraneous-conductive-parts that can be touched by livestock;

- metallic grids laid in the floor for supplementary bonding are connected to the protective conductors of the installation.

## 9.3.9.9 Locations with increased risk of shock

For installations and locations with increased risk of shock (saunas, bathrooms and agricultural/horticultural premises etc.) certain additional measures may be required, such as:

- automatic disconnection of supply by means of an RCD with a rated residual operating current not exceeding 30 mA;
- supplementary equipotential bonding;
- a reduction in the maximum fault clearance time.

In these cases, test to confirm that for circuits supplying fixed equipment that is outside the earthed equipotential zone (and which have exposed-conductive-parts that could be touched by a person who has direct contact directly with Earth) the Earth fault loop impedance ensures that disconnection occurs within the time stated in Table 9.6.

**Table 9.6  Maximum disconnection times for TN systems**

| Installation nominal voltage, $U_o$ (V) | Maximum disconnection time, t (s) |
|---|---|
| 120 | 0.8 |
| 230 | 0.4 |
| 400 | 0.2 |
| >400 | 0.1 |

Test, measure and confirm that:

- if the installation is part of a TT system, all socket-outlet circuits have been protected by an RCD;
- automatic disconnection using an RCD has **not** been applied to a circuit incorporating a PEN conductor;
- installations that provide protection against indirect contact by automatically disconnecting the supply have a circuit protective conductor run to (and terminated at) each point in the wiring and at each accessory;

 Except suspended lampholders that have no exposed-conductive-parts.

- one or more of the following types of protective device has been used:

  o   an overcurrent protective device;
  o   an RCD;

- where an RCD is used in a TN-C-S system, a PEN conductor has not been used on the load side;

- the protective conductor to the PEN conductor is on the source side of the RCD;
- the maximum disconnection times to a circuit supplying socket-outlets and to other final circuits that supply portable equipment intended for manual movement during use, or hand-held Class I equipment does not exceed those shown in Table 9.7;

**Note:** This requirement does not apply to a final circuit supplying an item of stationary equipment connected by means of a plug and socket-outlet where precautions are already taken to prevent the use of the socket-outlet for supplying hand-held equipment, nor to reduced low-voltage circuits.

- where a fuse is used to satisfy this disconnection requirement, maximum values of Earth fault loop impedance ($Z_s$) corresponding to a disconnection time of 0.4 s are as stated in Table 9.8 for a nominal voltage to Earth ($U_o$) of 230 V;
- for a distribution circuit and a final circuit supplying only stationary equipment, the maximum disconnection time of 5 s is not exceeded:

**Table 9.7 Maximum Earth fault loop impedance ($Z_s$) for fuses, for a 0.4 s disconnection time with $U_o$ of 230 V (see Regulation 413-02-10)**

| General purpose (gG) fuses to BS 88–2.1 and BS 88–6 | | | | | | | |
|---|---|---|---|---|---|---|---|
| Rating (A) | 6 | 10 | 16 | 20 | 25 | 32 | 40 | 50 |
| $Z_s$ ($\Omega$) | 8.89 | 5.33 | 2.82 | 1.85 | 1.5 | 1.09 | 0.86 | 0.63 |

**Notes:**

1. the circuit loop impedances given in the table above should not be exceeded when the conductors are at their normal operating temperature. If the conductors are at a different temperature when tested, then the reading should be adjusted accordingly.
2. see appropriate British Standard for types and rated currents of fuses other than those mentioned in Table 9.6 on p. 597.

### 9.3.9.10 Protection by separation of circuits

Ensure that:

- the separation of circuits shall be verified for protection by:
  - SELV;
  - PELV;
  - electrical separation;
  - the separation of live parts from those of other circuits, and those of other circuits from Earth is verified by measuring that the insulation resistance in accordance with values show in Table 9.8;
- FELV circuits meet all the test requirements for low-voltage circuits.

**Table 9.8 Minimum values of insulation resistance**

| Circuit nominal voltage (V) | Test voltage d.c. (V) | Minimum insulation resistance (MΩ) |
|---|---|---|
| SELV and PELV | 250 | 0.25 |
| Up to and including 500V (with the exception of the above systems) | 500 | 0.5 |
| Above 500V | 1000 | 1.0 |

## 9.3.9.11 Polarity

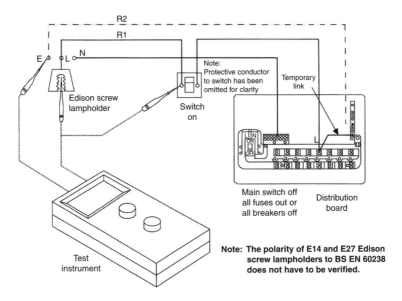

**Figure 9.7** Polarity test on a lighting circuit. (Courtesy of IET.)

Complete a polarity test (Figure 9.7) to verify that:

- fuses and single-pole control and protective devices are only connected in the phase conductor;
- circuits (other than BS EN 60238 E14 and E27 lampholders) that have an earthed neutral conductor centre contact bayonet (and Edison screw lampholders) have their outer or screwed contacts connected to the neutral conductor;
- wiring has been correctly connected to socket-outlets and similar accessories.

## 9.3.9.12 Use of Class II equipment

Class II equipment (often referred to as 'double insulated equipment') is typical of modern equipment intended to be connected to the fixed electrical installation (e.g. household appliances, portable tools and similar loads), and all live parts are insulated so as to prevent a fault in the basic insulation causing a dangerous voltage to appear on the exposed metalwork of electrical equipment.

To verify compliance, confirm that:

- circuits supplying Class II equipment have a circuit protective conductor that is run to (and terminated at) each point in the wiring and each accessory;

Except suspended lampholders that have no exposed-conductive-parts.

- the metalwork of exposed Class II equipment is mounted so that it is not in electrical contact with any part of the installation that is connected to a protective conductor;
- when Class II equipment is used as the sole means of protection against indirect contact, the installation or circuit concerned is under effective supervision while in normal use.

This form of protection **shall not** be used for circuits that include socket-outlets or where a user can change items of equipment without authorization.

### 9.3.10 Additional tests with the supply connected

Other than insulation tests, the following tests are to be completed with the supply connected:

- re-check of polarity;
- Earth electrode resistance;
- Earth fault loop impedance;
- prospective fault current.

### 9.3.10.1 Polarity

Repeat the polarity test shown in Section 9.3.9.11 to verify that:

- fuses and single-pole control and protective devices are only connected in the phase conductor;
- circuits (other than BS EN 60238 E14 and E27 lampholders) that have an earthed neutral conductor centre contact bayonet (and Edison screw lampholders) have their outer or screwed contacts connected to the neutral conductor;
- wiring has been correctly connected to socket-outlets and similar accessories.

### 9.3.10.2 Earth electrode resistance

If the earthing system incorporates an Earth electrode as part of the installation, measure the electrode resistance to Earth (using test equipment similar to that shown in Annex 9A).

If the electrode under test is being used in conjunction with an RCD protecting an installation, test (prior to energizing the remainder of the installation) between the phase conductor at the origin of the installation and the Earth electrode with the test link open.

**Note:** The resulting impedance reading (i.e. the electrode resistance) should then be added to the resistance of the protective conductor for the protected circuits.

### 9.3.10.3 Earth fault loop impedance

This is an extremely important test to ensure that, under Earth fault conditions, overcurrent devices disconnect fast enough to reduce the risk of electric shock. The Regulations stipulate that:

> *"If the protective measures employed require a knowledge of Earth fault loop impedance, then the relevant impedances must be measured."*

Where a fuse is used, maximum values of Earth fault loop impedance ($Z_s$) corresponding to a disconnection time of 0.4 s are stated in Table 9.9 for a nominal voltage to Earth ($U_o$) of 230 V.

**Note:** See the appropriate British Standard for types and rated currents of fuses other than those mentioned in Table 9.9.

Using test equipment (similar to that listed in Annex 9A), complete the following tests:

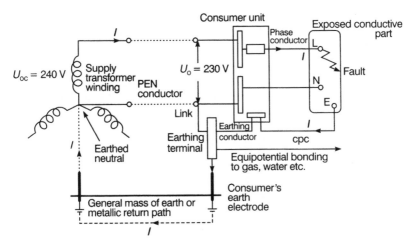

**Figure 9.8** Testing Earth fault loop impedance.

**Table 9.8 British Standards for fuse links (courtesy Scaddon)**

| Standard | | Current rating | Voltages rating | Breaking capacity | Notes |
|---|---|---|---|---|---|
| 1 | BS 2950 | Range 0.05–25A | Range 1000 V (0.05 A) to 32 V (25 A) a.c. and d.c. | Two or three times current rating | Cartridge fuse links for telecommunication and light electrical apparatus. Very low breaking capacity |
| 2 | BS 646 | 1, 2, 3 and 5 A | Up to 250 V a.c. and d.c. | 1000 A | Cartridge fuse intended for fused plugs and adapters to BS 546: 'roundpin' plugs |
| 3 | BS 1362 cartridge | 1, 2, 3, 5, 7, 10 and 13 A | Up to 250 V a.c. | 6000 A | Cartridge fuse primarily intended for BS 1363: 'flat pin' plugs |
| 4 | BS 1361 HRC cut-out fuses | 5, 15, 20, 30, 45 and 60 A | Up to 250 V a.c. | 16,500 A 33,000 A | Cartridge fuse intended for use in domestic consumer units. The dimensions prevent interchangeability of fuse links which are not of the same current rating |
| 5 | BS 88 motors | Four ranges, 2–1200 A | Up to 660 V, but normally 250 or 415 V a.c. and 250 or 500 V d.c. | Ranges from 10,000 to 80,000 A in four a.c. and three d.c. categories | Part 1 of Standard gives performance and dimensions of cartridge fuse links, whilst Part 2 gives performance and requirements of fuse carriers and fuse bases designed to accommodate fuse links complying with Part 1 |
| 6 | BS 2692 | Main range from 5 to 200 A; 0.5 to 3 A for voltage transformer protective fuses | Range from 2.2 to 132 kV | Ranges from 25 to 750 MVA (main range) 50 to 2500 MVA (VT fuses) | Fuses for a.c. power circuits above 660 V |
| 7 | BS 3036 rewirable | 5, 15, 20, 30, 45, 60, 100, 150 and 200 A | Up to 250 V to earth | Ranges from 1000 to 12,000 A | Semi-enclosed fuses (the element is a replacement wire) for a.c. and d.c. circuits |
| 8 | BS 4265 | 500 mA to 6.3 A 32 mA to 2 A | Up to 250 V a.c. | 1500 A (high breaking capacity); 35 A (low breaking capacity) | Miniature fuse links for protection of appliances of up to 250 V (metric standard) |

- Ensure that all main equipotential bonding is in place; connect the test equipment to the phase, neutral and Earth terminals at the remote end of the circuit under test. Press to test and record.

 **Note:** the circuit loop impedances given in the Table 9.9 should not be exceeded when the conductors are at their normal operating temperature. If the conductors are at a different temperature when tested, the reading should be adjusted accordingly.

### 9.3.10.4 Prospective fault current

Verify, test and ensure that:

- prospective fault currents (under both short-circuit and Earth fault conditions):
  - have been assessed for each supply source;
  - are calculated at every relevant point of the complete installation, either by enquiry or by measurement;
  - are measured at the origin and at other relevant points in the installation;
- protection of wiring systems against overcurrent takes into account minimum and maximum fault current conditions;
- fault current protective devices are provided:
  - at the supply end of each parallel conductor where two conductors are in parallel;
  - at the supply and load ends of each parallel conductor where more than two conductors are in parallel;
- fault current protective devices are:
  - less than 3 m in length between the point where the value of current-carrying capacity is reduced, and the position of the protective device;
  - installed so as to minimize the risk of fault current;
  - installed so as to minimize the risk of fire or danger to persons;
- fault current protective devices are placed (on the load side) at the point where the current-carrying capacity of the installation's conductors is likely to be lessened owing to:
  - the method of installation;
  - the cross-sectional area;
  - the type of cable or conductor used;
  - inherent environmental conditions;
- conductors are capable to carrying fault current without overheating;

**Note:** A single protective device may be used to protect conductors in parallel against the effects of fault current occurring.

Ensure that:

- fault current protection devices are capable of breaking;
- the breaking capacity rating of each device is not less than the prospective short-circuit current or Earth fault current at the point at which the device is installed;

**Note:** A lower breaking capacity is permitted if another protective device is installed on the supply side.

- the characteristics of each device used for overload current and/or for fault current protection has been co-ordinated so that the energy let through (i.e. by the fault current protective device) does not exceed the limiting values of the overload current protective device;
- devices providing protection against both overload current and fault current are capable of breaking;

**Note:** Overload current protection devices may have a breaking capacity below the value of the prospective fault current at the point where the device is installed.

- circuit-breakers used as fault current protection devices:

    ○ are capable of making any fault current up to and including the prospective fault current;
    ○ break any fault current flowing before that current causes danger due to thermal or mechanical effects produced in circuit conductors or associated connections;
    ○ break and make up any overcurrent up to and including the prospective fault current at the point where the device is installed;

- safety services with sources that are incapable of operating in parallel are protected against electric shock and fault current.

### 9.3.11 Insulation tests

When an electrical installation fails an insulation test, the installation must be corrected and the test made again. If the failure influences any previous tests that were made, those tests must also be repeated.

### 9.3.11.1 Locations with a risk of fire due to the nature of processed and/or stored materials

Test that wiring systems (less those using mineral-insulated cables and busbar trunking arrangements) have been protected against Earth insulation faults as follows:

- in TN and TT systems, by RCDs having a rated residual operating current ($I_{An}$) not exceeding 300 mA;
- in IT systems, by insulation monitoring devices with audible and visible signals.

**Notes:**

1. Adequate supervision is required to facilitate manual disconnection as soon as is appropriate.
2. The disconnection time of the overcurrent protective device, in the event of a second fault, shall not exceed 5 s.

### 9.3.11.2 Protection against electric shock – insulation tests

Confirm, measure and test that:

- circuit conductors for each SELV system are physically separated from those of any other system, or (i.e. where this proves impracticable) confirm that SELV circuit conductors are:

  ○ insulated for the highest voltage present;
  ○ enclosed in an insulating sheath additional to their basic insulation;

- equipment is capable of withstanding all mechanical, chemical, electrical and thermal influences stresses normally encountered during service;

**Note:** Paint, varnish, lacquer or similar products are **not** generally considered to provide adequate insulation for protection against direct contact in normal service.

- exposed-conductive-parts that might attain different potentials through failure of the basic insulation of live parts have been arranged so that a person will not come into simultaneous contact with two exposed-conductive-parts, or an exposed-conductive-part and any extraneous-conductive-part;

This may be achieved if the location has an insulating floor and insulating walls and one or more of the following arrangements apply:

- o   the distance between any separated exposed-conductive-parts (and between exposed-conductive-parts and extraneous-conductive-parts) is not less than 2.5 m (1.25 m for parts out of arm's reach);
- o   if protective obstacles (that are not connected to Earth or to exposed-conductive-parts and which are made out of insulating material) are used between exposed-conductive-parts and extraneous-conductive-parts;
- o   the insulation is of acceptable electrical and mechanical strength;

- •   if the nominal voltage of an SELV system exceeds 25 V a.c. rms or 60 V ripple-free d.c., then protection against direct contact has been provided by one (or more) of the following:

- o   insulation capable of withstanding a type-test voltage of 500 V a.c. rms for 60 s; or
- o   a barrier (or an enclosure) capable of providing protection to at least IP2X or IPXXB;

- •   in IT systems, an insulation monitoring device has been provided so as to indicate the occurrence of a first fault from a live part to an exposed-conductive-part or to Earth;
- •   insulating enclosures:

- o   are not pierced by conductive parts (other than circuit conductors) likely to transmit a potential;
- o   do not contain any screws of insulating material – the future replacement of which by metallic screws could impair the insulation provided by the enclosure;
- o   do not adversely affect the operation of the equipment protected;

 **Note:** Where the insulating enclosure has to be pierced by conductive parts (e.g. for operating handles of built-in equipment and for screws), protection against indirect contact shall not be impaired.

- •   live parts are completely covered with insulation which:

- o   can only be removed by destruction;
- o   is capable of durably withstanding electrical, mechanical, thermal and chemical stresses normally encountered during service;

- •   the basic insulation of operational electrical equipment is at least the degree of protection IP2X or IPXXB;
- •   where insulation has been applied during the erection of the installation, the quality of the insulation has been verified.

 Where the risk of electric shock is increased by a reduction in body resistance and/or by contact with Earth potential, confirm that protection has been provided by insulation of live parts, protection by obstacles, protection by barriers or enclosures, or SELV.

### 9.3.11.3 Protection against electric shock – special installations or locations – verification tests

Where SELV or PELV is used (whatever the nominal voltage) in locations containing a bath, shower, hot-air sauna and/or in a restrictive conductive location, confirm that protection against direct contact has been provided by:

- insulation capable of withstanding a type-test voltage of 500 V a.c. rms for 1 minute; or
- barriers and/or enclosures providing protection to at least IP2X or IPXXB.

The above requirements do not apply to a location in which freedom of movement is not physically constrained.

### 9.3.11.4 RCDs and RCBOs – verification tests

Test (on the load side of the RCD and as near as is practicable to its point of installation and between the phase conductor of the protected circuit and their associated circuit protective conductor) that:

- general-purpose RCDs:
  - do not open with a leakage current flowing equivalency to 50% of the rated tripping current;
  - open in less than 200 ms with a leakage current flowing equivalency to 100% of the rated tripping current of the RCD;

- general-purpose RCCDs to BS EN 61008 or RCBOs (residual current operated circuit-breaker without integral overcurrent protection) to BS EN 61009:
  - do not open with a leakage current flowing equivalency to 50% of the rated tripping current of the RCD;
  - open in less than 300 ms with a leakage current flowing equivalency to 100% of the rated tripping current of the RCD (unless it is a Type S (or selective) device that incorporates an intentional time delay, in which case it should trip between 130 ms and 500 ms);

- RCD protected socket-outlets to BS 7288:
  - do not open with a leakage current flowing equivalency to 50% of the rated tripping current of the RCD;

- o  open in less than 200 ms with a leakage current flowing equivalency to 100% of the rated tripping current of the RCD.

### 9.3.11.5  Selection and erection of wiring systems

Test to confirm that:

- all electrical joints and connections meet stipulated requirements concerning conductance, insulation, mechanical strength and protection;
- cables that are run in a thermally insulated spaces are not covered by the thermal insulation;
- the current-carrying capacity of cables that are installed in thermally insulated walls or above a thermally insulated ceiling conforms with Appendix 4 to the Regulations;
- the current-carrying capacity of cables that are totally surrounded by thermal insulation for less than 0.5 m has been reduced according to the size of cable, length and thermal properties of the insulation;
- the insulation and/or sheath of cables connected to a bare conductor or busbar is capable of withstanding the maximum operating temperature of the bare conductor or busbar;
- wiring systems are capable of withstanding the highest and lowest local ambient temperature likely to be encountered – or are provided with additional insulation suitable for those temperatures;
- wiring systems have been selected and erected so as to minimize (i.e. during installation, use and maintenance) damage to the sheath and insulation of cables and to insulated conductors and their terminations.

### 9.3.11.6  Site-applied insulation

Test to confirm that:

- insulation applied on site to protect against direct contact is capable of withstanding, without breakdown or flashover, an applied test voltage as specified in the British Standard for similar type-tested equipment;
- supplementary insulation applied to equipment during erection (i.e. to protect against indirect contact):
  - o  protects to at least IP2X or IPXXB; and
  - o  is capable of withstanding, without breakdown or flashover, an applied test voltage as specified in the British Standard for similar type-tested equipment.

### 9.3.11.7  Supplies for safety services (IT systems)

In an IT system, confirm that continuous insulation monitoring has been provided to give audible and visible indications of a first fault.

## 9.3.12 Periodic inspections and tests

Periodic inspection and testing of all electrical installations shall be carried out to confirm that the installation is in a satisfactory condition for continued service, and this inspection shall consist of careful scrutiny of the installation (dismantled or otherwise) using appropriate tests.

The aim of periodic inspection and testing is to:

- confirm that the safety of the installation has not deteriorated and that the installation has not been damaged;
- ensure the continued safety of persons and livestock against the effects of electric shock and burns;
- identify installation defects and non-compliance with the requirements of the Regulations that may give rise to danger;
- protect property from being damaged by fire and heat caused by a defective installation.

Precautions shall be taken to ensure that inspection and testing does not cause:

- danger to persons or livestock;
- damage to property and equipment (even if the circuit is defective).

The frequency of periodic inspection and testing of installations will depend on:

- the type of installation, its use and operation;
- the frequency and quality of maintenance; and
- the external influences to which it is subjected.

Periodic inspection and testing of supervised installations may take the form of continuous monitoring and maintenance by skilled persons. Appropriate records shall be kept.

## 9.3.13 Verification tests

All completed installations (including additions and/or alteration to existing installations) shall be inspected and tested for conformance to the requirements of BS 7621 – as amended).

### 9.3.13.1 Accessibility of connections

Confirm that all connections and joints are accessible for inspection, testing and maintenance, unless:

- they are in a compound-filled or encapsulated joint;
- the connection is between a cold tail and a heating element;
- the joint is made by welding, soldering, brazing or compression tool.

### 9.3.13.2 Appliances producing hot water or steam

Confirm that electric appliances producing hot water or steam have been protected against overheating.

### 9.3.13.3 Cables and conductors for low voltage

Confirm that flexible and non-flexible cables, flexible cords (and conductors used as an overhead line) operating at low voltage comply with the relevant British or Harmonized Standard.

### 9.3.13.4 Electric surface heating systems

Confirm that the equipment, system design, installation and testing of all electric surface heating systems meet the requirements of BS 6351.

### 9.3.13.5 Emergency switching – verification tests

For any exterior installation, confirm that the switch is placed outside the building, adjacent to the equipment (where this is not possible, a notice showing the position of the switch shall be placed adjacent to the equipment and a notice fixed near the switch shall indicate its use).

### 9.3.13.6 Forced air heating systems

Confirm (by inspection and test) that electric heating elements of forced air heating systems (other than those of central-storage heaters):

- are incapable of being activated until the prescribed air flow has been established;
- deactivate when the air flow is reduced or stopped;
- do not have two, independent, temperature limiting devices;
- have frames and enclosures that are constructed of non-ignitable material.

### 9.3.13.7 Heating cables

Check that:

- heating conductors and cables that pass through (or are in close proximity to) to a fire hazard:
    - are enclosed in material with an ignitability characteristic P as specified in BS 476;
    - are protected from any mechanical damage;
- heating cables that have been laid (directly) in soil, concrete, cement screed, or other material used for road and building construction are:

- o    capable of withstanding mechanical damage,
- o    constructed of material that will be resistant to damage from damp-ness and/or corrosion;

- heating cables that have been laid (directly) in soil, a road, or the struc-ture of a building are installed so that they:

  - o    are completely embedded in the substance they are intended to heat;
  - o    are not damaged by movement (by the cables themselves or by the substance in which they are embedded);
  - o    comply in all respects with the maker's instructions and recommendations;

- the maximum loading of floor-warming cable under operating conditions is no greater than the temperatures shown in Table 9.10.

**Table 9.10 Maximum conductor operating temperatures for a floor-warming cable**

| Type of cable | Maximum conductor operating temperature ( C) |
|---|---|
| General-purpose PVC over conductor | 70 |
| Enamelled conductor, polychlorophene over enamel, PVC over all | 70 |
| Enamelled conductor, PVC over all | 70 |
| Enamelled conductor, PVC over enamel, lead-alloy E sheath over all | 70 |
| Heat-resisting PVC over conductor | 85 |
| Nylon over conductor, heat-resisting PVC over all | 85 |
| Synthetic rubber or equivalent elastomeric insulation over conductor | 85 |
| Mineral insulation over conductor, copper sheath over all | Temperature dependent on type of seal employed, outer covering etc. |
| Silicone-treated woven-glass sleeve over conductor | 180 |

## 9.3.13.8 Heating and ventilation systems

In locations where heating and ventilation systems containing heating ele-ments are installed and where there is a risk of fire due to the nature of proc-essed or stored materials, check to ensure that:

- the dust or fibre content and the temperature of the air does not present a fire hazard;
- temperature-limiting devices have a manual reset;
- heating appliances are fixed;

- heating appliances mounted close to combustible materials are protected by barriers to prevent the ignition of such materials;
- heat-storage appliances are incapable of igniting combustible dust and/or fibres;
- enclosures of equipment such as heaters and resistors do not attain higher surface temperatures than:
  - 90°C under normal conditions, and
  - 115°C under fault conditions.

### 9.3.13.9  Identification of conductors by letters and/or numbers

Test to confirm that all individual conductors and groups of conductors:

- have been identified by a label containing either letters or numbers that are clearly legible;
- have numerals that contrast, strongly, with the colour of the insulation;
- numerals <u>6</u> and <u>9</u> have been underlined (so as to avoid confusion).

### 9.3.13.10  Plugs and socket-outlets

Inspect and test to ensure that any plug and socket-outlet used in single-phase a.c. or two-wire d.c. circuits that does **not** comply with BS 1363, BS 546, BS 196 or BS EN 60309–2 has either been designed especially for that purpose or:

- the plug and socket-outlet used for an electric clock has been specially designed for that purpose and the plug has a fuse not exceeding 3 A and which complies with BS 646 or BS 1362;
- the plug and socket-outlet used for an electric shaver is either part of the shaver supply unit which complies with BS 3535 (as amended) or is in a room (other than a bathroom) that complies with BS 4573.

### 9.3.13.11  Water heaters

Confirm that water heaters (or boilers) having immersed and uninsulated heating elements are permanently connected to the electricity supply via a double-pole linked switch, which is either:

- separate from and within easy reach of the heater/ boiler; or
- part of the boiler/heater (provided that the wiring from the heater or boiler is directly connected to the switch without use of a plug and socket-outlet).

#### 9.3.13.11.1  Functional testing

The following are amongst the most important functional tests that should be completed:

- verify the effectiveness of RCDs providing protection against indirect contact (or supplementary protection against direct contact) by a test simulating a typical fault condition;
- functionally test assemblies (such as switchgear and controlgear assemblies, drives, controls and interlocks) to show that they are properly mounted, adjusted and installed in accordance with the Regulations.

### 9.3.14 Electrical connections

Of main concern are:

- connections between conductors and between a conductor and equipment;
- main earthing terminals or bars;
- final and distribution circuits.

Test and verify that:

- connections between conductors and between a conductor and equipment (and between a conductor and equipment) provide durable electrical continuity and adequate mechanical strength;
- the earthing conductor of main earthing terminals (or bars) is capable of being disconnected to enable the resistance of the earthing arrangements to be measured;

 For convenience (and if required) this may be combined with the main earthing terminal or bar.

- all joints:
  - o are capable of disconnection only by means of a tool;
  - o are mechanically strong;
  - o ensure the maintenance of electrical continuity;
- the wiring of final and distribution circuits to equipment with a protective conductor current exceeding 10 mA has a protective connection complying with one or more of the following:
  - o a single protective conductor with a cross-sectional area greater than 10 mm$^2$;
  - o a single (mechanically protected) copper protective conductor with a cross-sectional area greater than 4 mm$^2$;
  - o two individual protective conductors;
  - o a BS 4444 Earth monitoring system that will automatically disconnect the supply to the equipment in the event of a continuity fault;

- connection (i.e. of the equipment) to the supply is by means of a double-wound transformer, which has its secondary winding connected to the protective conductor of the incoming supply and the exposed-conductive-parts of the above.

### 9.3.15 Tests for compliance with the Building Regulations

As shown in Table 9.11, there are four types of installation that have to be inspected and tested for compliance with the Building Regulations.

Table 9.11 Types of installation

| Type of inspection | When is it used? | What should it contain? | Remarks |
|---|---|---|---|
| Electrical Installation Certificate | For the initial certification of a new installation or for the alteration of or addition to an existing installation where new circuits have been introduced | A schedule of inspections and test results as required by Part 6 (of BS 7671) A certificate, including guidance for recipients (standard form from Appendix 6 of BS 7671) | The original certificate shall be given to the person ordering the work, and a duplicate shall be retained by the contractor |
| Minor Electrical Installation Works Certificate | For additions and alterations to an installation such as an extra socket-outlet or lighting point to an existing circuit, the relocation of a light switch etc. | Relevant provisions of Part 6 of BS 7671 | This certificate may also be used for the replacement of equipment such as accessories or luminaires, but not for the replacement of distribution boards (or similar items) or the provision of a new circuit |
| Electrical Installation Certificate | For the inspection of an existing electrical installation | A schedule of inspections and a schedule of test results as required by Part 6 (of BS 7671) | For safety reasons, the electrical installation will need to be inspected at appropriate intervals by a competent person |
| Building Regulations Compliance Certificate | Confirmation that the work carried out complies with the Building Regulations | The basic details of the installation, the location, the completion date and the name of the installer | A purchaser's solicitor may request this document when you come to sell your property. Looking further ahead, it may be required as one of the documents that will make up your Home Information Pack |

Figure 9.9 indicates how to choose which type of inspection is required.

Part P applies only to fixed electrical installations that are intended to oper-
ate at low voltage or extra-low voltage, and which are not controlled by the
Electricity Supply Regulations 1988 as amended, or the Electricity at Work
Regulations 1989 as amended.

### 9.3.16  Additional tests required for special installations and locations

Part 7 of the Regulations contains additional requirements in respect of installations where the risk of electric shock is increased by a reduction in body
resistance or by contact with Earth potential (e.g. locations containing a bath
or shower, swimming pools, hot-air sauna, construction installations, agricultural and horticultural premises, caravans, motor caravans and highway power
supplies etc.).

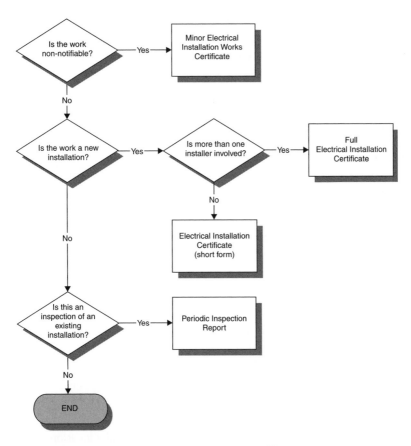

**Figure 9.9** Choosing the correct inspection certificate.

Additional tests and inspections for these special installations and locations have been listed at the end of the following sections in the form of check sheets.

### 9.3.16.1 Agricultural and horticultural premises

All fixed agricultural and horticultural installations (outdoors and indoors) and locations where livestock is kept (such as stables, chicken houses, piggeries, feed-processing locations, lofts and storage areas for hay, straw and fertilizers) shall be inspected to confirm that they comply with Part 705 of the Regulations. (See Section 7.3.1 for a list of inspections and tests that need to be completed.)

 **Note:** If these premises include dwellings that are intended solely for human habitation, then the dwellings are excluded from the scope of these particular Regulations.

### 9.3.16.2 Conducting locations with restricted movement

Fixed equipment in conducting locations (particularly where the movement of persons is restricted by the location) and the supplies to mobile equipment for use in such locations shall be inspected to confirm that they comply with Part 706 of the Regulations. (See Section 7.3.2 for a list of inspections and tests that need to be completed.)

### 9.3.16.3 Construction and demolition sites

Installations providing electricity supply for:

- new building construction;
- repairs, alterations or extensions to, or demolition of, existing buildings;
- engineering construction;
- earthworks;

shall be inspected to confirm that they comply with Part 704 of the Regulations. (See Section 7.3.3 for a list of inspections and tests that need to be completed.)

 These requirements do **not** apply to:

- construction site offices, cloakrooms, meeting rooms, canteens, restaurants, dormitories and toilets;
- installations covered by BS 6907.

### 9.3.16.4 Electrical installations in caravan/camping parks and similar locations

Electrical installations in caravan/camping parks and similar locations providing facilities for supplying leisure accommodation vehicles (including caravans) or tents shall be inspected to confirm that they comply with Part 708 of the Regulations. (See Section 7.3.4 for a list of inspections and tests that need to be completed.)

### 9.3.16.5 Electrical installations in caravans and motor caravans

All electrical installations in caravans and motor caravans shall be inspected to confirm that they comply with Part 721 of the Regulations. (see Section 7.3.5 for a list of inspections and tests that need to be completed.)

It should be noted that the requirements of this section do **not** apply to:

- electrical circuits and equipment covered by the Road Vehicles Lighting Regulations 1989;
- installations covered by BS EN 1648–1 and BS EN 1648–2;
- internal electrical installations of mobile homes, fixed recreational vehicles, transportable sheds and the like, temporary premises or structures.

### 9.3.16.6 Exhibitions, shows and stands

Temporary electrical installations in exhibitions, shows and stands (including mobile and portable displays and equipment) shall be inspected to confirm that they comply with Part 711 of the Regulations. (See Section 7.3.6 for a list of inspections and tests that need to be completed.)

### 9.3.16.7 Floor and ceiling heating systems

Electric floor and ceiling heating systems that are erected as either thermal storage heating systems or direct heating systems shall be inspected to confirm that they comply with Part 753 of the Regulations. (See Section 7.3.7 for a list of inspections and tests that need to be completed.)

### 9.3.16.8 Locations containing a bath or shower

All locations containing a bath or shower shall be inspected to confirm that they comply with Part 701 of the Regulations. (See Section 7.3.8 for a list of inspections and tests that need to be completed.)

Locations containing baths or showers for medical treatment, or for disabled persons, may have special requirements.

### 9.3.16.9 Marinas and similar locations

Circuits intended to supply pleasure craft or houseboats in marinas and similar locations shall be inspected to confirm that they comply with Part 709 of the Regulations. (See Section 7.3.9 for a list of inspections and tests that need to be completed.)

### 9.3.16.10 Medical locations

Section 710 is a new section specifically aimed at electrical installations in medical locations. Although it refers mainly to hospitals, private clinics, medical and dental practices, healthcare centres and dedicated medical rooms in the workplace, Section 710 also applies to electrical installations in locations designed for medical research and (where applicable) to veterinary clinics.

The International Electrotechnical Commission (IEC) and/or British Standards (BS) manufacturers' standards for medical electrical equipment consist of two types of testing: type testing and routine testing. (See Section 7.3.10 for a full list of inspections and tests that need to be completed.)

#### 9.3.16.10.1  Type testing

Type testing is carried out by an approved test house (under tightly specified and tightly controlled environmental conditions) on a single representative sample of a piece of equipment for which certification of compliance with a standard is being sought. These tests are not intended for routine use – indeed, it has been documented that repetition of many of the tests would certainly cause deterioration in the performance and safety of the equipment under test.

#### 9.3.16.10.2  Routine testing

Routine testing, on the other hand, is intended to provide an indication of the inherent safety of the equipment, without subjecting it to undue stress that would be liable to cause deterioration.

### 9.3.16.11 Mobile and transportable units

A vehicle and/or mobile (self-propelled or towed) or transportable structure (such as a container or cabin) in which all or part of an electrical installation is contained and which is provided with a temporary supply by means of, for example, a plug and socket-outlet, shall be inspected to confirm that they comply with Part 717 of the Regulations. (See Section 7.3.11 for a list of inspections and tests that need to be completed.)

### 9.3.16.12 Operating and maintenance gangways

Section 729 is a new (fairly small) addition to the standard. It centres on the operation and safe maintenance of switchgear and controlgear within areas that include gangways and where access is restricted to skilled or instructed persons. Access areas and the requirements for operating and maintenance gangways should comply with Part 729 of the Regulations. (See Section 7.3.12 for a list of inspections and tests that need to be completed.)

### 9.3.16.13 Rooms and cabins containing saunas

Installations supplying electricity for locations containing hot-air sauna heating equipment (in accordance with BS EN 60335–2–53) shall be inspected to confirm that they comply with Part 703 of the Regulations. (See Section 7.3.13 for a list of inspections and tests that need to be completed.)

### 9.3.16.14 Solar, photovoltaic (PV) power supply systems

Electrical installations of PV power supply systems (including subsystems with a.c. modules) shall be inspected to confirm that they comply with Part 712 of the Regulations. (See Section 7.3.14 for a list of inspections and tests that need to be completed.).

### 9.3.16.15 Swimming pools and other basins

Requirements applicable to basins of swimming pools, paddling pools and other basins, plus their surrounding zones, shall be inspected to confirm that they comply with Part 702 of the Regulations. (See Section 7.3.15 for a list of inspections and tests that need to be completed.)

### 9.3.16.16 Temporary electrical installations for structures, amusement devices and booths at fairgrounds, amusement parks and circuses

Electrical installation required for the safe design, installation and operation of temporarily erected mobile or transportable electrical machines and structures that incorporate electrical equipment shall be inspected to confirm that they comply with Part 740 of the Regulations. (See Section 7.3.16 for a list of inspections and tests that need to be completed.)

## 9.4 Identification and notices

For safety purposes, the Regulations require a number of notices and labels to be used for electrical installations, and these will need to be checked during initial and periodic inspections.

### 9.4.1 General

Verify that:

- labels (or other means of identification) indicate the purpose of each item of switchgear and controlgear;
- wiring is marked and/or arranged so that it can be quickly identified for inspection, testing, repair or alteration of the installation;
- unambiguous marking has been provided at the interface between conductors identified in accordance with these Regulations and conductors identified to previous versions of the Regulations;

**Note:** Appendix 7 of the Regulations provides guidance on how this can be achieved.

- orange-coloured conduits have been used to distinguish an electrical conduit from other services or other pipelines.

### 9.4.2 Conductors

Verify that:

- conductor cable cores are identified by colour and/or lettering and/or numbering;
- binding and sleeves used for identifying protective conductors comply with BS 3858;
- neutral or midpoint conductors are coloured **blue**;
- protective conductor cable cores are identifiable at all terminations (and preferably throughout their length);
- protective conductors are a bi-colour combination of **green-and-yellow**, and neither colour covers more than 70% of the surface being coloured;

 Verify that this combination of colours has **not** been used for any other purpose.

- single-core cables used as protective conductors are coloured **green-and-yellow** throughout their length;
- bare conductors or busbars used as protective conductors are identified by equal **green-and-yellow** stripes that are 15 mm to 100 mm wide;

 If adhesive tape is used, then it has been bicoloured.

- PEN conductors (when insulated) are either:
  - **green-and-yellow** throughout their length, with **blue** markings at the terminations, or
  - **blue** throughout their length, with **green-and-yellow** markings at the terminations;
- bare conductors are painted or identified by a coloured tape, sleeve or disc as per Table RLT-9.27.

**ALL other conductors** (including those used to identify conductor switchboard busbars and conductors) **shall** be coloured as shown in accordance with Table 9.12.

 The colour <u>green</u> shall **not** be used on its own.

**Table 9.12 Identification of conductors**

| Function | Alphanumeric | Colour |
|---|---|---|
| Protective conductors | | Green-and-yellow |
| Functional earthing conductor | | Cream |
| a.c. power circuit (Note 1) | | |
| Phase of single-phase circuit | L | Brown |
| Neutral of single- or three-phase circuit | N | Blue |
| Phase 1 of three-phase a.c. circuit | L1 | Brown |
| Phase 2 of three-phase a.c. circuit | L2 | Black |
| Phase 3 of three-phase a.c. circuit | L3 | Grey |
| Two-wire unearthed d.c. power circuit | | |
| Positive of two-wire circuit | L+ | Brown |
| Negative of two-wire circuit | L– | Grey |
| Two-wire earthed d.c. power circuit | | |
| Positive (of negative earthed) circuit | L+ | Brown |
| Negative (of negative earthed) circuit (Note 2) | M | Blue |
| Positive (of positive earthed) circuit (Note 2) | M | Blue |
| Negative (of positive earthed) circuit | L– | Grey |
| Three-wire d.c. power circuit | | |
| Outer positive of two-wire circuit derived from three-wire system | L+ | Brown |
| Outer negative of two-wire circuit derived from three-wire system | L– | Grey |
| Positive of three-wire circuit | L+ | Brown |
| Mid-wire of three-wire circuit (Notes 2 and 3) | M | Blue |
| Negative of three-wire circuit | L– | Grey |
| Control circuits, ELV and other applications | | |
| Phase conductor | L | Brown, black, red, orange, yellow, violet, grey, white, pink or turquoise |
| Neutral or mid-wire (Note 4) | N or M | Blue |

Notes:

1. Power circuits include lighting circuits.
2. M identifies either the middle wire of a three-wire d.c. circuit, or the earthed conductor of a two-wire earthed d.c. circuit.
3. Only the middle wire of three-wire circuits may be earthed.
4. An earthed PELV conductor is **blue**.

Further information on cable identification colours for extra-low-voltage and d.c. power circuits is available from the IET website at http://www.iet.org.

### 9.4.3  Identification of conductors by letters and/or numbers

Where letters and/or numbers are used to identify conductors, check to confirm that:

- individual conductors and/or groups of conductors are identified by either letters or numbers that are clearly legible;
- all numerals contrast, strongly, with the colour of the insulation;
- numerals <u>6</u> and <u>9</u> are underlined;
- protective devices are arranged and identified so that the circuit protected is easily recognizble;
- protective conductors coloured **green-and-yellow** are not numbered other than for the purpose of circuit identification;
- the alphanumeric numbering system is in accordance with Table 9.12.

### 9.4.4  Omission of identification by colour or marking

Colour or marking is **not** required for:

- concentric conductors of cables;
- metal sheath or armour of cables (when used as a protective conductor);
- bare conductors (where permanent identification is impracticable);
- extraneous-conductive-parts used as a protective conductor;
- exposed-conductive-parts used as a protective conductor.

### 9.4.5  Diagrams

Verify that all available diagrams, charts, tables or schedules that have been used indicate:

- the type and composition of each circuit (points of utilization served, number and size of conductors, type of wiring); and
- the method used for protection against indirect contact;
- the data (where appropriate) and characteristics of each protective device used for automatic disconnection;
- the identification (and location) of all protection, isolation and switching devices;
- the circuits or equipment that are susceptible to a particular test.

 Verify that distribution board schedules have been provided within (or adjacent to) each distribution board.

 All symbols used shall comply with BS EN 60617 (see Annex A to this book).

### 9.4.6 Warning notices

Check to confirm that the following warning notices are appropriate to the situation, correctly positioned and contain the right information.

#### 9.4.6.1 Earth-free local equipotential bonding

Where Earth-free local equipotential bonding has been used, confirm that an appropriate notice has been fixed in a prominent position adjacent to every point of access to the location concerned.

#### 9.4.6.2 Earthing and bonding connections

Confirm that a permanent label with the words shown in Figure 9.10 has been permanently fixed at or near:

- the connection point of every earthing conductor to an Earth electrode;
- the connection point of every bonding conductor to an extraneous-conductive-part;
- the main Earth terminal (when separated from the main switchgear).

> **Safety Electrical Connection – Do Not Remove**

**Figure 9.10** Earthing and bonding notice.

Where protection is by Earth-free local equipotential bonding or electrical separation, confirm that the warning notice reads as per Figure 9.11.

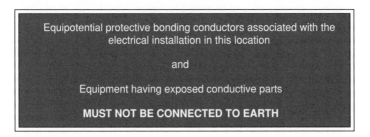

Equipotential protective bonding conductors associated with the electrical installation in this location

and

Equipment having exposed conductive parts

**MUST NOT BE CONNECTED TO EARTH**

**Figure 9.11** Earthing, bonding and electrical separation notice.

#### 9.4.6.3 Electrical installations in caravans, motor caravans and caravan parks

Confirm that:

- a notice has been fixed on (or near) the electrical inlet recess of the caravan or motor caravan installation containing details concerning the:

- o   nominal (design) voltage and frequency,
- o   current rating;

- a notice, worded as shown in Figure 9.12 has been permanently fixed near the main isolating switch.

---

### INSTRUCTIONS FOR ELECTRICITY SUPPLY

**TO CONNECT**

1.  Before connecting the caravan installation to the mains supply, check that:

    (a)   the supply available at the caravan pitch supply point is suitable for the caravan electrical installation and appliances, and
    (b)   the voltage and frequency and current ratings are suitable, and
    (c)   the caravan main switch is in the OFF position.

Also, prior to use, examine the supply flexible cable to ensure there is no visible damage or deterioration.

2.  Open the cover to the appliance inlet provided at the caravan supply point, if any, and insert the connector of the supply flexible cable.
3.  Raise the cover of the electricity outlet provided on the pitch supply point and insert the plug of the supply cable.

**THE CARAVAN SUPPLY FLEXIBLE CABLE MUST BE FULLY UNCOILED TO AVOID DAMAGE BY OVERHEATING**

4.  Switch on at the caravan main isolating switch.
5.  Check the operation of residual current devices (RCDs) fitted in the caravan by depressing the test button(s) and reset.

**IN CASE OF DOUBT OR, IF AFTER CARRYING OUT THE ABOVE PROCEDURE THE SUPPLY DOES NOT BECOME AVAILABLE, OR IF THE SUPPLY FAILS, CONSULT THE CARAVAN PARK OPERATOR OR THE OPERATOR'S AGENT OR A QUALIFIED ELECTRICIAN.**

**TO DISCONNECT**

6.  Switch off at the caravan main isolating switch, unplug the cable first from the caravan pitch supply point and then from the caravan inlet connector.

### PERIODIC INSPECTION

Preferably not less than once every three years and annually if the caravan is used frequently, the caravan electrical installation and supply cable should be inspected and tested and a report on their condition obtained as prescribed in BS 7671 Requirements for Electrical Installations published by the Institution of Engineering and Technology and BSI.

---

**Figure 9.12** Electrical safety notice.

## 9.4.6.4 Emergency switching

For any exterior installation for emergency switching, confirm that the switch is placed outside the building, adjacent to the equipment (where this is not possible, a notice showing the position of the switch shall be placed adjacent to the equipment and a notice fixed near the switch shall indicate its use).

## 9.4.6.5 Final-circuit distribution boards

Confirm that distribution boards for socket-outlet final circuits have a notice that clearly indicates circuits that have a high protective conductor current, and that this information is positioned so as to be plainly visible to a person employed in modifying or extending the circuit.

## 9.4.6.6 Inspection and testing

Confirm that:

* a notice has been fixed in a prominent position at or near the origin of every installation upon completion of the work (e.g. initial verification, alterations and additions to an installation, and periodic inspection and testing);
* the notice has been inscribed in indelible characters and reads as shown in Figure 9.13;

> **IMPORTANT**
> This installation should be periodically inspected and tested and a report on its condition obtained, as prescribed in BS 7671:2008 Requirements for Electrical Installations.
>
> Date of last inspection......................
>
> Recommended date of next inspection..................

**Figure 9.13** Inspection and testing notice.

* if an installation includes an RCD then it has a notice (fixed in a prominent position) that reads as shown in Figure 9.14.

> This installation, or part of it, is protected by a device that automatically switches off the supply if an earth fault develops.
>
> Test quarterly by pressing the button marked 'T' or 'Test'. The device should switch off the supply and should then be switched on to restore the supply.
>
> If the device does not switch off the supply when the button is pressed, seek expert advice.

**Figure 9.14** Inspection and testing notice.

Confirm that, following initial verification, an Electrical Installation Certificate (together with a schedule of inspections and a schedule of test results) was given to the person ordering the work, and that:

- the schedule of test results identified every circuit and its related protective device(s), and recorded the results of the appropriate tests and measurements;
- the certificate took account of the respective responsibilities for the safety of that installation and the relevant schedules;
- defects or omissions revealed during inspection and testing of the installation work covered by the certificate were rectified before the certificate is issued.

Confirm that:

- an Electrical Installation Certificate (containing details of the installation, together with a record of the inspections made and the test results) or a Minor Electrical Installation Works Certificate (for all minor electrical installations that do not include the provision of a new circuit) was provided for all alterations or additions to electrical circuits; and
- any defects found in an existing installation were recorded on an Electrical Installation Certificate or a Minor Electrical Installation Works Certificate.

### 9.4.6.7 Isolation

Verify that a notice is fixed in each position where there are live parts that are not capable of being isolated by a single device.

 The location of each isolator shall be indicated, unless there is no possibility of confusion.

If an installation is supplied from more than one source, confirm that:

- a main switch is provided for each source of supply;
- a notice has been placed warning operators that more than one switch needs to be operated.

Unless an interlocking arrangement is provided, confirm that a notice has been provided warning people that they will need to use the appropriate isolating devices if an equipment or enclosure that contains live parts cannot be isolated by a single device.

### 9.4.6.8 Voltage

Verify that:

- items of equipment (or enclosures) within which a nominal voltage exceeding 230 V exists (and where the presence of such a voltage would not normally be expected) are arranged so that, before access is gained to a live part, a warning of the maximum voltage present is clearly visible;

- where terminals (or other fixed live parts between which a nominal voltage exceeding 230 V exists) are housed in separate enclosures (or items of equipment which, although separated, can be reached simultaneously by a person) a notice has been secured in a position such that anyone, before gaining access to such live parts, is warned of the maximum voltage that exists between those parts;
- means of access to all live parts of switchgear and other fixed live parts where different nominal voltages exist are marked to indicate the voltages present.

### 9.4.6.9 Warning notice – non-standard colours

If an installation that was wired to a previous version of the Regulations is partially altered or rewired according to the current Regulations (i.e. in accordance with Table 9.12), confirm that a warning notice (Figure 9.15) has been placed at (or near) the appropriate distribution board.

> **WARNING**
> This installation has wiring colours to two versions of BS 7671.
> Great care should be taken before undertaking extension, alteration or repair that all conductors are correctly identified.

**Figure 9.15** Non-standard colours.

## 9.5 What type of certificates and reports are there?

There are three types of certificate associated with electrical installations. These are:

| | |
|---|---|
| Electrical Installation Certificate | For the design, construction, inspection and testing of the work |
| Minor Electrical Installation Works Certificate | For the design, construction, inspection and testing of minor work |
| Electrical Installation Condition Report | For the regular inspection and testing of an electrical installation |

All certificates need to be made out and signed (or otherwise authenticated) by a competent person or persons. If the design, construction and inspection and testing is the responsibility of one person, then a certificate (similar to that shown in Figure 9.16) may be used as a replacement for the multiple signatures section of the model form.

> **FOR DESIGN, CONSTRUCTION, INSPECTION & TESTING**
>
> I being the person responsible for the Design, Construction, Inspection & Testing of the electrical installation (as indicated by my signature below), particulars of which are described above, having exercised reasonable skill and care when carrying out the Design, Construction, Inspection & Testing, hereby CERTIFY that the said work for which I have been responsible is to the best of my knowledge and belief in accordance with BS 7671:2008 except for the departures, if any, detailed as follows:
>
> . . . . . . . . . . . . . . . . . . . . . . . . . . . . . . . . . . . . . . . . . . . . . . . . . . . . . . . . . . . . . . . . .
>
> . . . . . . . . . . . . . . . . . . . . . . . . . . . . . . . . . . . . . . . . . . . . . . . . . . . . . . . . . . . . . . . . .
>
> . . . . . . . . . . . . . . . . . . . . . . . . . . . . . . . . . . . . . . . . . . . . . . . . . . . . . . . . . . . . . . . . .
>
> Name:. . . . . . . . . . . . . . . .   Signature:. . . . . . . . . . . . . . . .   Date:. . . . . . . .

**Figure 9.16** Single-signature declaration form.

Schedules of inspections and test results shall be issued along with the associated Electrical Installation Certificate or Electrical Installation Condition Report (see Figures 9.17 and 9.18).

### 9.5.1 Electrical installation certificate

The Electrical Installation Certificate (containing details of the installation together with a record of the inspections made and the test results – see Figures 9.17 and 9.18) is **only** used for the initial certification of a new installation, or for an alteration or addition to an existing installation where new circuits have been introduced.

**Notes:**

1. The Electrical Installation Certificate is **not** to be used for a periodic inspection (for which an Electrical Installation Condition Report form should be used).
2. For an alteration or addition that does not involve the introduction of a new circuit(s), a Minor Electrical Installation Works Certificate should be used.

The Electrical Installation Certificate is **only** valid if accompanied by the Schedule of Inspections and the Schedule(s) of Test Results.

**Note:** The original certificate shall be given to the person ordering the work, and a duplicate shall be retained by the contractor.

**SCHEDULE OF INSPECTIONS**

**Methods of protection against electric shock**

**(a) Protection against both direct and indirect contact:**

☐ (i)   SELV

☐ (ii)  Limitation of discharge of energy

**(b) Protection against direct contact:**

☐ (i)    Insulation of live parts

☐ (ii)   Barriers or enclosures

☐ (iii)  Obstacles

☐ (iv)   Placing out of reach

☐ (v)    PELV

☐ (vi)   Presence of RCD for supplementary protection

**(c) Protection against indirect contact:**

☐ (i)   EEBADS including:

☐      Presence of earthing conductor

☐      Presence of circuit protective conductors

☐      Presence of main equipotential bonding conductors

☐      Presence of supplementary equipotential bonding conductors

☐      Presence of earthing arrangements for combined protective and functional purpose

☐      Presence of adequate arrangements for alternative source(s), where applicable

☐      Presence of residual current device(s)

☐ (ii)   Use of Class II equipment or equivalent insulation

☐ (iii)  Non-conducting location: Absence of protective conductors

☐ (iv)   Earth-free equipotential bonding: Presence of earth-free equipotential bonding conductors

☐ (v)    Electrical separation

**Prevention of mutual detrimental influence**

☐ (a)   Proximity of non-electrical services and other influences

☐ (b)   Segregation of band I and bandl circuits or band II insulation used

☐ (c)   Segregation of safety circuits

**Identification**

☐ (a)   Presence of diagrams, instructions, circuit charts and similar information

☐ (b)   Presence of danger notices and other warning notices

☐ (c)   Labelling of protective devices, switches and terminals

☐ (d)   Identification of conductors

**Cables and conductors**

☐ (a)   Routing of cables in prescribed zones or within mechanical protection

☐ (b)   Connection of conductors

☐ (c)   Erection methods

☐ (d)   Selection of conductors for current-carrying capacity and voltage drop

☐ (e)   Presence of fire barriers, suitable seals and protection against thermal effects

**General**

☐ (a)   Presence and correct location of appropriate devices for isolation and switching

☐ (b)   Adequacy of access to switchgear and other equipment

☐ (c)   Particular protective measures for special installations and locations

☐ (d)   Connection of single-pole devices for protection or switching in phase conductors only

☐ (e)   Correct connection of accessories and equipment

☐ (f)   Presence of undervoltage protective devices

☐ (g)   Choices and setting of protective and monitoring devices for protection against indirect contact and/or overcurrent

☐ (h)   Selection of equipment and protective measures appropriate to external influences

☐ (i)   Selection of appropriate functional switching devices

Inspected by ....................................................   Date ...............................................................

**Notes:**

✓   to indicate an inspection has been carried out and the result is satisfactory

✗   to indicate an inspection has been carried out and the result was unsatisfactory

N/A  to indicate  the inspection is not applicable

LIM  to indicate that, exceptionally, a limitation agreed with the person ordering the work prevented the inspection or test being carried out.

**Figure 9.17** Schedule of inspections (for new installation work only).

The notice shown in Figure 9.19 shall be attached to the certificate (Figure 9.20).

The Electrical Installation Certificate may be produced in any durable medium, including written and electronic media.

# SCHEDULE OF TEST RESULTS

Contractor: ....................................

Test Date: ....................................

Signature: ....................................

Method of protection against indirect contact: ....................................

Equipment vulnerable to testing: ....................................

Address/Location of distribution board: ....................................

....................................

Type of Supply: TN-S/TN-C-S/TT

$Z_e$ at origin: ..........ohms

PFC: ..........kA

*Instruments*

loop impedance: ....................

continuity: ....................

insulation: ....................

RCD tester: ....................

Description of Work: ....................................

| Circuit Description | Overcurrent Device Short-circuit capacity: ......kA | | Wiring Conductors | | | Continuity | | | Test Results | | | | | | | Remarks |
|---|---|---|---|---|---|---|---|---|---|---|---|---|---|---|---|---|
| | | | | | | | | | Insulation Resistance | | Polarity | Earth Loop Imped-ance $Z_s$ | Functional Testing | | | |
| | Type | Rating $I_n$ | Live | CPC | | $(R_1 + R_2)$* | $R_2$* | *$R_{in}$ | Live/Live | Live/Earth | | | RCD time | Other | | |
| | | | $mm^2$ | $mm^2$ | | $\Omega$ | $\Omega$ | $\Omega$ | $M\Omega$ | $M\Omega$ | | $\Omega$ | ms | | | |
| 1 | 2 | 3 | 4 | 5 | | 6 | 7 | 8 | 9 | 10 | 11 | 12 | 13 | 14 | | 15 |
| | | | | | | | | | | | | | | | | |
| | | | | | | | | | | | | | | | | |
| | | | | | | | | | | | | | | | | |
| | | | | | | | | | | | | | | | | |
| | | | | | | | | | | | | | | | | |
| | | | | | | | | | | | | | | | | |
| | | | | | | | | | | | | | | | | |
| | | | | | | | | | | | | | | | | |
| | | | | | | | | | | | | | | | | |
| | | | | | | | | | | | | | | | | |
| | | | | | | | | | | | | | | | | |

Deviations from Wiring Regulations and special notes:

* Complete column 6 or 7.

**Figure 9.18** Generic Schedule of Test Results.

GUIDANCE FOR RECIPIENTS

This safety Certificate has been issued to confirm that the electrical installation work to which it relates has been designed, constructed and inspected and tested in accordance with British Standard 7671:2008 (the IEEE Wiring Regulations).

You should have received an original Certificate and the contractor should have retained a duplicate Certificate.

The 'original' Certificate should be retained in a safe place and be shown to any person inspecting or undertaking further work on the electrical installation in the future. If you later vacate the property, this Certificate will demonstrate to the new owner that the electrical installation complied with the requirements of British Standard 7671:2008 at the time the Certificate was issued. The Construction (Design and Management) Regulations require that for a project covered by those Regulations, a copy of this Certificate, together with schedules, is included in the project health and safety documentation.

For safety reasons, the electrical installation will need to be inspected at appropriate intervals by a competent person. The maximum time interval recommended before the next inspection is stated on Page 1 under 'Next Inspection'.

This Certificate is intended to be issued only for a new electrical installation or for new work associated with an alteration or addition to an existing installation. It should not have been issued for the inspection of an existing electrical installation. A 'Periodic Inspection Report' should be issued for such a periodic inspection.

**Figure 9.19** Guidance notice to accompany the Electrical Installation Certificate.

 The Electrical Installation Certificate shall be compiled and signed by a competent person(s).

## 9.5.2 Minor Electrical Installation Works Certificate

A Minor Electrical Installation Works Certificate is used for additions and alterations to an installation, such as an extra socket-outlet or lighting point to an existing circuit, the relocation of a light switch etc. This certificate may also be used for the replacement of equipment such as accessories or luminaires, but **not** for the replacement of distribution boards (or similar items) or the provision of a new circuit.

The notice shown in Figure 9.21 shall be attached to the certificate (Figure 9.2).

Minor Electrical Installation Works Certificates:

- may be produced in any durable medium, including written and electronic media;
- shall be compiled and signed by a competent person(s).

Form 2                                                    Form No  /2

## ELECTRICAL INSTALLATION CERTIFICATE (notes 1 and 2)

(REQUIREMENTS FOR ELECTRICAL INSTALLATIONS - BS 7671 [IEE WIRING REGULATIONS])

**DETAILS OF THE CLIENT** (note 1) ........................................................................................................
........................................................................................................
........................................................................................................

**INSTALLATION ADDRESS**
........................................................................................................
........................................................................................................
...........................................Postcode..........................................

**DESCRIPTION AND EXTENT OF THE INSTALLATION** Tick boxes as appropriate
(note 1)

Description of installation: ...........................................

| | |
|---|---|
| Extent of installation covered by this Certificate: ........................................ | New installation ☐ |
| ........................................................................ | Addition to an existing installation ☐ |
| ........................................................................ | Alteration to an existing installation ☐ |

**FOR DESIGN**

I/We being the person(s) responsible for the design of the electrical installation (as indicated by my/our signatures below), particulars of which are described above, having exercised reasonable skill and care when carrying out the design, hereby CERTIFY that the design work for which I/our have been responsible is to the best of my/our knowledge and belief in accordance with BS 7671: ........., amended to.......... (date) except for the departures, if any, detailed as follows:

> Details of departures from BS 7671 (Regulations 120-01-03, 120-02):

The extent of liability of the signatory or the signatories is limited to the work described above as the subject of this Certificate.
For the DESIGN of the installation.                    **(Where there is mutual responsibility for the design)

Signature: ................................... Date ................   Name (BLOCK LETTERS): ........................................ Designer No. 1
Signature: ................................... Date ................   Name (BLOCK LETTERS): ........................................ Designer No. 2**

**FOR CONSTRUCTION**

I/We being the person(s) responsible for the construction of the electrical installation (as indicated by my/our signatures below), particulars of which are described above, having exercised reasonable skill and care when carrying out the construction, hereby CERTIFY that the construction work for which I/we have been responsible is to the best of my/our knowledge and belief in accordance with BS 7671: ...........amended to .............(date) except for the departures, If any, detailed as follows:

> Details of departures from BS 7671 (Regulations 120-01-03, 120-02):

The extent of liability of the signatory is limited to the work described above as the subject of this Certificate.
For CONSTRUCTION of the installation:

Signature: ................................................................. Date ................

Name (BLOCK LETTERS): ......................................................................................................... Constructor

**FOR INSPECTION & TESTING**

I/We being the person(s) responsible for the inspection & testing of the electrical installation (as indicated by my/our signatures below), particulars of which are described above, having exercised reasonable skill and care when carrying out the inspection & testing, hereby CERTIFY that the work for which I/we have been responsible is to the best of my/our knowledge and belief in accordance with BS 7671:........., amended to ........... (date) except for the departures, if any, detailed as follows:

> Details of departures from BS 7671 (Regulations 120-01-03, 120-02):

The extent of liability of the signatory is limited to the work described above as the subject of this Certificate.
For INSPECTION & TEST of the installation:                    **(Where there is mutual responsibility for the design)

Signature: ................................................................. Date ................

Name (BLOCK LETTERS): ......................................................................................................... Inspector

**NEXT INSPECTION (notes 4 and 7)**

I/We the designer(s) recommend that this installation is further inspected and tested after an interval of not more than............ years/months

**Figure 9.20** Electrical Installation Certificate.

**PARTICULARS OF THE SIGNATORIES TO THE ELECTRICAL INSTALLATION CERTIFICATE (note 3)**

**Designer (No. 1)**
Name: ...................................... Company: .........................................
Address: ..............................................................................................
........................................... Postcode: ................... Tel No.: .................

**Designer (No. 2)**
(if applicable) Name: ...................................... Company: .........................................
Address: ..............................................................................................
........................................... Postcode: ................... Tel No.: .................

**Constructor**
Name: ...................................... Company: .........................................
Address: ..............................................................................................
........................................... Postcode: ................... Tel No.: .................

**Inspector**
Name: ...................................... Company: .........................................
Address: ..............................................................................................
........................................... Postcode: ................... Tel No.: .................

**SUPPLY CHARACTERISTICS AND EARTHING ARRANGEMENTS** Tick boxes and enter details, as appropriate

| Earthing arrangements | Number and Type of Live Conductors | Nature of Supply Parameters | Supply Protective Device Characteristics |
|---|---|---|---|
| TN-C ☐ | a.c. ☐   d.c. ☐ | Nominal voltage, $U/U_0^{(1)}$ ............................. V | |
| TN-S ☐ | 1-line, 2-wire ☐   2-pole ☐ | Nominal frequency, $f^{(1)}$ ............................ Hz | Type: ......................... |
| TN-C-S ☐ | 1-line, 3-wire ☐   3-pole ☐ | Prospective fault current, $I_{pf}^{(2)}$ (note 6)....... kA | ................................. |
| TT ☐ | 2-line, 3-wire ☐   other ☐ | External loop impedance, $Z_e^{(2)}$...................$\Omega$ | |
| IT ☐ | 3-line, 3-wire ☐ | (Note: (1) by enquiry, (2) by enquiry or by measurement) | Nominal current rating |
| | 3-line, 4-wire ☐ | | .................................A |
| Alternative source ☐ of supply (to be detailed on attached schedules) | | | |

**PARTICULARS OF INSTALLATION REFERRED TO IN THE CERTIFICATE** Tick boxes and enter details, as appropriate

**Means of Earthing**

Distributor's facility ☐

**Maximum Demand**

Maximum demand (load) ..................................................................Amps per phase

**Details of Installation Earth Electrode** (where applicable)

Installation earth electrode ☐

Type (e.g. rod(s), tape etc.)    Location    Electrode resistance to earth

.................................    .................................    .................................$\Omega$

**Main Protective Conductors**

Earthing conductor: material ................................. csa ..........................mm$^2$  connection verified ☐

Main equipotential bonding conductors: material ................................. csa ..........................mm$^2$  connection verified ☐

To incoming water and/or gas service ☐    To other elements .................................

**Main Switch or Circuit-breaker**

BS. Type ............................. No. of poles ........................ Current rating ........................A  Voltage rating ........................V

Location ............................................................................  Fuse rating or setting ....................A

Rated residual operating current $I_{\Delta n}$ = ...... mA, and operating time of .................................ms (at $I_{\Delta n}$)
(applicable only where an RCD is suitable and is used as a main circuit-breaker)

**COMMENTS ON EXISTING INSTALLATION:** (In the case of an alteration or addition see Section 743)

..........................................................................................................................
..........................................................................................................................
..........................................................................................................................
..........................................................................................................................

**SCHEDULES** (note 2)
The attached Schedules are part of this document and this Certificate is valid only when they are attached to it.
............ Schedules of Inspections and ............ Schedules of Test Results are attached. (Enter quantities of schedules attached)

**Figure 9.20** *(Continued)*

**MINOR ELECTRICAL INSTALLATION WORKS CERTIFICATE GUIDANCE FOR RECIPIENTS**

This Certificate has been issued to confirm that the electrical installation work to which it relates has been designed, constructed and inspected and tested in accordance with British Standard 7671:2008 (IEE Wiring Regulations).

You should have received an 'original' Certificate and the contractor should have retained a duplicate. If you were the person ordering the work, but not the owner of the installation, you should pass this Certificate, or a copy of it, to the owner. A separate Certificate should have been received for each existing circuit on which minor works have been carried out. This Certificate is not appropriate if you requested the contractor to undertake more extensive installation work, for which you should have received an Electrical Installation Certificate.

The Certificate should be retained in a safe place and be shown to any person inspecting or undertaking further work on the electrical installation in the future. If you later vacate the property, this Certificate will demonstrate to the new owner that the minor electrical installation work carried out complied with the requirements of British Standard 7671:2008 at the time the Certificate was issued.

**Figure 9.21** Guidance notice to accompany the Minor Electrical Installation Certificate.

### 9.5.3 Electrical Installation Condition Report

An Electrical Installation Condition Report is used for reporting on the condition of an existing installation, and shall include schedules of both the inspection and the test results.

The notice shown in Figure 9.23 shall be attached to the certificate (Figure 9.24) on completion of the inspection.

Electrical Installation Condition Reports:

- may be produced in any durable medium, including written and electronic media;
- shall be compiled and signed by a competent person(s).

 An installation that was designed to an earlier edition of the Regulations, and which does not fully comply with the current edition, is not necessarily unsafe for continued use and does not necessarily require upgrading. Only damage, deterioration, defects, dangerous conditions and non-compliance with the requirements of the Regulations – which may give rise to danger – should be recorded.

## MINOR ELECTRICAL INSTALLATION WORKS CERTIFICATE
### (REQUIREMENTS FOR ELECTRICAL INSTALLATIONS - BS 7671 [IEE WIRING REGULATIONS])

To be used only for minor electrical work which does not include the provision of a new circuit

---

**PART 1:  Description of minor works**

1. Description of the minor works: ......................................................................................................................

2. Location/Address: ......................................................................................................................................

3. Date minor works completed: ...................................................................................................................

4. Details of departures, if any, from BS 7671

   ...........................................................................................................................................................

   ...........................................................................................................................................................

   ...........................................................................................................................................................

---

**PART 2:  Installation details**

1. System earthing arrangement:                       TN-C-S ☐   TN-S ☐   TT ☐

2. Method of protection against indirect contact: ...............................................................................................

3. Protective device for the modified circuit:        Type BS ............................   Rating ...............................A

4. Comments on existing installation, including adequacy of earthing and bonding arrangements:
   (see Regulation 130-07) .............................................................................................................................

   ...........................................................................................................................................................

   ...........................................................................................................................................................

   ...........................................................................................................................................................

---

**PART 3:  Essential Tests**

1. Earth continuity: satisfactory   ☐

2. Insulation resistance:

              Phase/neutral .................................................M$\Omega$

              Phase/earth ....................................................M$\Omega$

              Neutral/earth ..................................................M$\Omega$

3. Earth fault loop impedance:  ...................................................$\Omega$

4. Polarity: satisfactory          ☐

5. RCD operation (if applicable): Rated residual operating current $I_{\Delta n}$ .......mA and operating time of ..........ms (at $I_{\Delta n}$)

---

**PART 4:  Declaration**

1. I/We CERTIFY that the said works do not impair the safety of the existing installation, that the said works have been designed, constructed, inspected and tested in accordance with BS 7671: ............. (IEE Wiring Regulations), amended to ...................................... and that the said works, to the best of my/our knowledge and belief, at the time of my/our inspection, complied with BS 7671 except as detailed in Part 1.

2. Name: ........................................................................          3. Signature: ...............................................................

   For and on behalf of: ....................................................          Position: ...................................................................

   Address: ......................................................................

   ....................................................................................          Date: ........................................................................

   ....................................................................................

   ................................................Postcode: ...................

---

**Figure 9.22** Minor Electrical Installation Certificate.

---

**CONDITION REPORT**

**This Report is an important and valuable document which should be retained for future reference.**

The purpose of this Condition Report is to confirm, so far as reasonably practicable, whether or not the electrical installation is in a satisfactory condition for continued service (see Section E).

The Report should identify any damage, deterioration, defects and/or conditions which may give rise to danger (see Section K).

The person ordering the Report should have received the 'original' Report and the inspector should have retained a duplicate.

The 'original' Report should be retained in a safe place and be made available to any person inspecting or undertaking work on the electrical installation in the future. If the property is vacated, this Report will provide the new owner /occupier with details of the condition of the electrical installation at the time the Report was issued.

Where the installation incorporates a residual current device (RCD) there should be a notice at or near the device stating that it should be tested quarterly. **For safety reasons it is important that this instruction is followed.**

Section D (Extent and Limitations) should identify fully the extent of the installation covered by this Report and any limitations on the inspection and testing. The inspector should have agreed these aspects with the person ordering the Report and with other interested parties (licensing authority, insurance company, mortgage provider and the like) before the inspection was carried out.

Some operational limitations such as inability to gain access to parts of the installation or an item of equipment may have been encountered during the inspection. The inspector should have noted these in Section D.

For items classified in Section K as C1 ('Danger present'), **the safety of those using** the installation **is at** risk, and it is recommended that a competent person undertakes the necessary remedial work immediately.

For items classified in Section K as C2 ('Potentially dangerous'), **the safety of those using the installation may be at risk** and it is recommended that a competent person undertakes the necessary remedial work as a matter of urgency.

Where it has been stated in Section K that an observation requires further investigation the inspection has revealed an apparent deficiency which could not, due to the extent or limitations of the inspection, be fully identified. Such observations should be investigated as soon as possible. A further examination of the installation will be necessary, to determine the nature and extent of the apparent deficiency (see Section F).

For safety reasons, the electrical installation should be re-inspected at appropriate intervals by a competent person. The recommended date by which the next inspection is due is stated in Section F of the Report under 'Recommendations' and on a label at or near to the consumer unit / distribution board.

---

**Figure 9.23** Guidance notice to accompany the Electrical Installation Condition Report.

**PERIODIC INSPECTION REPORT FOR AN ELECTRICAL INSTALLATION**
(REQUIREMENTS FOR ELECTRICAL INSTALLATIONS - BS 7671 [IEE WIRING REGULATIONS])

**DETAILS OF THE CLIENT**

Client: ...............................................................................................................................................

Address: ............................................................................................................................................

Purpose for which this Report is required: ....................................................................................

**DETAILS OF THE INSTALLATION** Tick boxes as appropriate

Occupier: ..........................................................................................................................................

Installation: .......................................................................................................................................

Address: ............................................................................................................................................

Description of Premises:        Domestic ☐         Commercial ☐  Industrial ☐

Other ☐

Estimated age of the Electrical        .....................years
Installation:

Evidence of Alterations or Additions:        Yes ☐         No ☐         Not apparent ☐

If 'Yes', estimate age:  ............................ years

Date of last inspection: ...............................        Records available ?        Yes ☐        No ☐

**EXTENT AND LIMITATIONS OF THE INSPECTION**

Extent of electrical installation covered by this report: ...................................................................

..........................................................................................................................................................

..........................................................................................................................................................

Limitations: .......................................................................................................................................

..........................................................................................................................................................

..........................................................................................................................................................

This inspection has been carried out in accordance with BS 7671: 2001 (IEE Wiring Regulations), amended to ......
Cables concealed within trunking and conduits, or cables and conduits concealed under floors, in roof spaces and
generally within the fabric of the building or underground have not been inspected.

**NEXT INSPECTION**

I/We recommend that this installation is further inspected and tested after an interval of not more than ........
months/years, provided that any observations 'requiring urgent attention' are attended to without delay.

**DECLARATION**

INSPECTED AND TESTED BY

Name: ..................................................................        Signature: ...................................................

For and on behalf of: ...........................................        Position: .....................................................

Address: ...............................................................

..............................................................................        Date: ...........................................................

..............................................................................

**SUPPLY CHARACTERISTICS AND EARTHING ARRANGEMENTS** Tick boxes and enter details, as appropriate

| Earthing arrangements | | Number and Type of Live Conductors | | | | Nature of Supply Parameters | Supply Protective Device Characteristics |
|---|---|---|---|---|---|---|---|
| TN-C | ☐ | a.c. | ☐ | d.c. | ☐ | Nominal voltage, $U/U_0$ [1] .................V | Type: ................ |
| TN-S | ☐ | 1-line, 2-wire | ☐ | 2-pole | ☐ | Nominal frequency, $f$ [1] .................Hz | |
| TN-C-S | ☐ | 2-line, 3-wire | ☐ | 3-pole | ☐ | Prospective fault current, $I_{pf}$ [2] ........kA | Nominal current rating .............A |
| TT | ☐ | 3-line, 3-wire | ☐ | other | ☐ | External loop impedance, $Z_e$ [2] ........$\Omega$ | |
| IT | ☐ | 3-line, 4-wire | ☐ | | | (Note: (1) by enquiry, (2) by enquiry or by measurement) | |

**PARTICULARS OF INSTALLATION REFERRED TO IN THE REPORT** Tick boxes and enter details, as appropriate

| Means of Earthing | | Details of Installation Earth electrode (where applicable) | | |
|---|---|---|---|---|
| | | Type | Location | Electrode resistance |
| Distributor's facility | ☐ | | | |
| Installation earth electrode | ☐ | (e.g. rod(s), tape etc) ................................ | ................................ | to earth ............................ $\Omega$ |

**Figure 9.24** Electrical Installation Condition Report.

**Main Protective Conductors**

Earthing conductor:                    material ............................ csa ...............................

Main equipotential bonding
conductors                             material ............................ csa ...............................

To incoming water service ☐   To incoming gas service ☐   To incoming oil service ☐   To structural steel ☐
To lightning protection ☐     To other incoming service(s) ☐ (state details ...........................................................)

**Main Switch or Circuit-breaker**

BS, type and number of poles ...............................   Current rating ...............A          Voltage rating ...............V

Location .............................................          Fuse rating or setting .................A

Rated residual operating current $I_{\Delta n}$ = ............ mA, and operating time of ........ ms (at $I_{\Delta n}$) (applicable only where an RCD is suitable and is used as a main circuit-breaker)

| OBSERVATIONS AND RECOMMENDATIONS Tick boxes as appropriate | Recommendations detailed below |
|---|---|

Referring to the attached Schedule(s) of Inspection and Test Results, and subject to the limitations specified at
the Extent and Limitations of the Inspection section
☐ No remedial work is required          ☐ The following observations are made:

...........................................................................................................................................

...........................................................................................................................................

...........................................................................................................................................

...........................................................................................................................................

...........................................................................................................................................

...........................................................................................................................................

One of the following number, as appropriate, is to be allocated to each of the observations made above to indicate to the person(s)
responsible for the installation the action recommended.

☐1 requires urgent attention     ☐2 requires improvement     ☐3 requires further investigation

☐4 does not comply with BS 7671: 2001 amended to .................... This does not imply that the electrical installation inspected is
unsafe.

**SUMMARY OF THE INSPECTION**

Date(s) of the inspection: .........................................................................................................................

General condition of the installation: ...........................................................................................................

...........................................................................................................................................

...........................................................................................................................................

Overall assessment: Satisfactory/Unsatisfactory

**SCHEDULE(S)**
The attachment Schedules are part of this document and this Report is valid only when they are attached to it.
................ Schedules of Inspections and ............... Schedules of Test Results are attached.
(Enter quantities of schedules attached).

**Figure 9.24** (*Continued*)

# 9.6 Test requirements specific for compliance to the Building Regulations

## 9.6.1 Mandatory requirements

### 9.6.1.1 Part P – Electrical safety

Confirm that:

- reasonable provision has been made in the design, installation, inspection and testing of electrical installations to protect persons from fire or injury;
- sufficient information has been provided so that persons wishing to operate, maintain or alter an electrical installation can do so with reasonable safety.

### 9.6.1.2 Part M – Access and facilities for disabled people

Confirm that (in addition to the requirements of the Disability Discrimination Act 1995) precautions have been taken to ensure that:

- new non-domestic buildings and/or dwellings (e.g. houses and flats used for student living accommodation etc.);
- extensions to existing non-domestic buildings;
- non-domestic buildings that have been subject to a material change of use (e.g. so that they become a hotel, boarding house, institution, public building or shop);

are capable of allowing people, regardless of their disability, age or gender to:

- gain access to buildings;
- gain access within buildings;
- be able to use the facilities of the buildings (both as visitors and as people who live or work in them);
- use sanitary conveniences in the principal storey of any new dwelling.

 **Note:** From 1 October 2010, the Equality Act replaced most of the Disability Discrimination Act 1995 (DDA). However, the Disability Equality Duty in the DDA continues to apply.

### 9.6.1.3  Part L – Conservation of fuel and power

- Confirm that energy-efficiency measures have been provided that ensure that lighting systems utilize energy-efficient lamps with:
  - manual switching controls; or
  - automatic switching (in the case of external lighting fixed to the building); or
  - both manual and automatic switching controls;

so that the lighting systems can be operated effectively with regard to the conservation of fuel and power.

- Confirm that building occupiers have been supplied with sufficient information (including results of performance tests carried out during the works) to show how the heating and hot water services can be operated and maintained.

## 9.6.2  Inspection and test

Verify that all electrical installations have been inspected and tested during and at the end of installation, and before they are taken into service, and that they:

- are reasonably safe and that they comply with BS 7671:2008 (Incorporating Amendment No. 1);
- meet the relevant equipment and installation standards.

Confirm that all components that are part of an electrical installation have been inspected (during installation as well as on completion) to verify that the components have been:

- selected and installed in accordance with the IET Wiring Regulations;
- made in compliance with appropriate British or Harmonized European Standards;
- evaluated against external influences (such as the presence of moisture);
- checked to see that they have not been visibly damaged (or are defective) so as to be unsafe;
- tested to check satisfactory performance with respect to continuity of conductors, insulation resistance, separation of circuits, polarity, earthing and bonding arrangements, Earth fault loop impedance and functionality of all protective devices, including RCDs;
- inspected for conformance with the IET Wiring Regulations;
- had their test results recorded;
- had their test results compared with the relevant performance criteria to confirm compliance.

 **Note**: Inspections and testing of DIY work should **also** meet the above requirements.

### 9.6.3 Extensions, material alterations and material changes of use

Where any electrical installation work is classified as an extension, a material alteration or a material change of use, confirm that:

- the existing fixed electrical installation in the building is capable of supporting the amount of additions and alterations that will be required;
- the earthing and bonding systems are satisfactory and meet the requirements;
- the mains supply equipment is suitable and can carry the additional loads envisaged;
- any additions and alterations to the circuits that feed them comply with the requirements of the Regulations;
- the protective measures required meet the requirements;
- the rating and the condition of existing equipment (belonging to both the consumer and the electricity distributor) is sufficient.

### 9.6.4 Design

Confirm that electrical installations have been designed and installed so that they:

- are suitably enclosed (and appropriately separated) to provide mechanical and thermal protection;
- do not present an electric shock or fire hazard to people;
- meet the requirements of the Building Regulations;
- provide adequate protection for persons against the risks of electric shock, burn or fire injuries;
- provide adequate protection against mechanical and thermal damage.

### 9.6.5 Electricity distributors' responsibilities

Prior to starting works, confirm that the electricity distributor has:

- accepted responsibility for ensuring that the supply is mechanically protected and can be safely maintained;
- evaluated and agreed the proposal for a new (or significantly altered) installation;
- installed the cut-out and meter in a safe location.

Confirm that the distributor has:

- provided an earthing facility for new connections;
- maintained the supply within defined tolerance limits;
- provided certain technical and safety information to the consumer to enable him to design his installation(s);
- ensured that the distributor's equipment on consumers' premises:

    o  is suitable for its purpose;
    o  is safe in its particular environment;
    o  clearly shows the polarity of the conductors.

### 9.6.6 Consumer units

Ensure that accessible consumer units have been fitted with a child-proof cover or installed in a lockable cupboard.

### 9.6.7 Earthing

Inspect and confirm that:

- electrical installations have been properly earthed;
- lighting circuits include a circuit protective conductor;
- socket-outlets that have a rating of 32 A or less and that may be used to supply portable equipment for use outdoors are protected by an RCD;
- the distributor has provided an earthing facility for all new connections;

- new or replacement, non-metallic light fittings, switches or other components do not require earthing (e.g. non-metallic varieties) unless new circuit protective (earthing) conductors have been provided;
- socket-outlets that will accept unearthed (two-pin) plugs do not use supply equipment that needs to be earthed.

### 9.6.8 Electrical installations

Verify (by inspection and test) that during installation, at the end of installation and before they are taken into service, all electrical installations:

- have been designed and installed (suitably enclosed and appropriately separated) to provide mechanical and thermal protection;
- provide adequate protection for persons against the risks of electric shock, burn or fire injuries;
- meet the requirements of the Building Regulations.

Confirm that all electrical installation work:

- has been carried out professionally;
- complies with the Electricity at Work Regulations 1989 as amended;
- has been carried out by persons who are competent to prevent danger and injury while doing it, or who are appropriately supervised.

### 9.6.9 Wiring and wiring systems

Confirm that

- cables concealed in floors and walls have (if required):
  - an earthed metal covering; or
  - are enclosed in steel conduit; or
  - have some form of additional mechanical protection;
- cables to an outside building (e.g. garage or shed), if run underground, have been routed and positioned so as to give protection against electric shock and fire as a result of mechanical damage to a cable;
- heat-resisting flexible cables (if required) have been supplied for the final connections to certain equipment (see maker's instructions).

### 9.6.9.1 Equipotential bonding conductors

Confirm that:

- main equipotential bonding conductors for water service pipes, gas installation pipes, oil supply pipes plus certain other 'earthy' metalwork have been provided;

- where there is an increased risk of electric shock (e.g. in bathrooms and shower rooms), supplementary equipotential bonding conductors have been installed;
- the minimum size of supplementary equipotential bonding conductors (without mechanical protection) is $4\,mm^2$.

## 9.6.10  Socket-outlets

Confirm by inspection that:

- older types of socket-outlet designed non-fused plugs are not connected to a ring circuit;
- RCD protection has been provided for all socket-outlets that have a rating of 32 A or less and that may be used to supply portable equipment for use outdoors;
- switched socket-outlets indicate whether they are ON;
- socket-outlets that will accept unearthed (two-pin) plugs are not used to supply equipment that needs to be earthed;
- the requirements in Table 9.13 for wall sockets have been met.

**Table 9.13  Building Regulations requirements for wall sockets**

| Type of wall | Requirement |
| --- | --- |
| Timber framed | Power points:<br><br>• that have been set in the linings have a similar thickness of cladding behind the socket box<br>• have not been placed back to back across the wall |
| Solid masonry | Deep sockets and chases have not been used in separating walls<br><br>The position of sockets has been staggered on opposite sides of the separating wall |
| Cavity masonry | The position of sockets has been staggered on opposite sides of the separating wall<br><br>Deep sockets and chases have not been used in a separating wall<br><br>Deep sockets and chases in a separating wall have not been placed back to back |
| Framed walls with absorbent material | Sockets have:<br><br>• been positioned on opposite sides of a separating wall<br>• not been connected back to back<br>• been staggered a minimum of 150 mm edge to edge |

Confirm (by inspection and testing) that all socket-outlets used for lighting:

- all socket-outlets are wall-mounted;
- are easily reachable;

- have been installed between 450 mm and 1200 mm from the finished floor level;
- socket-outlets are located no nearer than 350 mm from room corners;
- switched socket-outlets indicate whether they are 'ON';

### 9.6.11 Switches

Ensure that all controls and switches:

- are easy to operate, visible and free from obstruction;
- have been located between 750 mm and 1200 mm above the floor;
- do not require the simultaneous use of both hands (unless necessary for safety reasons) to operate;

and that:

- mains and circuit isolator switches clearly indicate whether they ON or OFF;
- individual switches on panels and on multiple socket-outlets have been well separated;
- front plates contrast visually with their backgrounds;
- controls that need close vision (e.g. thermostats) have been located between 1200 mm and 1400 mm above the floor;
- the operation of switches, outlets and controls does not require the simultaneous use of both hands (unless necessary for safety reasons);
- where possible, light switches with large push pads shave been used in preference to pull cords;
- the colours **red** and **green** have not been used in combination as indicators of ON and OFF for switches and controls;
- **all** switches used for lighting:
  - are easily reachable;
  - have been installed between 450 mm and 1200 mm from the finished floor level.

Confirm that light switches:

- have large push pads (in preference to pull cords);
- align horizontally with door handles;
- are within the 900 mm to 1100 mm from the entrance door opening;
- are located between 750 mm and 1200 mm above the floor;
- are not coloured **red** and **green** (i.e. as a combination) as indicators for ON and OFF.

## 9.6.12 Telephone points and TV sockets

Confirm that all telephone points and TV sockets have been located between 400 mm and 1000 mm above the floor (or 400 mm and 1200 mm above the floor for permanently wired appliances).

## 9.6.13 Equipment and components

### 9.6.13.1 Emergency alarms

Test and inspect to ensure that:

- emergency assistance alarm systems have:

  ○ visual and audible indicators to confirm that an emergency call has been received;
  ○ a reset control reachable from a wheelchair, WC or a shower/changing seat;
  ○ a signal that is distinguishable visually and audibly from the fire alarm;

- emergency alarm pull cords should be:

  ○ coloured **red**;
  ○ located as close to a wall as possible;
  ○ have two **red** 50 mm diameter bangles;

- front plates contrast visually with their backgrounds;
- the colours **red** and **green** have not been used (in combination) to indicate ON and OFF for switches and controls.

### 9.6.13.2 Fire alarms

Verify (by test and inspection) that fire-detection and fire-warning systems have been properly designed, installed and maintained, and that:

- all buildings have arrangements for detecting fire;
- all buildings have been fitted with a suitable (electrically operated) fire-warning system (in compliance with BS 5839) or have means of raising an alarm in case of fire (e.g. rotary gongs, handbells or shouting **'FIRE'**);
- fire alarms emit an audio and visual signal to warn occupants with hearing or visual impairments;
- the fire-warning signal is distinct from other signals that may be in general use;
- in premises that are used by the general public (e.g. large shops and places of assembly) a staff alarm system (complying with BS 5839) has been used.

### 9.6.13.3 Heat emitters

Check that heat emitters:

- are either screened or have their exposed surfaces kept at a temperature below 43°C;
- that are located in toilets and bathrooms do not restrict:
  - ○ the minimum clear wheelchair-manoeuvring space,
  - ○ the space beside a WC used to transfer from the wheelchair to the WC.

### 9.6.13.4 Portable equipment for use outdoors

Verify that RCDs have been provided for all socket-outlets that have a rating of 32 A or less and that may be used to supply portable equipment for use outdoors.

### 9.6.13.5 Power-operated entrance doors

Confirm that all power-operated doors have been provided with:

- safety features to prevent injury to people who are struck or trapped (such as a pressure-sensitive door edge that operates the power switch);
- a readily identifiable (and accessible) stop switch;
- a manual or automatic opening device in the event of a power failure where and when necessary for health or safety.

Confirm that:

- all doors to accessible entrances have been provided with a power-operated door opening and closing system if a force greater than 20 N is required to open or shut a door;
- once open, all doors to accessible entrances are wide enough to allow unrestricted passage for a variety of users, including wheelchair users, people carrying luggage, people with assistance dogs, and people with pushchairs and small children;
- power-operated entrance doors:
  - ○ have a sliding, swinging or folding action controlled manually (by a push pad, card swipe, coded entry or remote control), or automatically controlled by a motion sensor or proximity sensor such as a contact mat;
  - ○ open towards people approaching the doors;
  - ○ provide visual and audible warnings that they are operating (or about to operate);
  - ○ incorporate automatic sensors to ensure that they open early enough (and stay open long enough) to permit safe entry and exit;

- o incorporate a safety stop that is activated if the doors begin to close when a person is passing through;
- o revert to manual control (or fail safe) in the open position in the event of a power failure;
- o when open, do not project into any adjacent access route;

- its manual controls are:

  - o located between 750 mm and 1000 mm above floor level;
  - o operable with a closed fist;
  - o set back 1400 mm from the leading edge of the door when fully open;
  - o clearly distinguishable against the background;
  - o contrast visually with the background.

## 9.6.14 Thermostats

Check that all controls that need close vision (e.g. thermostats) are located between 1200 mm and 1400 mm above the floor.

### 9.6.14.1 Smoke alarms – dwellings

Confirm by test and inspection that smoke alarms have been positioned:

- in the circulation space within 7.5 m of the door to every habitable room;
- in the circulation spaces between sleeping spaces and in places where fires are most likely to start (e.g. kitchens and living rooms);
- on every storey of a house (including bungalows).

Confirm by test and inspection that:

- kitchen areas that are not separated from the stairway or circulation space by a door have been equipped with an additional heat detector in the kitchen that is interlinked to the other alarms;
- if more than one smoke alarm has been installed in a dwelling, they have been linked so that if a unit detects smoke it will operate the alarm signal of all the smoke detectors.

Verify by inspection that smoke alarms:

- have ideally been mounted, 25 mm and 600 mm below the ceiling (25–150 mm in the case of heat detectors) and at least 300 mm from walls and light fittings;
- have not been fixed over a stair shaft or any other opening between floors;
- have not been fitted:

- o   in places that get very hot (such as a boiler room);
- o   in a very cold area (such as an unheated porch);
- o   in bathrooms, showers, cooking areas or garages, or any other place where steam, condensation or fumes could give false alarms;
- o   next to or directly above heaters or air-conditioning outlets;
- o   on surfaces that are normally much warmer or colder than the rest of the space.

Test, inspect and confirm that the power supply for a smoke alarm system:

- •   has been derived from the dwelling's mains electricity supply via a single independent circuit at the dwelling's main distribution board (consumer unit);
- •   includes a standby power supply that will operate during mains failure;
- •   is not (preferably) protected by an RCD.

### 9.6.15  Lighting

#### 9.6.15.1  External lighting fixed to the building

Confirm (by test and inspection) that all external lighting (including lighting in porches, but not lighting in garages and carports):

- •   automatically extinguishes when there is enough daylight, and when not required at night;
- •   has sockets that can only be used with lamps having an efficiency greater than 40 lumens per circuit-watt (such as fluorescent or compact fluorescent lamp types, and **not** GLS tungsten lamps with bayonet-cap or Edison-screw bases).

#### 9.6.15.2  Fittings, switches and other components

Confirm that new or replacement non-metallic light fittings, switches or other components that require earthing (e.g. non-metallic varieties) have been provided with new circuit protective (earthing) conductors.

#### 9.6.15.3  Fixed lighting

Ensure that in locations where lighting can be expected to have most use, fixed lighting (e.g. fluorescent tubes and compact fluorescent lamps – but not GLS tungsten lamps with bayonet-cap or Edison-screw bases) with a luminous efficacy greater than 40 lumens per circuit-watt have been made available.

#### 9.6.15.4  Lighting circuits

Verify that all lighting circuits include a circuit protective conductor.

### 9.6.16  Lecture/conference facilities

In lecture hall and conference facilities, confirm that artificial lighting has been designed to:

• give good colour rendering of all surfaces;
• be compatible with other electronic and radio-frequency installations.

### 9.6.17  Cellars or basements

Ensure that liquid petroleum gas (LPG) storage vessels and LPG-fired appliances that are fitted with automatic ignition devices or pilot lights have not been installed in cellars or basements (J-3.5i).

## 9.7  What about test equipment

**Figure 9.25** A selection of test equipment.

Although BS 7671 lays great emphasis on the requirements for inspection and testing (Chapter 6 of the Regulations), the only reference (as far as I can tell!) to actual test equipment concerning insulation monitoring devices is a statement which says that:

• *"an insulation monitoring device shall be so designed or installed that it can only be possible to modify the setting with the use of a key or a tool;*
• *in an IT system, an insulation monitoring device shall be provided so as to indicate the occurrence of a first fault from a live part to an exposed-conductive-part, or to Earth;*

- *installations forming part of an IT system can make use of an insulation monitoring device."*

Of course, the actual choice of test equipment that the electrician chooses to use will normally be based on personal preference and experience. Nevertheless, it is essential that any piece of test equipment (including software) that is used when installing or inspecting electrical installations for compliance with the Regulations can be relied on to produce accurate results.

ISO 9001:2008 (i.e. the internationally recognized standard for quality management) specifies the requirements for the control of test equipment (although it actually refers to them as 'measuring and monitoring devices') as shown in Figure 9.26.

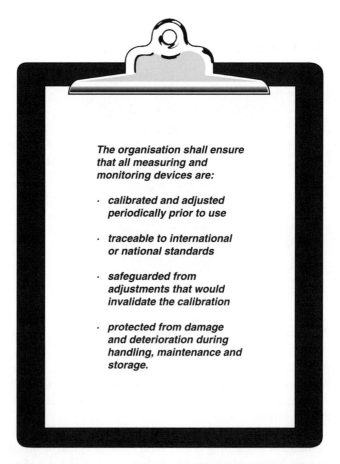

**The organisation shall ensure that all measuring and monitoring devices are:**

· *calibrated and adjusted periodically prior to use*

· *traceable to international or national standards*

· *safeguarded from adjustments that would invalidate the calibration*

· *protected from damage and deterioration during handling, maintenance and storage.*

**Figure 9.26** Mandatory requirements from the Building Regulations.

| Proof | The controls that an organization has in place to ensure that equipment (including software) used for conformance to specified requirements is properly maintained |
|---|---|
| **Likely documentation** | Equipment records of maintenance and calibration<br>Work instructions |

Although the majority of electricians probably work on an individual basis and the requirement to operate as an accredited and registered ISO 9001:2008 company does not really apply, following the recommendations of this standard can only help to improve the quality of any organization – no matter what its size.

In general, therefore:

- all measuring and test equipment that is used by an electrician needs to be well maintained, in good condition and capable of safe and effective operation within a specified tolerance of accuracy;
- all measuring and test equipment should be regularly inspected and/or calibrated to ensure that it is capable of accurate operation (and, where necessary, by comparison with external sources traceable back to national standards);
- any electrostatic protection equipment that is utilized when handling sensitive components should be regularly checked to ensure that it remains fully functional;
- the control of measuring and test equipment (whether owned by the electrician, on loan, hired or provided by the customer) should always include a check that the equipment:

  - is exactly what is required;
  - has been initially calibrated before use;
  - operates within the required tolerances;
  - is regularly recalibrated; and
  - facilities exist (either within the organization or via a third party) to adjust, repair or recalibrate it as necessary.

If the measuring and test equipment is used to verify process outputs against a specified requirement, the equipment needs to be maintained and calibrated against national and international standards, and the results of any calibrations carried out **must** be retained and the validity of previous results reassessed if they are subsequently found to be out of calibration.

### 9.7.1 Control of inspection, measuring and test equipment

Measuring and test equipment should always be stored correctly and be satisfactorily protected between uses (to ensure the correct bias and precision), and should be verified and/or recalibrated at appropriate intervals.

### 9.7.2 Computers

Special attention should be paid to computers if they are used for controlling processes, and particular attention should be paid to the maintenance and accreditation of any related software.

### 9.7.3 Software

Software used for measuring, monitoring and/or testing specified requirements should be validated prior to use.

### 9.7.4 Calibration

Without exception, all measuring instruments can be subject to damage, deterioration or just general wear and tear when they are in regular use. The electrician should, therefore, take account of this fact and ensure that all of his test equipment is regularly calibrated against a known working standard.

The accuracy of an instrument will depend very much on what items it is going to be used to test and the frequency of use of the test instrument, and the electrician will have to decide on the maximum tolerance of accuracy for each item of test equipment.

Of course, calibrating against a 'working standard' is pretty pointless if that particular standard cannot be relied upon, and so the workshop standard must also be calibrated, on a regular basis, at either a recognized calibration centre or at the UK Physical Laboratory against one of the national standards.

The electrician will, therefore, have to make allowances for:

- the calibration and adjustment of all measuring and test equipment that can affect the quality of his inspection and/or test;
- the documentation and maintenance of calibration procedures and records;
- the regular inspection of all measuring or test equipment to ensure that it is capable of the accuracy and precision that is required;
- the environmental conditions being suitable for the calibrations, inspections, measurements and tests to be completed.

If an instrument is found to be outside of its tolerance of accuracy, any items previously tested with the instrument must be regarded as suspect. In these circumstances, it would be wise to review the test results obtained using the individual instrument. This could be achieved by compensating for the extent of inaccuracy to decide if the acceptability of the tested item would be reversed.

### 9.7.5 Calibration methods

There are various possibilities, such as:

- sending all working equipment to an external calibration laboratory;
- sending one of each item (i.e. a 'workshop standard') to a calibration laboratory, and then sub-calibrating each working item against the workshop standard;
- testing by attributes – i.e. take a known 'faulty' product, and a known 'good' product and then test each one to ensure that the test equipment can identify the faulty and the good product correctly.

### 9.7.6  Calibration frequency

The calibration frequency depends on how much the instrument is used, its ability to retain its accuracy and how critical the items are that are being tested. Infrequently used instruments are often only calibrated prior to their use, while frequently used items would normally be checked and recalibrated at regular intervals, again depending on product criticality, cost, availability etc. Normally 12 months is considered about the maximum calibration interval.

### 9.7.7  Calibration ideals

- Each instrument should be uniquely identified, allowing it to be traced.
- The calibration results should be clearly indicated on the instrument.
- The calibration results should be retained for reference.
- The instrument should be labelled to show the next 'calibration-due' date in order to easily avoid its use outside of the period of confidence.
- Any means of adjusting the calibration should be sealed, allowing easy identification if it has been tampered with (e.g. a label across the joint of the casing).

 **Note:** Examples of test equipment normally used by electricians is shown in Annex 9A.

## 9.8  Requirements from the Regulations – Testing

Every electrical connection and joint shall be accessible      WR-526.3
for testing and maintenance, except for the following:

- a joint designed to be buried in the ground;
- a compound-filled or encapsulated joint;
- a connection between a cold tail and the heating element, as in ceiling heating, floor heating or a trace heating system;

- a joint made by welding, soldering, brazing or appropriate compression tool;
- a joint or connection made in equipment by the manufacturer of the product and not intended to be inspected and/or tested;
- equipment complying with BS 7671:200 (Incorporating Amendment No. 1) for a maintenance-free accessory and marked with the symbol.

When undertaking testing in a potentially explosive atmosphere, appropriate safety precautions in accordance with BS EN 60079–17 and BS EN 61241–17 are necessary.

| | |
|---|---|
| The tests in Part 6 of the Regulations, where relevant, shall be carried out, in that order, before the installation is energized. Where the installation incorporates an Earth electrode, the test given in Regulation 612.7 shall also be carried out before the installation is energized. | WR-612.1 |

If any test indicates a failure to comply, that test and any preceding test, the results of which may have been influenced by the fault indicated, shall be repeated after the fault has been rectified.

### 9.8.1 Protective systems

#### 9.8.1.1 SELV

| | |
|---|---|
| The separation of the live parts from those of other circuits shall be confirmed by a measurement of the insulation resistance. The resistance values obtained shall be in accordance with Table 9.14. | WR-612.4.1 |

Table 9.14  Minimum values of insulation resistance (data from BS 7671:2008)

| Circuit nominal voltage (V) | Test voltage d.c. (V) | Minimum insulation resistance (MΩ) |
|---|---|---|
| SELV and PELV | 250 | >0.5 |
| Up to and including 500V, with the exception of the above systems | 500 | >1.0 |
| Above 500V | 1000 | >1.0 |

## 9.8.1.2 PELV

The separation of the live parts from other circuits shall    WR-612.4.2
be confirmed by a measurement of the insulation resistance. The resistance values obtained shall be in accordance with Table 9.14.

## 9.8.1.3 FELV

FELV circuits shall meet all the test requirements for    WR-612.4.4
low-voltage circuits.

## 9.8.1.4 Basic protection by a barrier or an enclosure provided during erection

Where basic protection is intended to be afforded by a    WR-612.4.5
barrier or an enclosure provided during erection, it shall
be verified by test that each barrier or enclosure affords
a degree of protection not less than IP2X or IPXXB, or
IP4X or IPXXD, as appropriate, where that Regulation
so requires.

## 9.8.1.5 Protection by automatic disconnection of the supply

Where RCDs are also used for protection against fire, the    WR-612.8
conditions for protection by automatic disconnection of
the supply shall be verified.

The verification of the effectiveness of the measures for fault protection by automatic disconnection of supply depends on the type of system used, and is as follows:

### 9.8.1.5.1 TN system

Compliance shall be verified by:

• measurement of the Earth fault loop impedance;
• verification of the characteristics and/or the effectiveness of the associated protective device.

### 9.8.1.5.2 TT system

Compliance shall be verified by:

- measurement of the resistance of the Earth electrode for exposed-conductive-parts of the installation;
- verification of the characteristics and/or effectiveness of the associated protective device.

### 9.8.1.5.3 IT system

Compliance shall be verified by calculation or measurement of the current ($I_d$) in case of a first fault at the line conductor or at the neutral.

Where conditions that are similar to the conditions of a TT system occur, in the event of a second fault in another circuit verification shall be made according to a TT system.

Where conditions that are similar to the conditions of a TN system occur, in the event of a second fault in another circuit verification shall be made according to a TN system.

### 9.8.1.6 Protection by electrical separation

| | |
|---|---|
| The separation of the live parts from those of other circuits and from Earth shall be confirmed by a measurement of the insulation resistance. The resistance values obtained shall be in accordance with Table 9.14. | WR-612.4.3 |

### 9.8.1.7 Additional protection

| | |
|---|---|
| The verification of the effectiveness of the measures applied for additional protection is fulfilled by visual inspection and test. | WR-612.10 |
| Where RCDs are required for additional protection, the effectiveness of automatic disconnection of supply by RCDs shall be verified using suitable test equipment according to BS EN 61557–6 to confirm that the relevant requirements are met. | WR-612.10 |

### 9.8.1.8 Earth fault loop impedance

| | |
|---|---|
| Where protective measures are used that require knowledge of the Earth fault loop impedance, the relevant impedances shall be measured, or determined by an alternative method. | WR-612.9 |

**Note:** Further information on the measurement of Earth fault loop impedance can be found in Appendix 14 of BS 7671:2008.

### 9.8.1.9 Functional testing

| | |
|---|---|
| Where fault protection and/or additional protection is to be provided by an RCD, the effectiveness of any test facility incorporated in the device shall be verified. | WR-612.13.1 |
| Equipment, such as switchgear and controlgear assemblies, drives, controls and interlocks, shall be subjected to a functional test to show that it is properly mounted, adjusted and installed in accordance with the relevant requirements of the Regulations. | WR-612.13.2 |

## 9.8.2 Other tests

### 9.8.2.1 Continuity of protective conductors, including main and supplementary equipotential bonding

| | |
|---|---|
| A continuity test shall be made. It is recommended that the test be carried out with a supply having a no-load voltage between 4 V and 24 V, d.c. or a.c., and a short-circuit current of not less than 200 mA. | WR-612.2.1 |

### 9.8.2.2 Continuity of ring final circuit conductors

| | |
|---|---|
| A test shall be made to verify the continuity of each conductor, including the protective conductor, of every ring final circuit. | WR-612.2.2 |

### 9.8.2.3 Earth electrode resistance

| | |
|---|---|
| Where the earthing system incorporates an Earth electrode as part of the installation, the electrode resistance to Earth shall be measured. | WR-612.7 |

### 9.8.2.4 Insulation resistance

| | |
|---|---|
| The insulation resistance shall be measured between live conductors and between live conductors and the protective conductor connected to the earthing arrangement. | WR-612.3.1 |
| The insulation resistance measured with the test voltages indicated in Table 9.14 shall be considered satisfactory if the main switchboard and each distribution circuit tested separately, with all its final circuits connected but with current-using equipment disconnected, has an insulation resistance **not less than** the appropriate value given in Table 9.14. | WR-612.3.2 |

More stringent requirements are applicable for the wiring of fire alarm systems in buildings (see BS 5839–1).

| | |
|---|---|
| Where a surge protective device or other equipment is likely to influence the verification test, or be damaged, such equipment shall be disconnected before carrying out the insulation resistance test. | WR-612.3.2 |
| In locations exposed to a fire hazard, a measurement of the insulation resistance between the live conductors should be applied. | WR-612.3.2 |

Insulation resistance values are usually much higher than those in Table 9.14.

| | |
|---|---|
| Where the circuit includes electronic devices that are likely to influence the results, or be damaged, then only a measurement between the live conductors connected together and the earthing arrangement shall be made. | WR-612.3.3 |

### 9.8.2.5 Insulation resistance/impedance of floors and walls

| | |
|---|---|
| In a non-conducting location at least three measurements shall be made in the same location, one of these measurements being approximately 1 m from any accessible extraneous-conductive-part in the location. | WR-612.5.1 |

The other two measurements shall be made at greater distances. The measurement of resistance/impedance of insulating floors and walls is carried out with the system voltage to Earth at nominal frequency. The above series of measurements shall be repeated for each relevant surface of the location.

 **Note:** Further information on measurement of the insulation resistance/impedance of floors and walls can be found in Appendix 13 of BS 7671: 2008.

Any insulation or insulating arrangement of extraneous-conductive-parts:                                          WR-612.5.2

*   when tested at 500 V d.c. shall be not less than 1 MΩ; and
*   shall be able to withstand a test voltage of at least 2 kV a.c. rms; and
*   shall not pass a leakage current exceeding 1 mA in normal conditions of use.

### 9.8.2.6 Polarity

A test of polarity shall be made, and it shall be verified that:                                                       WR-612.6

*   every fuse and single-pole control and protective device is connected in the line conductor only; and
*   except for E14 and E27 lampholders to BS EN 60238, in circuits having an earthed neutral conductor, centre-contact bayonet and Edison-screw lampholders have the outer or screwed contacts connected to the neutral conductor; and
*   wiring has been correctly connected to socket-outlets and similar accessories.

### 9.8.2.7 Phase sequence

For multi-phase circuits it shall be verified that the phase sequence is maintained.                                           WR-612.12

### 9.8.2.8 Prospective fault current

| | |
|---|---|
| The prospective short-circuit current and prospective Earth fault current shall be measured, calculated or determined by another method, at the origin and at other relevant points in the installation. | WR-612.11 |

### 9.8.2.9 Verification of voltage drop

| | |
|---|---|
| When required, compliance with the Regulations may be confirmed by using the following options:<br><br>• the voltage drop may be evaluated by measuring the circuit impedance;<br>• the voltage drop may be evaluated by using calculations – for example, by diagrams or graphs showing maximum cable length versus load current for different conductor cross-sectional areas with different percentage voltage drops for specific nominal voltages, conductor temperatures and wiring systems. | WR-612.14 |

 **Note:** Verification of voltage drop is not normally required during initial verification.

## 9.8.3 Certification and reporting

Electrical Installation Certificates, Electrical Installation Compliance Reports and Minor Electrical Installation Works Certificates shall be compiled and signed or otherwise authenticated by a competent person or persons.

| | |
|---|---|
| Electrical Installation Certificates, Electrical Installation Compliance Reports and Minor Electrical Installation Works Certificates may be produced in any durable medium, including written and electronic media. | WR-631.5 |
| Regardless of the media used for original certificates, reports or their copies, their authenticity and integrity shall be verified by a reliable process or method. | WR-631.5 |
| The process or method shall also verify that any copy is a true copy of the original. | WR-631.5 |

An Electrical Installation Certificate or a Minor Electri-   WR-633.1
cal Installation Works Certificate (as appropriate) shall
apply to all work competed, including the additions or
alterations.

The contractor or other person responsible for the new   WR-633.2
work, or a person authorized to act on his behalf, shall
record on the Electrical Installation Certificate or the
Minor Electrical Installation Works Certificate, any
defects found, insofar as is reasonably practicable, in the
existing installation.

### 9.8.3.1 Initial verification

During erection, on completion of an installation, or   WR-134.2.1
addition or alteration to an installation (and before
being put into service), appropriate inspection and
testing shall be carried out by competent persons to
verify that the requirements of the IET Wiring
Regulations have been met.

 Certification shall be completed in accordance with Section 631 and 632.

The designer of the installation is responsible for recom-   WR-134.2.2
mending the interval to the first periodic inspection and
test, as detailed in Part 6 of BS 7671:2008 (Incorporating Amendment No. 1).

### 9.8.3.2 Electrical Installation Certificate

Upon completion of the verification of a new installation   WR-631.1
or changes to an existing installation, an Electrical Installation Certificate (based on the model shown in Figure A9.1),
together with a schedule of inspections and a schedule of
test results, shall be given to the person ordering the work.

The schedule of test results shall identify every circuit,   WR-632.2
including its related protective device(s), and shall record
the results of the appropriate tests and measurements.

> The person or persons responsible for the design, con-    WR-632.3
> struction, inspection and testing of the installation shall
> provide the person ordering the work with a certificate
> that takes account of their respective responsibilities for
> the safety of that installation, together with the schedules
> described in Regulation 632.1 of BS 7671:2008 (Incorpo-
> rating Amendment No. 1).

 Defects or omissions revealed during inspection and testing of the installation work covered by the Electrical Installation Certificate **shall** be made good
before the certificate is issued.

### 9.8.3.3 Periodic inspection and testing

> Where required, periodic inspection and testing of    WR-621.1
> every electrical installation shall be carried out (in
> accordance with Regulations 621.2 to 621.5 of BS
> 7671:2008 (Incorporating Amendment No. 1)) in
> order to determine, insofar as is reasonably practicable,
> whether the installation is in a satisfactory condition for
> continued service.
>
> The documentation arising from the initial certification    WR-621.1
> and any previous periodic inspection and testing shall be
> taken into account. Where no previous documentation
> is available, investigation of the electrical installation
> shall be undertaken prior to carrying out the periodic
> inspection and testing.

 **Note:** A generic list of examples of items requiring inspection is given in
Appendix 6 to BS 7671:2008 (Incorporating Amendment No. 1).

> Periodic inspections comprising a detailed examina-    WR-621.2
> tion of the installation shall be carried out without
> dismantling, or with partial dismantling as required, sup-
> plemented by appropriate tests to show that the require-
> ments for disconnection times for protective devices are
> complied with, to provide for:
>
> • safety of persons and livestock against the effects of
>    electric shock and burns;

- protection against damage to property by fire and heat arising from an installation defect;
- confirmation that the installation is not damaged or deteriorated so as to impair safety;
- the identification of installation defects and departures from the requirements of the Regulations that may give rise to danger.

Precautions shall be taken to ensure that the periodic inspection and testing shall not cause danger to persons or livestock, and shall not cause damage to property and equipment, even if the circuit is defective.                           WR-621.3

Measuring instruments and monitoring equipment and methods shall be chosen in accordance with relevant parts of BS EN 61557.                           WR-621.3

The extent and results of the periodic inspection and testing of an installation, or any part of an installation, shall be recorded.                           WR-621.4

 Periodic inspection and testing shall be undertaken by a competent person.

Upon completion of the periodic inspection and testing of an existing installation, an Electrical Installation Report, based on the model show in Figure A9.6, shall be provided.                           WR-631.2

Such documentation shall include details of the extent of the installation and limitations of the inspection and testing.                           WR-631.2

A copy of the Electrical Installation Condition Report, together with a schedule of inspections and a schedule of test results, shall be given by the person carrying out the inspection, or a person authorized to act on his behalf, to the person ordering the inspection.                           WR-634.1

Any damage, deterioration, defects, dangerous conditions and non-compliance with the requirements of the Regulations, which may give rise to danger, together with any significant limitations of the inspection and testing, including the reasons, shall be recorded.                           WR-634.2

**Notes:**

1. The frequency of periodic inspection and testing of an installation will depend on the type of installation and equipment, its use and operation, the frequency and quality of maintenance, and the external influences to which it is subjected.
2. The results and recommendations of the previous report, if any, shall be taken into account.
3. In the case of an installation under an effective management system for preventive maintenance in normal use, periodic inspection and testing may be replaced by an adequate regime of continuous monitoring and maintenance of the installation and all its constituent equipment, by skilled persons competent in such work.
4. Appropriate records shall be kept.

### 9.9.9.1 9.8.3.4 Minor Electrical Installation Works Certificate

Where minor electrical installation work does not include the provision of a new circuit, a Minor Electrical Installation Works Certificate, based on the model given in Figure 9.22, may be provided for each circuit altered or extended, as an alternative to an Electrical Installation Certificate.    WR-631.3

The Minor Electrical Installation Works Certificates shall be compiled and signed or otherwise authenticated by a competent person or persons.

# Annex 9A – Examples of test equipment used to test electrical installations

The following are examples of instruments that are required to test electrical installations for compliance with the requirements of BS 7671.

Quite a lot of test equipment manufacturers now produce dual or multi-functional instruments, and so it is quite common to find an instrument that can be used to undertake a number of different types of test – for example, continuity and insulation resistance, loop impedance and prospective fault current. It is, therefore, wise to carry out a little research before purchasing!

## 9A.1 Continuity tester

All protective and bonding conductors must be tested to ensure that they are electrically safe and correctly connected. Low-resistance ohmmeters and simple multi-meters are normally used for continuity testing. Ideally, these should have a no-load voltage of between 4 V and 24 V, be capable of producing an a.c. or d.c. short-circuit voltage of not less than 200 mA, and have a resolution of at least 0.01 mΩ.

| TM INS1600 Digital insulation/ continuity tester By TLC: http://www. tlc-direct. co.uk | A compact, easy to read, battery-operated insulation and continuity tester, covering a wide measurement range (up to 2000 MΩ/1000 V), a.c. voltage (up to 600 V) and with a continuity beeper<br><br>Capable of testing all requirements of the 17th edition and Guidance Notes | • Easy to read, 0.65 inch LCD display<br>• Data-hold switch<br>• Power ON lock facility for hands-free operation<br>• Rotary switch for easy range selection<br>• Overload protection<br>• Dimensions: 165 × 100 × 57 mm<br>• a.c. voltage range: 600 V<br>• Batteries: 6× A<br>• Sampling rate: 2.5 times/s<br>• Weight: 500 g |

**Figure A 9.1**
Continuity testing.
(Courtesy of TLC.)

## 9A.2 Insulation resistance tester

A low resistance between the phase and neutral conductors, or between the live conductors to Earth, will cause a leakage current that will cause a weaken-

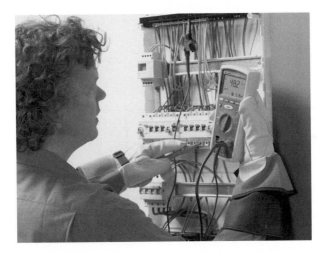

**Figure A 9.2** Insulation tester.

ing of insulation, as well as involving a waste of energy, which would increase the running costs of the installation. To overcome this problem, the resistance between poles or to Earth must never be less than 0.5 mΩ for the usual supply voltages.

## 9A.3 Loop impedance tester

Loop testing is a quick, convenient and highly specific method of testing an electrical circuit for its ability to engage protective devices (circuit-breakers, fuses etc.) by simulating a fault from live to Earth or from live to neutral (short-circuit). The tester first measures the unloaded voltage, and then connects a known resistance between the conductors, thereby simulating a fault. The voltage drop is measured across the known resistor, in series with the loop, and the proportion of the supply voltage that appears across the resistor will be dependent on the impedance of the loop.

---

KMP4120DL
By Robin
Electronics:
http://www.
robinelectronics.
com

**Figure A 9.3**
Loop impedance
tester. (Courtesy of
Robin Electronics.)

An Earth loop impedance tester with a 0.01 Ω resolution, capable of performing loop tests without tripping most passive RCDs

- Three pre-selectable loop impedance ranges (20, 200 and 2000 Ωs)
- Three prospective short-circuit (PSC) ranges (200, 2000 and 20 kA)
- Ability to test the majority of passive RCDs without tripping
- A lock-down test button allowing hands-free operation

---

## 9A.4 RCD tester

The standard method for protecting electrical installations is to make sure that an Earth fault results in a fault current that is high enough to operate the protective device quickly, so that fatal shock is prevented. However, there are cases where the impedance of the Earth fault loop, or the impedance of the fault itself, are too high to enable enough fault current to flow. In such a case, either:

- the current will continue to flow to Earth, perhaps generating enough heat to start a fire; or
- metalwork that can be touched may be at a high potential relative to Earth, resulting in severe shock danger.

Either or both of these possibilities can be removed by the installation of an RCD.

 **Note:** RCDs are also, sometimes, referred to as:

RCCD    residual current operated circuit-breaker
SRCD    socket-outlet incorporating an RCD
PRCD    portable RCD, usually an RCD incorporated in a plug
RCBO    an RCCD that includes overcurrent protection
SRCBO   a socket-outlet incorporating an RCBO and RCD tester allows a selection of out-of-balance currents to flow through the RCD and cause its operation.

 The RCD tester should not be operated for longer than 2 s.

| | | |
|---|---|---|
| **TM-RC-70 RCD Tester** By TLC: http://www. tlc-direct.co.uk  **Figure A 9.4** RCD tester. (Courtesy of TLC.) | A compact and simple instrument for monitoring 10 30, 100, 300 and 500 mA RCD devices, as well as selected 100, 300 and 500 mA RCDs | • Display: LCD • RCD current: 10, 30, 100 and 300 mA • Power supply: 230 V ± 10% • Measurement range: 10–300 mA • Test sequence: 10–300 mA • Dimensions: 235 × 103 × 70 mm • Weight: approx. 700 g |

## 9A.5 Prospective fault current tester

A prospective fault current (PFC) tester is used to measure the prospective phase neutral fault current.

| | | |
|---|---|---|
| **Profitest Master** By Gossen Metrawatt: techinfo@ gossen-metrawatt.com | Designed to test the PFC, as well as other insulation and impedance tests | • Overcurrent protection devices<br>• Loop and line impedance<br>• Earth resistance, Earth leakage resistance<br>• Insulation resistance<br>• Phase sequence indicator |

**Figure A 9.5**
Prospective fault current tester. (Courtesy of Gossen Metrawatt.)

It is usual to find that PFC testers are part of a combined PFC/loop impedance tester.

## 9A.6  Test lamp or voltage indicator

These types of tester (often referred to as a 'tetrascope' or 'neon screwdriver') are frequently used by electricians, and might look like the example shown in Figure A9.6.

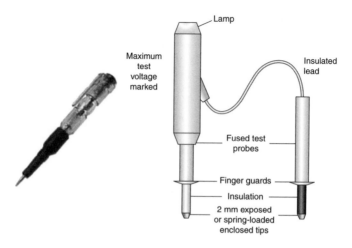

**Figure A 9.6**  Typical test lamp and voltage indicators. (Courtesy of TLC.)

These compact screwdriver multi-testers are normally water and impact resistant, and provide a.c. voltage test, contact test 70–250 V a.c., non-contact 100–1000 V a.c., polarity test 1.5–36 V d.c., continuity check 0–5 Ω and auto-power ON/OFF.

## 9A.7  Earth electrode resistance

The Earth electrode (when used), is the means of making contact with the general mass of Earth and should be regularly tested to ensure that good contact is made. In all cases, the aim is to ensure that the electrode resistance is not so high that the voltage from earthed metalwork to Earth exceeds 50 V.

 **Note:** Acceptable electrodes are rods, pipes, mats, tapes, wires, plates and structural steelwork buried or driven into the ground. The pipes of other services such as gas and water must **not** be used as Earth electrodes (although they must be bonded to Earth).

**Fluke 1620 Series GEO
Earth Ground Testers**
By Fluke (UK) Ltd: http://
www.fluke.co.uk

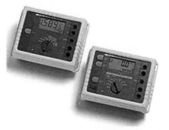

**Figure A 9.7**  Earth electrode
resistance tester. (Courtesy
of Fluke (UK) Ltd.)

# 10

# Installation, maintenance and repair

According to studies recently completed by the Comité Européen de Normalisation (CEN) and the Comité Européen de Normalisation Electrotechnique (CENELEC), the installation and maintenance engineer is the primary cause of reliability degradations during the in-service stage of most electrical installations. Therefore, the problems associated with poorly trained, poorly supported and/or poorly motivated personnel with respect to reliability and dependability require careful assessment and quantification.

The most important factor, however, that affects the overall reliability of a modern product (system or installation) is the increased number of individual components that is required in a product. As most system failures are actually caused by the failure of a single component (equipment or sub-assembly), the reliability of each individual component must be considerably better than the reliability of the overall system.

Because of this, quality standards for the installation, maintenance, repair and inspection of in-service products have had to be laid down in engineering standards, handbooks and local operating manuals (written for specific items and equipment). These publications are used by maintenance engineers, and should always include the most recent amendments. It is essential that assurance personnel use the same procedures for their inspections as were used for the installation.

 **Note:** Schedules for installing wiring systems (together with guidance for selecting the appropriate size of cable, current ratings and so on) are shown in Appendix 4 Tables 4A1 and 4A2 of the Regulations.

Although this final chapter is a comparatively small chapter (compared with some of the others in this book!), it is nevertheless extremely important, as it provides some guidance on the requirements for installation, maintenance and repair. It also lists the requirements of the Wiring Regulations for maintenance etc. with respect to electrical installations – but, as per the previous chapters, which contain similar lists, please remember that this is **only** the author's impression of the most important aspects of the Regulations, and electricians should **always** consult BS 7671 to satisfy compliance. Finally, in Annex 10A there is a comprehensive set of checklists for the quality control of electrical equipment and electrical installations.

The Regulations devote a complete part to inspection and testing (i.e. Part 6), and emphasize the need for continual improvement by stating:

> *"Every installation (or alteration to an existing installation) shall, during erection and on completion before being put into service, be inspected and tested to verify, so far as is reasonably practicable, that the requirements of the Regulations have been met.*
>
> *The verification shall be made by a competent person and on completion of the verification, a certificate shall be prepared."*

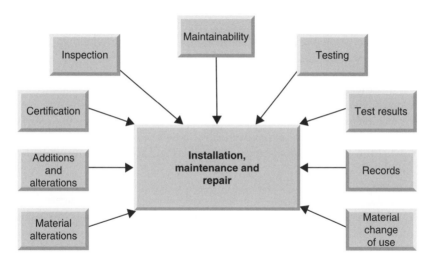

**Figure 10.1** Installation, repair and maintenance.

## 10.1 General

As previously shown, the requirements of BS 7671:2008 (Incorporating Amendment No. 1) apply to the design, erection and verification of electrical installations, such as those of:

- agricultural and horticultural premises;
- caravans, caravan parks and similar sites;
- commercial premises;
- construction sites, exhibitions, shows, fairgrounds and other installations for temporary purposes, including professional stage and broadcast applications;

- domestic buildings;
- external lighting and similar installations;
- industrial premises;
- highway equipment and street furniture;
- low-voltage generating sets;
- marinas;
- medical locations;
- mobile or transportable units;
- photovoltaic systems;
- prefabricated buildings;
- public premises;
- residential premises.

**Note:** 'Premises' covers the land **and** all facilities, including buildings, belonging to it.

The Regulations do **not** apply to any of the following installations:

- aircraft equipment;
- electrical equipment of machines covered by BS EN 60204;
- equipment for mobile and fixed offshore installations;
- equipment for motor vehicles;
- equipment on board ships covered by BS 8450;
- lightning protection systems for buildings and structures covered by BS EN 62305;
- systems for the distribution of electricity to the public;
- railway traction equipment, rolling stock and signalling equipment;
- those aspects of lift installations covered by relevant parts of BS 5655 and BS EN 81–1;
- those aspects of mines and quarries specifically covered by Statutory Regulations;
- radio-interference suppression equipment (except insofar as it affects safety of the electrical installation).

## 10.2 Installation

Many requirements and recommendations for the installation of electrical equipment are to be found in BS 7671:2008 (Incorporating Amendment No. 1). Most of these have already been mentioned in other chapters of this book, and the intention here is to bring to your attention some of the most important ones. It should **not**, however, be viewed as a complete list!

**Note:** See also Chapter 7 of this book for specific requirements concerning special installations and locations, and Chapter 9 for inspections and tests.

## 10.2.1 Cables

| | |
|---|---|
| A cable buried in the ground (i.e. not installed in a conduit or duct) shall incorporate an earthed armour or metal sheath or both, suitable for use as a protective conductor. | WR-522.8.10 |
| A cable installed under a floor or above a ceiling shall be run in such a position that it is not liable to be damaged by contact with the floor or the ceiling or their fixings. | WR-522.6.100 |
| A cable should preferably not be installed in a location where it is liable to be covered by thermal insulation. | WR-523.9 |
| Cables complying with the requirements of BS EN 60332–1–2 may be installed without special precautions. | WR-527.1.3 |
| Where multi-core cables are installed in parallel, each cable shall contain one conductor of each line. | WR-521.8.1 |
| The line and neutral conductors of each final circuit shall be electrically separate from those of every other final circuit, so as to prevent the indirect energizing of a final circuit intended to be isolated. | WR-521.8.2 |

## 10.2.2 Conductors

 A bare live conductor **shall** always be installed on insulators.

| | |
|---|---|
| A device protecting a conductor against overload may be installed along the run of that conductor, provided that part of the run (i.e. between the point where a change occurs and the position of the protective device) has neither branch circuits nor outlets for connection of current-using equipment, and is protected against fault current; or:<br><br>• its length does not exceed 3 m;<br>• it is installed so as to reduce the risk of fault to a minimum; and<br>• it is installed so as to reduce to a minimum the risk of fire or danger to persons. | WR-433.2.2 |

> A device protecting a conductor may be installed on        WR-434.2.2
> the supply side of the point where a change occurs,
> provided that it possesses an operating characteristic
> such that it protects the wiring situated on the load side
> against fault current.

**Note:** The number and type of live conductors (e.g. single-phase two-wire a.c., three-phase four-wire a.c.) are established based on the source of energy and the type of circuit(s) used within the installation.

### 10.2.3 Design

Two of the most important requirements for the installation of electrical equipment are that:

- installed equipment and its connections must be accessible for operational, inspection and maintenance purposes; and
- equipment must be arranged to allow easy access for periodic inspection, testing and maintenance.

> Every installation shall be divided into circuits, as neces-   WR-314.1
> sary, to:
>
> - avoid danger and minimize inconvenience in the
>   event of a fault;
> - ensure safe inspection, testing and maintenance;
> - prevent the indirect energizing of a circuit that is
>   intended to be isolated;
> - reduce the possibility of unwanted tripping of
>   residual current devices (RCDs) due to excessive
>   protective conductor (PE) currents produced by
>   equipment;
> - reduce the effects of electromagnetic interference
>   (EMI);
> - take account of any danger that may arise from the
>   failure of a single circuit (e.g. a lighting circuit).
>
> Separate circuits shall be provided for parts of the        WR-314.2
> installation that need to be separately controlled.
>
> The electrical installation shall be designed to provide    WR-132.1
> for:

- the protection of persons, livestock and property;
- the proper functioning of the electrical installation for the intended use;
- the maximum demand of an installation.

| | |
|---|---|
| The electrical installation shall be arranged in such a way that no mutual detrimental influence will occur between electrical installations and non-electrical installations. | WR-132.11 |
| Equipment shall be selected and installed to provide for the safety and proper functioning for the intended use of the installation. | WR-530.3 |
| Equipment installed shall be appropriate to the external influences foreseen. | WR-530.3 |
| Where the use of a new material or invention leads to departures from the Regulations, the resulting degree of safety of the installation shall be not less than that obtained by compliance with the Regulations. | WR-133.5 |

### 10.2.4 Electromagnetic comparability

| | |
|---|---|
| All electrical installations and equipment shall be in accordance with the current and relevant electromagnetic compatibility (EMC) regulations and the relevant EMC standard. | WR-332.1 |
| Consideration shall be given by the planner and designer of the electrical installation to measures reducing the effect of induced voltage disturbances and EMI. | WR-332.2 |
| The installation design shall take into consideration the anticipated electromagnetic emissions, generated by the installation or the installed equipment. | WR-131.6.4 |

### 10.2.5 Electromechanical stresses

Every conductor or cable shall have adequate strength, and be so installed as to withstand the electromechanical forces that may be caused by any current, including fault current, it may have to carry in service.

## 10.2.6 Installation of equipment

| | |
|---|---|
| Electrical equipment not installed on or in a combustible wall shall be enclosed with a suitable thickness of non-flammable material. | WR-422.4.101 |
| Electrical equipment shall be arranged so that: | WR-132.12 |
| • there is sufficient space for the initial installation and later replacement of individual items of electrical equipment;<br>• the equipment is accessible for operation, inspection, testing, fault detection, maintenance and repair. | |
| Electrical equipment shall be installed so that design temperatures are not exceeded. | WR-134.1.5 |
| Electrical joints and connections shall be properly constructed with regard to conductance, insulation, mechanical strength and protection. | WR-134.1.4 |
| If an item of equipment (e.g. a capacitor) is installed behind a barrier or in an enclosure (and that equipment could retain a dangerous electrical charge after it has been switched off) a warning label shall be provided. | WR-416.2.5 |
| The characteristics of the electrical equipment shall not be impaired by the process of erection. | WR-134.1.2 |
| Switchgear or controlgear shall be installed outside the location, unless: | WR-422.3.3 |
| • it is suitable for the location; or<br>• it is installed in an enclosure providing a degree of protection of at least IP4X, or (in the presence of dust) IP5X, or (in the presence of electrically conductive dust) IPX6. | |

## 10.2.7 Power supplies

The type of earthing system to be used for the installation should be determined taking into consideration the characteristics of the source of energy and (in particular) any earthing facilities.

| | |
|---|---|
| Electrical sources for safety services shall be installed as fixed equipment, and in such a manner that they cannot be adversely affected by failure of the normal source. | WR-560.6.2 |
| In a TN, TT or IT system, an RCD with a rated residual operating current of not more than 30 mA shall be installed to protect every circuit. | WR-551.4.4.2 |
| Stationary batteries shall be installed so that they are accessible **only** to skilled or instructed persons. | WR-551.8.1 |
| Where danger or damage is expected to arise due to an interruption of supply, suitable provisions shall be made in the installation or installed equipment. | WR-131.7 |
| An interrupting device shall be installed in such a way that it can be easily recognized, and effectively and rapidly operated case of danger where there is a necessity for immediate interruption of supply. | WR-132.9 |

## 10.2.8 Protective devices

| | |
|---|---|
| A device for protection against overload shall be installed at the point where a reduction occurs in the value of the current-carrying capacity of the conductors of the installation. | WR-433.2.1 |
| A device protecting a conductor against overload may be installed along the run of that conductor, provided that part of the run (i.e. between the point where a change occurs and the position of the protective device) has neither branch circuits nor outlets for connection of current-using equipment and is protected against fault current; or | WR-433.2.2 |

- its length does not exceed 3 m;
- it is installed so as to reduce the risk of fault to a minimum; and
- it is installed so as to reduce to a minimum the risk of fire or danger to persons.

A device protecting a conductor may be installed on          WR-434.2.2
the supply side of the point where a change occurs
provided that it possesses an operating characteristic
such that it protects the wiring situated on the load
side against fault current.

A device providing protection against fault current          WR-434.2
shall be installed at the point where a reduction
in the cross-sectional area or other change causes
a reduction in the current-carrying capacity of the
conductors of the installation (except installations situated in locations presenting a fire risk or
risk of explosion, and where the requirements for
special installations and locations specify different
conditions).

A device (such as a circuit-breaker with a short-circuit     WR-432.3
release, or a fuse) providing protection against fault
current shall only be installed where overload protection is achieved by other means.

Except where a protective device is installed to             WR-411.6.3.2
interrupt the supply in the event of the first Earth
fault, a residual current monitor or an insulation
fault location system shall be provided to indicate
the occurrence of a first fault from a live part to an
exposed-conductive-part or to Earth.

The installation of equipment (e.g. fixing, connection       WR-412.2.3.1
of conductors) shall not affect the protection afforded
by an enclosure.

## 10.2.9 Thermal effects

Electrical installations shall be so arranged that:          WR-131.3.1

- the risk of ignition of flammable materials due to
  high temperature or electric arc is minimized;
- during normal operation of the electrical equipment, there shall be minimal risk of burns to persons or livestock.

## 10.2.10 Wiring systems

Electrical installations shall be so arranged that:

- the risk of ignition of flammable materials due to high temperature or electric arc is minimized;
- during normal operation of the electrical equipment, there shall be minimal risk of burns to persons or livestock.

 **Note:** Where an electrical service is located in close proximity to one or more non-electrical services, it shall meet the following conditions:

| | |
|---|---|
| • the wiring system shall be suitably protected against the hazards likely to arise from the presence of the other services in normal use;<br>• fault protection shall be provided by means of automatic disconnection of supply. | WR-528.3.4 |

| | |
|---|---|
| The choice of the type of wiring system and the method of installation shall include consideration of the following:<br><br>• the nature of the location;<br>• the nature of the structure supporting the wiring;<br>• accessibility of wiring to persons and livestock;<br>• voltage;<br>• the electromechanical stresses likely to occur due to short-circuit and Earth fault currents;<br>• EMI;<br>• other external influences (e.g. mechanical, thermal and those associated with fire) to which the wiring is likely to be exposed during the erection of the electrical installation or in service. | WR-132.7 |
| The installation of wiring systems will meet the requirements if:<br><br>• the rated voltage of the cable(s) is not less than the nominal voltage of the system and at least 300/500 V; and | WR-412.2.4.1 |

- adequate mechanical protection of the basic insulation is provided by (one or more) of the following:

  ○ the non-metallic sheath of the cable;
  ○ non-metallic trunking or ducting (complying with the BS EN 50085);
  ○ non-metallic conduit (complying with the BS EN 61386).

A wiring system that passes through the location but is not intended to supply electrical equipment in the location shall:                                              WR-422.3.5

- have no connection or joint within the location, unless the connection or joint is installed in an enclosure; and
- be protected against overcurrent; and
- **not** use bare live conductors.

A wiring system shall be installed so that the general structural performance and fire safety of the building are not reduced.                                              WR-527.1.2

A wiring system **shall not** be installed in the vicinity of services that produce heat, smoke or fumes likely to be detrimental to the wiring, **unless** it is protected from harmful effects by shielding arranged so as not to affect the dissipation of heat from the wiring.                  WR-528.3.1

### 10.2.11 Initial verification

During erection, on completion of an installation, addition or alteration to an installation (and before being put into service), appropriate inspection and testing shall be carried out by competent persons to verify that the requirements of BS 7671:2008 (Incorporating Amendment No. 1) have been met.                          WR-134.2.1

## 10.3 Maintenance, inspection and repair

The aim of periodic inspection and testing is to:

- confirm that the installation is in a satisfactory condition for continued service;

- confirm that the safety of the installation has not deteriorated or that the installation has not been damaged;
- ensure the continued safety of persons and livestock against the effects of electric shock and burns;
- identify installation defects and non-compliance with the requirements of the Regulations that may give rise to danger;
- protect property from being damaged by fire and heat caused by a defective installation.

The Regulations are clear in the requirements that maintenance inspections shall consist of a careful scrutiny of the installation (dismantled or otherwise) using the appropriate tests described in Chapter 6 of the Regulations.

The frequency of maintenance inspections will depend on:

- the type of installation, its use and operation;
- the frequency and quality of maintenance; and
- the external influences to which it is subjected.

Precautions shall be taken to ensure that maintenance inspections do not cause:

- danger to persons or livestock;
- damage to property and equipment (even if the circuit is defective).

The maintenance inspection shall be made to verify that the installed electrical equipment:

- complies with the requirements of the Regulations and the appropriate National Standards and European Harmonized Directives;
- is correctly selected and erected in accordance with the Regulations;
- is not visibly damaged or defective so as to impair safety.

The inspection shall include the following items, where relevant:

- access to switchgear and equipment;
- cable routing in safe zones;
- choice and setting of protective and monitoring devices;
- connection of conductors;
- connection of single-pole devices for protection or switching in phase conductors only;
- correct connection of accessories and equipment;
- erection methods;
- identification of conductors;
- labelling of protective devices, switches and terminals;
- presence of danger notices and other warning signs;
- presence of diagrams, instructions and similar information;

- presence of fire barriers and suitable seals;
- presence of isolation and switching devices (and their correct location);
- prevention of mutual (i.e. detrimental) influence;
- presence of undervoltage protective devices;
- protection against electric shock (direct and indirect contact);
- protection against external influences;
- protection against direct and indirect contact;
- protection against mechanical damage;
- protection against overcurrent;
- protection against thermal effects selection of conductors for current-carrying capacity and voltage drop.

 **Note:** Generally speaking, the following CEN/CENELEC recommendations are relevant for all installed electrical and/or electronic equipment.

For ease of maintenance, all equipment provided should have:

- easily accessible test points to facilitate fault location;
- modules that have been constructed so as to facilitate the connection of test equipment (e.g. logic analysers, emulators and test ROMs);
- fault location provision to allow functional areas within each module or equipment to be isolated.

Under workshop conditions it should be possible to gain access to all circuitry while operating, with a minimum of effort required, in order to partially dismantle the module concerned, with the assumption that there will be a minimum of risk to the components or the testing maintenance staff.

Special connecting leads, printed wiring extension boards and any other special items required for maintenance purposes, together with the mating half of all necessary connectors, will probably have to be obtained from the manufacturer.

Equipment is normally expected to have been designed to have a useful life of not less than 20 years. 'Useful life' normally means 'the period for which the equipment will continue to operate with the specified level of reliability'.

No components, modules or equipment should, therefore, be used (insofar as can be ascertained at the time of manufacture) for which spares cannot be fully guaranteed to be available throughout the life of the equipment.

An assessment shall be made of the frequency and quality    WR-341.1
of maintenance that the installation can reasonably be
expected to receive during its intended life, and to ensure
that:

- any periodic inspection and testing, maintenance and repairs likely to be necessary during the intended life can be readily and safely carried out; and
- the effectiveness of the protective measures for safety during the intended life shall not diminish; and
- the reliability of equipment for proper functioning of the installation is appropriate to the intended life.

## 10.3.1 Accessibility of electrical equipment

| | |
|---|---|
| Provision shall be made for safe and adequate access to all parts of a wiring system that may require maintenance. | WR-529.3 |

| | |
|---|---|
| Electrical equipment shall be arranged so that: | WR-132.12 |

- there is sufficient space for the initial installation and later replacement of individual items of electrical equipment;
- the equipment is accessible for operation, inspection, testing, fault detection, maintenance and repair.

| | |
|---|---|
| Every item of equipment shall be arranged so as to enable its operation, inspection and maintenance and access to each connection. | WR-513.1 WR-543.3.2 |

| | |
|---|---|
| Every installation shall be divided into circuits, as necessary, to: | WR-314.1 |

- avoid hazards and minimize inconvenience in the event of a fault;
- ensure safe inspection, testing and maintenance;
- prevent the indirect energizing of a circuit that is intended to be isolated;
- reduce the possibility of unwanted tripping of RCDs due to excessive protective conductor currents produced by equipment;
- reduce the effects of EMI;
- take account of any danger that may arise from the failure of a single circuit (e.g. a lighting circuit).

> The selection and erection of equipment for solar and photovoltaic (PV) power supply systems shall enable safe maintenance, and shall not adversely affect provisions made by the manufacturer of the PV equipment to enable maintenance or service work to be carried out safely.
>
> WR-712.513.1

### 10.3.1.1 Operating and maintenance gangways

Section 729 is a new (fairly small) addition to the standard that centres on the operation and safe maintenance of switchgear and controlgear within areas that included gangways and where access is restricted to skilled or instructed persons.

#### 10.3.1.1.1 Restricted access areas

> Restricted access areas shall:
>
> WR-729.3
>
> - be clearly and visibly marked by appropriate signs;
> - not provide access to unauthorized persons; and
> - provide closed doors that, although normally in the closed position, nevertheless will allow easy evacuation in case of danger, without the use of a key, tool or any other device that is not part of the opening mechanism.

#### 10.3.1.1.2 Requirements for operating and maintenance gangways

The width of gangways and access areas shall be adequate for work, operational access, emergency access, emergency evacuation and for transport of equipment.

> Gangways shall permit at least a 90° opening of equipment doors or hinged panels (see Figure RLT-7-1)
>
> WR. 729.513.2

#### 10.3.1.1.3 Restricted access areas where basic protection is provided by barriers or enclosures

Where basic protection is provided by barriers or enclosures, the following minimum dimensions apply (Figure 10.2):

| Gangway | Dimensions |
|---|---|
| Gangway width between barriers or enclosures and switch handles or circuit-breakers:<br><br>• in the most onerous position; and<br>• in the most onerous position and the wall | 700 mm |
| Gangway width between barriers or enclosures or other barriers or enclosures and the wall | 700 mm |
| Height of gangway to barrier or enclosure above floor | 2000 mm |
| Live parts placed out of reach | 2500 mm |

 **Note:** Where additional workspace is needed (e.g. for special switchgear and controlgear assemblies), larger dimensions may be required.

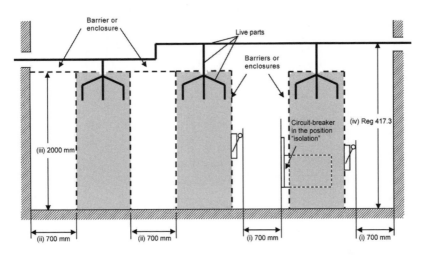

**Figure 10.2** Gangways in installations with protection by barriers or enclosures. (Courtesy of the BSI.)

 **Note:** The above dimensions apply after barriers and enclosures have been fixed and with circuit-breakers and switch handles in the most onerous position, including 'isolation'.

### 10.3.1.1.4 Restricted access areas where the protective measure of obstacles is applied

Where basic protection is provided by obstacles, the following minimum dimensions apply (Figure 10.3):

| Gangway | Dimensions |
|---|---|
| Gangway width between obstacles and switch handles or circuit-breakers: | 700 mm |
| • in the most onerous position; and<br>• in the most onerous position and the wall | |
| Gangway width between obstacles or other obstacles and the enclosures and the wall | 700 mm |
| Height of gangway to obstacles, barrier or enclosure above floor | 2000 mm |
| Live parts placed out of reach | 2500 mm |

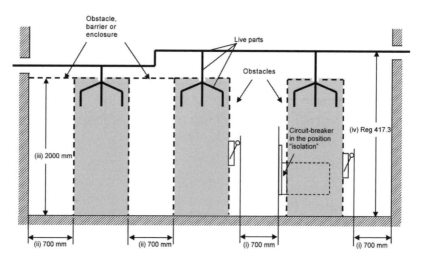

**Figure 10.3** Gangways in installations with protection by obstacles. (Courtesy of the BSI.)

## 10.3.1.1.5 Access of gangways

| | |
|---|---|
| Gangways longer than 10 m shall be accessible from both ends. | WR-729.513.2.3 1 |

 Figure 729.3 of BS 7671:2008 (Incorporating Amendment No. 1) also provides examples of positioning of doors in closed restricted access.

## 10.3.2  Connections

Every connection shall be accessible for inspection, test-   WR-526.3
ing and maintenance, except for the following:

- a joint designed to be buried in the ground;
- a compound-filled or encapsulated joint;
- a connection between a cold tail and the heating
  element, as in ceiling heating, floor heating or a
  trace heating system;
- a joint made by welding, soldering, brazing or
  appropriate compression tool;
- a joint (or connection) made in the equipment by
  the manufacturer and not intended to be inspected
  or maintained;
- a joint forming part of the equipment complying
  with the appropriate product standard.

## 10.3.3  Frequency of inspection and testing

The frequency of periodic inspection and maintenance   WR-622.1
of an installation will depend on the type of installation
and equipment, its use and operation, the frequency and
quality of maintenance and the external influences to
which it is subjected.

The results and recommendations of the previous report,   WR-622.1
if any, shall be taken into account.

In the case of an installation under an effective   WR-622.2
management system for preventive maintenance in normal use, periodic inspection and testing may be
replaced by an adequate regime of continuous monitoring and maintenance of the installation and all its constituent equipment by skilled persons competent in
such work.

 Full records of these inspections and tests **must**
be retained.

## 10.3.4 Switching off for mechanical maintenance

| | |
|---|---|
| The capability of switching off for mechanical maintenance shall be provided where mechanical maintenance could involve a risk of physical injury. | WR-537.3.1.1 |
| Suitable means shall be provided to prevent electrically powered equipment from becoming unintentionally reactivated during mechanical maintenance. | WR-537.3.1.2 |

### *10.3.4.1 Devices for switching off for mechanical maintenance*

A device such as a:                                    WR-537.3.2.1

- multi-pole switch;
- circuit-breaker;
- control and protective switching device (CPS);
- control switch operating a contactor;
- plug and socket-outlet;

may be inserted in the main supply circuit for switching off for mechanical maintenance.

A device for switching off for mechanical maintenance (or a control switch for such a device) shall:          WR-537.3.2.2

- require manual operation;

- be designed and/or installed so as to prevent inadvertent or unintentional switching on;          WR-537.3.2.3

- be so placed and durably marked so as to be readily identifiable and convenient for the intended use;          WR-537.3.2.4

- ensure that the open position of the contacts of the device shall be visible or be clearly and reliably indicated by the use of the symbols O and I to indicate the open and closed positions, respectively.          WR-537.3.2.2

A plug and socket-outlet or similar device of rating not exceeding 16 A may be used as a device for switching off for mechanical maintenance.          WR-537.3.2.6

Where a switch is used as a device for switching off for mechanical maintenance, it shall be capable of cutting off the full load current of the relevant part
of the installation.

### 10.3.5 Insulating enclosure

 An insulating enclosure **shall not** contain any screws or other fixing means
that might need to be removed (e.g. during installation and maintenance) and
that *could* be replaced by metallic screws or some other type of fixing that could
affect the insulation of the enclosure.

### 10.3.6 Low-power supply sources

| | |
|---|---|
| The power output of a low-power supply system should be inspected so as to ensure that it is limited to 500 W for a 3–hour duration or 1500 W for a 1–hour duration. | WR-560.6.10 |

 However, the batteries may be of the gastight or valve-regulated maintenance-
free type, and the minimum design life of the batteries shall be 5 years.

### 10.3.7 Multi-phase sequence

| | |
|---|---|
| For multi-phase circuits, it shall be verified that the phase sequence is maintained. | WR-612.12 |

### 10.3.8 Safety protection

| | |
|---|---|
| Disconnecting devices shall be provided so as to allow electrical installations, circuits or individual items of equipment to be switched off or isolated for the purposes of operation, inspection, fault detection, testing, maintenance and repair. | WR-132.10 |
| Where it is necessary to remove a protective measure in order to carry out maintenance, provision shall be made so that the protective measure can be reinstated without a reduction in the degree of protection originally intended. | WR-529.2 |

The protective measures of placing out of reach and obstacles **shall not** be used except where:

WR-559.10.1

- the maintenance of equipment is restricted to skilled persons who are specially trained;
- items of street furniture are within 1.5 m of a low-voltage overhead line.

### 10.3.9 Environmental aspects

The wiring system shall be selected and erected so:

WR-522.3.1

- that no damage is caused by condensation or ingress of water during installation, use and maintenance;
- as to minimize the damage arising from mechanical stress (e.g. by impact, abrasion, penetration, tension or compression) during installation, use or maintenance;

WR-522.6.1

- as to avoid during installation, use or maintenance, damage to the sheath or insulation of cables and their terminations.

WR-522.8.1

 Notes:

1. The degree of protection of electrical equipment shall be maintained after installation of the cables and conductors.
2. A wiring system buried in a floor shall be sufficiently protected to prevent damage caused by the intended use of the floor.

 The use of any lubricants that could have a detrimental effect on the cable or wiring system is **not** permitted.

### 10.3.10 Disconnecting devices

Disconnecting devices shall be provided so as to allow electrical installations, circuits or individual items of equipment to be switched off or isolated for the purposes of maintenance and repair.

WR-132.10

## 10.4 Certification and reporting

As previously explained in Chapter 9 of this book, the following certificates shall be completed where appropriate:

 Please note that all Electrical Installation Certificates, Electrical Installation Condition Reports and Minor Electrical Installation Works Certificates **MUST** be compiled and signed (or otherwise authenticated) by a competent person or persons.

| | | |
|---|---|---|
| Electrical Installation Certificate | Containing details of the installation, together with a record of the inspections made and the test results, shall be provided for new installations and changes to existing installations | WR-631.1 |
| Electrical Installation Condition Report | Containing details of the extent of the installation and limitations of the inspection and testing covered by the report – together with records of inspection, the results of testing and a recommendation of when the next periodic inspection should occur | WR-631.2 |
| Minor Electrical Installation Works Certificate | For all minor electrical installation work that does not include the provision of a new circuit | WR-631.3 |

Although alterations and repairs may be completed during an installation, normal repairs are completed following a periodic inspection, and reports of these repairs shall be compiled and signed or otherwise authenticated by a competent person or persons.

| | |
|---|---|
| Following the periodic inspection, an Electrical Installation Report, together with a schedule of inspections and a schedule of test results, shall be given by the person carrying out the inspection, or a person authorized to act on his behalf, to the person ordering the inspection. | WR-634.1 |

| | |
|---|---|
| Any damage, deterioration, defects, dangerous conditions and non-compliance with the requirements of the Regulations, which may give rise to danger, together with any significant limitations of the inspection and testing, including the reasons, shall be recorded. | WR-634.2 |
| Reports may be produced in any durable medium, including written and electronic media. | WR-631.5 |
| Regardless of the media used for original certificates, reports (or their copies) shall have their authenticity and integrity verified by a reliable process or method. | WR-631.5 |
| The process or method shall also verify that any copy is a true copy of the original. | WR-631.5 |

 Records are an important part of the management of any electrical installation, as they provide objective evidence of activities performed and/or results achieved, and the following requirements come from the Regulations.

## 10.5 Additions and alterations to an installation

The Regulations require every electrical installation to be inspected and tested during erection and on completion before being put into service to verify that the requirements of BS 7671:2008 (Incorporating Amendment No. 1) have been met. By definition, this requirement applies to alterations and/or additions to an existing installation, as well as to entirely new installations.

The following is a summary of the relevant requirements for additions and alterations to an installation.

| | |
|---|---|
| No addition or alteration, either temporary or permanent, shall be made to an existing installation:<br><br>• unless it has been ascertained that the rating and the condition of any existing equipment (including that of the distributor) will be adequate for the altered circumstances;<br>• the earthing and bonding arrangements used as a protective measure for the safety of the addition or alteration are adequate. | WR-132.16 |

If wiring additions or alterations are made to an installation such that some of the wiring complies with the current Regulations but there is also wiring to previous versions of these Regulations, a warning notice shall be affixed at or near the appropriate distribution board with the wording shown in Figure 10.4.

WR-514.14.1

**CAUTION**

This installation has wiring colours to two versions of BS 7671. Great care should be taken before undertaking extension, alteration or repair that all conductors are correctly identified.

**Figure 10.4** Warning notice – non-standard colours.

An Electrical Installation Certificate or a Minor Electrical Installation Works Certificate (as appropriate) shall apply to all work competed, including the additions or alterations.

WR-633.1

The contractor or other person responsible for the new work, or a person authorized to act on his behalf, shall record on the Electrical Installation Certificate or the Minor Electrical Installation Works Certificate any defects found, insofar as is reasonably practicable, in the existing installation.

WR-633.2

## 10.5.1 Warning notice - alternative supplies

Where an installation includes alternative or additional sources of supply, warning notices shall be affixed at the following locations in the installation:

- at the origin of the installation;
- at the meter position, if remote from the origin;
- at the consumer unit or distribution board to which the alternative or additional sources are connected;
- at all points of isolation of all sources of supply.

WR-514.15.1

The warning notice shall have the wording shown in Figure 10.5.

> **WARNING – MULTIPLE SUPPLIES**
>
> Isolate all electrical supplies before carrying out work.
>
> Isolate the mains supply at ..................................
>
> Isolate the alternative supplies at .......................

**Figure 10.5** Warning notice – alternative or additional sources of supply.

 **Note:** In some circumstances (particularly when an historic building is undergoing a material change of use and where the special characteristics of the building need to be recognized) it may **not** be practical to improve sound insulation to the standards set out in Part E1 or resistance to contaminants and water as set out in Part C. In these cases, the aim should be to improve the insulation and resistance where it is practicably possible – always provided that the work does not prejudice the character of the historic building, or increase the risk of long-term deterioration to the building fabric and/or fittings.

## 10.6  Material changes of use

Where there is a material change of use of a building, any work carried out shall ensure that the building complies with the applicable requirements of the following paragraphs of Schedule 1 of the Building Act 1984:

*(a)  In all cases:*

- *means of warning and escape (B1);*
- *internal fire spread - linings (B2);*
- *internal fire spread - structure (B3);*
- *external fire spread (B4);*
- *access and facilities for the fire service (B5);*
- *resistance to moisture (C1)(2);*
- *dwelling houses and flats formed by material change of use (E4);*
- *ventilation (F1);*
- *sanitary conveniences and washing facilities (G1);*
- *bathrooms (G2);*
- *foul water drainage (H1);*
- *solid waste storage (H6);*
- *combustion appliances (J1, J2 & J3);*
- *conservation of fuel and power – dwellings (L1);*
- *conservation of fuel and power - buildings other than dwellings (L2);*
- *electrical safety (P1, P2);*

*(b)  In other cases:*

**Table 10.1  Building Act requirements**

| Material change of use | Requirement | Approved Document |
|---|---|---|
| The building is used as a dwelling, where previously it was not | Resistance to moisture | C2<br>E1, E2, E3 |
| The public building consists of a new school | Acoustic conditions in schools | E4 |
| The building contains a flat, where previously it did not | Resistance to the passage of sound | E1, E2, E3 |
| The building is used as an hotel or a boarding house, where previously it was not | Structure | A1, A2, A3<br>E1, E2, E3 |
| The building is used as an institution, where previously it was not | Structure | A1, A2, A3 |
| The building is used as a public building, where previously it was not | | A1, A2, A3<br>E1, E2, E3 |
| The building is not a building described in Classes I to VI in Schedule 2, where previously it was not | Structure | A1, A2, A3 |
| The building, which contains at least one room for residential purposes, contains a greater or lesser number of dwellings than it did previously | Structure | A1, A2, A3<br>E1, E2, E3 |
| The building, which contains at least one dwelling, contains a greater or lesser number of dwellings than it did previously | Resistance to the passage of sound | E1, E2, E3 |

# Annex 10A – Example stage audit checks

## Design stage

| | Item | | Related item | Remark |
|---|---|---|---|---|
| 1 | Requirements | 1.1 | Information | Has the customer fully described his requirement? |
| | | | | Has the customer any mandatory requirements? |
| | | | | Are the customer's requirements fully understood by all members of the design team? |
| | | | | Is there a need to have further discussions with the customer? |
| | | | | Are other suppliers or sub-contractors involved?<br>If yes, who is the prime contractor? |

*(continued)*

| Item | Related item | Remark |
|---|---|---|
| | 1.2 Standards | What international standards need to be observed? <br> Are they available? |
| | | What national standards need to be observed? <br> Are they available? |
| | | What other information and procedures are required? <br> Are they available? |
| | 1.3 Procedures | Are there any customer-supplied drawings, sketches or plans? <br> Have they been registered? |
| 2 Quality procedures | 2.1 Procedures manual | Is one available? |
| | | Does it contain detailed procedures and instructions for the control of all drawings within the drawing office? |
| | 2.2 Planning implementation and production | Is the project split into a number of work packages? <br> If so: |
| | | • Are the various work packages listed? <br> • Have work package leaders been nominated? <br> • Is their task clear? <br> • Is their task achievable? |
| | | Is a time plan available? <br> Is it up to date? <br> Regularly maintained? <br> Relevant to the task? |
| 3 Drawings | 3.1 Identification | Are all drawings identified by a unique number? |
| | | Is the numbering system strictly controlled? |
| | 3.2 Cataloguing | Is a catalogue of drawings maintained? <br> Is this catalogue regularly reviewed and up to date? |
| | 3.3 Amendments and modifications | Is there a procedure for authorizing the issue of amendments, changes to drawings? |
| | | Is there a method for withdrawing and disposing of obsolete drawings? |
| 4 Components | 4.1 Availability | Are complete lists of all the relevant components available? |
| | 4.2 Adequacy | Are the selected components currently available and adequate for the task? <br> If not, how long will they take to procure? <br> Is this acceptable? |
| | 4.3 Acceptability | If alternative components have to be used, are they acceptable to the task? |
| 5 Records | 5.1 Failure reports | Has the design office access to all records, failure reports and other relevant data? |

(continued)

| Item | Related item | Remark |
|---|---|---|
| | 5.2 Reliability data | Are reliability data correctly stored, maintained and analysed? |
| | 5.3 Graphs, diagrams, plans | In addition to drawings, is there a system for the control of all graphs, tables, plans etc.? Are CAD facilities available? (If so, go to 6.1) |
| 6 Reviews and audits | 6.1 Computers | If a processor is being used: <br>• Are all the design office personnel trained in its use? <br>• Are regular back-ups taken? <br>• Is there an anti-virus system in place? |
| | 6.2 Manufacturing Division | Is a close relationship being maintained between the design office and the manufacturing division? |
| | 6.3 | Is notice being taken of the manufacturing division's exact requirements, their problems and their choices of components etc.? |

## Installation stage

| Item | Related item | Remark |
|---|---|---|
| 1 Degree of quality | 1.1 Quality control procedures | Are quality control procedures available? <br>Are they relevant to the task? <br>Are they understood by all members of the manufacturing team? <br>Are they regularly reviewed and up to date? <br>Are they subject to control procedures? |
| | 1.2 Quality control checks | What quality checks are being observed? Are they relevant? <br>Are there laid-down procedures for carrying out these checks? <br>Are they available? <br>Are they regularly updated? |
| 2 Reliability of product design | 2.1 Statistical data | Is there a system for predicting the reliability of the product's design? <br>Are sufficient statistical data available to be able to estimate the actual reliability of the design, before a product is manufactured? <br>Are the appropriate engineering data available? |
| | 2.2 Components and parts | Are the reliability ratings of recommended parts and components available? |

(continued)

| Item | Related item | Remark |
|------|--------------|--------|
| | | Are probability methods used to examine the reliability of a proposed design? |
| | | If so, have these checks revealed design deficiencies such as: |
| | | • Assembly errors? |
| | | • Operator learning, motivational, or fatigue factors? |
| | | • Latent defects? |
| | | • Improper part selection? |
| | | (Note: If necessary, use additional sheets to list actions taken.) |

## Acceptance stage

| Item | | Related item | Remark |
|------|--|--------------|--------|
| 1 | Product performance | | Does the product perform to the required function? |
| | | | If not, what has been done about it? |
| 2 | Quality level | 2.1 Workmanship | Does the workmanship of the product fully meet the level of quality required or stipulated by the user? |
| | | 2.2 Tests | Is the product subjected to environmental tests? |
| | | | If so, which ones? |
| | | | Is the product field tested as a complete system? |
| | | | If so, what were the results? |
| 3 | Reliability | 3.1 Probability function | Are individual components and modules environmentally tested? |
| | | | If so, how? |
| | | 3.2 Failure rate | Is the product's reliability measured in terms of probability function? |
| | | | If so, what were the results? |
| | | | Is the product's reliability measured in terms of failure rate? |
| | | | If so, what were the results? |
| | | 3.3 Mean time between failures | Is the product's reliability measured in terms of mean time between failures? |
| | | | If so, what were the results? |

## In-service stage

| Item | | Related item | Remark |
|------|--|--------------|--------|
| 1 | System reliability | 1.1 Product basic design | Are statistical methods being used to prove the product's basic design? |
| | | | If so, are they adequate? |
| | | | Are the results recorded and available? |

(continued)

| Item | | Related item | Remark |
|------|--|--------------|--------|
| | | | What other methods are used to prove the product's basic design? |
| | | | Are these methods appropriate? |
| 2 | Equipment reliability | 2.1   Personnel | Are there sufficient trained personnel to carry out the task? |
| | | | Are they sufficiently motivated? |
| | | | If not, what is the problem? |
| | | 2.1.1  Operators | Have individual job descriptions been developed? |
| | | | Are they readily available? |
| | | | Are all operators capable of completing their duties? |
| | | 2.1.2  Training | Do all personnel receive appropriate training? |
| | | | Is a continuous on-the-job training programme available to all personnel? If not, why not? |
| | | 2.2   Product dependability | What proof is there that the product is dependable? |
| | | | How is product dependability proved? Is this sufficient for the customer? |
| | | 2.3   Component reliability | Has the reliability of individual component been considered? |
| | | | Does the reliability of individual components exceed the overall system reliability? |
| | | 2.4   Faulty operating procedures | Are operating procedures available? |
| | | | Are they appropriate to the task? |
| | | | Are they regularly reviewed? |
| | | 2.5   Operational abuses | Are there any obvious operational abuses? |
| | | | If so, what are they? How can they be overcome? |
| | | 2.5.1  Extended duty cycle | Do the staff have to work shifts? If so, are they allowed regular breaks from their work? |
| | | | Is there a senior shift worker? If so, are his duties and responsibilities clearly defined? |
| | | | Are computers used? If so, are screen filters available? Do the operators have keyboard wrist rests? |
| | | 2.5.2  Training | Do the operational staff receive regular on-the-job training? |
| | | | Is there any need for additional in-house or external training? |
| 3 | Design capability | 3.1   Faulty operating procedures | Are there any obvious faulty operating procedures? |
| | | | Can the existing procedures be improved upon? |

# Annex A

# Symbols used in electrical installations

## SYMBOLS

| Symbol | Description |
|---|---|
| | Socket-outlet |
| | Switched socket-outlet |
| | Switch |
| | Two-way switch, single-pole |
| | Intermediate switch |
| | Pull switch, single-pole |
| | Lighting outlet position |
| | Fluorescent luminaire |
| | Wall mounted luminaire |
| | Emergency lighting luminaire (or special circuit) |
| | Self-contained emergency lighting luminaire |
| | Push button |
| | Clock |
| | Bell |
| | Buzzer |
| | Horn |
| | Telephone handset |
| | Microphone |
| | Loudspeaker |
| | Antenna |
| | Machine<br>* Function<br>M = Motor<br>G = Generator |
| G | Generator |
| | Indicating instrument<br>* function<br>V = Voltmeter<br>A = Ammeter |
| | Integrating instrument or Energy meter<br>* function<br>Wh = Watt-hour<br>VArh = Volt ampere reactive hour |
| | Load<br>*details |
| | Motor starter<br>*indicates type |
| | Class II appliance |
| III | Class III appliance |
| | Safety isolating transformer |
| | Isolating transformer |
| | Fuse link, rated current in amperes |
| | Operating device (coil) |
| | Make contact - normally open |
| | Break contact - normally closed |
| | Manually operated switch |
| | Three-phase winding - delta |
| | Three-phase winding - Star |
| | Changer, Converter |
| | Rectifier |
| | Invertor |
| | Primary cell - longer line positive, shorter line negative |
| | Battery |
| | Transformer - general symbol |

$10^9$ giga   G
$10^6$ mega   M
$10^3$ kilo   k
$10^{-3}$ milli   m
$10^{-6}$ micro   μ
$10^{-9}$ nano   n

# Annex B

# List of electrical and electromechanical symbols

| SYMBOL | DESCRIPTION |
|---|---|
| $\beta°$ | tube oscillating angle |
| $°C$ | degrees Celsius |
| $\Omega$ | ohm |
| $\mu g$ | microgram |
| $\mu g/m^3$ | micrograms per cubic metre |
| $\mu m$ | micrometre |
| $\mu s$ | microsecond |
| a | amplitude |
| A | ampere |
| A/m | amperes per metre |
| am | attometre |
| atm | standard atmosphere |
| C | coulomb |
| cd | candela |
| $cd/m^2$ | candelas per square metre |
| dB | decibels |
| dB(A) | decibel amps |
| dBm | decibel metres |
| $dm^3$ | cubic decimetre |
| $dm^3/mm$ | cubic decimetres/millimetre – flow |
| Em | exametre |
| eV | electronvolt |
| f | frequency |
| F | farad |
| fm | femtometre |
| ft | foot |
| g | gram |
| G | gauss |
| G | shock |
| $g^2/Hz$ | accelerated spectral density |
| GHz | Giga Hertz – frequency |
| Gm | gigametre |
| $g/m^3$ | grams per cubic metre |
| $g_n$ | peak acceleration |
| $G_s$ | setting value of a characteristic quantity |
| h | hour |
| H | henry |
| ha | hectare |
| hp | horsepower |

(*continued*)

| SYMBOL | DESCRIPTION |
|---|---|
| hr(s) | hour(s) – alternative to h |
| Hz | Hertz |
| I | amps |
| $I^2R$ | power |
| in | inch |
| J | joule |
| k | constant of the relay |
| K | kelvin |
| kA | kiloamps |
| kA/µs | kiloamps per micro second |
| kg | kilogram |
| $kg/m^3$ | kilograms per cubic metre |
| kgf | kilogram force |
| kHz | kilohertz |
| kPa | kilo Pascal – pressure |
| ks | kilosecond |
| kV | kilovolts |
| kW | kilowatt |
| $kW/m^2$ | kilowatts per square metre – irradiance |
| l | litre |
| lb | pound |
| lb/in | pounds per square inch |
| m | metre |
| m/s | metres per second |
| $m/s^2$ | metres per second per second – amplitude |
| $m^2$ | square metres |
| $m^3$ | cubic metres |
| mbar | millibar – pressure |
| MHz | Mega Hertz |
| min | minute |
| mm | millimetre |
| Mm | megametre |
| mm/h | millimetres per hour |
| $mm/m^2$ | millimetres/square metre – exposure |
| mol | mole |
| ms | millisecond |
| mV | millivolts |
| MVA | megavolt amps |
| N | newton |
| $N/m^2$ | newtons per square metre |
| NaCl | sodium chloride |
| nF | nano Farad |
| nm | nanometre |
| pH | alkalinity/acidity value |
| pm | picometre |
| Pm | petametre |
| $R$ | intensity of dropfield in mm/h |
| R | resistance |
| rad/s | radians per second |
| s | second |
| S | siemens |
| t | tonne |
| T | time |
| T | tesla |
| Tm | terametre |

(continued)

| SYMBOL | DESCRIPTION |
|---|---|
| $\hat{u}$ | amplitude of voltage surge |
| $U_n$ | nominal voltage |
| V | volt |
| V/$\mu$s | volts per microsecond |
| V/km | volts per kilometre |
| Vm | volts per metre |
| W | watt |
| Wb | weber |
| W/m$^2$ | watts per square metre – irradiance |
| yd | yard |
| ym | yocotmetre |
| Ym | Yottametre |
| zm | zeptometre |
| Zm | zettametre |

# Annex C

# SI units for existing technology

As Gregor M. Grant explained in his article published in the April/May 1997 issue of ElectroTechnology, the *Système International d'Unités* (SI) was a child of the 1960s, a creation of the 11th General Conference on Weights and Measures, *Conférence Générale des Poids et Mesures* (CGPM). This assembly endorsed the Italian physicist Professor Giovanni Giorgi's MKS (i.e. metre-kilogram-second) system of 1901 and decided to base the SI system on it. Seven basic units were adopted, as shown in Table A, each of which was harmonised to a standard value.

Of the seven units, only the kilogram (kg) is represented by a physical object, namely a cylinder of platinum-iridium kept at the International Bureau of Weights and Measures at Sèvres, near Paris, with a duplicate at the US Bureau of Standards.

The metre (m), on the other hand, "is the length of the path travelled by light in a vacuum during a time interval of 1/299,792,458 of a second".

The second (s) has been defined as "the duration of 9,192,631,770 periods of radiation corresponding to the energy-level change between the two hyperfine levels of the ground state of caesium-133 atom".

The ampere (A) is "that constant current which, if maintained in two straight parallel conductors of infinite length, of negligible circular cross section and placed 1m apart in vacuum, would produce between these conductors a force equal to $2 \times 10^{-7}$ newtons per metre length".

The unit of temperature is the kelvin (K), which is a thermodynamic measurement as opposed to one based on the properties of real material. Its origin is at absolute zero and there is a fixed point where the pressure and temperature of water, water vapour and ice are in equilibrium, which is defined as 273.16K.

The mole (mol) is "that quantity of substance of a system which contains as many elementary entities as there are atoms in 0.012kg of carbon-12". For definition purposes the entities *must* be specified, (e.g. atoms, electrons, ions or any other particles or groups of such particles).

Finally there is the candela (cd), the unit of light intensity. This is defined as "the luminous intensity, in the perpendicular direction, of a surface of $1/600,000\text{m}^2$ of a black body at the temperature of freezing platinum under a pressure of $101,325 \text{ N/m}^2$".

Two years before the creation of SI units, another international agreement had made the prefixes mega and micro official and introduced some new ones, such as the nano whose name derives from the Greek "nanos" meaning dwarf (see Table B). Its symbol is n, and its mathematical representation is $10^{-9}$, indicating the number of *digits* to the right of the decimal point, in this case 0.000 000 001.

Even these minute quantities, however, soon became inadequate and, by 1962, it was decided that a thousandth of a picometre be designated a femtometre and one thousandth of this new measurement be termed an attometre. Later on, the zeptometre and yoctometre were introduced.

# Basic SI units

Many SI units are named after people but when these units are written in full, they do not necessarily require initial capital letters, e.g. amperes, coulombs, newtons, siemens.

All the above examples are expressed in the plural, but note that siemens does not drop the final 's' in the singular as this was derived from a person's name (i.e. Siemens) thus we have one newton, but one siemens.

**Table A  Basic SI units**

| SI nomenclature | Abbreviation | Quantity |
|---|---|---|
| metre | m | length |
| kilogram | kg | mass |
| second | s | time |
| ampere | A | electrical current |
| kelvin | K | temperatures |
| mole | mol | amount of substance |
| candela | cd | luminous intensity |

# Small number SI prefixes

Within the SI units there is a distinction between a quantity and a unit. Length is a quantity, but metres (abbreviated to m) is a unit.

**Table B  Small number SI units**

| Measurement | Symbol | Equivalent to |
|---|---|---|
| millimetre | mm | 0.001m or $10^{-3}$m |
| micrometre | μm | 0.000 001m or $10^{-6}$m |
| nanometre | nm | 0.000 000 001m or $10^{-9}$m |
| picometre | pm | 0.000 000 000 001m or $10^{-12}$m |
| femtometre | fm | 0.000 000 000 000 001m or $10^{-15}$m |
| attometre | am | 0.000 000 000 000 000 001m or $10^{-18}$m |
| zeptometre | zm | 0.000 000 000 000 000 000 001m or $10^{-21}$m |
| yoctometre | ym | 0.000 000 000 000 000 000 000 001m or $10^{-24}$m |

# Large number SI prefixes

Table C  Large number SI prefixes

| Measurement | Symbol | Equivalent to |
|---|---|---|
| megametre | Mm | 1,000,000m or $10^6$m |
| gigametre | Gm | 1,000,000 000m or $10^9$m |
| terametre | Tm | 1,000,000,000,000m or $10^{12}$m |
| petametre | Pm | 1,000,000,000,000,000m or $10^{15}$m |
| exametre | Em | 1,000,000,000,000,000,000m or $10^{18}$m |
| zettametre | Zm | 1,000,000,000,000,000,000,000 or $10^{21}$m |
| yottametre | Ym | 1,000,000,000,000,000,000,000,000 or $10^{24}$m |

# Deprecated prefixes

Some non-SI fractions and multiples are occasionally used (see below), but they are not encouraged.

Table D  Deprecated prefixes

| Fractions | Prefix | Abbreviation | Multiple | Prefix | Abbreviation |
|---|---|---|---|---|---|
| $10^{-1}$ | deci | d | 10 | deka | da |
| $10^{-2}$ | centi | c | $10^2$ | hecto | h |

# Derived units

Some units, derived from the basic SI units, have been given special names, many of which originate from a person's name (i.e. Siemens).

Table E  Derived units

| Quantity | Name of unit | Abbreviation (symbol) | Expression in terms of other SI units |
|---|---|---|---|
| energy | joule | J | Nm |
| force | newton | N | – |
| power | watt | W | J/s |
| electric charge | coulomb | C | As |
| potential difference (voltage) | volt | V | W/A |
| electrical resistance (or reactance or impedance) | ohm | Ω | V/A |
| electrical capacitance | farad | F | C/V |
| magnetic flux | weber | Wb | Vs |
| Inductance (note that the plural of henry is henrys) | henry | H | Wb/A |

| | | | |
|---|---|---|---|
| magnetic flux density | tesla | T | Wb/m$^2$ |
| admittance (electrical conductance) | siemens | S | A/V$(=\Omega^{-1})$ |
| frequency | hertz | Hz | cycles per second (or events per second) |

# Units without special names

Other derived units, without special names, are listed below.

**Table F  Units without special names**

| QUANTITY | Unit | Abbreviation |
|---|---|---|
| area | square metres | m$^2$ |
| volume | cubic metres | m$^3$ |
| density | kilograms per cubic metre | kg/m$^3$ |
| velocity | metres per second | m/s |
| angular velocity (angular frequency) | radians per second | rad/s |
| acceleration | metres per second per second | m/s$^2$ |
| pressure | newtons per square metre | N/m$^2$ |
| electric field strength | volts per metre | Vm |
| magnetic field strength | amperes per metre | A/m |
| luminance | candelas per square metre | cd/m$^2$ |

# Tolerated units

Some non-SI units are tolerated in conjunction with SI units.

**Table G  Tolerated units**

| Quantity | Unit | Abbreviation (symbol) | Definition |
|---|---|---|---|
| area | hectare | ha | $10^4$m$^2$ |
| volume | litre | l | $10^{-3}$m$^3$ |
| pressure | standard atmosphere | atm | 101,325Pa |
| mass | tonne | t | $10^3$kg(Mg) |
| energy | electronvolt | eV | $1.6021 \times 10^{19}$J |
| magnetic | gauss | G | $10^{-4}$T |

## Obsolete units

For historical interest, (as well as for completeness), the following table gives a list of obsolete units.

**Table H  Obsolete units**

| Quantity | Unit | Abbreviation (symbol) | Definition |
|---|---|---|---|
| length | inch | in | 0.0254m |
| | foot | ft | 0.3048m |
| | yard | yd | 0.9144m |
| | mile | mi | 1.60394km |
| mass | pound | lb | 0.4539237kg |
| force | dyne | dyn | $10^{-5}$N |
| | poundal | pdl | 0.138255N |
| | pound force | lbf | 4.44822N |
| | kilogram force | kgf | 9.80665N |
| pressure | atmosphere | atm | 101.325kN/m$^2$ |
| | torr | torr | 133.322N/m$^2$ |
| | pounds per square inch | lb/in | 6894.76N/m$^2$ |
| energy | erg | erg | $10^{-7}$J |
| power | horsepower | hp | 745.700W |

# Annex D

# Acronyms and Abbreviations

| | |
|---|---|
| a.c. | alternating current |
| ACS | assembly for construction sites |
| ADP | automatic data processing |
| BEC | British Electrotechnical Committee |
| BRE | Building Research Establishment Ltd |
| BS | British Standards |
| BSI | British Standards Institution |
| CAD | computer aided design |
| CBN | common bonding network |
| CE | Conformity Europe |
| CECC | CENELEC Electronic Components Committee |
| CEN | Comité Européen de Normalisation |
| CENELEC | Comité Européen de Normalisation Electrotechnique |
| CNE | combined neutral and earth |
| CPC | circuit protective conductor |
| CPS | control and protective switching device |
| d.c. | direct current |
| DCL | device for connecting a luminaire |
| DDA | Disability Discrimination Act |
| DIY | do it yourself |
| EBADS | equipotential bonding and automatic disconnection of supplies |
| ECA | Electrical Contractors Association |
| EEBAD | earthed equipotential bonding and automatic disconnection |
| EEC | European Economic Commission |
| ELECSA | Fenestration Self-Assessment Scheme |
| ELV | extra-low voltage |
| EMC | electromagnetic compatibility |
| EMF | electromotive force |
| EMI | electromagnetic interference |
| EN | European Normalisation |
| ESQCR | Electricity Safety, Quality and Continuity Regulations 2002 |
| EU | European Union |

(continued)

| | |
|---|---|
| EU | European Community |
| FE | Functional Earth |
| FELV | functional extra-low voltage |
| GLS | as in tungsten lights |
| HBES | home and building electronic systems |
| HD | Harmonized Directive |
| HELA | Health and Safety Executive/Local Authorities |
| HEMP | high altitude electromagnetic pulse |
| HSE | Health and Safety Executive |
| HV | high voltage |
| I/O | input/output |
| ICM | insulation current monitoring device |
| IEC | International Electrotechnical Commission |
| IEE | Institution of Electrical Engineers (who later changed their name to the Institution of Engineering and Technology – IET) |
| IET | Institution of Engineering and Technology |
| IIE | Institution on of Incorporated Engineers |
| ILU | integrated logistic unit |
| IMD | insulation monitoring device |
| $I_a$ | current causing automatic operation of the disconnecting device within the time specified in Table 3.3 |
| $I_n$ | residual operating current |
| $I_{imp}$ | impulse current |
| $I_{nspd}$ | nominal discharge current |
| $I_{sc\ STC}$ | short circuit currents (Standard Test Conditions for the PV industry) |
| IPC | implant point of coupling |
| ISM | industrial, scientific and medical |
| ISO | International Standards Organization |
| IT | information technology |
| ITCZ | international conveyance zone |
| ITE | information technology equipment |
| LPG | liquid petroleum gas |
| LSC | luminaire supporting coupler |
| LUR | logical user requirement |
| LV | low voltage |
| MDD | Medical Devices Directive |
| ME | medical electrical |
| MET | main earthing terminal |
| MKS | metre-kilogram-second |
| MMI | Man–machine interface |
| MTBF | mean time between failures |
| N | neutral |
| NAPIT | National Association of Professional Inspectors and Testers |
| NICEIC | National Inspection Council for Electrical Installation Counselling |
| NSO | National Standards Organisation |

(continued)

| | |
|---|---|
| OCPD | overcurrent protective device |
| OFTEC | Oil Firing Technical Association |
| OJT | on-the-job-training |
| PCB | printed circuit board |
| PE | protective Earth |
| PELV | protective extra-low voltage |
| PEN | combined protective and neutral (conductor) |
| PME | protective multiple earthing |
| PRCD | portable RCD, usually an RCD incorporated into a plug |
| prEN | European draft standards |
| PV | photovoltaic |
| PVC | polyvinyl chloride |
| QA | quality assurance |
| QC | quality control |
| QMS | quality management system |
| QP | quality procedure |
| rms | root mean square |
| $R_A$ | sum of the resistances of the Earth electrode and the protective conductor connecting it to the exposed-conductive-parts |
| RAH | relative air humidity |
| RAM | reliability, availability and maintainability |
| RCBO | residual current operated circuit-breaker without integral overcurrent protection |
| RCCB | residual current operated circuit-breaker with integral overcurrent protection |
| RCCD | residual current operated circuit-breaker |
| RCD | residual current device |
| RCM | residual current monitor |
| RF | radio frequency |
| RH | relative humidity |
| S/N | signal to noise ratio |
| SELV | separated extra-low voltage |
| SI | Statutory Instrument |
| SI | Système International d'Unités |
| SPD | surge protective device |
| SRCBO | socket outlet incorporating an RCBO and RCD tester |
| SRCD | socket outlet incorporating an RCD |
| SSEG | Small-Scale Embedded Generators |
| STC | Standard Test Conditions (for the PV industry) |
| t | time |
| TDS | time delay switches |
| TLV | threshold limit values |
| TOV | temporary overvoltage |
| TQM | total quality management |
| $U_o$ | nominal a.c. rms or d.c. line voltage to Earth |

*(continued)*

| $U_{oc\ STC}$ | open circuit voltage (Standard Test Conditions for the PV industry) |
| UPS | uninterruptible power system |
| VDU | visual display unit |
| VSD | variable speed drive |
| WAUILF | workplace applied uniform indicated low frequency (application) |
| WI | work instruction |
| YFR | yearly forecast rationale |
| $Z_s$ | Earth fault loop impedance |

# Annex E

# British Standards currently used with the Wiring Regulations

By the time you read this edition of *Wiring Regulations in Brief*, it is quite possible that some of the standards and directives listed in this book **and** the IET Wiring Regulations will have been reviewed and updated, and, although the ones listed in this annex are still relevant, the latest edition of these documents should **always** be taken into account.

For this reason, the reader is recommended to always have a quick check via Google (or some other search engine) to make sure that he or she is using the most up-to-date standard, regulation or recommendation.

## E.1 British Standards currently used with the Wiring Regulations – listing by standard

| BS or EN number | Title |
| --- | --- |
| BS 67:1987 (1999) | Specification for ceiling roses |
| BS 88 | Cartridge fuses for voltages up to and including 1000 V a.c. and 1500 V d.c. |
| BS 88–2:2007 | Low-voltage fuses. Supplementary requirements for fuses for use by authorized persons (fuses mainly for industrial application). Examples of standardized systems of fuses A to I |
| BS 88–2.2:1988 | Specification for fuses for use by authorized persons (mainly for industrial application). Additional requirements for fuses with fuse-links for bolted connections |
| BS 88–6:1988 | Specification of supplementary requirements for fuses of compact dimensions for use in 240/415 V a.c. industrial and commercial electrical installations |
| BS 196:1961 | Specification for protected-type non-reversible plugs, socket-outlets cable-couplers and appliance-couplers with earthing contacts for single phase a.c. circuits up to 250 volts |

**E.1  British Standards** (*continued*)

| BS or EN number | Title |
| --- | --- |
| BS 476 | Fire tests on building materials and structures |
| BS 476–4:1970 | Non-combustible test for materials |
| BS 476–12:1991 | Method of test for ignitability of products by direct flame impingement |
| BS 546:1950 (1988) | Specification. Two-pole and earthing-pin plugs, socket-outlets and socket-outlet adaptors |
| BS 559:1998 (2005) | Specification for design, construction and installation of signs |
| BS 646:1958 (1991) | Specification. Cartridge fuse-links (rated up to 5 amperes) for a.c. and d.c. service |
| BS 951:1999 | Electrical earthing. Clamps for earthing and bonding. Specification |
| BS 1361:1971 (1986) | Specification for cartridge fuses for a.c. circuits in domestic and similar premises |
| BS 1362:1973 (1992) | Specification for general purpose fuse links for domestic and similar purposes (primarily for use in plugs) |
| BS 1363 | 13 A plugs, socket-outlets, connection units and adaptors |
| BS 1363–1:1995 | Specification for rewirable and non-rewirable 13 A fused plugs |
| BS 1363–2:1995 | Specification for 13 A switched and unswitched socket-outlets |
| BS 1363–3:1995 | Specification for adaptors |
| BS 1363–4:1995 | Specification for 13 A fused connection units switched and unswitched |
| BS 3036:1958 (1992) | Specification. Semi-enclosed electric fuses (ratings up to 100 amperes and 240 volts to earth) |
| BS 3676 | Switches for household and similar fixed electrical installations. Specification for general requirements |
| BS 3858:1992 (2004) | Specification for binding and identification sleeves for use on electric cables and wires |
| BS 4177:1992 | Specification for cooker control units |
| BS 4444:1989 (1995) | Guide to electrical earth monitoring and protective conductor proving |
| BS 4573:1970 (1979) | Specification for 2–pin reversible plugs and shaver socket-outlets |
| BS 4662:2006 | Boxes for flush mounting of electrical accessories. Requirements and test methods and dimensions |
| BS 4727 | Glossary of electrotechnical power, telecommunications, electronics, lighting and colour terms |
| BS 5266 | Emergency lighting |
| BS 5467:1997 | Electric cables. Thermosetting insulated, armoured cables for voltages of 600/1000 V and 1900/3300 V |

**E.1 British Standards** (*continued*)

| BS or EN number | Title |
| --- | --- |
| BS 5499 | Graphical symbols and signs. Safety signs, including fire safety signs |
| BS 5655 | Lifts and service lifts |
| BS 5655–1:1986 | Safety rules for the construction and installation of electric lifts [Applicable only to the modernization of existing lift installations.] |
| 13S 5655–2:1988 | Safety rules for the construction and installation of hydraulic lifts [Applicable only to the modernization of existing lift installations.] |
| BS 5655–11:2005 | Code of practice for the undertaking of modifications to existing electric lifts [Applicable only to the modernization of existing lift installations.] |
| BS 5655–12:2005 | Code of practice for the undertaking of modifications to existing hydraulic lifts [Applicable only to the modernization of existing lift installations.] |
| BS 5733:1995 | Specification for general requirements for electrical accessories |
| BS 5803–5:1985 | Thermal insulation for use in pitched roof spaces in dwellings. Specification for installation of man-made mineral fibre and cellulose fibre insulation |
| BS 5839 | Fire detection and fire alarm systems for buildings |
| BS 5839–1:2002 | Code of practice for system design, installation, commissioning and maintenance |
| BS 6004:2000 (2006) | Electric cables. PVC insulated, non-armoured cables for voltages up to and including 450/750 V, for electric power, lighting and internal wiring |
| BS 6007:2006 | Electric cables. Single core unsheathed heat resisting cables for voltages up to and including 450/750 V, for internal wiring |
| BS 6220:1983 (1999) | Electric cables. Single core PVC insulated flexible cables of rated voltage 600/1000 V for switchgear and controlgear wiring |
| BS 6231:2006 | Electric cables. Single core PVC insulated flexible cables of rated voltage 600/1000 V for switchgear and controlgear wiring |
| BS 6346:1997 (2005) | Electric cables. PVC insulated, armoured cables for voltages of 600/1000 V and 1900/3300 V |
| BS 6351 | Electric surface heating |
| BS 6351–1:1983 (2007) | Specification for electric surface heating devices |
| BS 6351–2:1983 (2007) | Guide to the design of electric surface heating systems |
| BS 6351–3:1983 (2007) | Code of practice for the installation, testing and maintenance of electric surface heating systems |

**E.1 British Standards** (*continued*)

| BS or EN number | Title |
| --- | --- |
| BS 6500:2000 (2005) | Electric cables. Flexible cords rated up to 300/500 V, for use with appliances and equipment intended for domestic, office and similar environments |
| BS 6701:2004 | Telecommunications equipment and telecommunications cabling. Specification for installation, operation and maintenance |
| BS 6724:1997 (2007) | Electric cables. Thermosetting insulated, armoured cables for voltages of 600/1000 V and 1900/3300 V, having low emission of smoke and corrosive gases when affected by fire |
| BS 6907 | Electrical installations for open-cast mines and quarries |
| BS 6972:1988 | Specification for general requirements for luminaire supporting couplers for domestic, light industrial and commercial use |
| BS 6991:1990 | Specification for 6/10 A, two-pole weather-resistant couplers for household, commercial and light industrial equipment |
| BS 7001:1988 | Specification for interchangeability and safety of a standardized luminaire supporting coupler |
| BS 7211:1998 (2005) | Electric cables. Thermosetting insulated, non-armoured cables for voltages up to and including 450/750 V, for electric power, lighting and internal wiring, and having low emission of smoke and corrosive gases when affected by fire. |
| BS 7361:1991 | Cathodic protection. Code of practice for land and marine applications [Current, but partially replaced by BS EN 15112:2006 and BS EN 13636:2004.] |
| BS 7375:1996 | Code of practice for distribution of electricity on construction and building sites |
| BS 7430:1998 | Code of practice for earthing |
| BS 7454:1991 (2003) | Method for calculation of thermally permissible short-circuit currents, taking into account non-adiabatic heating effects |
| BS 7629–1:1997 (2007) | Specification for 300/500 V fire resistant electric cables having low emission of smoke and corrosive gases when affected by fire. Multi-core cables |
| BS 7697:1993 (2004) | Nominal voltages for low voltage public electricity supply systems |
| BS 7698–12:1998 | Reciprocating internal combustion engine driven alternating current generating sets. Emergency power supply to safety devices |
| BS 7769 | Electric cables. Calculation of the current rating |
| BS 7769–1.1:1997 | Has been superseded/withdrawn and replaced by BS IEC 60287–1–1:2006 |
| BS 7769–1.2:1994 (2005) | Current rating equations (100% load factor) and calculation of losses. Sheath eddy current loss factors for two circuits in flat formation |

**E.1 British Standards** (*continued*)

| BS or EN number | Title |
|---|---|
| BS 7769–2.2:1997 (2005) | Thermal resistance. A method for calculating reduction factors for groups of cables in free air, protected from solar radiation |
| BS 7769–2–2.1: 1997 (2006) | Thermal resistance. Calculation of thermal resistance. Section 2.1: Calculation of thermal resistance |
| BS 7769–3.1:1997 (2005) | Sections on operating conditions. Reference operating conditions and selection of cable type |
| BS 7846:2000 (2005) | Electric cables. 600/1000V armoured fire-resistant cables having thermosetting insulation and low emission of smoke and corrosive gases when affected by fire |
| BS 7889:1997 | Electric cables. Thermosetting insulated, unarmoured cables for a voltage of 600/1000V |
| BS 7909 | Code of practice for design and installation of temporary distribution systems delivering a.c. electrical supplies for lighting, technical services and other entertainment related purposes |
| BS 7919:2001 (2006) | Electric cables. Flexible cables rated up to 450/750V, for use with appliances and equipment intended for industrial and similar environments |
| BS 8436:2004 | Electric cables. 300/500V screened electric cables having low emission of smoke and corrosive gases when affected by fire, for use in walls, partitions and building voids. Multi-core cables |
| BS 8450:2006 | Code of practice for installation of electrical and electronic equipment in ships |
| BS 61535:2006 | Installation couplers intended for permanent connection in fixed installations |
| BS AU 149a:1980 (1987) | Specification for electrical connections between towing vehicles and trailers with 6V or 12V electrical equipment: type 12 N (normal) |
| BS AU 177a:1980 (1987) | Specification for electrical connections between towing vehicles and trailers with 6V or 12V electrical equipment: type 12 S (supplementary) |
| BS EN 81 | Safety rules for the construction and installation of lifts |
| BS EN 81–1:1998 | Electric lifts [Also known as: BS 5655–1:1986 Lifts and service lifts ... etc.] |
| BS EN 1648 | Leisure accommodation vehicles |
| BS EN 1648–1:2004 | 12V direct current extra low voltage electrical installations. Caravans |
| BS EN 1648–2:2005 | 12V direct current extra low voltage electrical installations. Motor caravans |
| BS EN 6100–1 | Glossary of building and civil engineering terms |
| BS EN 50085 | Cable trunking and cable ducting systems for electrical installations |

**E.1 British Standards** (*continued*)

| BS or EN number | Title |
| --- | --- |
| BS EN 50085–1: 1999 (2005) | General requirements [BS EN 50085–1:1999 remains current.] |
| BS EN 50085–2–1: 2006 | Cable trunking systems and cable ducting systems intended for mounting on walls and ceilings |
| BS EN 50085–2–3: 2001 | Particular requirements for slotted cable trunking systems intended for installation in cabinets. Section 3: Slotted in cabinets |
| BS EN 50086 | Specification for conduit systems for cable management |
| BS EN 50086–1: 1994 | General requirements [Replaced by BS EN 61386–1:2004, but remains current.] |
| BS EN 50086–2–1: 1996 | Particular requirements. Rigid conduit systems [Replaced by BS EN 61386–21:2004, but remains current.] |
| BS EN 50086–2–2: 1996 | Particular requirements. Pliable conduit systems [Replaced by BS EN 61386–22:2004, but remains current.] |
| BS EN 50086–2–3: 1996 | Particular requirements. Flexible conduit systems [Replaced by BS EN 61386–23:2004, but remains current.] |
| BS EN 50086–2–4: 1994 | Particular requirements. Conduit systems buried underground |
| BS EN 50107 | Signs and luminous-discharge-tube installations operating from a no-load rated output voltage exceeding 1 kV but not exceeding 10 kV |
| BS EN 50107–1: 2002 | General requirements |
| BS EN 50107–2: 2005 | Requirements for earth-leakage and open-circuit protective devices |
| BS EN 50171:2001 | Central power supply systems |
| BS EN 50174 | Information technology – Cabling installation |
| BS EN 50200:2006 | Method of test for resistance to fire of unprotected small cables for use in emergency circuits |
| BS EN 50266 | Common test methods for cables under fire conditions. Test for vertical flame spread of vertically-mounted bunched wires or cables. |
| BS EN 50266–1: 2001 (2006) | Apparatus |
| BS EN 50266–2–1: 2001 (2006) | Procedures. Category A F/R |
| BS EN 50266–2–2: 2001 (2006) | Procedures. Category A |
| BS EN 50266–2–3: 2001 (2006) | Procedures. Category B |
| BS EN 50266–2–4: 2001 (2006) | Procedures. Category C |

**E.1  British Standards** (*continued*)

| BS or EN number | Title |
|---|---|
| BS EN 50266–2–5: 2001 (2006) | Procedures. Small cables. Category D |
| BS EN 50281 | Electrical apparatus for use in the presence of combustible dust |
| BS EN 50281–1–1: 1999 | Electrical apparatus protected by enclosures. Construction and testing [Replaced by BS EN 60241–0:2006 and BS EN 61241–1:2004, but remains current.] |
| BS EN 50281–1–2: 1999 | Electrical apparatus protected by enclosures. Selection, installation and maintenance [Partially replaced by BS EN 61241–14:2004 and BS EN 61241–17: 2005.] |
| BS EN 50281–2–1: 1999 | Test methods. Methods of determining minimum ignition temperatures |
| BS EN 50362:2003 | Method of test for resistance to fire of larger unprotected power and control cables for use in emergency circuits |
| BS EN 50438 | Requirements for the connection of micro-cogenerators in parallel with public low-voltage distribution networks |
| BS EN 60079 | Electrical apparatus for explosive gas atmospheres |
| BS EN 60079–10: 2003 | Classification of hazardous areas |
| BS EN 60079–14: 2003 | Electrical installations in hazardous areas (other than mines) |
| BS EN 60079–17: 2003 | Inspection and maintenance of electrical installations in hazardous areas (other than mines) |
| BS EN 60092–507: 2000 | Electrical installations in ships – Pleasure craft |
| BS EN 60146–2: 2000 | Semiconductor convertors. General requirements and line commutated convertors. Self-commutated semiconductor converters including direct d.c. converters |
| BS EN 60204 | Safety of machinery. Electrical equipment of machines |
| BS EN 60204–1: 2006 | General requirements |
| BS EN 60228:2005 | Conductors of insulated cables |
| BS EN 60238:1999 (2004) | Edison screw lampholders [BS EN 60238:1999 remains current.] |
| BS EN 60255–22–1: 2005 | Electrical relays. Electrical disturbance tests for measuring relays and protection equipment. 1 MHz burst immunity tests |
| BS EN 60269 | Low-voltage fuses |
| BS EN 60269–1: 2007 | General requirements |

**E.1 British Standards** (*continued*)

| BS or EN number | Title |
| --- | --- |
| BS EN 60269–2: 1995 | Supplementary requirements for fuses for use by authorized persons (fuses mainly for industrial application) [Replaced by BS 88–2:2007 and BS EN 60269–1:2007, but remains current.] |
| BS EN 60269–3: 1995 | Supplementary requirements for fuses for use by unskilled persons (fuses mainly for household and similar applications) [Replaced by BS 88–3:2007 and BS EN 60269–1:2007, but remains current.] |
| BS EN 60309: | Plugs, socket-outlets and couplers for industrial purposes |
| BS EN 60309–1: 1999 | General requirements |
| BS EN 60309–2: 1999 | Dimensional interchangeability requirements for pin and contact-tube accessories |
| BS EN 60320–1: 2001 | Appliance couplers for household and similar general purposes. General requirements |
| BS EN 60332–1–2: 2004 | Tests on electric and optical fibre cables under fire conditions. Test for vertical flame propagation for a single insulated wire or cable. Procedure for 1 kW pre-mixed flame |
| BS EN 60335–1: 2002 | Household and similar electrical appliances. Safety. General requirements |
| BS EN 60335–2–29: 2004 | Particular requirements for battery chargers |
| BS EN 60335–2–41: 2003 | Particular requirements for pumps |
| BS EN 60335–2–53: 2003 | Particular requirements for sauna heating appliances |
| BS EN 60335–2–71: 2003 | Particular requirements for electrical heating appliances for breeding and rearing animals |
| BS EN 60335–2–76: 2005 | Particular requirements for electric fence energizers |
| BS EN 60335–2–96: 2002 | Particular requirements for flexible sheet heating elements for room heating |
| BS EN 60439 | Low-voltage switchgear and controlgear assemblies. |
| BS EN 60439–1: 1999 | Type-tested and partially type-tested assemblies |
| BS EN 60439–2: 2000 | Particular requirements for busbar trunking systems (busways) |
| BS EN 60439–3: 1991 | Particular requirements for low-voltage switchgear and controlgear assemblies intended to be installed in places where unskilled persons have access to their use. Distribution boards |
| BS EN 60439–4: 2004 | Particular requirements for assemblies for construction sites (ACS) |

**E.1 British Standards** (*continued*)

| BS or EN number | Title |
| --- | --- |
| BS EN 60445:2000 | Basic and safety principles for man-machine interface, marking and identification. Identification of equipment terminals and of terminations of certain designated conductors, including general rules for an alphanumeric system |
| BS EN 60446:2000 | Basic and safety principles for man-machine interface, marking and identification. Identification of conductors by colours or numerals |
| BS EN 60529:1992 (2004) | Specification for degrees of protection provided by enclosures (IP code) |
| BS EN 60570:2003 | Electrical supply track systems for luminaires [Replaces BS EN 60570:1997 and BS EN 60570–2–1:1995, which remain current.] |
| BS EN 60598 | Luminaires |
| BS EN 60598–1: 2004 | Luminaires. General requirements and tests |
| BS EN 60598–2–18: 1994 | Particular requirements. Luminaires for swimming pools and similar applications |
| BS EN 60598–2–23: 1997 | Particular requirements. Extra-low voltage lighting systems for filament lamps |
| BS EN 60598–2–24: 1999 | Particular requirements. Luminaires with limited surface temperatures |
| BS EN 60664–1: 2003 | Insulation coordination for equipment within low-voltage systems. Principles, requirements and tests |
| BS EN 60669 | Switches for household and similar fixed electrical installations |
| BS EN 60669–1: 2000 | General requirements |
| BS EN 60669–2–1: 2004 | Particular requirements. Electronic switches |
| BS EN 60669–2–2: 2006 | Particular requirements. Electromagnetic remote-control switches (RCS) |
| BS EN 60669–2–3: 2006 | Particular requirements. Time Delay Switches (TDS) |
| BS EN 60669–2–4: 2005 | Particular requirements. Isolating switches |
| BS EN 60670 | Boxes and enclosures for electrical accessories for household and similar fixed electrical installations |
| BS EN 60670–1: 2005 | General requirements |
| BS EN 60670–22: 2006 | Particular requirements for connecting boxes and enclosures |
| BS EN 60684 | Flexible insulating sleeving |
| BS EN 60702–1: 2002 | Mineral insulated cables and their terminations with a rated voltage not exceeding 750 V. Cables |

**E.1 British Standards** (*continued*)

| BS or EN number | Title |
| --- | --- |
| BS EN 60721 | Classification of environmental conditions |
| BS EN 60721–3–3: 1995 (2005) | Classification of groups of environmental parameters and their severities. Stationary use at weather protected locations |
| BS EN 60721–3–4: 1995 (2005) | Classification of groups of environmental parameters and their severities. Stationary use at non-weather protected locations |
| BS EN 60898:1991 | Specification for circuit-breakers for overcurrent protection for household and similar installations [Replaced by BS EN 60898–1:2003, but remains current.] |
| BS EN 60898–1: 2003 | Circuit breakers for a.c. operation |
| BS EN 60898–2: 2001 | Circuit-breakers for a.c. and d.c. operation [BS EN 60898–2:2001 remains current. (It was withdrawn in error and has been reinstated.)] |
| BS EN 60904–3: 1993 | Photovoltaic devices. Measurement principles for terrestrial Photovoltaic (PV) solar devices with reference spectral irradiance data |
| BS EN 60947 | Low-voltage switchgear and control gear |
| BS EN 60947–2: 2006 | Circuit-breakers |
| BS EN 60947–3: 1999 | Switches, disconnectors, switch-disconnectors and fuse-combination units |
| BS EN 60947–4–1: 2001 | Contactors and motor starters — Electromechanical contactor and motor starters |
| BS EN 60947–5–1: 2004 | Control circuit devices and switching elements — Electromechanical control circuit devices |
| BS EN 60947–6–1: 2005 | Multiple function equipment — Transfer switching equipment |
| BS EN 60947–6–2: 2003 | Multiple function equipment — Control and protective switching devices (or equipment) (CPS) |
| BS EN 60947–7 | Specification for low-voltage switchgear and controlgear |
| BS EN 60947–7–1: 2002 | Ancillary equipment — Terminal blocks for copper conductors |
| BS EN 60947–7–2: 2002 | Ancillary equipment — Protective conductor terminal blocks for copper conductors |
| BS EN 60998 | Connecting devices for low-voltage circuits for household and similar purposes |
| BS EN 60998–2–1: 2004 | Particular requirements for connecting devices as separate entities with screw-type clamping units |
| BS EN 60998–2–2: 2004 | Particular requirements for connecting devices as separate entities with screwless-type clamping units |
| BS EN 61000 | Electromagnetic compatibility (EMC) |

**E.1 British Standards** (*continued*)

| BS or EN number | Title |
|---|---|
| BS EN 61008–1: 1995 (2004) | Residual current operated circuit-breakers without integral overcurrent protection for household and similar uses (RCCBs). General rules<br>[BS EN 61008–1:1995 remains current.] |
| BS EN 61009–1: 1995 (2004) | Electrical accessories. Residual current operated circuit-breakers with integral overcurrent protection for household and similar uses (RCBOs)<br>General rules.<br>BS EN 61009–1:1995 remains current |
| BS EN 61034–2: 2005 | Measurement of smoke density of cables burning under defined conditions. Test procedure and requirements |
| BS EN 61095:1993 | Specification for electromechanical contactors for household and similar purposes |
| BS EN 61140:2002 | Protection against electric shock. Common aspects for installation and equipment |
| BS EN 61184:1997 | Bayonet lampholders |
| BS EN 61215:2005 | Crystalline silicon terrestrial photovoltaic (PV) modules. Design qualification and type approval |
| BS EN 61241 | Electrical apparatus for use in the presence of combustible dust |
| BS EN 61241–17: 2005 | Inspection and maintenance of electrical installations in hazardous areas (other than mines) |
| BS EN 61347 | Lamp controlgear |
| BS EN 61347–1: 2001 | General and safety requirements |
| BS EN 61347–2–2: 2001 | Particular requirements for d.c. or a.c. supplied electronic step-down convertors for filament lamps |
| BS EN 61386 | Conduit systems for cable management |
| BS EN 61386–1: 2004 | General requirements |
| BS EN 61386–21: 2004 | Particular requirements. Rigid conduit systems |
| BS EN 61386–22: 2004 | Particular requirements. Pliable conduit systems |
| BS EN 61386–23: 2004 | Particular requirements. Flexible conduit systems |
| BS EN 61534 | Powertrack systems |
| BS EN 61534–1: 2003 | General requirements |
| BS EN 61534–21: 2006 | Particular requirements for powertrack systems intended for wall and ceiling mounting |

**E.1 British Standards** (*continued*)

| BS or EN number | Title |
| --- | --- |
| BS EN 61537: 2002 (2007) | Cable tray systems and cable ladder systems for cable management [BS EN 61537:2002 remains current.] |
| BS EN 61557 | Electrical safety in low voltage distribution systems up to 1000 V a.c. and 1500 V d.c. Equipment for testing, measuring or monitoring of protective measures. General requirements |
| BS EN 61557–2: 2007 | Insulation resistance |
| BS EN 61557–6: 1998 | Residual current devices (RCD) in TT, TN and IT systems |
| BS EN 61557–8: 1997 | Insulation monitoring devices for IT systems |
| BS EN 61557–9: 2000 | Equipment for insulation fault location in IT systems |
| BS EN 61558–1: 1998 (2005) | Safety of power transformers, power supply units and similar. General requirements and tests [BS EN 61558–1:1998 remains current.] |
| BS EN 61558–2–4: 1998 | Particular requirements for isolating transformers for general use |
| BS EN 61558–2–5: 1998 | Particular requirements for shaver transformers and shaver supply units |
| BS EN 61558–2–6: 1998 | Particular requirements for safety isolating transformers for general use |
| BS EN 61558–2–23: 2001 | Particular requirements for transformers for construction sites |
| BS EN 62020:1999 | Electrical accessories. Residual current monitors for household and similar uses (RCMs) |
| BS EN 62040 | Uninterruptible power systems (UPS) |
| BS EN 62208:2003 | Empty enclosures for low-voltage switchgear and controlgear assemblies. General requirements |
| BS EN 62262:2002 | Degrees of protection provided by enclosures for electrical equipment against external mechanical impacts (IK code) |
| BS EN 62305 | Protection against lightning |
| BS EN 62305–1: 2006 | General requirements |
| BS EN 62305–2: 2006 | Risk management |
| BS EN 62305–3: 2006 | Physical damage to structures and life hazard |
| BS EN 62305–4: 2006 | Electrical and electronic systems within structures |
| BS EN ISO 11446:2004 | Road vehicles. Connectors for the electrical connection of towing and towed vehicles. 13–pole connectors for vehicles with 12 V nominal supply voltage |

# E.2 British Standards currently used with the Wiring Regulations – listing by subject

| Title | BS or EN number |
| --- | --- |
| 13A fused plugs (Specification for rewirable and non-rewirable) | BS 1363–1:1995 |
| 13A plugs, socket-outlets, connection units and adaptors | BS 1363 |
| 13A switched and unswitched socket-outlets | BS 1363–2:1995 |
| 13A fused connection units switched and unswitched | BS 1363–4:1995 |
| 2–pin reversible plugs and shaver socket-outlets | BS 4573:1970 (1979) |
| Adaptors | BS 1363–3:1995 |
| Ancillary equipment — Protective conductor terminal blocks for copper conductors | BS EN 60947–7–2:2002 |
| Ancillary equipment — Terminal blocks for copper conductors | BS EN 60947–7–1:2002 |
| Appliance couplers for household and similar general purposes. General requirements | BS EN 60320–1:2001 |
| Assemblies for construction sites (ACS) | BS EN 60439–4:2004 |
| Basic and safety principles for man-machine interface, marking and identification. Identification of equipment terminals and of terminations of certain designated conductors, including general rules for an alphanumeric system | BS EN 60445:2000 |
| Basic and safety principles for man-machine interface, marking and identification. Identification of conductors by colours or numerals | BS EN 60446:2000 |
| Battery chargers | BS EN 60335–2–29:2004 |
| Bayonet lampholders | BS EN 61184:1997 |
| Binding and identification sleeves for use on electric cables and wires | BS 3858:1992 (2004) |
| Boxes and enclosures for electrical accessories for household and similar fixed electrical installations | BS EN 60670 |
| Boxes for flush mounting of electrical accessories. Requirements and test methods and dimensions | BS 4662:2006 |
| Busbar trunking systems (busways) | BS EN 60439–2:2000 |
| Cable tray systems and cable ladder systems for cable management | BS EN 61537:2002 (2007) |
| Cable trunking and cable ducting systems for electrical installations | BS EN 50085 |
| Cable trunking systems and cable ducting systems intended for mounting on walls and ceilings | BS EN 50085–2–1:2006 |
| Caravans – 12V direct current extra low voltage electrical installations | BS EN 1648–1:2004 |
| Cathodic protection. Code of practice for land and marine applications [Current, but partially replaced by BS EN 15112:2006 and BS EN 13636:2004.] | BS 7361:1991 |

**E.2 British Standards** (*continued*)

| Title | BS or EN number |
|---|---|
| Ceiling roses | BS 67:1987 (1999) |
| Central power supply systems | BS EN 50171:2001 |
| Circuit breakers for a.c. operation | BS EN 60898–1:2003 |
| Circuit-breakers | BS EN 60947–2:2006 |
| Circuit-breakers for a.c. and d.c. operation [BS EN 60898–2:2001 remains current. (It was withdrawn in error and has been reinstated.)] | BS EN 60898–2:2001 |
| Circuit-breakers for overcurrent protection for household and similar installations [Replaced by BS EN 60898–1:2003, but remains current.] | BS EN 60898:1991 |
| Classification of environmental conditions | BS EN 60721 |
| Classification of groups of environmental parameters and their severities. Stationary use at weather protected locations | BS EN 60721–3–3:1995 (2005) |
| Classification of groups of environmental parameters and their severities. Stationary use at non-weather protected locations | BS EN 60721–3–4:1995 (2005) |
| Classification of hazardous areas | BS EN 60079–10:2003 |
| Conductors of insulated cables | BS EN 60228:2005 |
| Conduit systems buried underground | BS EN 50086–2–4:1994 |
| Conduit systems for cable management | BS EN 61386 |
| Conduit systems for cable management | BS EN 50086 |
| Connecting boxes and enclosures | BS EN 60670–22:2006 |
| Connecting devices as separate entities with screwless-type clamping units | BS EN 60998–2–2:2004 |
| Connecting devices as separate entities with screw-type clamping units | BS EN 60998–2–1:2004 |
| Connecting devices for low-voltage circuits for household and similar purposes | BS EN 60998 |
| Contactors and motor starters — Electromechanical contactor and motor starters | BS EN 60947–4–1:2001 |
| Control and protective switching devices (or equipment) (CPS) | BS EN 60947–6–2:2003 |
| Control circuit devices and switching elements — Electromechanical control circuit devices | BS EN 60947–5–1:2004 |
| Cooker control units | BS 4177:1992 |
| Crystalline silicon terrestrial photovoltaic (PV) modules. Design qualification and type approval | BS EN 61215:2005 |
| Current rating equations (100% load factor) and calculation of losses. Sheath eddy current loss factors for two circuits in flat formation | BS 7769–1.2:1994 (2005) |
| Degrees of protection provided by enclosures for electrical equipment against external mechanical impacts (IK code) | BS EN 62262:2002 |

**E.2 British Standards** (*continued*)

| Title | BS or EN number |
| --- | --- |
| Design and installation of temporary distribution systems delivering a.c. electrical supplies for lighting, technical services and other entertainment related purposes | BS 7909 |
| Design, construction and installation of signs | BS 559:1998 (2005) |
| Dimensional interchangeability requirements for pin and contact-tube accessories | BS EN 60309–2:1999 |
| Distribution of electricity on construction and building sites | BS 7375:1996 |
| Earthing | BS 7430:1998 |
| Earthing-pin plugs, socket-outlets and socket-outlet adaptors | BS 546:1950 (1988) |
| Edison screw lampholders | BS EN 60238:1999 (2004) |
| Electric cables. 300/500 V screened electric cables having low emission of smoke and corrosive gases when affected by fire, for use in walls, partitions and building voids. Multi-core cables | BS 8436:2004 |
| Electric cables. 600/1000 V armoured fire-resistant cables having thermosetting insulation and low emission of smoke and corrosive gases when affected by fire | BS 7846:2000 (2005) |
| Electric cables. Calculation of the current rating | BS 7769 |
| Electric cables. Flexible cables rated up to 450/750 V, for use with appliances and equipment intended for industrial and similar environments | BS 7919:2001 (2006) |
| Electric cables. Flexible cords rated up to 300/500 V, for use with appliances and equipment intended for domestic, office and similar environments | BS 6500:2000 (2005) |
| Electric cables. PVC insulated, armoured cables for voltages of 600/1000 V and 1900/3300 V | BS 6346:1997 (2005) |
| Electric cables. PVC insulated, non-armoured cables for voltages up to and including 450/750 V, for electric power, lighting and internal wiring | BS 6004:2000 (2006) |
| Electric cables. Single core PVC insulated flexible cables of rated voltage 600/1000 V for switchgear and controlgear wiring | BS 6220:1983 (1999) |
| Electric cables. Single core PVC insulated flexible cables of rated voltage 600/1000 V for switchgear and controlgear wiring | BS 6231:2006 |
| Electric cables. Single core unsheathed heat resisting cables for voltages up to and including 450/750 V, for internal wiring | BS 6007:2006 |
| Electric cables. Thermosetting insulated, armoured cables for voltages of 600/1000 V and 1900/3300 V | BS 5467:1997 |
| Electric cables. Thermosetting insulated, armoured cables for voltages of 600/1000 V and 1900/3300 V, having low emission of smoke and corrosive gases when affected by fire | BS 6724:1997 (2007) |

**E.2  British Standards** (*continued*)

| Title | BS or EN number |
|---|---|
| Electric cables. Thermosetting insulated, non-armoured cables for voltages up to and including 450/750 V, for electric power, lighting and internal wiring, and having low emission of smoke and corrosive gases when affected by fire | BS 7211:1998 (2005) |
| Electric cables. Thermosetting insulated, unarmoured cables for a voltage of 600/1000 V | BS 7889:1997 |
| Electric fence energizers | BS EN 60335–2–76:2005 |
| Electric lifts<br>[Also known as: BS 5655–1:1986 Lifts and service lifts, etc.] | BS EN 81–1:1998 |
| Electric surface heating devices | BS 6351–1:1983 (2007) |
| Electric surface heating systems | BS 6351–2:1983 (2007) |
| Electric surface heating | BS 6351 |
| Electrical accessories | BS 5733:1995 |
| Electrical accessories. Residual current monitors for household and similar uses (RCMs) | BS EN 62020:1999 |
| Electrical accessories. Residual current operated circuit-breakers with integral overcurrent protection for household and similar uses (RCBOs). General rules<br>[BS EN 61009–1:1995 remains current.] | BS EN 61009–1:1995 (2004) |
| Electrical and electronic systems within structures | BS EN 62305–4:2006 |
| Electrical apparatus for explosive gas atmospheres | BS EN 60079 |
| Electrical apparatus for use in the presence of combustible dust | BS EN 50281 |
| Electrical apparatus for use in the presence of combustible dust | BS EN 61241 |
| Electrical apparatus protected by enclosures. Construction and testing<br>[Replaced by BS EN 60241–0:2006 and BS EN 61241–1:2004, but remains current.] | BS EN 50281–1–1:1999 |
| Electrical apparatus protected by enclosures. Selection, installation and maintenance<br>[Partially replaced by BS EN 61241–14:2004 and BS EN 61241–17: 2005.] | BS EN 50281–1–2:1999 |
| Electrical connections between towing vehicles and trailers with 6 V or 12 V electrical equipment: type 12 N (normal) | BS AU 149a:1980 (1987) |
| Electrical connections between towing vehicles and trailers with 6 V or 12 V electrical equipment: type 12 S (supplementary) | BS AU 177a:1980 (1987) |
| Electrical earth monitoring and protective conductor proving | BS 4444:1989 (1995) |
| Electrical earthing. Clamps for earthing and bonding. Specification | BS 951:1999 |
| Electrical heating appliances for breeding and rearing animals | BS EN 60335–2–71:2003 |

**E.2 British Standards** (*continued*)

| Title | BS or EN number |
|---|---|
| Electrical installations for open-cast mines and quarries | BS 6907 |
| Electrical installations in hazardous areas (other than mines) | BS EN 60079–14:2003 |
| Electrical installations in ships – Pleasure craft | BS EN 60092–507:2000 |
| Electrical relays. Electrical disturbance tests for measuring relays and protection equipment. 1 MHz burst immunity tests | BS EN 60255–22–1:2005 |
| Electrical safety in low voltage distribution systems up to 1000 V a.c. and 1500 V d.c. Equipment for testing, measuring or monitoring of protective measures. General requirements | BS EN 61557 |
| Electrical supply track systems for luminaires [Replaces BS EN 60570:1997 and BS EN 60570–2–1: 1995, which remain current.] | BS EN 60570:2003 |
| Electromagnetic compatibility (EMC) | BS EN 61000 |
| Electromagnetic remote-control switches (RCS) | BS EN 60669–2–2:2006 |
| Electromechanical contactors for household and similar purposes | BS EN 61095:1993 |
| Electronic switches | BS EN 60669–2–1:2004 |
| Emergency lighting | BS 5266 |
| Empty enclosures for low-voltage switchgear and controlgear assemblies. General requirements | BS EN 62208:2003 |
| Enclosures (IP code) | BS EN 60529:1992 (2004) |
| Equipment for insulation fault location in IT systems | BS EN 61557–9:2000 |
| Extra-low voltage lighting systems for filament lamps | BS EN 60598–2–23:1997 |
| Filament lamps (d.c. or a.c. supplied electronic step-down convertors) | BS EN 61347–2–2:2001 |
| Fire detection and fire alarm systems for buildings | BS 5839 |
| Fire resistant electric cables having low emission of smoke and corrosive gases when affected by fire. Multi-core cables | BS 7629–1:1997 (2007) |
| Fire tests on building materials and structures | BS 476 |
| Flexible conduit systems [Replaced by BS EN 61386–23:2004, but remains current.] | BS EN 50086–2–3:1996 |
| Flexible conduit systems | BS EN 61386–23:2004 |
| Flexible insulating sleeving | BS EN 60684 |
| Fuse links for domestic, and similar purposes (primarily for use in plugs) | BS 1362:1973 (1992) |
| Fuse-links (rated up to 5 amperes) for a.c. and d.c. service | BS 646:1958 (1991) |
| Fuses – semi-enclosed electric fuses (ratings up to 100 amperes and 240 volts to earth) | BS 3036:1958 (1992) |
| Fuses for a.c. circuits in domestic and similar premises | BS 1361:1971 (1986) |
| Fuses for use by authorized persons (fuses mainly for industrial application) [Replaced by BS 88–2:2007 and BS EN 60269–1:2007, but remains current.] | BS EN 60269–2:1995 |

**E.2  British Standards** (*continued*)

| Title | BS or EN number |
|---|---|
| Fuses for use by authorized persons (mainly for industrial application). Additional requirements for fuses with fuse-links for bolted connections | BS 88–2.2:1988 |
| Fuses for use by unskilled persons (fuses mainly for household and similar applications) [Replaced by BS 88–3:2007 and BS EN 60269–1:2007, but remains current.] | BS EN 60269–3: 1995 |
| Fuses for voltages up to and including 1000 V a.c. and 1500 V d.c. | BS 88 |
| General and safety requirements | BS EN 61347–1:2001 |
| Glossary of building and civil engineering terms | BS EN 6100–1 |
| Glossary of Electrotechnical power, telecommunications, electronics, lighting and colour terms | BS 4727 |
| Graphical symbols and signs. Safety signs, including fire safety signs | BS 5499 |
| Household and similar electrical appliances. Safety. General requirements | BS EN 60335–1:2002 |
| Information technology – Cabling installation | BS EN 50174 |
| Inspection and maintenance of electrical installations in hazardous areas (other than mines) | BS EN 60079–17:2003 |
| Inspection and maintenance of electrical installations in hazardous areas (other than mines) | BS EN 61241–17:2005 |
| Installation couplers intended for permanent connection in fixed installations | BS 61535:2006 |
| Installation of electrical and electronic equipment in ships | BS 8450:2006 |
| Installation, testing and maintenance of electric surface heating systems | BS 6351–3:1983 (2007) |
| Insulation coordination for equipment within low-voltage systems. Principles, requirements and tests | BS EN 60664–1:2003 |
| Insulation monitoring devices for IT systems | BS EN 61557–8:1997 |
| Insulation resistance | BS EN 61557–2:2007 |
| Isolating switches | BS EN 60669–2–4:2005 |
| Isolating transformers for general use | BS EN 61558–2–4:1998 |
| Lamp controlgear | BS EN 61347 |
| Leisure accommodation vehicles | BS EN 1648 |
| Lifts and service lifts | BS 5655 |
| Low voltage public electricity supply systems | BS 7697:1993 (2004) |
| Low-voltage fuses | BS EN 60269 |
| Low-voltage fuses. Supplementary requirements for fuses for use by authorized persons (fuses mainly for industrial application). Examples of standardized systems of fuses A to I | BS 88–2:2007 |
| Low-voltage switchgear and controlgear | BS EN 60947 |
| Low-voltage switchgear and controlgear | BS EN 60947–7 |

## E.2 British Standards (continued)

| Title | BS or EN number |
|---|---|
| Low-voltage switchgear and controlgear assemblies intended to be installed in places where unskilled persons have access to their use. Distribution boards | BS EN 60439–3:1991 |
| Low-voltage switchgear and controlgear assemblies | BS EN 60439 |
| Luminaire supporting coupler (Specification for interchangeability and safety of a standardized luminaire supporting coupler) | BS 7001:1988 |
| Luminaire supporting couplers for domestic, light industrial and commercial use | BS 6972:1988 |
| Luminaires | BS EN 60598 |
| Luminaires for swimming pools and similar applications | BS EN 60598–2–18:1994 |
| Luminaires with limited surface temperatures | BS EN 60598–2–24:1999 |
| Luminaires. General requirements and tests | BS EN 60598–1:2004 |
| Measurement of smoke density of cables burning under defined conditions. Test procedure and requirements | BS EN 61034–2:2005 |
| Method for calculation of thermally permissible short-circuit currents, taking into account non-adiabatic heating effects | BS 7454:1991 (2003) |
| Method of test for ignitability of products by direct flame impingement | BS 476–12:1991 |
| Method of test for resistance to fire of larger unprotected power and control cables for use in emergency circuits | BS EN 50362:2003 |
| Method of test for resistance to fire of unprotected small cables for use in emergency circuits | BS EN 50200:2006 |
| Mineral insulated cables and their terminations with a rated voltage not exceeding 750 V. Cables | BS EN 60702–1:2002 |
| Modifications to existing electric lifts [Applicable only to the modernization of existing lift installations.] | BS 5655–11:2005 |
| Modifications to existing hydraulic lifts [Applicable only to the modernization of existing lift installations.] | BS 5655–12:2005 |
| Motor caravans – 12 V direct current extra low voltage electrical installations | BS EN 1648–2:2005 |
| Non-combustible test for materials | BS 476–4:1970 |
| Operating conditions and selection of cable type | BS 7769–3.1:1997 (2005) |
| Photovoltaic devices. Measurement principles for terrestrial photovoltaic (PV) solar devices with reference spectral irradiance data | BS EN 60904–3:1993 |
| Physical damage to structures and life hazard | BS EN 62305–3:2006 |
| Pliable conduit systems | BS EN 61386–22:2004 |
| Pliable conduit systems. pliable conduit systems [Replaced by BS EN 61386–22:2004, but remains current.] | BS EN 50086–2–2:1996 |
| Plugs, socket-outlets and couplers for industrial purposes | BS EN 60309: |
| Powertrack systems | BS EN 61534 |

**E.2 British Standards** (*continued*)

| Title | BS or EN number |
| --- | --- |
| Powertrack systems intended for wall and ceiling mounting | BS EN 61534–21:2006 |
| Procedures. Category A | BS EN 50266–2–2:2001 (2006) |
| Procedures. Category A F/R | BS EN 50266–2–1:2001 (2006) |
| Procedures. Category B | BS EN 50266–2–3:2001 (2006) |
| Procedures. Category C | BS EN 50266–2–4:2001 (2006) |
| Procedures. Small cables. Category D | BS EN 50266–2–5:2001 (2006) |
| Protected-type non-reversible plugs, socket-outlets cable-couplers and appliance-couplers with earthing contacts for single phase a.c. circuits up to 250 volts | BS 196:1961 |
| Protection against electric shock. Common aspects for installation and equipment | BS EN 61140:2002 |
| Protection against lightning | BS EN 62305 |
| Pumps | BS EN 60335–2–41:2003 |
| Reciprocating internal combustion engine driven alternating current generating sets. Emergency power supply to safety devices | BS 7698–12:1998 |
| Requirements for earth-leakage and open-circuit protective devices | BS EN 50107–2:2005 |
| Requirements for the connection of micro-cogenerators in parallel with public low-voltage distribution networks [This document currently at DPC stage. (Expired 2004/11/30.)] | BS EN 50438 |
| Residual current devices (RCD) in TT, TN and IT systems | BS EN 61557–6:1998 |
| Residual current operated circuit-breakers without integral overcurrent protection for household and similar uses (RCCBs). General rules [BS EN 61008–1:1995 remains current.] | BS EN 61008–1:1995 (2004) |
| Residual current operated circuit-breakers with integral overcurrent protection | BS EN 61009–1:1995 (2004) |
| Rigid conduit systems [Replaced by BS EN 61386–21:2004, but remains current.] | BS EN 50086–2–1:1996 |
| Rigid conduit systems | BS EN 61386–21:2004 |
| Risk management | BS EN 62305–2:2006 |
| Road vehicles. Connectors for the electrical connection of towing and towed vehicles. 13–pole connectors for vehicles with 12V nominal supply voltage | BS EN ISO 11446:2004 |
| Rom heating (particular requirements for flexible sheet heating elements) | BS EN 60335–2–96:2002 |
| Safety isolating transformers for general use | BS EN 61558–2–6:1998 |

**E.2 British Standards** (*continued*)

| Title | BS or EN number |
|---|---|
| Safety of machinery. Electrical equipment of machines | BS EN 60204 |
| Safety of power transformers, power supply units and similar General requirements and tests [BS EN 61558–1:1998 remains current.] | BS EN 61558–1:1998 (2005) |
| Safety rules for the construction and installation of electric lifts [Applicable only to the modernization of existing lift installations.] | BS 5655–1:1986 |
| Safety rules for the construction and installation of hydraulic lifts [Applicable only to the modernization of existing lift installations.] | 13S 5655–2:1988 |
| Safety rules for the construction and installation of lifts | BS EN 81 |
| Sauna heating appliances | BS EN 60335–2–53:2003 |
| Semiconductor convertors. General requirements and line commutated convertors. Self-commutated semiconductor converters including direct d.c. converters | BS EN 60146–2:2000 |
| Shaver transformers and shaver supply units | BS EN 61558–2–5:1998 |
| Signs and luminous-discharge-tube installations operating from a no-load rated output voltage exceeding 1 kV but not exceeding 10 kV | BS EN 50107 |
| Slotted cable trunking systems intended for installation in cabinets. Section 3: Slotted in cabinets | BS EN 50085–2–3:2001 |
| Specification of supplementary requirements for fuses of compact dimensions for use in 240/415 V a.c. industrial and commercial electrical installations | BS 88–6:1988 |
| Switches for household and similar fixed electrical installations. Specification for general requirements | BS 3676 |
| Switches for household and similar fixed electrical installations | BS EN 60669 |
| Switches, disconnectors, switch-disconnectors and fuse-combination units | BS EN 60947–3:1999 |
| System design, installation, commissioning and maintenance | BS 5839–1:2002 |
| Telecommunications equipment and telecommunications cabling. Specification for installation, operation and maintenance | BS 6701:2004 |
| Test methods for cables under fire conditions. Test for vertical flame spread of vertically-mounted bunched wires or cables | BS EN 50266 |
| Test methods. Methods of determining minimum ignition temperatures | BS EN 50281–2–1:1999 |
| Tests on electric and optical fibre cables under fire conditions. Test for vertical flame propagation for a single insulated wire or cable. Procedure for 1 kW pre-mixed flame | BS EN 60332–1–2:2004 |

**E.2 British Standards** (*continued*)

| Title | BS or EN number |
|---|---|
| Thermal insulation for use in pitched roof spaces in dwellings. Specification for installation of man-made mineral fibre and cellulose fibre insulation | BS 5803–5:1985 |
| Thermal resistance. A method for calculating reduction factors for groups of cables in free air, protected from solar radiation | BS 7769–2.2:1997 (2005) |
| Thermal resistance. Calculation of thermal resistance. Section 2.1: Calculation of thermal resistance | BS 7769–2–2.1:1997 (2006) |
| Time delay switches (TDS) | BS EN 60669–2–3:2006 |
| Transfer switching equipment | BS EN 60947–6–1:2005 |
| Transformers for construction sites | BS EN 61558–2–23:2001 |
| Type-tested and partially type-tested assemblies | BS EN 60439–1:1999 |
| Uninterruptible Power Systems (UPS) | BS EN 62040 |
| Weather-resistant couplers for household, commercial and light industrial equipment | BS 6991:1990 |

# E.3 Other standards to which reference is made in the Regulations

## E.3.1 IEC and ISO

| | |
|---|---|
| IEC 60038–am 2 Ed 6 | IEC standard voltages |
| IEC 60364 | Low-voltage electrical installations |
| IEC 60364–5–51 | Electrical installations of buildings – Part 5–51: Selection and erection of electrical equipment – Common rules |
| IEC 60449–am 1 Ed 1 | Voltage bands for electrical installations of buildings |
| IEC 60502–1 Ed 2 | Power cables with extruded insulation and their accessories for rated voltages from 1 kV (Urn = 1,2 kV) up to 30 kV (Urn = 36 kV) – Part 1: Cables for rated voltages of 1 kV (Urn = 1,2 kV) and 3 kV (Urn = 3,6 kV) |
| IEC 60755–am 2 | General requirements for residual current operated protective devices |
| IEC 60884 Ed 3.1 | Plugs and socket-outlets for household and similar purposes – Part 1: General requirements |
| IEC 60906 | IEC system of plugs and socket-outlets for household and similar purposes |
| IEC 61201:1992 | Extra-low voltage (ELV). Limit values [Also known as PD 6536.] |

**E.3 Other Standards** (*continued*)

| | |
|---|---|
| IEC 61386 | Conduit systems for cable management [BS EN 61386 series.] |
| IEC 61386–24 Ed 1 | Particular requirements – Conduit systems buried underground |
| IEC 61662 TR2 Ed 1 | Assessment of the risk of damage due to lightning |
| IEC 61936–1 Ed 1 | Power installations exceeding 1 kV a.c. – Part 1: Common rules |
| IEC 61995–1 Ed 1 | Devices for the connection of luminaires for household and similar purposes – Part 1: General requirements |
| IEC/TS 62081 Ed 1 | Arc welding equipment. Installation and use |
| ISO 8820 | Road vehicles. Fuse-links |

## E.3.2  CENELEC Harmonized Documents

BS 7671 Requirements for Electrical Installations takes account of the technical substance of agreements reached in CENELEC. In particular, the technical intent of the following CENELEC Harmonization Documents is included.

### E.3.2.1  CENELEC Harmonized Documents – listed by subject

| | |
|---|---|
| Agricultural and horticultural premises | HD 60364–7–705:2007 |
| Application of measures for protection against overcurrent | HD 384.4.473 A1:1980 |
| Caravan parks, camping parks and similar locations | HD 384.7.708:2005 |
| Caravan parks, camping parks and similar locations | HD 60364–7–708:2009 |
| Conducting locations with restricted movement | HD 60364–7–706:2007 |
| Construction and demolition site installations | HD 60364–7–704:2007 |
| Devices for protection against overvoltage | HD 60364–5–534:2008 |
| Earthing arrangements | HD 60364–5–54:2007 |
| Earthing arrangements, protective conductors and protective bonding conductors | prHD 60364–5–54:2004 |
| Electrical installations in caravans and motor caravans | HD 60364–7–721:2009 |
| Electrical installations in caravans and motor caravans | prHD 60364–7–721:2007 |
| Exhibitions, shows and stands | HD 384.7.711:2003 |
| External influences | HD 60364–5–51:2006 |
| Extra-low-voltage lighting installations | HD 60364–7–715:2005 |
| Fundamental principles, assessment of general characteristics and definitions | HD 60364–1:2008 |
| Identification of cores in cables and flexible cords | HD 308 S2:2001 |
| Initial verification | HD 384.6.61 S2:2003 |

**listed by subject** (*continued*)

## E.3.2.2  CENELEC Harmonized Documents – listed by Directive

| | |
|---|---|
| FprHD 60364–4–444:200X | Measures against electromagnetic disturbances |
| FprHD 60364–7–702:2009 | Swimming pools and other basins |
| FprHD 60364–7–710:2010 | Medical locations |
| FprHD 60364–7–717:2009 | Mobile or transportable units |
| HD 308 S2:2001 | Identification of cores in cables and flexible cords |
| HD 384.4.41 S2/A1:2002 | Protection against electric shock |
| HD 384.4.42 S1 A2:1994 | Protection against thermal effects |
| HD 384.4.43 S2:2001 | Protection against overcurrent |
| HD 384.4.443 S1:2000 | Protection against overvoltages |
| HD 384.4.473 A1:1980 | Application of measures for protection against overcurrent |
| HD 384.4.482 51:1997 | Protection against fire where particular risks or danger exist |
| HD 384.5.551:1997 | Low-voltage generating sets |
| HD 384.6.61 S2:2003 | Initial verification |
| HD 384.7.702 S2:2002 | Swimming pools and other basins |
| HD 384.7.703:2005 | Rooms and cabins containing sauna heaters |
| HD 384.7.708:2005 | Caravan parks, camping parks and similar locations |
| HD 384.7.711:2003 | Exhibitions, shows and stands |
| HD 384.7.714 51:2000 | Outdoor lighting installations |
| HD 60364–1:2008 | Fundamental principles, assessment of general characteristics and definitions |
| HD 60364–4–41:2007 | Protection against electric shock |
| HD 60364–4–42:2001 | Protection against fire |
| HD 60364–4–42;2001 | Protection against thermal effects |
| HD 60364–4–43:2008 | Protection against overcurrent |
| HD 60364–4–43:2008 | Measures against overcurrent |
| HD 60364–4–442:1997 | Protection of low voltage installations against temporary overvoltages |
| HD 60364–4–443:2006 | Protection against overvoltages |
| HD 60364–5–51:2006 | Selection and erection – Common rules |
| HD 60364–5–51:2006 | External influences |
| HD 60364–5–534:2008 | Devices for protection against overvoltage |
| HD 60364–5–54:2007 | Earthing arrangements |
| HD 60364–5–559:2005 | Outdoor lighting installations |
| HD 60364–6:2007 | Initial verification |
| HD 60364–7–701:2007 | Locations containing a bath or shower |
| HD 60364–7–703:2005 | Sauna heaters |
| HD 60364–7–704:2007 | Construction and demolition site installations |
| HD 60364–7–705:2007 | Agricultural and horticultural premises |
| HD 60364–7–706:2007 | Conducting locations with restricted movement |
| HD 60364–7–708:2009 | Caravan parks, camping parks and similar locations |
| HD 60364–7–709:2009 | Marinas and similar locations |

**listed by Directive** (*continued*)

| | |
|---|---|
| HD 60364–7–712:2005 | Solar photovoltaic (PV) power supply systems |
| HD 60364–7–715:2005 | Extra-low-voltage lighting installations |
| HD 60364–7–717:2004 | Mobile or transportable units |
| HD 60364–7–721:2009 | Electrical installations in caravans and motor caravans |
| HD 60364–7–729:2009 | Operating and maintenance gangways |
| HD 60364–7–740:2006 | Temporary electrical installations for structures, amusement devices |
| IEC 60364–4–44:2008 | Introduction to voltage and electro disturbances |
| prHD 60364–5–51:2003 | Selection and erection of equipment – Common rules |
| prHD 60364–5–54:2004 | Earthing arrangements, protective conductors and protective bonding conductors |
| prHD 60364–7–709:2007 | Marinas and similar locations |
| prHD 60364–7–721:2007 | Electrical installations in caravans and motor caravans |
| prHD 60364–7–740:2006 | Temporary electrical installations for structures, amusement devices and booths at fairgrounds, amusement parks and circuses |

Where the above Harmonized Documents contain UK special national conditions, those conditions have been incorporated within BS 7671. If BS 7671 is applied in other countries, the above Harmonized Documents should be consulted to confirm the status of a particular regulation.

 BS 7671 will continue to be amended from time to time to take account of the publication of new or amended CENELEC standards.

# Annex F

# Books by the same author

| Title | Extracts from book reviews | Publisher and ISBN |
|---|---|---|
| *Building Regulations in Brief* (7th edition)  | The most popular and trusted guide to the Building Regulations, *Building Regulations in Brief* is updated regularly to reflect constant changes. Now in its 7th edition, it has sold over 28,000 copies since its first publication in 2003. This new edition includes the latest on all the significant amendments to the Building Regulations, Planning Permission and the Approved Documents that occurred in October 2010, and includes changes to Parts F and L, as well as Approved Documents A, C and J. It also contains changes reflecting the consolidation of the Building Regulations. | Routledge ISBN 978–0415809696 |
| *Water Regulations in Brief*  | *Water Regulations in Brief* is a unique reference book, providing all the information needed to comply with the Regulations, in an easy to use, full-colour format. Crucially (unlike other titles on this subject) this book doesn't just cover the Water Regulations – it also clearly shows how they link in with the Building Regulations and the Wiring Regulations. | Routledge ISBN 978–1856176286 |
| *ISO 9001:2008 for Small Businesses* (4th edition) | The new edition of this top-selling quality management handbook contains a full description of the ISO 9001:2008 standard plus detailed information on quality control and quality assurance. Fully updated following 11 years of practical field experience of the standard, it includes a sample quality manual (that can be customised to suit individual requirements) and on-line assistance on self-certification etc. | Routledge ISBN 978–1856178617 |

(continued)

| Title | Extracts from book reviews | Publisher and ISBN |
|---|---|---|
| *ISO 9001:2008 Quality Manual & Audit Checksheets* (2nd edition)  | A CD containing a soft copy of the generic Quality Management System featured in *ISO 9001 for Small Businesses* (4th edition) plus a soft copy of all the check sheets and example audit forms contained in *ISO 9001 Audit Procedure* (2nd edition). A comprehensive CD containing all the vital documentation, information and guidance needed to develop a full quality management system. | Herne European Consultancy Ltd ISBN 978–0954864798 |
| *Quality Management System for ISO 9001:2000* (2nd edition)  | This book, together with the accompanying CD, provides probably the most comprehensive set of ISO 9001:2000 compliant documents available worldwide. Fully customisable, they can be used as basic templates for any organization wishing to work in compliance with, or gain registration to, ISO 9001. | Herne European Consultancy Ltd ISBN 978–0954864743 |
| *ISO 9001:2000 Audit Procedures* (2nd edition) | This book usefully describes methods for completing management reviews and quality audits. It contains a complete set of audit check sheets and explanations to assist quality managers and auditors in completing internal, external and third-party audits of ISO 9001 quality management systems. | Routledge ISBN 978–0750666152 |

*(continued)*

| Title | Extracts from book reviews | Publisher and ISBN |
|---|---|---|
| *Auditing Quality Management Systems* (4th edition)  | This book is the result of over a decade's experience of all major international standards for integrated quality management systems. It includes a comprehensive CD containing all the major audit check sheets and forms that are required to conduct either a simple internal audit or an external assessment of an organization against the formal requirements of ISO 9001:2008. Now fully updated to include checklists for project management and health & safety in the workplace, it also includes background notes for auditors. | Herne European Consultancy Ltd ISBN 978–0954864774 |
| *ISO 9001:2000 in Brief* (2nd edition)  | Revised and expanded, this new edition of this easy to understand guide provides practical information on how to set up a cost-effective ISO 9001 compliant quality management system. | Routledge ISBN 978–0750666169 |
| *MDD Compliance using Quality Management Techniques*  | The Medical Device Directive (MDD) is difficult to understand and interpret, but this book covers the subject superlatively. The book is a good reference for understanding the requirements of the MDD, and will aid companies of all sizes in adding these requirements to an existing quality management system. | Routledge ISBN 978–0750644419 |
| *Quality and Standards in Electronics*  | A manufacturer or supplier of electronic equipment or components needs to know the precise requirements for component certification and quality conformance to meet the demands of the customer. This book ensures that the professional is aware of all the UK, European and international necessities, knows the current status of the regulations and standards, and where to obtain them. | Newnes ISBN 978–0750625319 |

*(continued)*

| Title | Extracts from book reviews | Publisher and ISBN |
|---|---|---|
| *Environmental Requirements for Electromechanical and Electronic Equipment* | This is the definitive reference containing all the background guidance, typical ranges, details of recommended test specifications, case studies and regulations covering the environmental requirements on designers and manufacturers of electrical and electromechanical equipment worldwide. | Newnes ISBN 978–0750639026 |
| *CE Conformity Marking* | CE marking can be regarded as a product's trade passport for Europe. It is a mandatory European marking for certain product groups to indicate conformity with the essential health and safety requirements set out in the European Directive. This practical and easy to understand book contains essential information for any manufacturer or distributor wishing to trade in the European Union. | Routledge ISBN 978–0750648134 |
| *Optoelectronics and Fibre Optic Technology: A Practical Guide* | Students, technicians and professional readers could benefit from this introduction to the fascinating technology of fibre optics. Simply written in an easily accessible style that does not put the reader off, the book covers all the basic topics in an appropriate and logical order. Topical areas such as optoelectronics in LANs and WANs, cable TV systems, and the global fibre-optic highway make this book essential reading for anyone who needs to keep up with the technology of modern data communications. | Newnes ISBN 978–0750653701 |

And for those who would like to relax with some cooking recipes – based on cyder and apples why not try . . .

| *The Cyder Book* | A unique combination of an historical overview of cider making through the ages, the cider making process and a collection of recipes using cider and cider apples. | Herne European Consultancy Ltd ISBN 978–0954864767 |
|---|---|---|

# Index